ANNUAL REVIEW OF NUCLEAR SCIENCE

EDITORIAL COMMITTEE (1976)

ANNUAL REVIEW OF NUCLEAR SCIENCE

EMILIO SEGRÈ, *Editor*
University of California, Berkeley

J. ROBB GROVER, *Associate Editor*
Brookhaven National Laboratory

H. PIERRE NOYES, *Associate Editor*
Stanford University

VOLUME 26

1976

ANNUAL REVIEWS INC.   4139 EL CAMINO WAY   PALO ALTO, CALIFORNIA 94306

ANNUAL REVIEWS INC.
Palo Alto, California, USA

International Standard Book Number: 0-8243-1526-X
Library of Congress Catalog Card Number: 53-995

Annual Reviews Inc. and the Editors of its publications assume no responsibility
for the statements expressed by the contributors to this *Review*.

REPRINTS

The conspicuous number aligned in the margin with the title of each article in this
volume is a key for use in ordering reprints. Available reprints are priced at the
uniform rate of $1 each postpaid. The minimum acceptable reprint order is 10
reprints and/or $10.00, prepaid. A quantity discount is available.

FILMSET BY TYPESETTING SERVICES LTD, GLASGOW, SCOTLAND
PRINTED AND BOUND IN THE UNITED STATES OF AMERICA

CONTENTS

Emilio Segrè

PREFACE

This is the last volume of the *Annual Review of Nuclear Science* that will appear under my editorship. I have served for 26 years on the Editorial Committee, and since 1957 as Editor; it is now time to relinquish the post to a member of a younger generation.

This is an appropriate occasion to comment on the history and purpose of the *Annual Review of Nuclear Science* during the period of my service.

The first volume of the series appeared in 1952 and its preface, by A. F. Thompson, chairman of the Subcommittee on Publications and Information of the Committee on Nuclear Science of the National Research Council, stated:

> Early in 1950, the Committee on Nuclear Science of the National Research Council decided to sponsor the preparation of a series of annual review volumes covering the most important developments in the fields of nuclear science each year. The Committee members agreed that a collection of critical appraisals of each year's progress in the various fields of nuclear research in a single annual volume would be of great value to all of those interested in this rapidly developing field. . . .

World War II had ended five years earlier and had left a vast amount of new technical knowledge and data in the nuclear field. An important fraction of this was in Manhattan Project literature that was slowly being declassified. Scientists returning from war projects to peaceful pursuits had problems in deciding how to handle technical information that was still classified and in choosing what to teach the eager new students. Much of the prewar literature had become obsolete; new books were being prepared, but were not yet ready. All this contributed to the establishment of the *Annual Review of Nuclear Science* and several of its first promoters were connected with the former Manhattan Project or with the then-new Atomic Energy Commission and the bureau in charge of declassification.

I joined the Editorial Committee in 1952. J. G. Beckerley was Editor, and the National Research Council was still providing financial assistance. The third volume, published in 1953, was the first for which the Editorial Committee, then composed of J. G. Beckerley, Editor; M. D. Kamen, D. Mastick, and L. I. Schiff, Associate Editors; and C. D. Coryell, L. F. Curtiss, E. Segrè, and R. E. Zirkle, Members, was entirely responsible. By the time we reached Volume 5, we had a clearer idea of what we were doing. The preface to Volume 5 shows the results of our experience:

> . . . Therefore it seems appropriate to note here some of the basic problems which the Editorial Committee has faced in determining the content of these volumes and to record the method by which these problems have been resolved.

One important question has been, "To whom are the individual reviews directed?" We have concluded that a variety of readers must be considered, ranging from the specialists in the field under review, who probably will not learn very much new, to the scientists in neighboring fields, who will probably profit substantially. Other readers will profit mainly from those chapters dealing with common phenomena of broad scientific interest and those written by authors particularly adept in expressing novel and complex ideas in simple terms. Because of these realities the individual chapters will have a varying usefulness to those working in fields remote from the ones under review.

Another difficult question is, "What is the proper scope of the volumes?" We have been unable to find a rigid definition of "Nuclear Science"; in fact, we are not certain any can exist. An expanding field has by its nature undefined boundaries. We feel certain that nuclear physics, pure and applied, experimental and theoretical, belongs here. But, for example, radiation chemistry—does this belong to these volumes, or is it properly a matter for the *Annual Review of Physical Chemistry?* ...

It has also been difficult in defining the scope of these volumes to decide the extent of coverage appropriate to radiobiology. . . .

Of the many other problems faced by the Editorial Committee in planning each volume, one stands out in difficulty. This is the question of timing. Complete coverage in any one volume is clearly impossible. Consequently we have to be concerned with when to include a review on any particular subject. Is the topic "ripe" for reviewing? Has anything of consequence happened since the last review? Have journals or books appeared (or are they about to appear) which would make a review at this time inappropriate? There seems to be no simple formula for obtaining the answers to such questions. It is a matter of predicting each year the activity and accomplishments expected during the next year and a half or more (the interval between the invitation to the author and the appearance of the corresponding volume). . . .

Of comparable difficulty to the timing question is the space allocation. This is, of course, related to the scope of the review and to the intended audience. Even with these latter items fixed, the optimum number of pages to be available for each chapter is still debatable. Too few pages tends to stimulate a review which is essentially an annotated reference list; too many pages encourages inclusion of material of secondary interest. . . .

In spite of these and other problems, the Editorial Committee believes the results to date have been well worth the effort. The steadily increasing proliferation of research papers in the literature of nuclear science steadily increases the value of such integrating activities as these volumes represent.

It should not be construed from the above comments that the Editorial Committee considers its actions as creative. The real creators of each Annual Review *are the contributing authors. To them belongs the credit, and to them we express our appreciation.* [Italics added.]

The Editorial Committee and the Editor have been concerned with clarity and readability. We have always urged our authors to be intelligible, at least in the opening sections of an article, to specialists outside their field. For a while, E. P. Wigner, a person of uncommon erudition and intelligence, was on our Editorial Committee. After serving his term he wrote a review of Volume 22, in which he says ". . . a friend, himself a contributor to the book, and deeply interested in reducing the specialization of his colleagues to some narrow area of nuclear physics, asked me whether I could conclude that this book would contribute to the broadening of the interest of nuclear physicists. I read the book, therefore, not only to learn some of its contents, but also to arrive at some answer to my friend's question.

"I am glad to report that the answer arrived at is rather positive, although it is possible to criticize almost every one of the 12 articles . . . ." He then reviews the single articles and concludes "the articles one can understand are rewarding reading."

The choice of authors is a subject of pride. We succeeded in having the collaboration of a certain number of celebrities, and, even more important, many authors have written for *Annual Reviews* before they achieved fame, and they have given us some truly memorable articles of permanent value. Indeed, I am tempted to mention some of the articles that impressed me most for their clarity, profundity, or both (even that is possible), but the choice would be too personal and possibly unfair. A better test would be to check the condition of Annual Reviews volumes in the library of an active institution. I expect that the pages of specific articles will be well worn.

Not surprisingly, the collection of *Annual Review of Nuclear Science* faithfully reflects the development of nuclear and particle physics over the years. Other aspects of nuclear science are also well covered, but perhaps less thoroughly. The volumes show the evolution of nuclear science in the last 25 years as on a theater screen. Among the tools we see the accelerators, from the first postwar innovations to modern colliding beams; the detection techniques, from photographic emulsions to bubble chambers and solid-state detectors. In nuclear physics, we see the shell model unifying many different phenomena, followed by the unraveling of beta decay, the clarification (as far as it goes) of nucleon–nucleon interaction, the introduction of polarization among the subjects of study, the heavy-ion reactions, the great efforts at mastering the problems of nuclear matter, the ever-surprising study of fission, and, not long ago, a sort of bird's-eye view of the whole field. In particle physics we start with the new particles as observed in cosmic rays, the first steps of pion physics, the clarification brought about on many subjects by isospin, the multiplication of the resonances and their classification by symmetry arguments, antiparticles, the nonconservation of parity, Regge poles, electron scattering, and several theoretical articles concerned with very fundamental aspects. At the beginning of the series, we find nuclear chemistry reviews on the first transuranics, and now on elements with $Z > 100$, the development of many chemical techniques and even applications far afield, such as in archaeology.

Many articles cover borderline fields between atomic or solid-state physics and nuclear physics. They are all interesting and reinforce the lesson of the interrelation of different branches of physics. We have in mind subjects such as positronium, muonium, exotic atoms, the Mössbauer effect, neutron optics, and passage of radiation through matter.

The reactors and nuclear energy are treated according to the prevailing opinions, and, in time, one perceives a change in mood on their practical application.

Fusion recurs repeatedly. Here we approach the geological and astrophysical applications, from geochronology to neutron stars.

Biological applications were treated more frequently in the early volumes. As time went on, the subject diverged too much from the main theme for systematic inclusion, although radiation protection is of paramount importance to all practicing nuclear and particle scientists.

The collection of the 26 volumes of the *Annual Review of Nuclear Science* thus forms a good encyclopedia on the subject. Perhaps it was not consciously planned that way, but the result is there and is of considerable importance.

This extremely brief mention of some of the contents of the series omits the majority of the articles, many of primary importance to the practitioner.

I can only close this preface with renewed thanks to the staff, the Editorial Committees, and above all to the authors. They have truly made the *Annual Review of Nuclear Science* what it is today and all the readers and users of the volumes owe them a debt of gratitude for their unselfish labors.

Emilio Segrè

Ann. Rev. Nucl. Sci. 1976. 26: 1–50

CP VIOLATION AND K^0 DECAYS

✕5570

Konrad Kleinknecht

Institut für Physik, Universität Dortmund, Dortmund, Germany

CONTENTS

1 INTRODUCTION

Symmetries and conservation laws have long played an important role in physics. The simplest examples of macroscopic relevance are the conservation of energy and momentum, which are due to the invariance of forces under translation in time and space, respectively.

In the domain of quantum phenomena, there are also conservation laws corresponding to discrete transformations. One of these is reflection in space ("parity operation") P (1). Invariance of laws of nature under P means that the mirror image of an experiment yields the same result in its reflected frame of reference as the original experiment in the original frame of reference. This means that "left" and "right" cannot be defined in an absolute sense.

Similarly, the particle-antiparticle conjugation C (see, for example, 2) transforms each particle into its antiparticle, whereby all additive quantum numbers change their sign. C invariance of laws again means that experiments in a world consisting mainly of antiparticles will give identical results to the ones in our world with the one exception that all names of particles are "anti" relative to ours. Here again it will be not possible to define in an absolute way whether a particle consists of antimatter or matter: an antiatom composed of antinucleons and positrons emits the same spectral lines as the corresponding atom.

A third transformation of this kind is time reversal T, which reverses momenta and angular momenta (3). This corresponds formally to an inversion of the direction of time. According to the CPT theorem of Lüders & Pauli (4, 5) there is a connection between these three transformations such that under rather weak assumptions in a local field theory all processes are invariant under the combined operation $C \cdot P \cdot T$.

For a long time it was assumed that all elementary processes are also invariant under the application of each of the three operations C, P, and T separately. However, the work of Lee & Yang (6) questioned this assumption, and the subsequent experiments demonstrated the violation of P and C invariance in weak decays of nuclei (7) and of pions and muons (8, 9). This violation can be visualized by the longitudinal polarization of neutrinos emerging from a weak vertex: they are left-handed when they are particles and right-handed when antiparticles. Application of P or C to a neutrino leads to an unphysical state.

The combined operation CP, however, transforms a left-handed neutrino into a right-handed antineutrino, thus connecting two physical states. CP invariance therefore was considered (10) to be replacing the separate P and C invariance of weak interactions.

One consequence of this postulated CP invariance for the neutral K mesons was predicted by Gell-Mann & Pais (11): there should be a long-lived partner to the known $V^0(K_1^0)$ particle of short lifetime (10^{-10} sec). According to this proposal these two particles are mixtures of two strangeness eigenstates, K^0 ($S = +1$) and $\overline{K^0}$ ($S = -1$) produced in strong interactions. Weak interactions do not conserve strangeness and the physical particles should be eigenstates of CP if the weak interactions are CP invariant. These eigenstates are (with $\overline{K^0} = CP\ K^0$)

$$CP\ K_1 = CP\ \left[(K^0 + \overline{K^0})/\sqrt{2}\right] = (\overline{K^0} + K^0)/\sqrt{2} = K_1,$$
$$CP\ K_2 = CP\ \left[(K^0 - \overline{K^0})/\sqrt{2}\right] = (\overline{K^0} - K^0)/\sqrt{2} = -K_2. \qquad 1.$$

Because of $CP\ (\pi^+\pi^-) = (\pi^+\pi^-)$ for π mesons in a state with angular momentum zero, the decay into $\pi^+\pi^-$ is allowed for the K_1 but forbidden for the K_2; hence the longer lifetime of K_2, which was indeed confirmed when the K_2 was discovered (12, 13).

In 1964, however, Christenson, Cronin, Fitch & Turlay (14) discovered that the long-lived neutral K meson also decays to $\pi^+\pi^-$ with a branching ratio of $\sim 2 \times 10^{-3}$, From then on the long-lived state was called K_L because it was no longer identical to the CP eigenstate K_2; similarly, the short-lived state was called K_S. The CP violation that manifested itself by the decay $K_L \to \pi^+\pi^-$ was confirmed by subsequent discoveries of the decay $K_L \to \pi^0\pi^0$ (15, 16), and of a charge asymmetry in the decays $K_L \to \pi^\pm e^\mp v$ and $K_L \to \pi^\pm \mu^\mp v$ (17, 18).

The question whether this CP violation (and T violation through the CPT theorem) is due to a fifth force, the "superweak" interaction (19), to a T-violating part of the weak interaction (20–28), or to interference of the T-invariant weak interaction with a T-violating part in the electromagnetic (29–32) or strong (33–35) interactions, has been studied by many subsequent experiments. Such experiments are described in Section 5.

Section 2 is devoted to the phenomenological description of the K^0 system, Section 3 to experimental techniques, and Section 4 to the experiments on CP violation parameters in K^0 decay.

Reviews on related subjects (36–43) include the classical ones by Lee & Wu (36) and by Bell & Steinberger (37).

2   PHENOMENOLOGICAL DESCRIPTION OF K^0 DECAYS

The formalism of particle mixture introduced by Gell-Mann & Pais (11) has been modified to describe a possible CP violation by Lee, Oehme & Yang (44). Further extensions were given by Sachs (45), Wu & Yang (46), Bell & Steinberger (37), and Lee & Wu (36). For simplicity, CPT invariance is assumed here.

2.1   Mass Matrix

Let the eigenstates of strangeness S produced in strong interactions be K^0 ($S = +1$) and $\overline{K^0}$ ($S = -1$), where $\overline{K^0} = CP\ K^0$. Weak interactions do not conserve strangeness, whereby K^0 and $\overline{K^0}$ can mix by second-order weak transitions through intermediate states like $2\pi, 3\pi, \pi\mu v, \pi e v$. The states obeying an exponential decay law are linear superpositions of K^0 and $\overline{K^0}$:

$$\alpha | K^0\rangle + \beta | \overline{K^0}\rangle = \begin{pmatrix} \alpha \\ \beta \end{pmatrix}. \qquad\qquad 2.$$

The time-dependent Schrödinger equation then becomes a matrix equation (47),

$$i\frac{d}{dt}\begin{pmatrix} \alpha \\ \beta \end{pmatrix} = \mathbf{X}\begin{pmatrix} \alpha \\ \beta \end{pmatrix}, \qquad\qquad 3.$$

where $X_{ik} = M_{ik} - i\Gamma_{ik}/2$, and M_{ik} and Γ_{ik} are Hermitian matrices, named mass matrix and decay matrix, respectively. The elements of the matrix $\mathbf{X}$ are

$$\mathbf{X}_{11} = \langle K^0|H| K^0\rangle, \quad \mathbf{X}_{22} = \langle \overline{K^0}|H| \overline{K^0}\rangle, \quad \mathbf{X}_{12} = \langle K^\circ |H| \overline{K^0}\rangle,$$
$$\mathbf{X}_{21} = \langle \overline{K^0}|H| K^0\rangle, \qquad\qquad 4.$$

where CPT invariance requires $\mathbf{X}_{11} = \mathbf{X}_{22}$. The solutions of the eigenvalue equations are $\mathbf{M}_L = \mathbf{X}_{11} + ipq$ and $\mathbf{M}_S = \mathbf{X}_{11} - ipq$ if one defines $p^2 = i\mathbf{X}_{12}$ and $q^2 = i\mathbf{X}_{21}$. The corresponding eigenstates are

$$|K_L\rangle = (p|K^0\rangle - q|\overline{K^0}\rangle)/\sqrt{|p^2| + |q^2|},$$
$$|K_S\rangle = (p|K^0\rangle + q|\overline{K^0}\rangle)/\sqrt{|p|^2 + |q|^2}.$$

5.

These are the observable particles for which from the Schrödinger equation

$$i\frac{d}{dt}|K_L\rangle_S = M_L\,|K_L\rangle_S$$

6.

follows an exponential decay law with an oscillating phase factor. The real part of the eigenvalue M_L is the mass m_L of the particle, and the imaginary part of M_L is half of the decay width Γ_L:

$$M_L = m_L - i\Gamma_L/2, \qquad M_S = m_S - i\Gamma_S/2.$$

7.

The time evolution of the states is given by

$$|K_L\rangle \rightarrow |K_L\rangle \exp(-iM_L\tau) = |K_L\rangle \exp(-\Gamma_L\tau/2)\exp(-im_L\tau)$$

8.

and analogously for K_S.

If one defines an asymmetry parameter

$$\varepsilon = (p-q)/(p+q),$$

then one obtains

$$\varepsilon = \frac{\operatorname{Im}\Gamma_{12}/2 + i\operatorname{Im}\mathbf{M}_{12}}{i(\Gamma_S - \Gamma_L)/2 - (m_S - m_L)}$$

9.

and

$$\langle K_L | K_S \rangle = 2\operatorname{Re}\varepsilon,$$

i.e. K_L and K_S are only orthogonal if CP is conserved, which implies $\varepsilon = 0$. The nondiagonal elements of the decay and mass matrix can be expressed by the first terms of a perturbation expansion:

$$\Gamma_{12} = 2\pi \sum_F \rho_F \langle \overline{K^0} | H_{WK} | F \rangle \langle F | H_{WK} | K^0 \rangle,$$

where the sum runs over all possible intermediate states F with phase space density ρ_F. Similarly,

$$\mathbf{M}_{12} = \langle \overline{K^0} | H_{WK} | K^0 \rangle + \sum_n \left(\langle \overline{K^0} | H_{WK} | n \rangle \langle n | H_{WK} | K^0 \rangle / (m_{K^0} - m_n) \right)$$

where the sum extends over all possible states n.

In terms of the parameter ε and of the pure CP eigenstates K_1 and K_2 of Section 1, the physical states can be written:

$$|K_L\rangle = (|K_2\rangle + \varepsilon|K_1\rangle)/\sqrt{1+|\varepsilon|^2}$$
$$= [(1+\varepsilon)|K^0\rangle - (1-\varepsilon)|\overline{K^0}\rangle]/\sqrt{2(1+|\varepsilon|^2)},$$

10.

$$|K_S\rangle = (|K_1\rangle + \varepsilon|K_2\rangle)/\sqrt{1+|\varepsilon|^2}$$
$$= [(1+\varepsilon)|K^0\rangle + (1-\varepsilon)|\bar{K}^0\rangle]/\sqrt{2(1+|\varepsilon|^2)}. \qquad 11.$$

CP violation in the decay matrix ($\text{Im } \Gamma_{12} \neq 0$) or in the mass matrix ($\text{Im } \mathbf{M}_{12} \neq 0$) therefore implies a CP impurity in the physical states measurable by the parameter ε.

If CPT invariance was not assumed, the equations above for K_L and K_S would contain admixture parameters ε_L and ε_S, respectively, and a CPT violation would manifest itself by $\varepsilon_L \neq \varepsilon_S$.

2.2   Isospin Decomposition

In $K_{S,L} \to 2\pi$ decays, the angular momentum of the pions vanishes. The spatial part of the wave function is therefore symmetric, and since pions are bosons, the isospin wave function must be symmetric too. The two symmetric combinations of two $I = 1$ states have $I = 0$ and $I = 2$, and the four existing transition amplitudes are

$$\langle 0|T|K_S\rangle, \quad \langle 2|T|K_S\rangle, \quad \langle 0|T|K_L\rangle, \quad \langle 2|T|K_L\rangle,$$

which can be reduced to three complex numbers by normalizing to the amplitude $\langle 0|T|K_S\rangle$:

$$\varepsilon_0 = \langle 0|T|K_L\rangle / \langle 0|T|K_S\rangle,$$
$$\varepsilon_2 = \frac{1}{\sqrt{2}} \langle 2|T|K_L\rangle / \langle 0|T|K_S\rangle, \qquad 12.$$
$$\omega = \langle 2|T|K_S\rangle / \langle 0|T|K_S\rangle.$$

The experimentally observable quantities are

$$\eta_{+-} = \langle \pi^+\pi^-|T|K_L\rangle / \langle \pi^+\pi^-|T|K_S\rangle,$$
$$\eta_{00} = \langle \pi^0\pi^0|T|K_L\rangle / \langle \pi^0\pi^0|T|K_S\rangle, \qquad 13.$$
$$\delta_L = [\Gamma(K_L \to \pi^- l^+ \nu) - \Gamma(K_L \to \pi^+ l^- \nu)]/[\Gamma(K_L \to \pi^- l^+ \nu) + \Gamma(K_L \to \pi^+ l^- \nu)].$$

Relating the isospin states to the physical 2π states

$$\langle 0| = \frac{1}{\sqrt{3}}\langle \pi^-\pi^+| - \frac{1}{\sqrt{3}}\langle \pi^0\pi^0| + \frac{1}{\sqrt{3}}\langle \pi^+\pi^-|,$$
$$\langle 2| = \frac{1}{\sqrt{6}}\langle \pi^-\pi^+| + \sqrt{\frac{2}{3}}\langle \pi^0\pi^0| + \frac{1}{\sqrt{6}}\langle \pi^+\pi^-|,$$

one obtains

$$\eta_{+-} = (\varepsilon_0 + \varepsilon_2)/(1+\omega/\sqrt{2}), \qquad \eta_{00} = (\varepsilon_0 - 2\varepsilon_2)/(1-\sqrt{2}\omega). \qquad 14.$$

Because of the validity of the $\Delta I = \frac{1}{2}$ rule for CP-conserving weak nonleptonic decays $\omega \ll 1$ and therefore can be neglected.

A suitable choice for the phase of the $K^0 \to 2\pi$ $(I = 0)$ amplitude is obtained by choosing this amplitude to be real except for final-state interactions leading to a phase shift δ_0:

$$\langle 0|T|K^0\rangle = \exp(i\delta_0)A_0 \quad \text{and} \quad A_0 \text{ real.}$$

Similarly

$$\langle 2|T|K^0\rangle = \exp(i\delta_2)A_2.$$

Then one gets

$$\varepsilon_0 = \varepsilon, \qquad \varepsilon_2 = i(2^{-1/2})(\text{Im}\,A_2/A_0)\exp[i(\delta_2-\delta_0)] = \varepsilon'.$$

Therefore, representing ε and ε' in the complex plane, one obtains the Wu-Yang triangle relations

$$\eta_{+-} = \varepsilon+\varepsilon', \qquad \eta_{00} = \varepsilon-2\varepsilon' \qquad\qquad 15.$$

and for the charge asymmetry

$$\delta_{\text{L}} = \frac{1-|x|^2}{|1-x|^2}\,2\,\text{Re}\,\varepsilon,$$

with $x = g/f$ being the ratio of $\Delta Q = -\Delta S$ to $\Delta Q = \Delta S$ amplitudes (see Section 2.5.2).

The decomposition of the observable decay amplitude into ε and ε' corresponds to a separation of the CP-violating effects due to the mass and decay matrices (ε), which are seen also in the impurity of the K_{L} and K_{S} states and of CP violation in the transition matrix element.

The phase of ε' is $\text{Arg}(\varepsilon') = (\pi/2)+(\delta_2-\delta_0)$. The $\pi\pi$ phase shifts have been measured very precisely in recent experiments. They obtain $\delta_2 = (-7.2\pm1.3)°$ (48) and $\delta_0 = (46\pm5)°$ (49). Therefore $\text{Arg}(\varepsilon') = (37\pm5)°$. The phase of ε is given by

$$\text{Arg}\,\varepsilon = \phi_{\text{D}}+\arctan(2\Delta m/\Gamma_{\text{S}}),$$

where $\Delta m = m_{\text{L}}-m_{\text{S}}$ and

$$\phi_{\text{D}} = -\arctan(\text{Im}\,\Gamma_{12}/2\,\text{Im}\,M_{12}).$$

If there is no strong CP violation in the channels $K \to 2\pi(I = 2)$, $K \to \pi l\nu$, and $K \to 3\pi$, ϕ_{D} is very small. Using the present experimental limits on CP violation in the leptonic decays (Section 5.4.3), we get

$$|\phi_{\text{D}}(K_{e3})| < 0.6°, \qquad |\phi_{\text{D}}(K_{\mu3})| < 1.2°;$$

in the same way the upper limit on the CP-violating $K_{\text{S}} \to \pi^+\pi^-\pi^0$ decay (50),

$$\Gamma(K_{\text{S}} \to \pi^+\pi^-\pi^0)/\Gamma(K_{\text{L}} \to \pi^+\pi^-\pi^0) < 0.12,$$

allows one to set the limit

$$|\phi_{\text{D}}(\pi^+\pi^-\pi^0)| < 1.3°.$$

Similarly, from the limit (51) $\Gamma(K_{\text{S}} \to 3\pi^0)/\Gamma(K_{\text{L}} \to 3\pi^0) < 1.2$ follows

$$|\phi_{\text{D}}(3\pi^0)| < 7.1°.$$

If we use the experimental value of Δm from Section 4.2, then

$$\arg \varepsilon = 43.8 \pm 0.2° \text{ for models with } \varepsilon' = 0, \qquad\qquad 16.$$

$$\arg \varepsilon = 44 \pm 8° \text{ for models with } \varepsilon' \neq 0. \qquad\qquad 17.$$

2.3   Models of CP Violation

Apart from early models that tried to avoid the consequence of a CP violation altogether by giving up the superposition principle (52) or by postulating a long-range cosmological field due to the local preponderance of matter over antimatter (53, 54), which were soon discarded by experiment, there are four classes of models. Since most of them preserve CPT conservation, we consider first the experimental evidence supporting CPT conservation.

2.3.1   CPT INVARIANCE   The principle of CPT invariance (4, 5, 44) requires masses, lifetime, and other static quantities for particle and antiparticle to be equal in magnitude. The best evidence for it is given by the value of the $K_L - K_S$ mass difference (Section 2.4),

$$(m_K - m_{\bar{K}})/m_K \lesssim (m_L - m_S)/m_K \approx 0.7 \times 10^{-14}.$$

This gives an upper limit of 10^{-14} for CPT-violating strong interactions. Since electromagnetic contributions to the K^0 mass of the order of $\alpha \cdot m_K$ can be expected, this limit allows at most a 10^{-12} contribution of a CPT-violating interaction. Another test of CPT invariance in electromagnetic interactions is obtained by comparing the magnetic moments of a particle and its antiparticle. In particular, the measurements on the g factor of positive and negative muons (55, 56) are very precise. The results are

$$g(\mu^+) - g(\mu^-) = (-7.2 \pm 4.9) \times 10^{-7};$$

for the electron (57, 58)

$$g(e^+) - g(e^-) = (16 \pm 22) \times 10^{-6};$$

and the corresponding limits on a CPT-violating part of the electromagnetic interaction are of the order of 10^{-6}.

Even for the question of whether CPT is conserved in weak interactions the $K_L - K_S$ mass difference gives the lowest upper limit. A first-order CPT-violating weak interaction could be at most of a strength of 10^{-7} relative to the CPT-invariant weak interaction in order not to contradict the experimental value.

Independent but less stringent evidence for CPT invariance in weak interactions with $\Delta S = 1$ or $\Delta S = 0$ comes from the experimental near equality of particle and antiparticle lifetimes, as determined for μ, π, and K mesons (59–61):

$$(\tau_{\mu^+} - \tau_{\mu^-})/(\tau_{\mu^+} + \tau_{\mu^-}) = (0.0 \pm 0.5) \times 10^{-3},$$

$$(\tau_{\pi^+} - \tau_{\pi^-})/(\tau_{\pi^+} + \tau_{\pi^-}) = (0.32 \pm 0.35) \times 10^{-3},$$

$$(\tau_{K^+} - \tau_{K^-})/(\tau_{K^+} + \tau_{K^-}) = (0.25 \pm 0.50) \times 10^{-3}.$$

All of these results agree with the predictions of *CPT* invariance. A test of *CPT* at a different level of sensitivity will be obtained from the study of *CP*-violating K^0 decays by checking whether the Wu-Yang triangle mentioned in Section 2.2 "closes," i.e. whether the complex numbers η_{+-}, η_{00}, and $3\varepsilon'$ really form a triangle. This is discussed in Section 4.

2.3.2 MODELS   Assuming now *CPT* invariance of all interactions, the observed *CP*-violating effects in K decay imply also T violation (the experimental data of Section 4 are even sufficient for proving T violation without *CPT* invariance).

In general, with *CPT* invariance there are four combinations of violations possible:

(a) T-conserving, C-violating, and P-violating;

(b) T-violating, C-conserving, and P-violating;

(c) T-violating, C-violating, and P-conserving;

(d) T-violating, C-violating, and P-violating.

Parity conservation in strong and electromagnetic interactions has been tested, e.g. by looking for a circular polarization of γ rays from nuclear transitions. The presence of a wrong parity admixture in one of the nuclear states involved will cause a small amplitude for a γ transition with abnormal multipolarity that can interfere with the dominant amplitude and cause such a circular polarization. In the experiments of Lobashov et al (62), polarizations of the order of 10^{-5} have been measured. These are consistent with being due to the two-nucleon force $np \rightarrow pn$ induced by the weak interaction [see (43) for a review].

From many experiments of similar nature one can infer that strong and electromagnetic interactions are not of type a, b, or d. Therefore if the source of the *CP*-violating phenomena is located in strong or electromagnetic interactions, there must be a part of those interactions belonging to class c, i.e. C- and T-violating, but P-conserving.

Proposed models can be included in the following four categories:

1. Millistrong *CP* violation models (33–35) postulate the existence of C- and T-violating terms of the order 10^{-3} in the strong interaction. The process $K_L \rightarrow \pi^+\pi^-$ is supposed to occur by interference of two amplitudes: first, the K_L decays via normal *CP*-conserving weak interaction with $\Delta S = 1$ into an intermediate state X, and this state decays into $\pi^+\pi^-$ by this T-violating strong interaction. The amplitude of the process is of order $G_F \cdot a$, where G_F is the Fermi coupling constant and a is the coupling of this *CP*-violating strong interaction. From the experimental value of $|\eta_{+-}|$ one concludes that $a \approx 10^{-3}$.

2. Electromagnetic *CP* violation models (29–32) require large parts of the electromagnetic interaction of hadrons to be C- and T-violating, but P-conserving. A two-step process $K_L \rightarrow x \rightarrow 2\pi$ could then occur through interference of a weak and an electromagnetic *CP*-violating amplitude. The product of G_F with the fine structure constant α is not too far from $G_F \times 10^{-3}$, as required by the magnitude of $|\eta_{+-}|$.

3. Milliweak models (20–28) assume that a part of the order of 10^{-3} in the weak interaction is CP-violating and responsible for the observed effects. The decay $K_L \to 2\pi$ is then a one-step process, and CP or T violations of the order of 10^{-3} should show up in other weak processes.

4. The superweak model (19) postulates a new $\Delta S = 2\, CP$-violating interaction that has a coupling (coupling constant g) smaller than second-order weak interaction. This interaction could induce a transition $K_L \to K_S$, with subsequent decay $K_S \to 2\pi$. More precisely, this interaction would cause a first-order transition matrix element

$$M_{Sw} = \langle \bar{K} | \mathscr{H}_{Sw} | K \rangle \sim gG_F.$$

The mass difference itself is related to the second-order weak matrix element

$$\mathbf{M}_{\bar{K}K} = \sum_n \langle \bar{K} | \mathscr{H}_w | n \rangle \langle n | \mathscr{H}_w | K \rangle / (E_K - E_n + i\varepsilon),$$

where n is an intermediate state with energy E_n and $\mathscr{H}_w$ is the weak Hamiltonian. In order that the CP-violating amplitude for $K_L \to 2\pi$ relative to the CP-conserving one be of the observed magnitude, the ratio $\mathbf{M}_{Sw}/\mathbf{M}_{\bar{K}K}$ must be of the order 10^{-3}. Since $\mathbf{M}_{Sw} \approx gG_F$ and $\mathbf{M}_{\bar{K}K} \sim G_F^2 m_p^2$, where the proton mass m_p is used as a cutoff in the integration, this yields $g \sim G_F m_p^2 10^{-3} \approx 10^{-8}$.

This superweak interaction can only be detected in the $K_L - K_S$ system because this is the only known pair of states with such a small difference in energy that it is sensitive to forces weaker than second-order weak interaction.

These four classes of models differ in their predictions. For K^0 decays the superweak model states $\varepsilon' = 0$ and consequently $\eta_{+-} = \eta_{00} = \varepsilon$. The Wu-Yang triangle collapses to a line at an angle $\phi_{+-} = \phi_{00} = \phi_\varepsilon = (43.8 \pm 0.2)°$. For the other classes, ε' should not vanish and could be of the order of the violation of the $\Delta I = \frac{1}{2}$ rule in nonleptonic decays, i.e. $|\varepsilon'|/|\varepsilon| \sim 5 \times 10^{-2}$.

In any case, for models other than the superweak one, violations of CP or T should manifest themselves in other reactions of particles or nuclei.

2.4 K_L-K_S Regeneration

The term *regeneration* has been introduced by Pais & Piccioni (63) to designate the creation of short-lived K_S mesons when long-lived K_L mesons traverse matter. It was predicted that this process would happen when the short- and long-lived K mesons are two different linear combinations of the K^0 and $\bar{K}^0$, the eigenstates of strangeness ($S = +1$ and $S = -1$). Because strong interactions conserve strangeness, several reactions that the $\bar{K}^0$ can undergo (e.g. $\bar{K}^0 p \to \Lambda \pi^+$) have no counterpart for the K^0. Therefore the total cross-sections of $\bar{K}^0$ on the nucleons p and n are both bigger than the corresponding total cross-sections of K^0 on the nucleons, and as a consequence, the total cross-sections of $\bar{K}^0$ on any nucleus N also obey the inequality $\sigma_T(\bar{K}^0 N) \geqq \sigma_T(K^0 N)$, which can become an equality in the asymptotic region. The optical theorem relates the total cross-section σ_T to the forward scattering amplitude $f(0)$, $\mathrm{Im}\, f(0) = (k/4\pi)\sigma_T$, where $k = p/h$ is the wave number of the K meson. Therefore, at any finite momentum we have $|\mathrm{Im}\, \bar{f}(0)| > |\mathrm{Im}\, f(0)|$ with $\bar{f}$ designating the $\bar{K}^0$ scattering amplitude.

Assuming for the moment that the real part of $f(0)$ is not correspondingly bigger than Re $f(0)$, it follows that $|\bar{f}(0)| > |f(0)|$. An incident long-lived neutral K is given by Equation 5.

By interaction of the K_L with matter, the two components of K_L, K^0 and $\bar{K}^0$, are altered differently and the state emerging is

$$\psi_f(\theta)\rangle = [f(\theta)pK^0\rangle - \bar{f}(\theta)q\bar{K}^0\rangle]/\sqrt{|p|^2 + |q|^2}$$
$$= \frac{f(\theta) - \bar{f}(\theta)}{2} K_S\rangle + \frac{f(\theta) + \bar{f}(\theta)}{2} K_L\rangle.$$

Now, if $f(\theta) \neq \bar{f}(\theta)$, then the emerging state contains a K_S component; this component is "regenerated" from the K_L beam. In particular, in the forward direction ($\theta = 0$) the amplitude of K_S regenerated by one scattering center is proportional to $f(0) - \bar{f}(0)$.

If one now considers two scattering centers, these two scatterers will act coherently if $d(p_S \cos\theta - p_L) \ll 1$, where p_S and p_L are the momenta of K_S and K_L respectively, d is the distance of two scattering centers along the K_L direction, and θ is the scattering angle between incoming K_L and outoing K_S (64, 75). Using the momentum transfer $p_S - p_L = \Delta m\, m_L/p_L$ (following from energy conservation) one obtains from this relation the coherence length, i.e. the distance d_{max} along the K_L direction at which two scattering centers can still interfere fully coherently for forward regeneration,

$$d_{max} \approx 1/(p_S - p_L) = p_L/(m_L\Delta m),$$

and the maximum scattering angle for coherent regeneration,

$$\theta^2 \lesssim 2(p_S - p_L)/p_S \approx 10^{-14}.$$

The coherence length is of the order of several K_S mean decay lengths and the momentum transfer of the order of 10^{-6} eV/c.

We therefore expect different classes of regeneration:

1. Coherent regeneration or "transmission regeneration" in the forward direction at angles $\theta \le 10^{-7}$ rad. Coherent addition of amplitudes from an extended region of several centimeters in length.
2. Elastic ("diffraction") regeneration from nuclei; incoherent addition of intensities from different nuclei, but coherent action of the nucleons inside the nucleus; and typical recoil momenta of $p^* \sim 50$ MeV/c and angular distribution Gaussian at small momentum transfers.
3. Inelastic regeneration: momentum transfers so big as to break up the nucleus or to transfer it to an excited state.

In our context, interest in the regeneration amplitudes is due to the fact that in the study of CP-violating effects, regeneration proves to be a powerful tool: it is possible to generate coherent mixtures of K_L and K_S waves and to study interference effects between the CP-conserving decay $K_S \to 2\pi$ from regenerated K_S and the CP-violating decay $K_L \to 2\pi$. This method has been used extensively (65–71). Also, the regeneration amplitudes measured by detecting $K_S \to \pi^+ \pi^-$ have been used to obtain an absolute normalization for the decay rate of $K_L \to \pi^0 \pi^0$ (15). Finally, the decay time distribution of $K \to \pi e v$ decays from a coherent K_L, K_S mixture has been used to measure the phase relation between outgoing K_S and incoming K_L, the "regeneration phase" (70, 72–74).

The connection between the "elementary" scattering amplitude $f(\theta) - \bar{f}(\theta)$ and the macroscopic observed regeneration probability for the coherent regeneration [class 1] can be expressed in the following way: one defines the state emerging from a block of matter of length L placed in a pure K_L beam as

$$\psi_f \rangle = K_L \rangle + \rho(L) K_S \rangle.$$

Then L is obtained by summing all contributions from individual scattering centers. The result is (75):

$$\rho = \pi i \, \frac{f(0) - \bar{f}(0)}{k} \, \Lambda_S N \, \frac{1 - \exp\left[(i\Delta m/\Gamma_S - \frac{1}{2})l\right]}{\frac{1}{2} - i\Delta m/\Gamma_S}, \qquad\qquad 18.$$

where

$f(0)$ and $\bar{f}(0)$ are, respectively, the K^0 and $\bar{K}^0$ scattering amplitudes in the forward direction;

$k = p_k/\hbar$ is the kaon wave number;

$\Lambda_S = \beta\gamma\tau_S$ is the mean decay length of K_S;

N is the density of scattering centers;

$\Delta m = m_L - m_S$ is the $K_L - K_S$ mass difference;

$\Gamma_S = \hbar/\tau_S$ is the K_S decay rate; and

$l = L/\Lambda_S$ is the length of the regenerator in units of K_S decay lengths.

This expression for ρ contains two factors: a nuclear part

$$\pi i \left\{ \frac{f(0) - \bar{f}(0)}{k} \right\} \Lambda_S N$$

and a geometrical part $G(L)$ depending on the length L of the regenerator. Therefore the phase of ρ is usually split into two parts:

$$\phi_\rho = \arg(\rho) = \phi_f + \phi_{\Delta m},$$

where

$$\phi_f = \arg\left[i\{f(0) - \bar{f}(0)\}\right]$$

is the nuclear regeneration phase and

$$\phi_{\Delta m} = \arg\left\{[1 - \exp(i\Delta m/\Gamma_S - \tfrac{1}{2})l]/[\tfrac{1}{2} - i\Delta m/\Gamma_S]\right\}$$

is the phase of the geometrical factor.

2.5 *Interference between Decay Amplitudes of K_L and K_S*

An arbitrary coherent mixture of K_L and K_S states will show interference phenomena when decaying into 2π or other common decay channels. According to Section 2.1

the eigentime development of K_L is

$$|K_L\rangle \rightarrow |K_L\rangle \exp(-iM_L\tau),$$

with $M_L = m_L - (i/2)\Gamma_L$, and correspondingly for K_S. An arbitrary mixture

$$|\psi(0)\rangle = a_S|K_S\rangle + a_L|K_L\rangle$$

will develop into

$$|\psi(\tau)\rangle = a_S \exp(-iM_S\tau)|K_S\rangle + a_L \exp(-iM_L\tau)|K_L\rangle.$$

2.5.1 2π DECAY   The 2π amplitude is therefore

$$\langle 2\pi | T | \psi(\tau)\rangle = a_S \exp(-iM_S\tau)\langle 2\pi | T | K_S\rangle + a_L \exp(-iM_L\tau)\langle 2\pi | T | K_L\rangle$$
$$= \langle 2\pi | T | K_S\rangle \{a_S \exp(-iM_S\tau) + a_L\eta \exp(-iM_L\tau)\},$$

where $\eta = \eta_{+-}$ for $\pi^+\pi^-$ decay and $\eta = \eta_{00}$ for $\pi^0\pi^0$ decay. The observed decay rate is proportional to

$$R(\tau) = |a_S|^2 \exp(-\Gamma_S\tau) + |a_L \cdot \eta|^2 \exp(-\Gamma_L\tau)$$
$$+ 2|a_S| \, |a_L| \, |\eta| \exp[-(\Gamma_L + \Gamma_S)\tau/2] \cos(\Delta m\tau + \phi) \qquad 19.$$

with $\phi = \arg(a_S) - \arg(\eta a_L)$. For different initial conditions of the mixture we obtain different results:

1. For an initially pure K^0 state ($a_S = 1 = a_L$),

$$R_1(\tau) = \exp(-\Gamma_S\tau) + |\eta|^2 \exp(-\Gamma_L\tau) + 2|\eta| \exp[-(\Gamma_L + \Gamma_S)\tau/2]$$
$$\cdot \cos(\Delta m\tau - \arg\eta).$$

2. For an initially pure $\bar{K}^0$ state the interference term changes sign.
3. For an incoherent mixture of K^0 (intensity N_K) and $\bar{K}^0$ (intensity $N_{\bar{K}}$), the interference term is multiplied by $(N_K - N_{\bar{K}})/(N_K + N_{\bar{K}})$. This is called the *vacuum interference method*.
4. For the coherent mixture behind a regenerator $a_S = \rho$, $a_L = 1$ and we obtain

$$R_2(\tau) = |\rho|^2 \exp(-\Gamma_S\tau) + |\eta|^2 \exp(-\Gamma_L\tau) + 2|\rho| \, |\eta| \exp[-(\Gamma_L + \Gamma_S)\tau/2]$$
$$\cdot \cos(\Delta m\tau + \phi_\rho - \arg\eta).$$

If the K_L beam passes through two regenerators characterized by their regeneration amplitudes ρ_1 and ρ_2 and if the downstream ends of these two slabs of matter are placed at a distance G along the beam direction, corresponding to a time in the K rest system of $\tau_G = Gm_K/(cp_K)$, where p_K is the kaon momentum, then the state behind the second regenerator is proportional to

$$|\psi\rangle = |K_L\rangle \exp(-iM_L\tau_G) + \rho_1|K_S\rangle \exp(-iM_S\tau_G) + \rho_2|K_S\rangle \exp(-iM_L\tau_G)$$

and the 2π decay amplitude directly behind the second regenerator is

$$\langle 2\pi | T | \psi\rangle \propto \rho_1 \exp(-iM_S\tau_G) + \rho_2 \exp(-iM_L\tau_G) + \eta \exp(-iM_L\tau_G).$$

Neglecting for simplicity the *CP*-violating terms, which is justified if $|\rho_1/\eta| \gg 1$ and $|\rho_2/\eta| \gg 1$ are chosen, the 2π decay rate becomes

$$I_{2\pi} = |\langle 2\pi | T | \psi \rangle|^2 \propto |\rho_1|^2 \exp(-\Gamma_S \tau_G) + |\rho_2|^2 + 2|\rho_1||\rho_2|\exp(-\Gamma_S \tau_G/2)$$
$$\cos(\Delta m \tau_G + \phi_{\rho 1} - \phi_{\rho 2}). \qquad 20.$$

2.5.2 SEMILEPTONIC DECAYS   The phenomenological Lagrangian for the semileptonic decay $K \to \pi l v$ is

$$L_{\Delta S = 1} = S_\lambda^* j_\lambda + S_\lambda j_\lambda^*,$$

where S_λ is the hadron current and j_λ the lepton current

$$j_\lambda = i \sum_{l=e,\mu} \psi_l^+ \gamma_4 \gamma_\lambda (1 + \gamma_5) \psi_{v_l}$$

in the notation of the article of Lee & Wu (36). The structure of S is a priori unknown. It may contain any of the five Lorentz-invariant operators for vector, axial-vector, scalar, tensor, or pseudoscalar interactions. Experiments are consistent with the absence of these interactions. The hadron current S_λ can then be decomposed unambiguously into a vector part S_λ^V and an axial vector part S_λ^A:

$$S_\lambda = S_\lambda^V + A_\lambda^A.$$

With the convention that the relative parity of K and π meson is $+1$, the K_{l3} decays are governed by the matrix element of S_λ^V, while $K_{\pi 2}$ is related to the axial vector part S_λ^A. For the vector part relevant for K_{l3} decay Lorentz invariance shows that only two form factors enter; these are denoted

f_+ and f_- for $K^0 \to \pi^- l^+ v_l$,

g_+ and g_- for $\overline{K^0} \to \pi^- l^+ v_l$,

$\overline{f_+}$ and $\overline{f_-}$ for $\overline{K^0} \to \pi^+ l^- \bar{v}_l$,

$\overline{g_+}$ and $\overline{g_-}$ for $K^0 \to \pi^+ l^- \bar{v}_l$,

where l denotes lepton (muon or electron). The matrix element for the first reaction is then given by

$$\langle \pi^- | S_\lambda(x) | K^0 \rangle = \tfrac{1}{2}(E_K E_\pi)^{1/2}[(K_\lambda + \pi_\lambda)f_+(q^2) + (K_\lambda - \pi_\lambda)f_-(g^2)] \exp(iq \cdot x),$$

where K_λ and π_λ are the four-momenta of kaon and pion, E_K and E_π their energies, and q^2 the square of the momentum transfer $q_\lambda = K_\lambda - \pi_\lambda$. Corresponding expressions hold for the three other reactions.

As discussed in (36),

1. CPT invariance requires $f_\pm = -\overline{f}_\pm^*$ and $\overline{g}_\pm = -g_\pm^*$.
2. T invariance allows choosing all four form factors to be real.
3. The $\Delta Q = \Delta S$ rule forbids the decays $\overline{K^0} \to \pi^- l^+ v_l$ and $K^0 \to \pi^+ l^- v$, i.e. $g_\pm = \overline{g}_\pm = 0$; the degree of violation of this rule is measured by the parameter $x = g_+/f_+$.
4. The $\Delta I = \tfrac{1}{2}$ rule relates the matrix elements of the two members of an isospin doublet by a Clebsch-Gordan coefficient:

$$2^{1/2} M(K^+ \to \pi^0 l^+ v) = M(K^0 \to \pi^- l^+ v).$$

From the Gell-Mann-Nishijima relation for hadrons,

$$Q = (B+S)/2+I_3,$$

where Q is the charge, B the baryon number, S the strangeness, and I_3 the third isospin component, one concludes that for transitions between mesons ($B = 0$),

$$\Delta Q = \Delta S/2 + \Delta I_3.$$

If $\Delta I = \frac{1}{2}$, then for a transition with $\Delta S = 1$, $\Delta Q = \Delta S$ also holds; in particular, this means $g_\pm = \overline{g}_\pm = 0$.

The decay amplitudes for $|K_S\rangle$ and $|K_L\rangle$ can be written, if we take for simplicity the electron decay, where the form factors f_- and g_- vanish:

$$\langle \pi^- e^+ v \,|\, T \,|\, K_S \rangle \propto (1+\varepsilon)\langle \pi^- e^+ v \,|\, T \,|\, K^0 \rangle + (1-\varepsilon)\langle \pi^- e^+ v \,|\, T \,|\, \overline{K^0} \rangle$$
$$= (1+\varepsilon)f + (1-\varepsilon)g,$$
$$\langle \pi^- e^+ v \,|\, T \,|\, K_L \rangle \propto (1+\varepsilon)f - (1-\varepsilon)g,$$
$$-\langle \pi^+ e^- \bar{v} \,|\, T \,|\, K_S \rangle \propto (1+\varepsilon)g^* + (1-\varepsilon)f^*,$$
$$-\langle \pi^+ e^- \bar{v} \,|\, T \,|\, K_L \rangle \propto (1+\varepsilon)g^* - (1-\varepsilon)f^*.$$

Taking an arbitrary coherent mixture of K_L and $K_S |\psi(\tau)\rangle$ as above, its decay amplitudes

$$A^\pm(\tau) = \langle \pi^\mp e^\pm v \,|\, T \,|\, \psi(\tau) \rangle$$

are given by

$$A^+(\tau) \propto a_S \exp(-iM_S\tau)[(1+\varepsilon)+(1-\varepsilon)x] + a_L \exp(-iM_L\tau) \cdot [(1+\varepsilon)-(1-\varepsilon)x],$$
$$-A^-(\tau) \propto a_S \exp(-iM_S\tau)[(1+\varepsilon)x^*+(1-\varepsilon)] + a_L \exp(-iM_L\tau)$$
$$\times [(1+\varepsilon)x^*-(1-\varepsilon)].$$

The corresponding decay rates for the two charge states are $N^\pm(\tau) = |A^\pm(\tau)|^2$, and if we call the ratio of the initial K_S and K_L components $R = a_S/a_L$, we obtain for the difference of charge states

$$N^+ - N^- \approx 4(1-|x|^2)\{\text{Re}\,\varepsilon[\exp(-\Gamma_L\tau)+|R|^2\exp(-\Gamma_S\tau)]$$
$$+ |R|\exp(-\overline{\Gamma}\tau)(1+|\varepsilon|^2)\cos(\Delta m\tau + \phi_R)\}$$

and

$$N^+ + N^- \approx 2\exp(-\Gamma_L\tau)[(1+|\varepsilon|^2)(1+|x|^2)-2\,\text{Re}\,x(1-|\varepsilon|^2)-4\,\text{Im}\,x\,\text{Im}\,\varepsilon]$$
$$+ 2|R|^2\exp(-\Gamma_S\tau)[(1+|\varepsilon|^2)(1+|x|^2)+2\,\text{Re}\,x(1-|\varepsilon|^2)$$
$$+ 4\,\text{Im}\,x\,\text{Im}\,\varepsilon]+8|R|\exp(-\overline{\Gamma}\tau)\{\text{Re}\,\varepsilon(1+|x|^2)\cos(\Delta m\tau + \phi_R)$$
$$- [2\,\text{Im}\,\varepsilon\,\text{Re}\,x-(1-|\varepsilon|^2)\,\text{Im}\,x]\sin(\Delta m\tau + \phi_R)\},$$

where $\overline{\Gamma} = \frac{1}{2}(\Gamma_S+\Gamma_L)$.

In the limit where $\text{Re}\,\varepsilon \ll 1$, $\text{Im}\,x \ll 1$, and $\tau > 3/\Gamma_S$, the expression for the charge asymmetry $\delta(\tau) = (N^+ - N^-)/(N^+ + N^-)$ can be simplified:

$$\delta(\tau) = 2\frac{(1-|x|^2)}{|1-x|^2}\{\text{Re}\,\varepsilon+|R|\exp[-(\Gamma_S-\Gamma_L/2)\tau]\cos(\Delta m\tau + \phi_R)\}. \qquad 21.$$

For an initially pure K_L beam ($R = 0$), the asymmetry is independent of decay time

$$\delta_L = 2\,\text{Re}\,\varepsilon[(1-|x|^2)/|1-x|^2],$$

while for an initial incoherent mixture of $K^0(N_K)$ and $\overline{K^0}(N_{\bar{K}})$ the quantity $|R|$ has to be replaced by $(N_K - N_{\bar{K}})/(N_K + N_{\bar{K}})$, i.e. by the same dilution factor as in 2π interference in a short-lived beam.

For the coherent mixture created by a regenerator, R is given by the regeneration amplitude ρ, and ϕ_R by the regeneration phase ϕ_ρ.

3   DETECTION OF K^0 DECAYS

The main decay modes originating from K^0's in a neutral beam and their respective branching ratios are (76)

$$
\begin{array}{lll}
K_L \to \pi^\pm e^\mp \nu & (39.0 \pm 0.6)\% & K_{e3}, \\
K_L \to \pi^\pm \mu^\mp \nu & (27.5 \pm 0.5)\% & K_{\mu 3}, \\
K_L \to \pi^+ \pi^- \pi^0 & (11.9 \pm 0.4)\% & K_{\pi 3}, \\
K_L \to \pi^0 \pi^0 \pi^0 & (21.3 \pm 0.6)\% & K_{\pi 3}, \\
K_S \to \pi^+ \pi^- & (68.77 \pm 0.26)\% & K_{\pi 2}, \\
K_S \to \pi^0 \pi^0 & (31.23 \pm 0.26)\% & K_{\pi 2}.
\end{array}
$$

The experimental problem is to detect the rare *CP*-violating decay modes $K_L \to \pi^+ \pi^-$ and $K_L \to \pi^0 \pi^0$ with branching ratios of 10^{-3} in this overwhelming background of other decays and to measure their decay rate and, by interference, their phase relation to *CP*-conserving decay amplitudes. In addition, the *CP* impurity in the K_L state can be obtained by measuring the charge asymmetry in the semileptonic decay modes.

3.1   *Charged Decay Modes*

The two charged decay products in $\pi^+ \pi^-$ and semileptonic decays were recorded in magnetic spectrometers consisting of a wide-aperture magnet and at least three layers of position-measuring detectors. The measurement of track position was done by optical spark chambers (14, 65–67), then by wire spark chambers with magnetostrictive readout (68), by counters only (17, 18), and by multiproportional wire chambers (71, 77, 81). As an example, Figure 1 shows the spectrometer of reference 81.

Here the magnet has an aperture of 240×60 cm^2. This permits observation of K decays with K momentum from 5 to 15 GeV/c over a 9-m-long decay path upstream of the magnet with geometrical acceptances around 25%, depending on the decay mode. Most important for the measurement of interference phenomena between K_S and K_L decays extending over at least 20 K_S mean lifetimes in the kaon rest frame is the fact that this geometric acceptance varies smoothly and by not more than 50% over 10K_S mean lifetimes. This is because the acceptance has

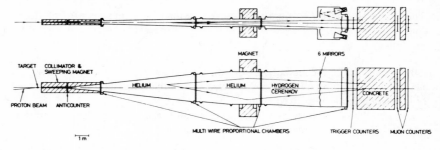

Figure 1   Magnetic spectrometer (81) for detecting two charged decay products of neutral
K mesons, side view (*upper*) and top view (*lower*). The spectrometer, consisting of a magnet
and 3×2 planes of proportional chambers, measures the momenta of the two charged
particles; the Cerenkov counter identifies electrons from K_{e3} decay and the muon hodoscopes
detect muons from $K_{\mu 3}$ decay.

to be computed by Monte Carlo methods, which is one of the ultimate limitations
of the experiment.

The vector momenta $\mathbf{p}_i$ ($i = 1, 2$) of the charged decay products are measured
by three multiwire proportional chambers, each equipped with a horizontal and a
vertical signal wire plane. The wire spacing is 2 mm, corresponding to a measurement
error of ± 0.7 mm; the wire diameter is 20 μm. The chambers can be used with a
time resolution of 40 nsec, thus permitting operation in regions of charged-particle
flux ten times higher than the ones sustainable by spark chambers with recovery
times of ~ 1 μsec. Furthermore, since the chambers are dc-operated, they do not
require additional scintillation counters, permitting a reduction of matter in the path
of detected particles down to 0.3 g cm^{-2} for the total apparatus. The readout
time of an electronically accepted event is 10 μsec, and more than 1000 events can
be recorded during one machine burst of 350 msec.

From the calculated vector momenta $\mathbf{p}_i$ one obtains the energies of the particles,
assuming their rest mass to be m_π,

$$E_i = (\mathbf{p}_i + m_\pi^2),$$

the invariant mass of the pair to be

$$m_{\pi\pi} = [(E_1 + E_2)^2 - (\mathbf{p}_1 + \mathbf{p}_2)^2]^{1/2},$$

and the kaon momentum to be $\mathbf{p}_K = \mathbf{p}_1 + \mathbf{p}_2$. The lifetime of the kaon from the
target (position along the beam Z_T) to the decay vertex (Z_V) in the kaon rest
system is given by $\tau = (Z_V - Z_T)m_K/(cp_Z)$, where m_K is the kaon mass, c the light
velocity, and p_Z the component of $\mathbf{p}_K$ along the beam line.

Two sets of information can be used to separate 2π, and leptonic decays: first,
the invariant mass $m_{\pi\pi}$ is required to be equal to m_K within the experimental
resolution for 2π decay (Figure 2), while leptonic decays show a broad $m_{\pi\pi}$ distribution, due to the wrong assignment of a pion mass to the lepton and to the
missing neutrino energy. Second, all experimenters use threshold gas Cerenkov

counters in order to identify electrons by their high velocity, while Cerenkov emission of pions is avoided by a suitable choice of refractive index n, such that the threshold momentum is $p\text{th} = m_\pi/\sqrt{2(n-1)} = 8.4\,\text{GeV}/c$ for hydrogen at atmospheric pressure. For the identification of muons, on the other hand, one uses their penetration through several (~ 8) interaction lengths of material in order to distinguish them from pions interacting in this absorber. For an absorber of 900 g/cm^2 of light concrete, the minimum momentum of a penetrating muon is 1.5 GeV/c, and the penetration probability of a pion through a hadronic cascade is 0.6% at 4 GeV/c momentum. In addition to this kind of misidentification of a pion, there is pion decay in flight.

In addition to the separation of decay modes, one has to know, in general, whether the K_S or K_L from which the decay products originate has undergone scattering on its way from its production to the decay point. In the case of a short-lived beam produced by protons interacting in a target near to the detector, this can be done by calculating the distance of the intercept of the reprojected kaon momentum p_K in the target plane from the target center ρ_T. Unscattered events cluster around $\rho_T = 0$, as shown in Figure 2. In the case of a long-lived beam, one uses the component of p_K transverse to the beam, ρ_T, or the angle θ between the kaon direction p_K and beam direction, in order to separate transmitted and coherently regenerated ($\theta = 0 = p_T$) kaons from events due to kaons having undergone scattering, or diffractive or inelastic regeneration.

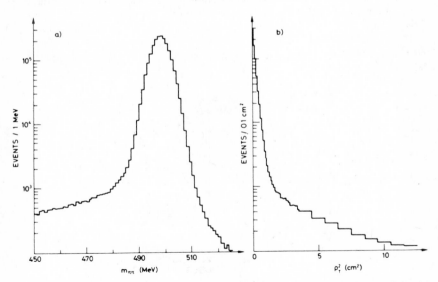

Figure 2 (*a*) Invariant $\pi^+\pi^-$ mass distribution of two-prong events recorded in the apparatus of Figure 1 after removing most of the three-body leptonic K_L decays identified through the Cerenkov or muon counters. (*b*) Distribution in ρ_T^2, the squared distance of the reprojected K momentum in the target plane from the target center (77).

This analysis is straightforward for 2π decays, but substantially more complicated for semileptonic decays, since the neutrino momentum escapes detection. Not only is it impossible for semileptonic events to separate coherently produced ones from incoherently produced ones, but there is also a twofold ambiguity in the kaon momentum. These difficulties can be solved by comparing the experimental distributions in momentum on a statistical basis with the corresponding distributions calculated by Monte Carlo methods from hypotheses on the "real" distributions.

3.2 Neutral Decay Modes

The detection of the neutral decay mode $K_L \to \pi^0\pi^0 \to 4\gamma$ is complicated by the presence of the decay $K_L \to 3\pi^0 \to 6\gamma$ with a 21% branching ratio that can simulate 4γ events if two γ rays are missed by the detector. Very specific kinematic features of the $2\pi^0$ decay must therefore be used in order to obtain a clean $K_L \to 2\pi^0$ signal (82–86).

In the apparatus (Figure 3) of the Princeton group (83), this is achieved by measuring with great precision the energy and direction of one γ ray of the four γ's from $2\pi^0$ decay. This photon can have a transverse momentum relative to the K_L direction of up to 240 MeV/c, since its parent π^0 has a momentum of exactly 209 MeV/c in the K_L rest system. The γ's from the background reaction $K_L \to 3\pi^0 \to 6\gamma$, however, are restricted to a transverse momentum below 167 MeV/c. Consequently, the apparatus of the Princeton group (83) included a pair spectrometer for the precise measurement of momentum, direction, and transverse momentum of one γ ray converted in a $\frac{1}{10}$ radiation length converter with standard deviations of 3%, 3 mrad, and 5 MeV/c respectively. In the spectrometer, positions of the e^+ and e^- trajectories bent by a $18'' \times 72''$ aperture magnet were recorded by five optical spark chambers. The remaining three γ rays were converted in a large optical spark chamber containing six radiation lengths of lead.

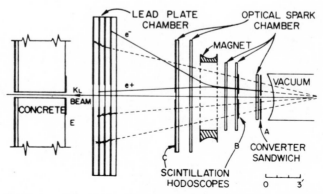

Figure 3 Apparatus of the Princeton group (83) for the detection of $K_L \to 2\pi^0$ decays. The rare $K_L \to 2\pi^0 \to 4\gamma$ decays are selected by requiring one γ ray converted in the converter A to have a transverse momentum $p_T > 170$ MeV/c, as measured in the magnetic pair spectrometer. The other three γ rays are detected in the lead-plate chamber.

Using the energy and direction of the spectrometer γ ray, the conversion points of the three other γ rays, and the decay vertex of the K_L computed from the spectrometer γ-ray direction and the beam line, the decay can be reconstructed with two constraints.

Alternatively, the design of the detector (Figure 4) of the Aachen-CERN-Torino (82) group has been guided by the idea of measuring accurately the energies of all four γ rays and the direction of at least two γ's of the four from $K_L \to \pi^0\pi^0 \to 4\gamma$ decay. The upstream part of the detector is the "direction-measuring" region totalling four spark chamber modules of four wire planes each. Every module is preceded by a lead converter foil of $\frac{1}{10}$ radiation length thickness, and the directions of the wires in the four planes are at $0°$, $45°$, $90°$, and $135°$ relative to the vertical in order to avoid ambiguities in the track reconstruction from sparks. The spatial resolution of the spark chambers is equal to the scattering error in a lead converter thickness of $\frac{1}{10}$ radiation length for a γ energy of 1 GeV, i.e. the average energy of γ's from $K_L \to 2\pi^0$ decay in their beam. In the downstream part of the detector the remaining γ rays are converted either in one of the two lead plates of $\frac{1}{4}$ radiation length thickness in front of two further spark chamber modules or in the array of 61 hexagonal lead-glass prisms of 13 radiation lengths thickness. These total absorption counters of lead-glass permit an energy measurement of the γ rays with a resolution (standard deviation) of $\sigma = 0.033/\sqrt{E\gamma}$, where $E\gamma$ is measured in GeV. A spark chamber trigger requires at least two channels in each of the vertical and horizontal counter arrays behind the fourth module; the energy deposited in at least three of the six outer sectors (see Figure 4) of the lead-glass counter array must exceed 450 MeV. The tracks of the e^+ e^- pairs from the γ converted in the first four modules are used to determine the decay point, and this position is used together with the conversion points of the other γ rays to determine their directions. Gamma rays are then paired such that the invariant mass of the pair is closest to the π^0 mass. Using the π^0 mass as a constraint, the 4γ invariant mass m, the angle θ

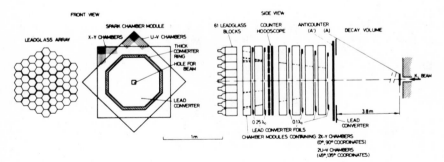

Figure 4 Apparatus of the Aachen-CERN-Torino group (82) for the measurement of the $K_L \to 2\pi^0$ decay rate. This detector consists of a direction-measuring part (four chamber modules upstream) where at least two γ rays are required to convert, and an energy-measuring part (two-chamber modules and lead-glass wall), where the energies of all four γ rays are recorded.

between the total visible momentum and the beam line, and p_t, the largest of the transverse momenta of the γ rays, can be used as constraints in the kinematical selection of $2\pi^0$ decays.

4   EXPERIMENTS ON CP VIOLATION IN K^0 DECAYS

CP violation in K^0 decays has been found in the decays $K_L \to \pi^+\pi^-$ (14) and $K_L \to \pi^0\pi^0$ (15, 16) and in the charge asymmetry δ_L of semileptonic K_{e3} (17) and $K_{\mu3}$ (18) decays. The measurable quantities are (Section 2.2) $|\eta_{+-}|$, ϕ_{+-}, $|\eta_{00}|$, ϕ_{00}, and δ_L. Since the interference of K_S and K_L decay amplitudes (Section 2.5) is used to determine the phase differences ϕ_{+-} and ϕ_{00}, the precise knowledge of two auxiliary quantities is required, the K_S lifetime τ_S and the $K_L - K_S$ mass difference $\Delta m = m_L - m_S$.

4.1 K_S Lifetime

There are three recent measurements with errors smaller than the 1971 world average (87). These are high-statistics measurements, one in a hydrogen bubble chamber by a CERN group (88) and the other two (77, 71) using proportional wire spectrometers to study the interference of $K_S \to \pi^+\pi^-$ with $K_L \to \pi^+\pi^-$ in a short-lived beam or behind a regenerator, respectively. The results are given in Table 1. The average of these results is $(4 \pm 1)\%$ higher than the former average in 1971 (87). The corrections in the hydrogen bubble chamber experiment due to the interference with $K_L \to \pi^+\pi^-$ and $K_L - K_S$ regeneration in hydrogen amount to $+1\%$. Therefore it is unlikely that ignoring these effects in the older experiments can be responsible for all of the discrepancy between the old and new results. One could conclude that the large number of events in the new experiments and the large acceptance allow a better study of systematic errors than before. The average for τ in Table 1 corresponds to $\Gamma_S = (1.120 \pm 0.003) \times 10^{10}$ sec^{-1}.

4.2 K_L-K_S Mass Difference

Measurements of the K_L-K_S mass difference $\Delta m = m_L - m_S$ make use of the different propagation of K_L and K_S waves in time (see Section 2.5). If τ is the eigentime in

Table 1   New measurements of the K_S lifetime

Group (Ref.)	Year	Technique	Events	Results
CERN (88)	1972	HBC	5×10^4	$(0.8958 \pm 0.0045) \times 10^{-10}$ sec
CERN-Heidelberg (77)	1974	PWC	6×10^6	$(0.8936 \pm 0.0048) \times 10^{-10}$ sec
Columbia (71)	1975	PWC	2×10^6	$(0.8913 \pm 0.0032) \times 10^{-10}$ sec

average (1975) = $(0.8930 \pm 0.0023) \times 10^{-10}$ sec
former average (1971) = $(0.862 \pm 0.006) \times 10^{-10}$ sec

ratio $\dfrac{\tau \text{ (new)}}{\tau \text{ (old)}} = 1.04 \pm 0.01$

the K rest frame, then K_L propagates with $\exp(-im_L\tau)$ and K_S with $\exp(-im_S\tau)$. The interference term of both waves is then proportional to $\exp(i\Delta m\tau)$, i.e. to $\cos(\Delta m\tau)$. All experiments measure such an interference term to determine Δm.

The traditional method is the varying gap method first used by the Princeton group (89). A K_L beam traverses two slabs of matter, thus regenerating coherent forward K_S amplitudes ρ_1 and ρ_2, and the intensity of K_S decays behind the second slab is measured as a function of the distance between the slabs; this intensity is given in Section 2.5.1. A severe problem in this experiment comparing $K \to 2\pi$ decay intensities for different distances τ_G is the monitoring of the K_L beam flux. This monitoring was done (90, 91) by recording K_{e3} decays or by recording neutrons in the beam or other charged particles from the proton target. However, corrections are needed because dead-time effects in the chambers vary with varying gap distance and possibly with time. The Aachen-CERN-Torino group (92) has avoided these problems by monitoring the K_L flux by recording on $K_S \to \pi^+\pi^-$ decays, i.e. by reserving a part of the K_L beam cross section for measuring the K_L flux multiplied with $|\rho_2|^2$ and another one for measuring $|\rho_1|^2 \exp(-\Gamma_S\tau_G)$. This fixed triple-gap method has given $\Delta m = (0.542 \pm 0.006) \times 10^{10}$ sec^{-1} and this result is independent of Γ_S.

A variant of this method has been used by the CERN-Heidelberg-Dortmund group (79). A quarter of the beam cross section is sacrificed for monitoring coherent regenerated $K_S \to \pi^+\pi^-$ decays. The intensities in two positions relative to the respective monitor events are measured. The ratio of exposure times in the two positions is $3:1$ to minimize the statistical error.

The fixed distance in space transforms (via the broad K_L momentum spectrum) to a distribution of gap flight times in the K^0 rest system; therefore the distribution $I_{2\pi}$ as a function of K_S momentum shows the typical interference pattern with destructive interference (see Figure 5). The result of this experiment is $\Delta m = (0.534 \pm 0.003) \times 10^{10}$ sec^{-1} for $\Gamma_S = 1.120 \times 10^{10}$ sec. The error is reduced to ± 0.00255 if the result is used in connection with the vacuum interference experiment done in the same apparatus (77).

A second measurement using a quite different method has been reported by the same group (80). Here they measure the charge asymmetry in K_{e3} and $K_{\mu3}$ decays at short distances from a K^0 production target. This charge asymmetry (see Section 2.5.2) is given by

$$\delta(\tau) = 2\,\frac{1-|x|^2}{|1-x|^2}\left[\exp(-\bar{\Gamma}\tau)\cdot\frac{N_K-N_{\bar{K}}}{N_K+N_{\bar{K}}}\cdot\cos(\Delta m\tau)+\mathrm{Re}\,\varepsilon\right].$$

Of the total data on leptonic decays, 6×10^6 K_{e3} decays and 2×10^6 $K_{\mu3}$ at times $\tau < 12.75 \times 10^{-10}$ sec were relevant to the measurement of the mass difference, while the rest of the data was used to determine the long-term asymmetry due to CP violation (see Section 4.7). The dilution factor was obtained from the size of the interference term in the vacuum interference experiment (77) (see Section 4.4). The results of the fit to the data displayed in Figures 6 and 7, including a $(0.45 \pm 0.10)\%$ shift due to radiative corrections in K_{e3}, are given in Table 2. These

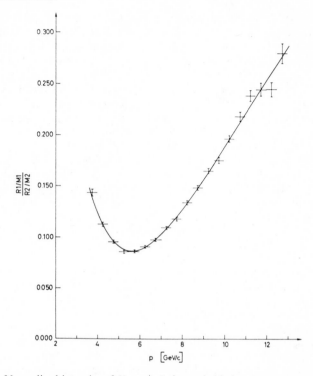

Figure 5 Normalized intensity of $K \to \pi^+\pi^-$ decays behind a two-regenerator setup as a function of the kaon momentum showing destructive interference between the two coherent K_S waves regenerated in the first and second slabs of copper (79).

results agree with those from the fixed-gap method, as shown in Table 2. The average of the results in Table 2 is $\Delta m = (0.536 \pm 0.002) \times 10^{10}$ sec^{-1}, while the CERN-Heidelberg data alone yields $\Delta m = (0.534 \pm 0.002) \times 10^{10}$ sec^{-1} if used in connection with data from the same apparatus, i.e. excluding the systematic error in the momentum scale.

4.3   *Rate of* $K_L \to \pi^+\pi^-$ *Decay*

This rate has been measured in two different ways. The first is by recording the number of $K_L \to \pi^+\pi^-$ decays in a certain decay volume together with leptonic decays $K_L \to \pi e v$ in the same volume. A Monte Carlo calculation of the relative detection efficiency of the apparatus for $K_{\pi 2}(\varepsilon_{\pi 2})$ and $K_{e3}(\varepsilon_{e3})$ decays is of crucial importance here; in particular, one needs to know the matrix element of K_{e3} decay, and the influence of this uncertainty will be more pronounced if the apparatus selects a small fraction of the Dalitz plot. Essentially all older measurements have been done in this way, and $|\eta_{+-}|$ is obtained by the relation

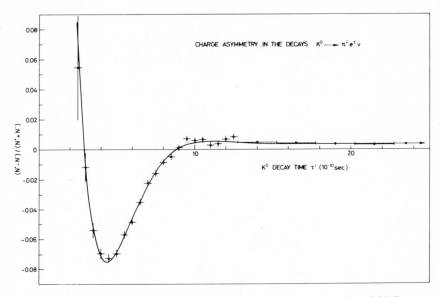

Figure 6   Eigentime dependence of charge asymmetry in K_{e3} decays from initially pure strangeness states (80) showing interference of K_S and K_L decay amplitudes.

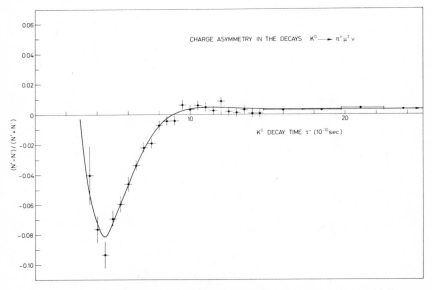

Figure 7   Eigentime dependence of charge asymmetry in $K_{\mu3}$ decays from initially pure strangeness states (80) showing interference of K_S and K_L decay amplitudes.

Table 2 Recent measurements of the $K_L - K_S$ mass difference

Group (Ref.)	Year	Method	Monitoring	Events	Results for $\Gamma_S = 1.120$
Aachen-CERN-Torino (92)	1970	fixed triple gap	$K_S \rightarrow 2\pi$ from fixed regenerator	3×10^4	0.542 ± 0.006
Chicago-Illinois (90)	1970	varying gap	neutron monitor and scatter monitor corrected for dead time	6×10^4	0.552 ± 0.006
Princeton (91)	1971	varying gap	K_{e3} flux corrected for scattering	2×10^4	0.533 ± 0.007
CERN-Heidelberg-Dortmund (79)	1974	fixed gap	$K_S \rightarrow 2\pi$ from fixed regenerator	3×10^5 ($+7 \times 10^5$ monitor)	0.534 ± 0.003
CERN-Heidelberg (80)	1974	K_{e3} charge asymmetry time distribution	not needed	6×10^6	0.5341 ± 0.0043
CERN-Heidelberg (80)	1974	$K_{\mu 3}$ charge asymmetry time distribution	not needed	2×10^6	0.529 ± 0.010

$$|\eta_{+-}|^2 = \frac{N(K_L \to \pi^+\pi^-) \cdot \varepsilon_{e3}}{N(K_L \to \pi ev) \cdot \varepsilon_{\pi 2}} \cdot \frac{\Gamma(K_L \to \pi ev)}{\Gamma(K_L \to \text{all})} \cdot \frac{\Gamma(K_L \to \text{all})}{\Gamma(K_S \to \text{all})} \cdot \frac{\Gamma_S}{\Gamma(K_S \to \pi^+\pi^-)}.$$

The second method can only be used in a short-lived K^0 beam. Strangeness eigenstates K^0 (and $\bar{K}^0$) are produced by an external proton beam hitting a nuclear target. The K_S and K_L components of the K^0 state propagate in time differently, but both decay into π^+ and π^-. Their two decay amplitudes interfere in the way described in Section 2.5.1. At decay times comparable to the K_S lifetime τ_S the resulting intensity of $K \to \pi^+ + \pi^-$ decays is mainly due to $K_S \to \pi^+\pi^-$ decay, and at late times ($\tau > 15\tau_S$) it is due to $K_L \to \pi^+\pi^-$ decay. The observation of such a time distribution will give therefore $|\eta_{+-}|^2$ as the ratio of the $\pi^+\pi^-$ rate at long times compared to the one extrapolated back to $\tau = 0$. Here also a Monte Carlo calculation is needed that allows comparison of the detection efficiency for the same two-body decay $K \to 2\pi$ at different points of the decay volume.

Four recent experiments have been reported:

1. The experiment on the vacuum interference by the CERN-Heidelberg group (77) has given: $|\eta_{+-}| = (2.30 \pm 0.035) \times 10^{-3}$ with $\Gamma_S = (1.119 \pm 0.006) \times 10^{10}$ sec^{-1} from the fit.
2. The same group reports (77) an independent measurement using K_{e3} decays as normalization with the result $|\eta_{+-}| = (2.30 \pm 0.06) \times 10^{-3}$ using $\Gamma_S = 1.119 \times 10^{10}$ sec^{-1}.
3. The Colorado-SLAC-Santa Cruz group (93) reported on a measurement in a K_L beam by normalizing in two ways:
   (a) comparing the $K_L \to \pi^+\pi^-$ rate to the $K_L \to \pi ev$ rate and
   (b) comparing it to $K_L \to \pi^+\pi^-\pi^0$ events.
   The results of both normalizations agree well and give together $|\eta_{+-}| = (2.23 \pm 0.05) \times 10^{-3}$ for $\tau_S = 0.862 \times 10^{10}$ sec. Using $\Gamma_S = 1.120 \times 10^{10}$ sec^{-1} this result is $|\eta_{+-}| = (2.27 \pm 0.05) \times 10^{-3}$.
4. A Dubna group (94) finds $|\eta_{+-}| = (2.05 \pm 0.11) \times 10^{-3}$ using K_{e3} as normalization.

Table 3 gives the results and the detection efficiencies for the decays used for normalization in the different experiments. The older experiments in general were designed to have a small detection efficiency for K_{e3} decays in order to get a small background K_{e3}-to-signal ($K_{\pi 2}$) ratio. The discrepancy between the pre-1967 and the post-1973 experiments cannot be negated. It is therefore not possible to take an average of the old and new data, but one must choose between them. The facts that the vacuum interference experiment depends less critically on Monte Carlo calculations and that the SLAC experiment has two independent normalizations that give consistent results favor the new group of results. The agreement between the CERN-Heidelberg vacuum interference experiment and the SLAC result using, along with other normalizations, the conventional normalization by K_{e3} also removes the possibility that there might be a deep physical reason for getting a different result from the two methods. The average from the new group of four experiments is

$$|\eta_{+-}| = (2.279 \pm 0.025) \times 10^{-3}.$$

Table 3 Measurements of $|\eta_{+-}|$

Group (Ref.)	Year	Decay used for flux normalization	Detection efficiency for normalizing decay ε_N (%)	$\dfrac{\varepsilon(K_L \to \pi^+\pi^-)}{\varepsilon_N}$	Result $|\eta_{+-}| \times 10^3$ for $\Gamma_S = 1.120 \times 10^{10}$ sec^{-1}
Princeton (14)	1964	$K_L \to$ all charged		4.4	1.96±0.20
Rutherford (95)	1965	$K_L \to \pi\mu\nu + \pi e\nu$	0.01	14	2.04±0.17
Saclay (96)	1966	$K_L \to$ leptonic	not given		1.88±0.20
CERN (97)	1966	$K_L \to \pi\mu\nu$		30	1.95±0.04
Princeton (98)	1967	$K_L \to \pi e\nu + \pi\mu\nu$	0.03	13	1.93±0.06
SLAC (93)	1973	a) $K_L \to \pi^+\pi^-\pi^0$	37	0.4	2.27±0.05
		b) $K_L \to$ all charged	22	0.7	
CERN-Heidelberg (77)	1974	$K_L \to \pi e\nu$	~2	2	2.30±0.06
CERN-Heidelberg (77)	1974	$K_S \to \pi^+\pi^-$	4	1	2.30±0.035
Dubna (94)	1974	$K_L \to \pi e\nu$			2.05±0.11

4.4   *Phase ϕ_{+-}*

The relative phase between the two amplitudes $a(K_L \rightarrow \pi^+\pi^-)$ and $a(K_S \rightarrow \pi^+\pi^-)$ has been measured by two distinct methods.

The first consists of measuring the interference of the $K_L \rightarrow \pi^+\pi^-$ amplitude

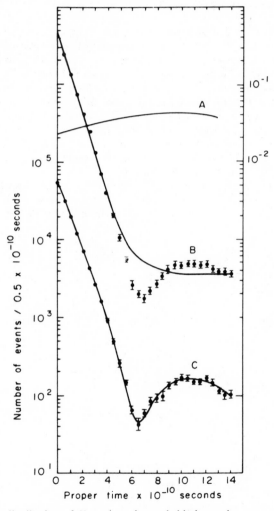

Figure 8   Time distribution of $K \rightarrow \pi^+\pi^-$ decays behind a carbon regenerator (71). (*a*) Detection efficiency for kaon momentum interval $5 < p_K < 6$ GeV/c. (*b*) Efficiency-corrected data summed over p_K; curve shows fit without interference. (*c*) Data for $5 < p_K < 6$ GeV/c with fit including interference.

with the coherently regenerated $K_S \to \pi^+\pi^-$ amplitude behind a slab of material (the regenerator). The experiments requires (a) the measurement of the $\pi^+\pi^-$ intensity as a function of K^0 eigentime behind the regenerator, which is given in Section 2.5.1, and (b) the measurement of the charge asymmetry δ in leptonic K^0 decays as a function of K eigentime τ behind the regenerator, given in Section 2.5.2. Experiment a gives the interference phase $\phi_{+-} - \phi_\rho$ and experiment b the phase of the regeneration amplitude ϕ_ρ.

The latest and most precise experiment in this series is the one of the Columbia group at Brookhaven (71, 74). Using a proportional wire spectrometer to detect $\pi^+\pi^-$ and leptonic decays at the same time, they obtained the time distribution of $\pi^+\pi^-$ events behind a carbon regenerator shown in Figure 8. The figure shows also the corresponding time distribution of events from the K^0 momentum interval $5 < p < 6$ GeV/c. There is a deep minimum from destructive interference at $\tau \sim 6.5 \times 10^{-10}$. The charge asymmetry of K_{e3} and $K_{\mu3}$ decays as a function of K^0 eigentime behind the regenerator is shown in Figure 9. The regeneration phase ϕ_ρ extracted from this asymmetry is used together with the combined phase $\phi_\rho - \phi_{+-}$ from the 2π interference to obtain the value $\phi_{+-} = (45.5 \pm 2.8)° + 120°$ $(\Delta m - 0.5348 \times 10^{10} \text{ sec}^{-1})/\Delta m$.

The other method for measuring ϕ_{+-} is the vacuum interference method mentioned above (Section 2.5.1), where one observes the $K \to \pi^+\pi^-$ distribution from an initially pure strangeness state. The information on ϕ_{+-} is contained

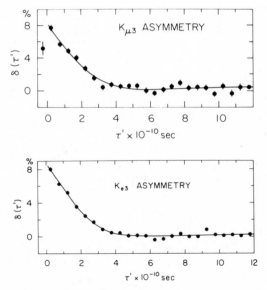

Figure 9 Charge asymmetry of semileptonic decays behind a carbon regenerator (74) for $K_{\mu3}$ (*upper*) and K_{e3} (*lower*) displaying interference of regenerated K_S wave with transmitted K_L wave.

in the interference term proportional to $\cos(\Delta m - \phi_{+-})$, and the time where the two interfering amplitudes are equal is $\tau \sim 12\tau_S$ such that the correlation of ϕ_{+-} with Δm is rather strong. This method requires the high precision of the experiments on the mass difference described in Section 4.2. An error of $\pm 0.002 \times 10^{10}$ sec^{-1} in Δm induces an uncertainty of $1.2°$ in ϕ_{+-}.

There are three experiments of this type (77, 99, 100). The analysis of the latest and most precise of those has been done by the CERN-Heidelberg group (77). The apparatus, situated in a 75-mrad short neutral beam, is described in Section 3.1. The time distribution of 6×10^6 $K_{S,L} \rightarrow \pi^+\pi^-$ decays is shown in Figure 10; curve a, together with the fitted distribution, is given in Section 2.5.1. The result of this fit is

$$\phi_{+-} = (49.4 \pm 1.0)° + 305°(\Delta m - 0.540 \times 10^{10} \text{ sec}^{-1})/\Delta m,$$

$$|\eta_{+-}| = (2.30 \pm 0.035) \times 10^{-3},$$

$$\Gamma_S = (1.119 \pm 0.006) \times 10^{10} \text{ sec}^{-1},$$

$$\chi^2 = 421 \text{ for } 444 \text{ degrees of freedom.}$$

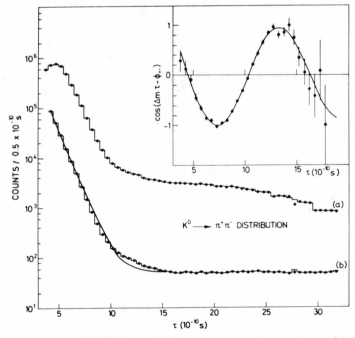

Figure 10 Time distribution of $K \rightarrow \pi^+\pi^-$ events from a coherent mixture of K_L and K_S produced in pure strangeness states (77). (*a*) Events (*histogram*) and fitted distribution (*dots*). (*b*) Events corrected for detection efficiency (*histogram*), fitted distribution with interference term (*dots*), and without interference term (*curve*). Insert: interference term as extracted from data (*dots*) and fitted term (*line*).

Table 4 Measurements of ϕ_{+-}

REGENERATION METHOD

	Interference phase measurement			Regeneration phase measurement			
Group (Ref.)	$\phi_{+-} - \phi_r$	Regenerator material	Group (Ref.)	ϕ_r (degrees)	ϕ_{+-} (degrees)	Δm used
CERN (67)	89.7±6.3	C	Opt. Model	−37 ±10	52.7±11.8	WA
CERN (66)	86.7±8	Cu	Columbia (72)	−51.9± 7.8	33.2±10.0	
			Princeton (73)	−55.3± 8.3		
CERN-Aachen-Torino (68)	88.1±5.5	Cu	Columbia (72)	−51.9± 7.8	36.2± 8.5	WA
			Princeton (73)	−55.3± 8.3		
ITEP (69)		Cu	ITEP	−43 ± 4	37 ±12	WA
Princeton (70)		Cu	Princeton		36.2± 6.1	WA
Columbia (71)		Cu	Columbia (74)		45.8± 2.8	WA

Average of regeneration measurements 43.3 ± 2.3

VACUUM INTERFERENCE METHOD

Group (Ref.)	ϕ_{+-} (degrees)	Δm	ϕ_{+-} (error from Δm not included)
CERN (99)	$(41 \pm 12) + 252(\Delta m - 0.526 \times 10^{10}\ \text{sec}^{-1})/\Delta m$	WA	45.8±12
Chicago-Illinois (100)	$(42.4 \pm 4) + 310(\Delta m - 0.538 \times 10^{10}\ \text{sec}^{-1})/\Delta m$	WA	41.3± 4
CERN-Heidelberg (77)	$(49.4 \pm 1) + 305(\Delta m - 0.540 \times 10^{10}\ \text{sec}^{-1})/\Delta m$	CERN-HD	45.9± 1

Average (error from Δm not included) 45.6 ± 1

Average of vacuum interference measurements 45.6 ± 1.6

Average 44.9 ± 1.3

Using the three mass-difference measurements (79, 80) obtained in the same apparatus (whereby systematic scale errors cancel) they obtain $\phi_{+-} = (45.9 \pm 1.6)$.

Using all the data listed in Table 4, the average comes out at $\phi_{+-} = (44.9 \pm 1.3)°$.

4.5   Rate of $K_L \rightarrow \pi^0\pi^0$ Decay

Early experiments including the first observations of this decay (16, 17) suffered from the difficulty of obtaining a reliable K_L flux measurement, of knowing the acceptance of the apparatus for this decay, and of extracting a background free signal. A compilation (87) gives an average for

$$|\eta_{00}| = (2.26 \pm 0.20) \times 10^{-3}.$$

The two recent precise measurements (82, 83) use the two apparatus described in Section 3.2. They both measure the ratio $|\eta_{00}/\eta_{+-}|$ by comparing the number N_1 of $K_L \rightarrow \pi^0\pi^0$ decays from a vacuum and the number N_2 of $K_S \rightarrow \pi^0\pi^0$ from coherently regenerated K_S. The K_L fluxes of the two exposures are obtained from beam monitors (83) or from $K_L \rightarrow 3\pi^0$ decays observed simultaneously (82). In addition, the ratio of $K_L \rightarrow \pi^+\pi^-$ decays (N_3) and of coherently regenerated $K_S \rightarrow \pi^+\pi^-$ (N_4) is measured in the same beam and the same regenerator. The ratio $(N_1 N_4)/(N_2 N_3)$ of the event rates normalized to the beam flux yields $|\eta_{00}/\eta_{+-}|^2$ where the detector efficiencies and the regeneration amplitudes cancel. The only remaining corrections are those for incoherent regeneration of K_S, which have to be applied to the $K_S \rightarrow \pi^0\pi^0$ rate behind the regenerator and, in the case of (82), to the $K_L \rightarrow 3\pi^0$ rate used for flux normalization.

Figure 11 shows the data obtained in (83), which display a good agreement

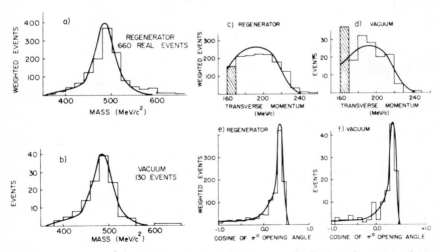

Figure 11 $K_{S,L} \rightarrow 2\pi^0$ results from (83): Invariant 4γ mass distribution of regenerated (*a*) and vacuum (*b*) events. Transverse momentum of spectrometer γ ray for regenerated (*c*) and vacuum (*d*) events. Opening angle of the two γ rays not paired with spectrometer γ ray for regenerated (*e*) and vacuum (*f*) events.

between the distributions for $K_L \to \pi^0\pi^0$ and regenerated $K_S \to \pi^0\pi^0$ events. The corresponding Figure 12 from (82) illustrates how it is possible to extract a clean signal of $K_L \to \pi^0\pi^0$ events in spite of the fact that $K_L \to 3\pi^0$ decays with four or five detected γ rays can simulate invariant masses near to the K^0 mass. The results of these two experiments are $|\eta_{00}/\eta_{+-}| = 1.00 \pm 0.06$ [Aachen-CERN-Torino (82)] and $\eta_{00}/\eta_{+-} = 1.03 \pm 0.07$ [Princeton (83)].

Using $|\eta_{+-}| = (2.279 \pm 0.026) \times 10^{-3}$ from Section 4.3, the old measurements can also be used (87) to get the ratio $|\eta_{00}|/|\eta_{+-}| = (0.992 \pm 0.088)$, in quite good agreement with the two direct measurements of the ratio. Therefore an average can be obtained: $|\eta_{00}|/|\eta_{+-}| = 1.008 \pm 0.041$, or $|\eta_{00}| = (2.30 \pm 0.10) \times 10^{-3}$.

4.6   Phase ϕ_{00}

This phase has been measured by studying the interference of $K_L \to \pi^0\pi^0$ and $K_S \to \pi^0\pi^0$ from coherently regenerated K_S (84–86), as in the case of $\pi^+\pi^-$ decay (Sections 2.5.1 and 4.4). However, the analysis of the eigentime distribution of $K \to 2\pi^0$ decays behind the regenerator is more difficult here because the background from $K_L \to 3\pi^0$ is higher and the spatial variation of the detection efficiency cannot be obtained from Monte Carlo calculations only. The latter point was

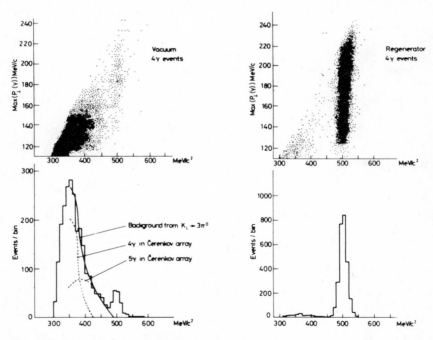

Figure 12 $K_{S,L} \to 2\pi^0$ data from (82): Invariant 4γ mass $m(4\gamma)$ vs p_T, the largest of the transverse momenta of the rays for regenerated K_S and vacuum K_L decays and projection of data on the $m(4\gamma)$ axis.

solved by measuring the spatial variation of the detection efficiency through $K_S \rightarrow 2\pi^0$ decays from a regenerator placed in different positions inside the decay volume.

From the resulting interference curve the Orsay-CERN group (84, 85) obtains the combined phase $\phi_{00} - \phi_f = 86° \pm 21°$, and using the nuclear regeneration phase $\phi_f = -48.2° \pm 3.5°$ from other measurements (101), they quote $\phi_{00} = 43° \pm 19°$.

The Aachen-CERN-Torino group (86) compares their interference curve to the corresponding one obtained earlier for the $\pi^+\pi^-$ decay mode (68), such that the difference of the two phases $\phi_{00} - \phi_{+-} = (7.6 \pm 18)°$ is obtained. Using ϕ_{+-} from Section 4.4, we get for the average of the two experiments $\phi_{00} = (48.0 \pm 13.1)°$ or $\phi_{00} - \phi_{+-} = (3.1 \pm 13.1)°$.

4.7 Charge Asymmetry in Semileptonic Decays

This asymmetry δ_L is the third manifestation of CP violation (see Section 2.2). It measures the CP impurity of the long-lived kaon state; in fact $\delta_L = 2\,\mathrm{Re}\,\varepsilon$ $(1 - |x|^2)/(|1 - x|^2)$, where x is the $\Delta S = \Delta Q$ violation parameter. Considerable improvement in precision has been achieved in the last three years: for the K_{e3} mode two experiments of the Princeton group (105) and of the CERN-Heidelberg group (78) have been reported; for the $K_{\mu 3}$ mode a new result comes from Stanford (109), one from CERN-Heidelberg (78), and a Brookhaven-Yale group (106) obtained a result for a mixture of both decay modes. Most remarkable are:

1. Event numbers of up to 34 million events in the K_{e3} mode and 15 million events in the $K_{\mu 3}$ mode.
2. Apparatus design such that the decay products (π and electron) traverse only minute amounts of matter (about 0.3–0.4 g cm^{-2}), thus diminishing corrections due to secondary interactions of these particles.
3. The precision of $K_{\mu 3}$ asymmetry measurements is now only a factor of 2 below the one for K_{e3}, making a comparison between the two possible. Table 5 gives the results and the average $\delta_L = (3.30 \pm 0.12) \times 10^{-3}$.

The charge asymmetries for K_{e3} and $K_{\mu 3}$ decays are equal to within 10%: $\delta_L^e / \delta_L^\mu = 1.089 \pm 0.116$. Assuming the validity of the $\Delta Q = \Delta S$ rule, which is supported by the present experiments (Section 5.4.3), one obtains $\mathrm{Re}\,\varepsilon = (1.64 \pm 0.06) \times 10^{-3}$.

4.8 Parameters of CP Violation

The results of all experiments on CP violation in K^0 decays are displayed in the Wu-Yang diagram (Figure 13) in the complex plane.

Comparing these results in particular to the superweak model that predicts $\eta_{+-} = \eta_{00} = \varepsilon$, we find the numbers quoted in Table 6.

All predictions of the superweak model are found to be in good agreement with these experiments. The experimental results are a fortiori compatible with a large class of milliweak models. If one wants to evaluate from experiment the ratio of CP violation in the transition matrix element (ε') relative to the one in the mass matrix (ε), the relation $\varepsilon' = \frac{1}{3}(\eta_{+-} - \eta_{00})$ can be used. For the component of ε' parallel to

Table 5 Charge asymmetry measurements in K_{l3} decays

Group (Ref.)	Year	Decay mode	Result ($\times 10^3$)	$\delta(\times 10^3)$	$\delta(\times 10^3)$
Columbia (102)	1969	K_{e3}	2.46 ± 0.59		
Columbia-Harvard-CERN (103)	1970	K_{e3}	3.46 ± 0.33	K_{e3} average	
San Diego-Berkeley (104)	1972	K_{e3}	3.6 ± 1.8	3.33 ± 0.14	
Princeton (105)	1973	K_{e3}	3.18 ± 0.38		
CERN-Heidelberg (78)	1974	K_{e3}	3.41 ± 0.18		
					$K_{\mu 3}$ and K_{e3} averaged
Brookhaven-Yale (106)	1973	$K_{e3} + K_{\mu 3}$	3.33 ± 0.50		Average 1976
SLAC-Berkeley (107)	1969	$K_{\mu 3}$	5.8 ± 1.7		3.30 ± 0.12
Berkeley (108)	1972	$K_{\mu 3}$	6.0 ± 1.4	$K_{\mu 3}$ average	
Stanford (109)	1972	$K_{\mu 3}$	2.78 ± 0.51	3.19 ± 0.24	
CERN-Heidelberg (78)	1974	$K_{\mu 3}$	3.13 ± 0.29		

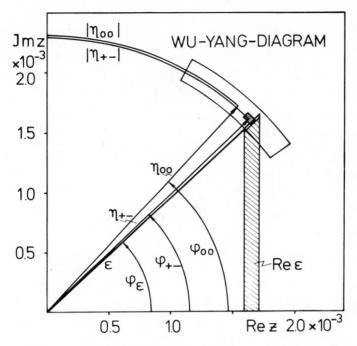

Figure 13 Wu-Yang diagram of *CP* violation parameters. ϕ_ε is drawn as given by the superweak model. The triangular relations are $\eta_{+-} = \varepsilon + \varepsilon'$ and $\eta_{00} = \varepsilon - 2\varepsilon'$.

$\varepsilon(\varepsilon_p')$, the measured moduli of η_{+-} and η_{00} yield $\varepsilon_p'/|\varepsilon| = 0.003 \pm 0.014$, while the difference of phases $\phi_{00} - \phi_{+-}$ determines the component ε_T' transverse to the ε direction: $\varepsilon_T'/|\varepsilon| \sim -0.02 \pm 0.08$. Using the phase of ε' derived from $\pi\pi$ phase shifts (Section 2.2) $\arg(\varepsilon') = (37 \pm 5)°$, as an additional constraint together with $\phi_{+-} = (44.9 \pm 1.3)$, we obtain

$$|\varepsilon_T'|/|\varepsilon| < 0.003,$$
$$|\varepsilon_p'|/|\varepsilon| < 0.012,$$
$$|\varepsilon| = (2.28 \pm 0.05) \times 10^{-3}.$$

Table 6 Comparison of experimental results with superweak model predictions

Quantity	Experiment average (1976)	Prediction of superweak model (SW)
ϕ_{+-}	$44.9° \pm 1.3°$	$\phi_{SW} = 43.8° \pm 0.2°$
ϕ_{00}	$48.0° \pm 13.1°$	$\phi_{SW} = 43.8° \pm 0.2°$
$|\eta_{00}|/|\eta_{+-}|$	1.008 ± 0.041	1.0
$\delta_L/|\eta_{+-}|$	1.448 ± 0.055	$2\cos\phi_{SW} = 1.443 \pm 0.005$

The experiments exclude therefore models with $|\varepsilon'/\varepsilon| \gtrsim 2 \times 10^{-2}$. Most of the milliweak models should predict values of $|\varepsilon'/\varepsilon|$ of the order of the violation of this $\Delta I = \frac{1}{2}$ rule, i.e. $|\varepsilon'/\varepsilon| \sim 0.05$. Therefore these experiments do not definitely exclude all milliweak models. It is not easy to improve substantially the experimental precision. A decision between superweak and milliweak models of CP violation will therefore probably have to come from other experimental information outside the K^0 system.

If, on the other hand, we use the experimental information to test CPT invariance by decomposing $\varepsilon_0 = 2\eta_{+-}/3 + \eta_{00}/3$ into a CPT-conserving, T-violating part $\tilde{\varepsilon}$ and a CPT-violating, T-conserving part $\tilde{\delta}$ following the analysis of Schubert et al (110), we obtain $\mathrm{Re}\,\tilde{\varepsilon} = (1.61 \pm 0.25) \times 10^{-3}$, $\mathrm{Im}\,\tilde{\varepsilon} = (1.40 \pm 0.25) \times 10^{-3}$, $\mathrm{Re}\,\tilde{\delta} = (-0.03 \pm 0.27) \times 10^{-3}$, and $\mathrm{Im}\,\tilde{\delta} = (-0.23 \pm 0.27) \times 10^{-3}$. This means that the observed CP violation is mainly due to a CPT-conserving, T-violating interaction. The strength of a CPT-violating interaction can be at most one quarter of the strength of a superweak interaction. This is probably the most sensitive test of CPT invariance at this time.

5   SEARCHES FOR C OR T VIOLATION

In order to distinguish between the different kinds of models of CP violation a whole series of experiments has been done. No evidence for C or T violation has been found.

5.1   Test of C and T Invariance in Strong Interactions

If, in a reaction $a + b \rightarrow c + d$, T and P are conserved by the interaction H, then the relevant matrix element for the transition from the initial state ψ_i to the final state ψ_f, $M_{if} = \langle \psi_f | H | \psi_i \rangle$, and the one for the inverse reaction, $M_{fi} = \langle \psi_i | H | \psi_f \rangle$, are related. Time reversal T exchanges in- and outgoing states and reverses momenta $\mathbf{p}_n$ and spins $\boldsymbol{\sigma}_n$ ($n = a, b, c, d$), while space inversion P reverses momenta and does not change spins. Therefore

$$PT \langle \psi_f(\boldsymbol{\sigma}_c, \boldsymbol{\sigma}_d, \mathbf{p}_c, \mathbf{p}_d) | H | \psi_i(\boldsymbol{\sigma}_a, \boldsymbol{\sigma}_b, \mathbf{p}_a, \mathbf{p}_b) \rangle$$
$$= \langle \psi_i(-\boldsymbol{\sigma}_a, -\boldsymbol{\sigma}_b, \mathbf{p}_a, \mathbf{p}_b) | H | \psi_f(-\boldsymbol{\sigma}_c, -\boldsymbol{\sigma}_d, \mathbf{p}_c, \mathbf{p}_d) \rangle.$$

Summation over all spin states gives the principle of detailed balance: $|M_{if}|^2 = |M_{fi}|^2$.

T invariance in strong interactions has been tested by checking the principle of detailed balance. The reactions studied experimentally were

$$d + {}^{24}\mathrm{Mg} \rightleftarrows p + {}^{25}\mathrm{Mg} \qquad (111),$$
$$\alpha + {}^{24}\mathrm{Mg} \rightleftarrows p + {}^{27}\mathrm{Al} \qquad (112),$$
$$d + {}^{16}\mathrm{O} \rightleftarrows \alpha + {}^{14}\mathrm{N} \qquad (113).$$

The comparison of the reactions with their inverse is in agreement with detailed balance in all cases. The upper limit on the contribution of T-violating amplitudes relative to the T-conserving one from these experiments is 3×10^{-3}.

C invariance in strong interactions has been tested by studying the annihilation of antiprotons in hydrogen into K and π mesons. One compares the annihilation into *C*-conjugate meson states

$$\bar{p}p \to 1+2+x, \qquad p\bar{p} \to \bar{1}+\bar{2}+\bar{x}.$$

If the angles of the two mesons 1 and 2 relative to the $\bar{p}$ direction as a polar axis are denoted by $\theta_1(\theta_2)$, the azimuth angle of particle 2 relative to the plane $(\bar{p}1)$ by ϕ_{12} and their momenta by $p_1(p_2)$, then the distributions in angle and momenta for the production of the two charge-conjugate states $1+2+x$ and $\bar{1}+\bar{2}+\bar{x}$ are related (114, 115):

$$W(p_1, \theta_1, p_2, \theta_2, \phi_{12}) = W(p_{\bar{1}}, \pi-\theta_{\bar{1}}, p_{\bar{2}}, \pi-\theta_{\bar{2}}, \phi_{\bar{1}\bar{2}}).$$

These relations have been tested by using a sample of ~ 4000 $\bar{p}$ annihilation events at 1.2 GeV/c in a hydrogen bubble chamber. The experiment allows to determine the relative amplitude of a *C*-violating strong interaction to be $(0.4 \pm 1.0) \times 10^{-2}$ of the *C*-conserving amplitude (115).

A similar experiment using $\bar{p}p$ annihilations at rest also obtained an upper limit of $\sim 1\%$ to such *C*-violating amplitudes (116).

5.2 Test of C Invariance in the Electromagnetic Interaction

C invariance in electromagnetic interactions can be tested in the same way as in strong interactions by searching for asymmetries between charge-conjugate states from the decay of an isospin singlet. In this context considerable effort has been devoted to the decays of the η and η' mesons, for which the effects of *C* violation were predicted to be of the order of 5% (141) if *CP* violation in K decays was due to the interference of an electromagnetic *C*-violating and the normal weak amplitude.

Experiments on the decays $\eta \to \pi^+\pi^-\pi^0$ and $\eta \to \pi^+\pi^-\gamma$ have been made with increasing precision from 1966 to 1974 as shown in table 7. In the more recent counter experiments η mesons are produced in the reaction $\pi^- p \to \eta n$ and identified by measuring the time of flight and direction of the neutron. The momenta of the charged decay products π^+ and π^- are then measured in a magnet around the hydrogen target, and the position of the events in the Dalitz plot is computed. Asymmetries are evaluated by comparing the event numbers in the charge-conjugate sextants of the Dalitz plot. No asymmetry is found, and upper limits to such an asymmetry are 3×10^{-3}.

A similar search for an asymmetry in the decay of the Regge recurrence of the η, $\eta'(958) \to \pi^+\pi^-\gamma$ was motivated by the fact that a kinematical suppression of a *C*-violating amplitude would be less pronounced in this decay than in $\eta \to \pi^+\pi^-\gamma$ because of the much higher available phase space. In fact, the ratio of sensitivity of $\eta' \to \pi^+\pi^-\gamma$ and $\eta \to \pi^+\pi^-\gamma$ decays to *C* violation has been estimated (128) to be 30.

In the experiment by an UCLA-LBL collaboration (129), 295 ± 31 events were recorded. An asymmetry of $A = -(6.9 \pm 7.8)\%$ was obtained. If the result is combined with the one of a former experiment (130), one obtains $A = -(2.9 \pm 6.2)\%$.

Table 7   Measurements of charge asymmetries in η decays

| Group (Ref.) | Year | $\eta \to \pi^+ \pi^- \pi^0$ | | $\eta \to \pi\pi\gamma$ | |
		Events (No.)	Charge asymmetry (%)	Events (No.)	Charge asymmetry (%)
Berkeley (117)	1966			33	-2 ± 17
Columbia-Berkeley-Purdue-Wisconsin-Yale (118)	1966	1300	$5.8 \ \pm 3.4$		
Columbia-Stony Brook (119)	1966	1441	$7.2 \ \pm 2.8$		
Saclay-Rutherford (120, 121)	1966	705	$-6.1 \ \pm 4.0$	160	$-4 \pm \ 8$
CERN-ETH-Saclay (122, 123)	1966	10,665	$0.3 \ \pm 1.0$	1620	$1.5 \pm \ 2.5$
Columbia-Brookhaven (124)	1968	36.800	$1.5 \ \pm 0.5$	6710	$2.4 \pm \ 1.4$
Columbia (125, 126)	1972	220,659	-0.05 ± 0.22	36,155	$0.5 \pm \ 0.6$
Rutherford-Westfield-Sussex (127)	1974	165,311	0.28 ± 0.26	34,680	$1.2 \pm \ 0.6$

A third test of *C* invariance in η decays is based on the equality of the *C* parities of η and π^0 mesons, both being $+1$ as deduced from their decays into two γ rays. A transition $\eta \to \pi^0 e^+ e^-$ with an electron-positron pair production mediated by a γ ray (with *C* parity $C_\gamma = -1$) is therefore forbidden by *C* invariance.

The experimental searches for this decay have been done in bubble chambers and upper limits for its branching ratio of 0.9×10^{-3} (131) and 3.7×10^{-4} (132) have been obtained.

In view of the fact that according to the original ideas a large part of the electromagnetic interaction had to be *C*-violating in order to account for the observed $K_L \to 2\pi$ decay rate, the experimental results do not support this proposal.

5.3   Test of *T* Invariance in the Electromagnetic Interaction

Here again the principle of detailed balance is used to test *T* invariance, but in addition there are other methods, including the very sensitive test on the electric dipole moment of the neutron.

5.3.1   DETAILED BALANCE EXPERIMENTS

Detailed balance in the reactions $\gamma + d \rightleftarrows n + p$ Here the cross sections for the photo disintegration of the deuteron in the region of the N^* (1236) resonance, i.e. around $E_\gamma \sim 300$ MeV had been measured with great precision by several groups (133–136). The inverse reaction at the corresponding center-of-mass (c.m.) energies has been measured by Bartlett et al (137) and by Schrock et al (138) using neutron beams of energies between 300 and 750 MeV. About 3×10^4 events have been registered in each experiment. The measured cross sections are analyzed in terms of a model of Barshay (139) and Austern (140), according to which the effect of a *T* violation should show up in the ratio A_2/A_0 when the differential cross sections are expressed as $d\sigma/d\Omega = A_0 + A_2 P_2(\cos \theta_{dn}^*)$, with θ_{dn}^* being the c.m. angle between deuteron and neutron and P_2 the Legendre polynomial of second order. A measure for *T* violation would be then the angle $\sin \phi \simeq 3[A_2/A_0(\gamma d \to np) - A_2/A_0(np \to \gamma d)]$ and *T* violation would give $\phi = 90°$. From the amplitude ratios A_2/A_0 for both reactions, a phase angle of $\phi = 4 \pm 10°$ at $T_n = 590$ MeV is deduced for one of the $np \to \gamma d$ experiments (138). No evidence for *T* violation is seen.

Detailed balance in the reactions $\gamma n \rightleftarrows \pi^- p$ The comparison of these two reactions is complicated by the necessity to extract the $\gamma n \to \pi^- p$ cross section from γd data (142–146) and by the difficult experimental problems in the radiative capture experiment $\pi^- p \to \gamma n$, e.g. the detection efficiency of neutron counters and the suppression of background from charge exchange $\pi^- p \to \pi^0 n$.

After some initial disagreement of the radiative capture data of the UCLA-LBL group (147, 148) with photoproduction data, the authors now conclude that the mutual agreement is acceptable, while a Lausanne-Munich-CERN group (149) finds good agreement with detailed balance. The latter group has evaluated the amount of a possible *T* violation by determining the phase Φ of a T-violating isovector $(M1^+)$ amplitude relative to the normal isoscalar and isovector amplitudes and finds $\Phi = (-0.5 \pm 2.5)°$. Again there is no evidence for *T* violation.

Detailed balance in the reactions $\gamma\ ^3He \rightleftarrows p+d$ These reactions have been studied by a UCSC-LBL group (150) in the γ energy interval between 200 and 600 MeV, i.e. around the energy necessary to excite one nucleon into the Δ (1236) resonance state. A comparison of $d\sigma/d\Omega$ was made at a proton kinetic energy of 462 MeV to the corresponding angular distribution of the photoproduction reaction, and the differential cross section at 90° in the c.m. at proton kinetic energies of 377, 462, and 576 MeV was compared to their photoproduction counterparts. Good agreement with detailed balance is found when comparing both angular distributions and absolute cross sections. The upper limit on a T-violating amplitude is 2%, using a multipole model for the evaluation.

5.3.2   TEST OF T INVARIANCE IN INELASTIC ELECTRON SCATTERING   Since Compton scattering from polarized protons has not been measured yet and since an apparent T violation in elastic electron-nucleon scattering would be a violation of conservation of the electromagnetic current at the same time, inelastic electron-scattering experiments are used to search for the existence of a term $\sigma_p \cdot (\mathbf{p}_i \times \mathbf{p}_f)$, where σ_p is the proton spin and $\mathbf{p}_i$ and $\mathbf{p}_f$ are the momenta of incident and outoing electrons, respectively. Such a term changes sign under T and must therefore vanish if T is conserved.

Inelastic electron scattering from polarized protons   The target polarization is perpendicular to the plane spanned by $\mathbf{p}_i$ and $\mathbf{p}_f$, and the cross sections for the inclusive reaction $e+p \rightarrow e+x$ are measured for proton spin up ($\sigma_\uparrow$) or down ($\sigma_\downarrow$). The asymmetry $A = (\sigma_\uparrow - \sigma_\downarrow)/(\sigma_\uparrow + \sigma_\downarrow)$ vanishes if T is conserved. Contributions from higher-order (α^3) electromagnetic effects to A change sign if e^- is replaced by e^+.

A contribution to A from T violation could occur (151) through the interference of amplitudes for transversely and longitudinally polarized photons. If the conjecture of Lee (152) and Okun (153) is assumed (that a T-violating current has to be an isoscalar), then the effect should show up in the excitation of the N^* (1512) and N^* (1688) resonance, but not with Δ (1236), because it is $I = \frac{3}{2}$. If the T-violating current is an isovector (30), then an asymmetry could also occur near the Δ (1236) resonance. Asymmetries of up to 35% could then occur, according to an estimate of the SLAC group (154).

The measured asymmetries at electron energies up to 18 GeV/c and momentum transfers up to 1 (GeV/c)^2 are consistent with $A = 0$. Taking the data near the excitation of the N^* (1512) resonance, e.g. one obtains $A = -(0.3\pm1.3)\%$ from the SLAC experiment (154). Similar but less precise data have been given by a CEA experiment (155), e.g. $A = (3.5\pm4.3)\%$ for excitation of the Δ (1236) resonance.

Measurement of vector polarization of deuterons in elastic e^-d scattering   In elastic electron-deuteron scattering, T violation is not coupled with violation of current conservation. Therefore a finite expectation value of the term $\sigma_d \cdot (\mathbf{p}_i \times \mathbf{p}_f)$ has been searched for in the experiment of Prepost et al (156) for an electron energy of 1 GeV and a recoil-deuteron four-momentum of 721 MeV/c. The recoil deuteron polarization was measured by recording the right-left asymmetry after scattering in a carbon absorber. The resulting deuteron polarization was found to be $|P| = 0.075 \pm 0.088$, in agreement with T invariance.

5.3.3 TESTS OF T INVARIANCE IN ELECTROMAGNETIC NUCLEAR TRANSITIONS   Since T invariance required coupling constants to be relatively real, i.e. the phases between interfering amplitudes to be either zero or $180°$, experiments have been done in order to measure the relative phase of two competing γ transitions of different multipolarity. These experiments have been reviewed by Henley (43) and more recently by Richter (157). The most precise experiments using Mössbauer resonant absorption obtain relative phases ϕ compatible with zero within a few mrad; e.g. the ^{99}Ru experiment (158a) gives $\phi = (0.0 \pm 1.7)$ mrad. The precision of these tests is remarkable; however, it is not clear whether T-violating effects would occur at such low energies.

5.3.4 ELECTRIC DIPOLE MOMENT OF THE NEUTRON   For an elementary particle, the only intrinsic definition of a direction is given by its spin σ, which is odd under C and T (inversion of "current") but—as an axial vector—is even under P. If the particle has a static electric dipole moment, this must be coupled to σ and its interaction with the electromagnetic field is given by $\sigma \cdot \mathbf{E}$. Since the electric field $\mathbf{E}$, as the gradient of a scalar field, is odd under P and odd under charge conjugation C, but even under T, the interaction Hamiltonian is odd under P and T, and even under C. Thus the existence of a nonvanishing electric dipole moment would be evidence for simultaneous T and P violation (10, 159–163).

If the source of the observed CP violation were in the electromagnetic interaction, such a dipole moment could be a combined effect of the T- and C-violating electromagnetic and the P- and C-violating weak interaction. The estimated size of such a moment for the neutron, D_n, would be: $D_n \sim e \times \lambda_c(n) \times G_F M^2$, where e is the elementary charge, $\lambda_c(n)$ the Compton wavelength of the neutron, and $G_F M^2$ the Fermi constant made dimension-free by multiplying with the square of the nucleon mass M. Then

$$D_n \sim eh/(Mc) \times 10^{-5} \simeq 10^{-19} \, e \cdot \text{cm}.$$

If it is inferred from the experiments on T violation in electromagnetic decays mentioned previously that the amount of T violation in the electromagnetic interaction is less than 10^{-2}, then this estimate is reduced to $10^{-21} \, e \cdot \text{cm}$. Estimates for the size of D_n from electromagnetic CP violation, milliweak, and superweak models have been given by Wolfenstein (163) and several other authors (164–168) and are compiled in Table 8, adapted from Wolfenstein (163).

Experiments on the neutron electric dipole moment have been made since 1950 with increasing sensitivity (169, 170, and references to earlier work quoted therein). They use a magnetic resonance spectrometer (Figure 14) placed in a slow (cold) neutron beam from a reactor. The neutrons are transversely polarized (and later analyzed) by reflection from magnetized ferromagnetic material. They are then allowed to pass through a long drift region with a transverse homogeneous magnetic field $\mathbf{B}$, where they precess with their Larmor frequency $v_L = \mu B/h$, where

Table 8   Values for the electric dipole moment of the neutron in $e \cdot cm$ from various models [from Wolfenstein (163)]

Source (Ref.)	Electromagnetic	Milliweak	Superweak
Rough estimate	10^{-20}	10^{-23}	10^{-29}
Model-independent calculation	10^{-21} to 10^{-23}	10^{-23}	10^{-30}
Salzman & Salzman (164)	10^{-20}	—	—
T. D. Lee (165)	—	10^{-23}	—
Mohapatra (166)	—	10^{-24}	—
Pais (167)	—	10^{-23}	—
Frenkel & Ebel (168)	—	10^{-29}	—
Experimental result (171)		$0.4 \pm 1.1 \times 10^{-24}$	

μ is their magnetic moment and $\hbar$ is Planck's constant. Two radio-frequency coils at the beginning and end of the drift region induce spin flip transitions when tuned to the resonance frequency $v = v_L$, thus causing depolarization of the neutron beam and loss of counting rate in the counter C behind the setup. This resonance behavior would be changed if the neutrons had an electric dipole moment precessing in addition in the electric field produced by the plates **E**. It is such a shift of resonance frequency that is searched for in the experiments. The most recent result comes from an improved version of the experiment of Dress et al (169) at the ILL reactor in Grenoble (171):

$$D_n = (0.4 \pm 1.1) \times 10^{-24} \, e \cdot \text{cm}.$$

This result, when compared to the calculations listed in Table 8, gives definite evidence against electromagnetic models of CP violation. Similarly, several milliweak models predicting $D_n \sim 10^{-23} \, e \cdot \text{cm}$ are hardly compatible with this result. An increase in experimental sensitivity by a factor of 10 or more (171, 172) could, if the electric dipole moment still failed to show up at this level, give definite evidence against most of the milliweak models.

5.4   Test of T Invariance in the Weak Interaction

The milliweak models of CP violation require a small (10^{-3}) fraction of the weak interaction to be T-violating. More precisely, they assume a finite transition

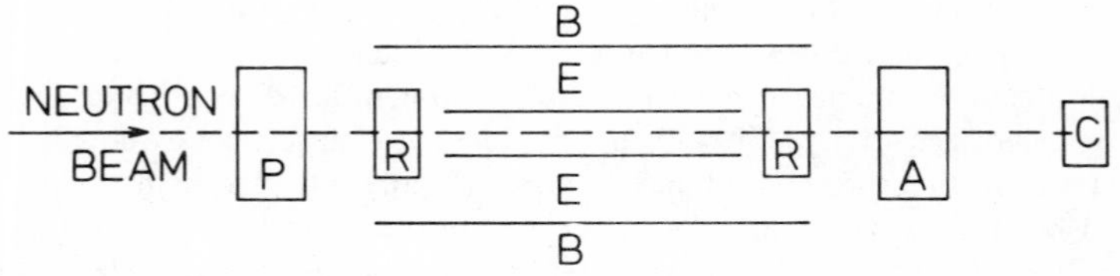

Figure 14   Schematic view of apparatus used in the search for an electric dipole moment of the neutron. P (spin polarizer), A (analyzer), B (pole pieces for magnetic guide field), E (plates for electric field), R (RF coils), and C (neutron counter).

amplitude from the pure $CP = -1$ state K_2^0 into 2π mesons ($CP = +1$). This milli-weak amplitude has $\Delta S = 1$, but searches for a small T-violating contribution in weak decays have been done for $\Delta S = 1$ as well as for $\Delta S = 0$ processes.

In general, triple-correlation products of two polar vectors (momenta) and one axial vector (spin) or of two spins and one momentum are searched for in these experiments. Because polar and axial vectors reverse their sign under the T operation, expectation values of these triple products must vanish if T invariance is valid.

5.4.1 BETA DECAY OF POLARIZED NEUTRONS AND POLARIZED ^{19}NE NUCLEI   Here the measured triple correlation is $\mathbf{P}(\mathbf{p}_e \times \mathbf{p}_\nu)$ with $\mathbf{P}$ the polarization vector of the decaying nucleus (nucleon) and $\mathbf{p}_e$ and $\mathbf{p}_{\bar{\nu}}$ the momenta of electron and antineutrino from the β decay. If the parent nucleus (nucleon) is at rest, the correlation is equivalent to $\mathbf{P}(\mathbf{p}_D \times \mathbf{p}_e)$, where $\mathbf{p}_D$ is the momentum of the daughter nucleus (nucleon). A practical way is to have the parent polarized along or opposite to its direction of flight and to observe the daughter and the electron at 90° relative to each other transverse to the beam direction.

In a weak decay mediated by V-A interaction, the magnitude D of the triple correlation is proportional to an interference term between axial (Gamow-Teller) and vector (Fermi) matrix elements. The results can be expressed therefore by calculating the relative phase ϕ between those two amplitudes (173).

Results of experiments are given in Table 9. The agreement with T invariance ($D = 0$, $\phi = 0$ or π) is perfect. T-violating amplitudes must be of order 10^{-2} or less in order to escape detection. The influence of final state interactions on these results has been discussed and found negligible at the present level of accuracy.

5.4.2 TEST OF T INVARIANCE IN Λ DECAY   The triple correlation here is $\mathbf{p}_p (\boldsymbol{\sigma}_p \times \boldsymbol{\sigma}_\Lambda)$, where $\mathbf{p}_p$ is the proton momentum and $\boldsymbol{\sigma}_p$ and $\boldsymbol{\sigma}_\Lambda$ are the spins of proton and Λ, respectively, in $\Lambda \to p\pi^-$ decay. A nonvanishing expectation value of the triple correlation would be induced by T violation through the interference of the s- and p-wave decay amplitudes. It would result in a shift Φ in the phase between these two amplitudes. This phase shift Φ is not zero in the absence of T violation because of final-state interactions, but is obtained from pion-nucleon scattering data to be $\Phi_{fs} = (6.5 \pm 1.5)°$ (178).

The measured phase shift between s and p waves is from two experiments (179–181)

$$\Phi = (7.5 \pm 3.9)°.$$

Table 9 Measurements of relative phase of A and V

Decay	Ref.	D	Φ (degrees)
$n \to p + e^- + \bar{\nu}$	(174)	-0.01 ± 0.01	181.3 ± 1.3
	(175)	-0.0011 ± 0.0017	180.14 ± 0.22
^{19}Ne $\to$ ^{19}F $+ e^+ + \nu$	(176)	$+0.002 \pm 0.014$	180.2 ± 1.6
	(177)	$+0.002 \pm 0.004$	180.2 ± 0.4

Therefore the deviation of the phase shift from Φ_{fs} is

$$\Delta\Phi = (1.0 \pm 4.0)^\circ.$$

Again there is agreement with the value $\Delta\Phi = 0$ required by T invariance.

5.4.3   SEARCH FOR SIMULTANEOUS VIOLATION OF THE $\Delta Q = \Delta S$ RULE AND T INVARIANCE IN SEMILEPTONIC K^0 DECAYS   If there were a $\Delta Q = -\Delta S$ amplitude at a phase Φ relative to the $\Delta Q = \Delta S$ with $\Phi \neq 0$ (mod. π), this would show up as an imaginary part $\text{Im } x \neq 0$ (see Section 2.5.2).

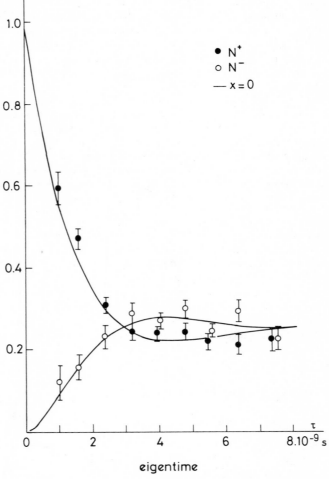

Figure 15   Time distribution of $K^0 \to \pi^- e^+ \nu(N^+)$ and $K^0 \to \pi^+ e^- \bar\nu(N^-)$ events from an initial K^0 state (182). The curves drawn assume validity of the $\Delta Q = \Delta S$ rule ($x = 0$).

Experiments aiming at Im x and Re x measure the eigentime distribution of leptonic K^0 decays from an initially pure eigenstate of strangeness. The time distribution of K^0 decays into the two states $e^+\pi^- v(N^+)$ and $e^-\pi^+ \bar{v}(N^-)$ has been given in Section 2.5.2. Essentially, the sum distribution $N^+ + N^-$ measures Im x, while the asymmetry is sensitive to Re x.

In the last three years the results of several new experiments on the $\Delta Q = \Delta S$ rule in K_{e3} and $K_{\mu3}$ decays were published [see (42) for a summary and for references]. The majority of these experiments were counter experiments. The errors on Re x_e in these new K_{e3} experiments range from ±0.07 to ±0.03, the ones on Im x_e from ±0.09 to ±0.05. All experiments are compatible with Re $x_e = 0 =$ Im x_e. As an example, Figure 15 shows the time distribution of N^+ and N^- obtained by the CERN-Orsay-Vienna group (182). In this experiment the K_0 momentum ambiguity in K_{e3} decay is resolved by measuring the production reaction products in $K^+p \to K^0p\pi^+$. From the 4724 reconstructed events the values Re $x_e = 0.04\pm0.03$ and Im $x_e = 0.06\pm0.05$ are deduced.

The experimental results on Re x_e and Re x_μ are summarized in Figure 16, the ones on Im x_e and Im x_μ in Figure 17, both taken from (42). There is mutual agreement between these results, and the averages computed from them are:

for K_{e3}:   Re $x_e = 0.021\pm0.014$ $\chi^2/DF = 19.8/16,$
Im $x_e = 0.016\pm0.022$ $\chi^2/DF = 15.4/16;$

for $K_{\mu3}$:   Re $x_\mu = 0.11\pm0.08$ $\chi^2/DF = 1.2/3,$
Im $x_\mu = 0.02\pm0.10$ $\chi^2/DF = 1.9/3.$

Within the experimental errors, the results for K_{e3} and $K_{\mu3}$ decays agree; for the real part of x there is supporting evidence for this from the measurements of the charge asymmetry δ_L in $K_L \to \pi l v$ decays (Section 4.7). In fact, we infer from

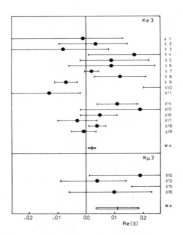

Figure 16 Summary of results on the parameter Re x measuring a violation of the $\Delta Q = \Delta S$ rule for K_{e3} and $K_{\mu3}$ decays (42).

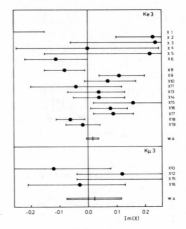

Figure 17 Summary of results on the parameter Im x measuring violation of the $\Delta Q = \Delta S$ rule and of T invariance in K_{e3} and $K_{\mu3}$ decays (42).

those measurements that

$$\delta_L^e/\delta_L^\mu = 1.089 \pm 0.116 = 1 + 2\,\mathrm{Re}\,(x_e - x_\mu) \quad \text{for} \quad |\mathrm{Im}\,x| \ll 1.$$

Therefore Re x_e − Re $x_\mu = 0.045 \pm 0.058$. Using Re x_e from the table above, this gives the independent result Re $x_\mu = -0.02 \pm 0.06$, which can be combined with the result above to give the value Re $x_\mu = 0.04 \pm 0.05$.

All these results demonstrate the absence of $\Delta Q = -\Delta S$ currents at a level of a few percentage points in amplitude relative to the $\Delta Q = \Delta S$ currents.

5.4.4 SEARCH FOR TRANSVERSE MUON POLARIZATION IN $K_{\mu3}$ DECAY   A triple correlation can be constructed for the decay $K_L \to \pi^\pm \mu^\mp \nu$ from the three vectors: muon polarization (σ_μ), muon momentum ($\mathbf{p}_\mu$), and pion momentum ($\mathbf{p}_\pi$): $\sigma_\mu(\mathbf{p}_\mu \times \mathbf{p}_\pi)$. A nonvanishing muon polarization $\mathbf{P}_T$ transverse to the decay plane spanned by $\mathbf{p}_\mu$ and $\mathbf{p}_\pi$ would indicate T violation. Such a term could be caused by the interference of the two hadronic amplitudes present in this decay: one proportional to the sum of the four-momenta of the kaon and pion, $p_K + p_\pi$, the other proportional to $p_K - p_\pi$ (183, 184). In two precise experiments (185, 186) transverse polarizations compatible with zero within 1–2% are found. The results can be expressed in terms of the imaginary part of the ratio of the two interfering amplitudes, Im ξ. They are Im $\xi = -0.02 \pm 0.08$ and Im $\xi = -0.060 \pm 0.045$. The real part of ξ is measured at the same time in these experiments and comes out Re $\xi = -1.81 \pm 0.50$ (185) and Re $\xi = -0.655 \pm 0.127$ (186), respectively. These values are barely compatible. Independent measurements of form factors and branching ratios in K_L decays favor the latter value. If the two experiments are nevertheless combined, one can extract the phase $\Phi = \arg \xi$ between the two amplitudes to be $\Phi = (4.3 \pm 3.3)°$, in agreement with $\Phi = 0$ required by T invariance. Final-state interactions are negligible here.

6   CONCLUSION

This review may be summarized as follows:

1. CP violation in decays of the long-lived K meson K_L is manifest in three ways: through the decay $K_L \to \pi^+ \pi^-$, the decay $K_L \to \pi^0 \pi^0$, and the charge asymmetry in the semileptonic K_L decays. All experimental information on amplitudes and phases of these decays gives a consistent picture in terms of a phenomenological analysis. If the observed amplitudes are divided up into a contribution from the CP impurity of the K_L state (ε) and a contribution from a CP-violating transition matrix element (ε'), then the experiments lead to the conclusion that $|\varepsilon'/\varepsilon| < 0.02$. This is in agreement with the superweak model ($\varepsilon' = 0$), but milliweak models cannot be excluded at this time.

2. If the observed CP-violating amplitudes are decomposed into a CPT-conserving and T-violating part $\tilde{\varepsilon}$ and a CPT-violating and T-conserving part $\tilde{\delta}$, then from experiment one finds $\tilde{\varepsilon} = (2.13 \pm 0.35) \times 10^{-3}$ and $|\tilde{\delta}/\tilde{\varepsilon}| < 0.25$. Therefore T violation is demonstrated independently of CPT invariance.

3. Searches for a T or C violation in strong, electromagnetic, and weak interactions have failed to detect any such symmetry violation down to levels of 10^{-2}–10^{-3} of the strength of strong, electromagnetic, or weak interactions, respectively. The most sensitive test is probably provided by the upper limit on the electric dipole moment of the neutron, which can be considered evidence against electromagnetic models of CP violation. Further improvements of this experiment may eventually facilitate the remaining choice between milliweak and superweak models of CP violation.

Literature Cited

1. Wigner, E. 1927. *Z. Phys.* 43:624
2. Wick, G. C. 1958. *Ann. Rev. Nucl. Sci.* 8:1
3. Wigner, E. 1932. *Nachr. Akad. Wiss. Göttingen* 31:546
4. Lüders, G. 1954. *Kgl. Danske Videnskab. Selskab, Matfys. Medd.* 28(5):1
5. Pauli, W. 1955. In *Niels Bohr and the Development of Physics*, ed. W. Pauli, p. 30. Oxford: Pergamon. 2nd ed.
6. Lee, T. D., Yang, C. N. 1956. *Phys. Rev.* 104:254
7. Wu, C. S. et al 1957. *Phys. Rev.* 105:1413
8. Garwin, R. L., Lederman, L. M., Weinrich, M. 1957. *Phys. Rev.* 105:1415
9. Friedman, J. I., Telegdi, V. L. 1957. *Phys. Rev.* 105:1681
10. Landau, L. D. 1957. *Nucl. Phys.* 3:127
11. Gell-Mann, M., Pais, A. 1955. *Phys. Rev.* 97:1387
12. Bardon, M. et al 1958. *Ann. Phys. NY* 5:156
13. Neagu, D. et al 1961. *Phys. Rev. Lett.* 6:552
14. Christenson, J. H., Cronin, J. W., Fitch, V. L., Turlay, R. 1964. *Phys. Rev. Lett.* 13:138
15. Gaillard, J. M. et al 1967. *Phys. Rev. Lett.* 18:20
16. Cronin, J. W. et al 1967. *Phys. Rev. Lett.* 18:25
17. Bennett, S. et al 1967. *Phys. Rev. Lett.* 19:993
18. Dorfan, D. et al 1967. *Phys. Rev. Lett.* 19:987
19. Wolfenstein, L. 1964. *Phys. Rev. Lett.* 13:562
20. Wolfenstein, L. 1966. *Nuovo Cimento* 42:17
21. Alles, W. 1965. *Phys. Lett.* 14:348
22. Glashow, S. L. 1965. *Phys. Rev. Lett.* 14:35
23. Zachariasen, F., Zweig, G. 1965. *Phys. Rev. Lett.* 14:794
24. Lotsoff, S. N. 1965. *Phys. Lett.* 14:344

48   KLEINKNECHT

25. Sachs, R. G. 1964. *Phys. Rev. Lett.* 13: 286
26. Mohapatra, R. N. 1972. *Phys. Rev. D* 6:2023
27. Pais, A. 1973. *Phys. Rev. D* 8:625
28. Lee, T. D. 1973. *Phys. Rev. D* 8:1226
29. Bernstein, J., Feinberg, G., Lee, T. D. 1965. *Phys. Rev. B* 139:1650
30. Barshay, S. 1965. *Phys. Lett.* 17:78
31. Salzman, F., Salzman, G. 1965. *Phys. Lett.* 15:91
32. Arbuzov, B. A., Filipov, A. T. 1966. *Phys. Lett.* 20:537, 21:771
33. Prentki, J., Veltman, M. 1965. *Phys. Lett.* 15:88
34. Okun, L. B. 1965. *Sov. J. Nucl. Phys.* 1:670
35. Lee, T. D., Wolfenstein, L. 1965. *Phys. Rev. B* 138:1490
36. Lee, T. D., Wu, C. S. 1966. *Ann. Rev. Nucl. Sci.* 16:511
37. Bell, J. S., Steinberger, J. 1966. *Proc. Int. Conf. Elem. Part., Oxford, 1965.* Chilton, Didcot, UK: Rutherford Lab.
38. Okun, L. B., Rubbia, C. 1968. *Proc. Int. Conf. Elem. Part., Heidelberg, 1967,* p. 301. Amsterdam: North-Holland
39. Cronin, J. W. 1968. *Proc. Int. Conf. High-Energy Phys., 14th, Vienna,* p. 281. Geneva: CERN
40. Steinberger, J. 1969, *Proc. Top. Conf. Weak Interactions,* Geneva: CERN 70-1:291
41. Winter, K. 1972. *Proc. Int. Conf. Elem. Part., Amsterdam, 1971,* p. 333. Amsterdam: North-Holland
42. Kleinknecht, K. 1974. *Proc. Int. Conf. High-Energy Phys., 17th, London, 1974,* p. III–23. Chilton, Didcot, UK: Sci. Res. Counc.
43. Henley, E. M. 1969. *Ann. Rev. Nucl. Sci.* 19:367
44. Lee, T. D., Oehme, R., Yang, C. N. 1957. *Phys. Rev.* 106:340
45. Sachs, R. G. 1963. *Ann. Phys. NY* 22:239
46. Wu, T. T., Yang, C. N. 1964. *Phys. Rev. Lett.* 13:380
47. Weisskopf, V. F., Wigner, E. P. 1930. *Z. Physik* 63:54, 65:18
48. Losty, M. F. et al 1974. *Nucl. Phys. B* 69:185
49. CERN-Munich (MPI) Collaboration. 1974. *Int. Conf. Meson Spectroscopy, 4th, Boston*
50. Metcalf, M. et al 1972. *Phys. Lett. B* 40:703
51. Barmin, V. V. et al 1973. *Phys. Lett. B* 46:465
52. Laurent, B., Roos, M. 1964. *Phys. Lett.* 13:269, 15:104
53. Bell, J. S., Perring, J. K. 1964. *Phys. Rev. Lett.* 13:348
54. Bernstein, J., Cabibbo, N., Lee, T. D. 1964. *Phys. Rev. Lett.* 12:146
55. Bailey, J. et al 1968. *Phys. Lett. B* 28:287
56. Bailey, J. et al 1975. *Phys. Lett. B* 55:420
57. Rich, A., Crane, H. R. 1966. *Phys. Rev. Lett.* 17:271
58. Wesley, J. C., Rich, A. 1971. *Phys. Rev. A* 4:1341
59. Meyer, S. L. et al 1963. *Phys. Rev.* 132:2693
60. Ayers, D. S. et al 1968. *Phys. Rev. Lett.* 21:261
61. Lobkowicz, F. et al 1966. *Phys. Rev. Lett.* 17:548
62. Lobashov, V. M. et al 1967. *Phys. Lett. B* 25:104
63. Pais, A., Piccioni, O. 1955. *Phys. Rev.* 100:1487
64. Kleinknecht, K. 1973. *Fortschr. Phys.* 21:57
65. Fitch, V. et al 1965. *Phys. Rev. Lett.* 15:73; *Phys. Rev.* 164:1711
66. Alff-Steinberger, C. et al 1966. *Phys. Lett.* 21:595
67. Bott-Bodenhausen, M. et al 1966. *Phys. Lett.* 23:277
68. Faissner, H. et al 1969. *Phys. Lett. B* 30:204
69. Balats, M. Ya. et al 1971. *Sov. J. Nucl. Phys.* 13:53
70. Carnegie, R. K. et al 1972. *Phys. Rev. D* 6:2335
71. Carithers, W. C. et al 1975. *Phys. Rev. Lett.* 34:1244
72. Bennett, S. et al 1968. *Phys. Lett. B* 27:239, 29:317
73. Strovink, M. W. 1970. PhD thesis, Tech. Rep. No. 6. Princeton Univ., Princeton, NJ
74. Carithers, W. C. et al 1975. *Phys. Rev. Lett.* 34:1240
75. Good, M. L. 1957. *Phys. Rev.* 106:591
76. Chaloupka, V. et al 1974. *Phys. Lett. B* 50:1
77. Geweniger, C. et al 1974. *Phys. Lett. B* 48:487
78. Geweniger, C. et al 1974. *Phys. Lett. B* 48:483
79. Geweniger, C. et al 1974. *Phys. Lett B* 52:108
80. Gjesdal, S. et al 1974. *Phys. Lett. B* 52:113
81. Dieperink, J. H. et al 1971. *Proc. Int. Conf. Instrument. High-Energy Phys., Dubna, 1970,* p. 251. Dubna: Joint Inst. Nucl. Res.
82. Holder, M. et al 1972. *Phys. Lett. B*

40:141
83. Banner, M. et al 1972. *Phys. Rev. Lett.* 28:1597
84. Wolff, B. et al 1971. *Phys. Lett. B* 36:517
85. Chollett, J. C. et al 1970. *Phys. Lett. B* 31:658
86. Barbiellini, G. et al 1973. *Phys. Lett. B* 43:529
87. Söding, P. et al 1972. *Phys. Lett. B* 39:1
88. Skjeggestad, O. et al 1972. *Nucl. Phys. B* 48:343
89. Christenson, J. H. et al 1965. *Phys. Rev. B* 140:74
90. Aronson, S. H. et al 1970. *Phys. Rev. Lett.* 25:1057
91. Carnegie, R. K. et al 1971. *Phys. Rev. D* 4:1
92. Cullen, M. et al 1970. *Phys. Lett. B* 32:523
93. Messner, R. et al 1973. *Phys. Rev. Lett.* 30:876
94. Genchev, V. I. et al 1974. Pap. No. 641. See Ref. 42
95. Galbraith, W. et al 1965. *Phys. Rev. Lett.* 14:383
96. Basile, P. et al 1965. *Proc. Balaton Conf.,* Balatonfüred, Hung., 1965
97. Bott-Bodenhausen, M. et al 1966. *Phys. Lett.* 23:277
98. Fitch, V. et al 1967. *Phys. Rev.* 164:1711
99. Böhm, A. et al 1969. *Nucl. Phys. B* 9:605
100. Jensen, D. A. et al 1969. *Phys. Rev. Lett.* 23:615
101. Darriulat, P. 1970. *Conf. Phys. Hautes Energies, Aix-en-Provence, 1970; Suppl. J. Phys.* 31 (11–12):C5–95
102. Saal, H. 1969. PhD thesis. Columbia Univ., New York; see also Ref. 17
103. Marx, J. et al 1970. *Phys. Lett. B* 32:219
104. Ashford, V. A. et al 1972. *Phys. Rev. Lett.* 31:47
105. Fitch, V. L. et al 1973. *Phys. Rev. Lett.* 31:1524
106. Williams, H. H. et al 1973. *Phys. Rev. Lett.* 31:1521
107. Paciotti, M. A. 1969. PhD thesis. UCRL-19446, Univ. Calif. Radiat. Lab., Berkeley; see also Ref. 18
108. McCarthy, R. L. et al 1973. *Phys. Rev. D* 7:687
109. Piccioni, R. et al 1972. *Phys. Rev. Lett.* 29:1412
110. Schubert, K. R. et al 1970. *Phys. Lett. B* 31:662
111. Weitkamp, W. G. et al 1968. *Phys. Rev.* 165:1233
112. Von Witsch, W. et al 1968. *Phys. Rev.* 169:923
113. Thornton, S. T. et al 1968. *Phys. Rev.*
Lett. 21:447
114. Pais, A. 1959. *Phys. Rev. Lett.* 3:242
115. Dobrzynski, L. et al 1966. *Phys. Lett.* 22:105
116. Baltay, C. et al 1965. *Phys. Rev. Lett.* 15:951
117. Crawford, F., Price, L. 1966. *Phys. Rev. Lett.* 16:333
118. Baltay, C. et al 1966. *Phys. Rev.* 149:1044
119. Baltay, C. et al 1966. *Phys. Rev. Lett.* 16:1224
120. Larribe, A. et al 1966. *Phys. Lett.* 23:600
121. Litchfield, P. J. et al 1967. *Phys. Lett. B* 24:486
122. Cnops, A. M. et al 1966. *Phys. Lett.* 22:546
123. Bowen, R. A. et al 1967. *Phys. Lett. B* 24:206
124. Gormley, M. et al 1968. *Phys. Rev. Lett.* 21:402, 21:339
125. Layter, J. G. et al 1972. *Phys. Rev. Lett.* 29:316
126. Thaler, J. J. et al 1972. *Phys. Rev. Lett.* 29:313
127. Jane, M. R. et al 1974. *Phys. Lett. B* 48:260, 48:265
128. Barrett, B., Tran Nguyen Truong 1966. *Phys. Rev.* 147:1161
129. Grigorian, A. et al 1974. Pap. No. 1075. See Ref. 42
130. Rittenberg, A., Kalbfleisch, G. 1965. *Phys. Rev. Lett.* 15:556
131. Baglin, C. et al 1967. *Phys. Lett. B* 24:637
132. Billing, K. D. et al 1967. *Phys. Lett. B* 25:435
133. Sober, D. I. et al 1969. *Phys. Rev. Lett.* 22:430
134. Buon, J. et al 1968. *Phys. Lett. B* 26:595
135. Anderson, R. L. et al 1969. *Phys. Rev. Lett.* 22:651
136. Kose, R. et al 1967. *Z. Phys.* 202:364
137. Bartlett, D. F. et al 1971. *Phys. Rev. Lett.* 27:881
138. Schrock, B. et al 1971. *Phys. Rev. Lett.* 26:1659
139. Barshay, S. 1965. *Phys. Rev. Lett.* 17:49
140. Austern, N. 1955. *Phys. Rev.* 100:1522
141. Lee, T. D. 1965. *Phys. Rev. B* 139:1415
142. Benz, P. et al 1973. *Nucl. Phys. B* 65:158
143. Rossi, V. et al 1973. *Nuovo Cimento A* 13:59
144. Boucrot, J. et al 1973. *Nuovo Cimento A* 18:635
145. von Holtey, G. et al 1974. *Nucl. Phys.*

B 70:379
146. Fujii, T. et al 1972. *Phys. Rev. Lett.* 28:1672
147. Berardo, P. A. et al 1974. *Phys. Rev. D* 9:621
148. Berardo, P. A. et al 1970. *Phys. Rev. Lett.* 24:419, 26:201
149. Guex, L. H. et al 1975. *Phys. Lett. B* 55:101, and private communication from C. Joseph, H. Schmitt
150. Heusch, C. A. et al 1974. Pap. No. 254, 255. Presented at *Proc. Int. Symp. Electron Photon Interactions High Energies, 6th,* Bonn, 1973. Amsterdam: North-Holland
151. Christ, N., Lee, T. D. 1966. *Phys. Rev.* 148:1520
152. Lee, T. D. 1965. *Phys. Rev. B* 140:959
153. Okun, L. B. 1966. *Phys. Lett.* 23:595
154. Rock, S. et al 1970. *Phys. Rev. Lett.* 24:748
155. Appel, J. A. et al 1970. *Phys. Rev. D* 1:1285
156. Prepost, R. et al 1968. *Phys. Rev. Lett.* 21:1271
157. Richter, A. 1975. *Proc. Symp. Interaction Stud. Nuclei, Mainz,* p. 191. Amsterdam: North-Holland
158. Kistner, O. C. 1967. *Phys. Rev. Lett.* 19:872
158a. Blume, M., Kistner, O. C. 1968. *Phys. Rev.* 171:417
159. Boulware, D. G. 1965. *Nuovo Cimento A* 40:1041
160. Feinberg, G. 1965. *Phys. Rev. B* 140:1402
161. Meister, N. T., Radha, T. K. 1964. *Phys. Rev. B* 135:769
162. Schwinger, J. 1964. *Phys. Rev. B* 136:1821
163. Wolfenstein, L. 1974. *Nucl. Phys. B*
77:375
164. Salzman, F., Salzman, G. 1966. *Nuovo Cimento A* 41:443
165. Lee, T. D. 1974. *Phys. Rep.* 9:143
167. Pais, A. 1973. *Phys. Rev. D* 8:1226
168. Frenkel, J., Ebel, M. E. 1973. Univ. Wis. Preprint
169. Dress, W. B., Miller, P. D., Ramsey, N. F. 1973. *Phys. Rev. D* 7:3147
170. Baird, J. K. et al 1969. *Phys. Rev.* 179:1285
171. Ramsey, N. F. 1976. *Bull. Am. Phys. Soc.* 21:61 and private communication
172. Mezei, F. 1972. *Z. Phys.* 255:146
173. Burgy, M. T. et al 1960. *Phys. Rev.* 120:1829
174. Erozolimsky, B. G. et al 1970. *Sov. J. Nucl. Phys.* 11:583
175. Steinberg, R. I. et al 1974. *Phys. Rev. Lett.* 33:41
176. Calaprice, F. P. et al 1969. *Phys. Rev.* 184:1117
177. Calaprice, F. P., Commins, E., Girvin, D. C. 1974. *Phys. Rev. D* 9:519
178. Barnes, S. W. et al 1960. *Phys. Rev.* 117:238
179. Overseth, O. E., Roth, R. F. 1967. *Phys. Rev. Lett.* 19:391
180. Cleland, W. E. et al 1967. *Phys. Lett. B* 26:45
181. Conforto, G. 1969. *Acta Phys. Austriaca Suppl.* 6:435
182. Niebergall, F. et al 1974. *Phys. Lett. B* 49:103
183. Sakurai, J. J. 1958. *Phys. Rev.* 109:980
184. Cabibbo, N., Maksymowicz, A., 1964. *Phys. Lett.* 9:352, 14:72
185. Longo, M. J. et al 1973. *Phys. Rev.* 181:1808
186. Sandweiss, J. et al 1973. *Phys. Rev. Lett.* 30:1002

Ann. Rev. Nucl. Sci. 1976. 26:51–87

GLOBAL CONSEQUENCES OF NUCLEAR WEAPONRY

×5571

J. Carson Mark

Los Alamos Scientific Laboratory, Los Alamos, New Mexico 87544

CONTENTS

INTRODUCTION

The first fission chain reaction of which we have knowledge occurred about two billion years ago as the result of the concentration of uranium by natural processes at the site of what is now the Oklo Mine in the Republic of Gabon in western Africa.[1] Man became aware of the possibility of such a reaction around 1940 and, by dint of great ingenuity, constructed an artful facsimile in Chicago in 1942. Happening as it did during a period of all-out war, the potential military or explosive use of the phenomenon was of paramount interest. The atomic bomb was the first result of which the public became aware.

The effects associated with the explosive release of the very large amounts of energy realized in nuclear weapons have been measured in connection with many test explosions; the results of such observations of the principal phenomena (blast wave, thermal radiation, and ionizing radiation) have been presented in detail (1) and in Volume 18 of the *Annual Review of Nuclear Science*. The consequences of the actual use of atomic weapons in 1945 at Hiroshima and Nagasaki have also been fully described. There is little new to say on these subjects.

Fortunately, we have as yet no experience with the use on actual targets of weapons very much larger than those employed in 1945 (15–20 kilotons) nor with the use of many nuclear weapons at one time. A necessarily hypothetical description of some of the possible consequences of events of this kind are given in the first sections of the following. The effects of the mere existence of nuclear weapons in conditioning our thought and behavior over the past thirty years are, however, not at all hypothetical. Aspects of these are considered in the final section.

[1] Added in proof: Cowan, G. A. 1976. *Sci. Am.* 235(1):36–47.

52

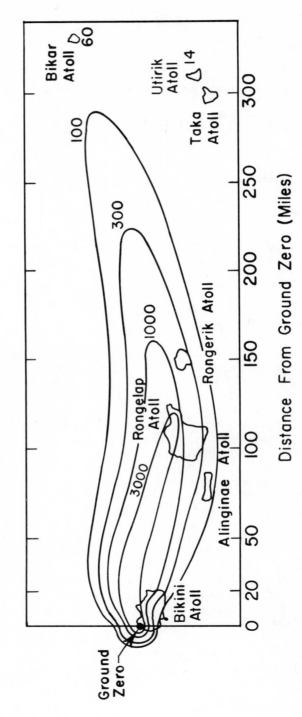

Figure 1 Isodose contours from Bikini test.

1   THE MAIN PROBLEM

On March 1, 1954, the United States tested a nuclear device—the Bravo shot of Operation Castle—on a coral reef at Bikini Atoll. It had a total yield of about 15 megatons. Within 24 hours, radioactive debris settling back to the surface had contaminated a long, oval-shaped area over 60 miles in width and extending more than 300 miles to the east in the downwind direction. Most of this area is open water, but there are several atolls in the region and 267 persons were on these atolls at the time. These people were evacuated about 2 days after the test. By then they had been exposed to levels of radiation ranging from a low of 14 r on Utirik Atoll 310 miles east of Bikini, to a high of 175 r on the southernmost part of Rongelap Atoll 105 miles from the test. Probable isodose contours are indicated in Figure 1 (1).

At a location downwind from such an explosion, fallout does not occur until the wind has moved the cloud of debris to a position over that location, so that 100 miles away in the direction of a 15 mph wind, radiation does not appear on the surface until after a lapse of 6 or 7 hours. Starting at that time, dust and granules falling and trailing from the cloud begin to accrete on the surface and the radioactive fission fragments they contain create a radiation field that increases as more particles arrive; that is, until the debris cloud (which is about 60 miles across in the case considered) should have passed by. Thereafter, the radiation rate on the surface (measured in r/hr) will decrease as $t^{-1.2}$, the decay rate applying to a mixture of fission fragments. Table 1 lists the dimensions of the regions in which an accumulated total exposure, equal to or larger than the level indicated, would be received in four days by an unprotected person; and also the time after the explosion at which the fallout might arrive at the extreme range.

Assuming that all the fallout were deposited just at the times indicated in Table 1 and that none of the material were subsequently removed by weather action, the curves of Figure 2 show the radiation rate and the accumulated exposure at later times. During the few hours while the fallout was in progress the values would differ from those shown, but they would not differ greatly at times after the

Table 1   Region exposed to various radiation levels

96-hr dose (roentgens γ, in air)	Length (miles)	Maximum width (miles)	Area (square miles)	Time of arrival (hrs)
($\geq$)3000[a]	~120	12	1,000	7.0
1000[b]	160	30	3,600	8.5
300[c]	220	45	8,000	12.0
120[d]	275	61	13,000	15.0
60	330	—	—	—

[a–d] See corresponding curves in Figure 2.

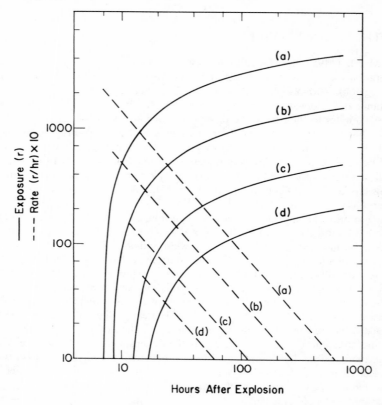

Figure 2   Accumulated exposure and radiation rate versus time for levels of Table 1.

fallout was complete. Generally similar curves, but shifted to earlier times, would apply at points closer to the explosion where the fallout would arrive sooner.

Of the 64 persons on Rongelap, about two thirds experienced symptoms of nausea and loss of appetite for a few days (2). None of these was incapacitated, and during the next 18 years no deaths among them could be attributed to radiation injury. Nevertheless, by a few weeks after the event their blood counts showed marked anomalies; these continued for about 15 years. In 1972 there was one death (of a 19-year-old youth) from leukemia, which was quite possibly related to the radiation exposure he had received in 1954. During the first 9 or 10 years no symptoms of thyroid malfunction were clinically apparent, but beginning in 1963 or 1964 trouble began to show up; by 1974, when 3 or 4 instances of thyroid lesions and impaired functions would have been expected in a similar unexposed group, 24 had occurred. Evidently, their exposure ($\sim$175 r), though not quite at the level where immediately obvious and serious effects would result, was a gravely damaging exposure. It is now assumed (3, 4) that of those who receive a total dose of radiation in the range 200–250 r, and who receive this total within a time as short as a few days, about half will require hospitalization for a week or so and about one in twenty will die. At a level of 400–450 r, all will require medical care and about 50% will die. No one has yet survived an acute exposure larger than 650 r.

Had this explosion occurred on the surface a few miles west of Elizabeth, New Jersey, the immediate effects on Manhattan or Long Island would not have been severe, though a lot of window glass would have been shattered, resulting in numerous injuries. Out to 20 miles from the scene (including, therefore, lower Manhattan and much of Brooklyn) many of the people on a clear line of sight to the fireball would have suffered second-degree burns (burns, that is, deep enough to cause blisters) on exposed portions of their skin. On the closest parts of the two islands some fires would probably have been started at places where the light from the explosion could shine directly on accumulations of combustible material. However, except possibly for rather flimsy sheds, buildings would have remained structurally intact and the amount of prompt nuclear radiation (neutrons and gammas) would have been too small to measure.

Gruesome disaster would have struck in New Jersey. There, in a series of rings around the point of burst, particular effects would have resulted in various kinds and degrees of damage. First, to a radius of about a third of a mile, there would be the crater. Though some hundreds of thousands of tons of the displaced material would have been vaporized and incorporated in the rising cloud, most of it would have been thrown into a ridge extending to a radius of about half a mile and having a maximum height of about 175 feet (~ 14 stories). Out to one and one half miles the fireball, which would have stopped growing at about that radius, would bathe the surface in an atmosphere of incandescent air with temperatures of a few thousand degrees centigrade for the first 15 seconds or so until it started to rise clear of the ground. Prompt neutrons and gammas would have provided a lethal exposure (> 1000 r) to an unprotected person at a radius of 2 miles; however, already at a distance of $2\frac{1}{2}$ miles and beyond, the prompt radiation exposure would not be serious (< 50 r), even for people without protection. [In fact, for weapons in this high- (megaton) yield class, injury from exposure to prompt radiation is not of much consequence, since the lethal ranges of blast and thermal effects for a person in the open are considerably larger. This is different from the situation at Hiroshima where, with a 20-kiloton weapon, the ranges of serious hazard from the effects of prompt radiation and blast were each about 1 mile.] Out to 5 miles almost all buildings would have been destroyed, or suffered considerable damage, and most of the occupants would have been injured—many fatally. Almost all persons in the open would have been killed, or have suffered fatal injuries—from flash burns, being thrown about, or being hit with flying or falling objects. In a further ring, to the 10-mile radius, many residential and apartment-type structures would have been severely damaged and many fires started (as also, of course, at closer locations) by the incendiary effects of thermal radiation, by electrical short circuits, and as a result of broken gas pipes. In the inner half of this ring many streets would be blocked with wreckage and overturned vehicles. Even near the outer part of this ring persons standing in the open, when struck by the blast wave sweeping out from the explosion, would be bodily impelled at such a velocity that collision with a wall or merely tumbling in the street would be extremely hazardous.

As already mentioned, there would also be many injuries in the region beyond ten miles; however, the greater the distance, the smaller the fraction of the population affected and the smaller, also, the fraction of the injuries that would be fatal. One possible effect at large distance is injury to the eyes. The range at which a person looking directly at the fireball would suffer burns on his retina depends on such things as the visibility at the time and whether his eye is bright- or dark-adapted, but in any case is larger than the range for second-degree burns on other parts of the body. However, in connection with a surface burst, a person at a considerable distance would have to be well above the ground (as up in a tall building) and looking directly at the spot, to experience such damage. Statistically speaking, this would be an unimportant class of injury. Temporary flash-blindness, on the other hand, whereby the eye is unable to receive new images for several seconds or longer, would occur at larger ranges than retinal burns, and, though not a very serious type of injury, would be experienced by a large fraction of those who happened to be outdoors, since the initial flash of light is so brilliant that it is not necessary to be viewing the fireball directly for the eye to be dazzled. Over a large area, then, most of those operating motor vehicles at the time would be affected and would be temporarily out of control. In moderate or heavy traffic, collisions and pile-ups would be certain to result; and the more heavily traveled roadways would be suddenly blocked at numerous points far outside the area affected by the blast.

The whole region suffering immediate effects would also be subjected to fallout, but for the most part only after an appreciable delay, since it requires some time for the fallout pattern already described to be established. The sequence of events is as follows: at first, all the residual radioactive material from the explosion, along with the other weapon material and very much larger amounts of material boiled off the surface, are present in the fireball in the form of extremely hot gas. Only as the fireball rises and expands and cools can these normally solid materials begin to condense into small droplets. At this time also, large quantities of dust and grains of material swept up from the surface are mixed into the cloud and provide condensation nuclei on which radioactive material is plated out. As it solidifies, by far the major portion of the radioactivity is lodged on particles so small ($\ll 1$ mm) that they continue to be carried upward as long as the cloud is rising. When the cloud approaches its ultimate height they can begin to fall back to earth under the force of gravity—the larger ones sooner and faster. In the present case, the cloud by then has a radius of about 30 miles. Its base (at the tropopause) will be at an altitude of 40 or 50 thousand feet, and its top somewhere above 100,000 feet. It takes about 10 minutes for the cloud to stabilize at its final height, and something like 20 minutes for the largest particles present in important numbers to fall back to the surface from the bottom of the cloud. Thus, in the area in which small particles from the cloud would reach the surface (which may be displaced laterally a few miles from the area directly below the cloud by the influence of the winds between the surface and the bottom of the cloud), radiation from fallout will not begin in earnest until about half an hour after the explosion. After that, however, the level will increase very rapidly towards the 1000-r/hr range; so that in this region a person without shelter would receive a dangerous exposure in a rather short time.

This brief period of grace from the onset of fallout radiation, which applies over most of the area within 20 or 30 miles of the burst point, is not similarly available in the area within about 5 miles of the explosion. Two processes are involved that are important only in this close-in area. In the case of a surface burst about half of the neutrons released are captured in the materials of the surface. Many of the capture products are harmless, such as that resulting from capture in ordinary hydrogen, but some are highly radioactive. Examples are captures in aluminum, sodium, and manganese, all of which are sufficiently common in soils and building materials that the residual induced radioactivity—though small on the scale of the total activity—would be quite significant when concentrated locally, as some of it will be. Much of this induced activity will be carried aloft with the cloud (from which it will contribute its small component to the general fallout); but some will remain where it was formed and some, though stirred up and carried about by the winds associated with the blast wave, will fall back to the surface without ever going very high or far. Since rather few neutrons penetrate as far as 2 miles from their source, the induced radioactivity on the surface is confined to a region a few miles across, but in that region a strong radiation field is established immediately. The other localized process has to do with the stem—the column of air rushing vertically upwards to fill the space vacated by the rising cloud. In the present case this column will be several miles in radius, and, for the first minute or so, the vertical wind velocity will be 200–300 miles per hour. By this means pebbles and even larger objects will be carried high enough to come in contact with parts of the cloud and have radioactive material deposited on them. The rate of rise of the cloud and the velocity of the vertical wind decrease rapidly, so that within a very few minutes, pebble-like objects are free to start back to the surface. As a result, while the fireball is still rising, an intense radiation field is already building up in the area directly below the stem. There is a continuous transition from this process to the general process of fallout from the stabilized cloud, so that downwind from the area under the stem there will be a spur along which some fallout will arrive relatively early.

With this pattern and timing of fallout, even an uninjured person in the five-mile zone could not safely leave the building in which he happened to be, except possibly for a single, brief excursion. Those in the five- to ten-mile zone who were not hurt would have a little time to get away from burning buildings or proceed to nearby shelter, but very little time to help others. An attempt on the part of a large fraction of the population to escape from this area would prove fatal for many. To embark on such a mission with any confidence, it would be necessary to know that not too many others were trying to do the same, which routes were open, and which would avoid the path of the cloud. This last depends on the wind direction at various altitudes, so the wind direction at the surface is not a fully reliable guide. One would also need to have enough gasoline since, with hand-operated pumps at service stations about as common as wood-burning locomotives, it is rather difficult to obtain gas in the event of a power outage.

The post-detonation predicament of those in this area would be much grimmer than that of those in the (smaller) corresponding area at Hiroshima. There, since an air burst was involved, there was no close-in fallout nor any significant residual radiation on the surface. People who could move about were able to get away from fires and to assist others without thereby being exposed to additional danger. People from outside the area of severe damage were able to come in and help some of the injured. Such fire-fighting equipment as had not been put out of action by the explosion went to fight some of the outlying fires. In the Japanese cities in the area corresponding (having similar blast pressure) to our five-mile zone, over 70% of those present were killed and half of the survivors were injured; in the area corresponding to our outer ring about a third were killed and half the survivors injured. In the present situation, with mobility sharply curtailed and outside assistance unavailable because of the extreme fallout hazard, the incidence of fatalities and injuries would be much higher. The particular region considered (Union, Essex, and Hudson counties in New Jersey, along with Staten Island) has a population density of about 7000 per square mile. The area of the inner zone is 80 square miles, and that of the outer, over 200, so that the number of persons exposed to the severe immediate effects described could be more than half a million in the inner circle and a million and a half in the outer ring. The number of fatalities in this phase of this single engagement could exceed the number of US battle deaths ($\sim 3/4$ million) in all wars from the American Revolution through Vietnam. Of course, many considerably "richer" targets—of larger population density—could easily be found.

As pictured, both Manhattan and Long Island, with their 8.5 million inhabitants, escaped with very little direct damage. However, the debris cloud when fully developed would overcast all of Manhattan, Brooklyn, and Queens, as well as the Bronx on the mainland. Within the hour fallout would begin to arrive over most of this area. On Long Island it would first appear and be most intense in the region where all the bridges are located. Only from the outlying suburbs on the mainland would evacuation to the north offer a reasonable prospect for any appreciable fraction of the population. Assuming a steady moderate west-by-south wind, the cloud would drift over Long Island. In about 5 or 6 hours it would have moved off to the east, but by that time enough fallout would have accumulated on Manhattan and the whole of Long Island to provide a radiation exposure greater than 1000 r in the 96-hour period following the explosion. Over much of this area the 96-hour exposure would exceed 3000 r (Figure 3).

To avoid radiation injury with serious early effects, a person must keep his total exposure during any period as short as a few days below 150 or 200 r. To manage this where the exposure in the open was headed towards 3000 r would essentially require that a person already be off the street (and off the beach, the parkway, and the LIRR) by the time the fallout arrived, and remain in a sufficiently

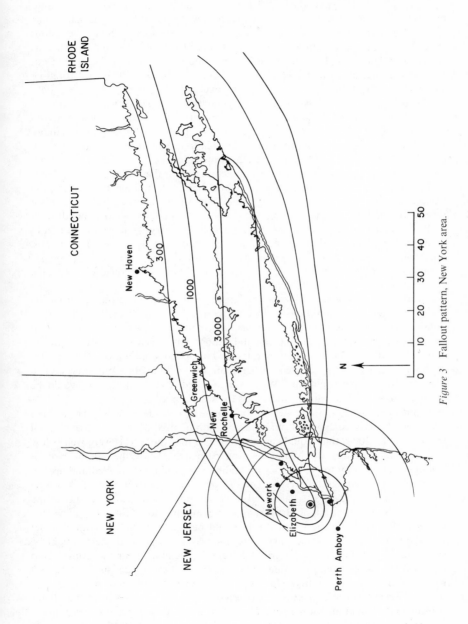

Figure 3   Fallout pattern, New York area.

sheltered place for several days. A basement without any exposed walls would probably provide adequate shielding; a basement with windows or some of the walls exposed might not; a mobile home would be completely useless. Since the fallout particles collect on flat surfaces such as streets and roofs, the central portions (away from windows) of the middle floors of a multistory building, being far from the sources of radiation, might serve. By some means or other the would-be survivor would need to arrange that he was exposed to no more than a twentieth of the outside level, or that he had an intensity-averaged shielding factor of ~ 20—though without a good radiation meter he would not know just how well he was doing. Over all or most of Long Island, through the first several days, attempts at evacuation, extended periods devoted to rescue work, repair of services, or police duty would be suicidal. In 1972, in the whole of New York City the fire department was required to cope with an actual fire on the average of once every $4\frac{1}{2}$ minutes. Through the first days, any fires that could not be handled by the persons and equipment already in the building would have to be left to burn themselves out. At standard rates, among the ~ 7 million persons huddled in shelters on Long Island, there would be one birth every $4\frac{1}{2}$ minutes, and from cardiovascular causes alone, one death or ominous attack every 15 minutes. Shattering decisions would be called for; to run the gauntlet of radiation (of unknown intensity) in order to reach hospital facilities and medical care, or take one's chance with what might be at hand.

A rational decision that it would be appropriate for some group to leave its shelter and get away from the Island would require information that could only be obtained outside the shelter. In particular, the results of a survey would be needed to develop an estimate of the incremental exposure involved in transit out of the contaminated area. With radiation rates at the 3000-r contour of 20 r/hr after 48 hours, such a survey started earlier than that would burn out survey crew members at a rather rapid rate. Whenever initiated, considerable time would be required to conduct the survey and assess the situation for each locality.

About the least restrictive criterion that could be considered would be that the incremental exposure expected during evacuation be itself a modest percentage of the amount (~ 200 r) that would cause trouble, so that those who were well below this level at the start would not become casualties. With the fallout pattern envisaged, evacuation to New Jersey or Manhattan would be excluded and, except for a few square miles at the western end of Long Island for which the Triborough Bridge might be the best alternative, all traffic would have to be funneled over the Throgs Neck and Whitestone Bridges and keep moving northwards for a few tens of miles after reaching the mainland. It would be necessary to allow for the fact that the radiation level would vary along the way and that some delay might be encountered at any point on the route. Except for locations near the Long Island end of these bridges, the minimum time base would have to be taken as several hours, even if an orderly dispatching of parties could be enforced. At the end of the first week it is quite likely that there would still be places within the 3000-r contour with radiation rates above 5 or 10 r/hr, so that evacuation (within an incremental exposure limit of 25 r) would not be feasible from all localities as early as that. By the end of the second week, with rates down by a further factor oi or more, evacuation might conceivably have been completed under the (rathₖ generous) criterion suggested.

By no means would all responses be of an orderly sort, or based on purely rational and technical considerations. A large number of isolated shelters would be in use whose very existence was unknown to any central authority. Not all would have radio equipment available and operating whereby they could receive information and instructions. Many persons would find themselves in makeshift or overcrowded shelters where problems of the supply of food—as well, possibly, as that of water, heat, light, and sanitary facilities—would soon become acute. Some would shortly come to the unshakeable conclusion that all hope for the future depended on getting away quickly, particularly when opposed only by a purely verbal construct of an impalpable and invisible hazard that was expiring, if not already expired. At least for several days neither police nor wardens could be deployed to patrol the streets effectively, and a large number of individual decisions would be made and implemented, whether judicious or not. With a general background of deep anxiety, mounting misgivings and discomfort, and at least some hysteria, it would not be long before the vaguest rumor that others had taken off without incident would bring whole subterranean contingents out onto the street, as might also the sight of someone displaying the well-known symptoms of nausea and vomiting, whether these were radiation-related or not.

From causes of this sort, along with more elementary accidents and errors such as not getting the word; not getting under cover soon enough; shelters inadequate to the need; or shelters rendered uninhabitable by fire, flooding, freezing, or whatever—from all the wealth of opportunity for fearful, headstrong, or even heroic actions, as well as pure misfortune, a very large number of serious overexposures to radiation would occur. Unless this overexposure were in the high range (> 500 r, say, where death in a short time is probable), after a day or so of discomfort (nausea and vomiting), there would usually be a latent period of 2 or 3 weeks (1) before clear symptoms became evident and those affected really knew they were in trouble.

The situation on Manhattan and in the Bronx would be essentially similar to that on Long Island, except that from being closer to areas free of fallout it would be easier to get away from the Bronx. As, over the first week or two, the several million people involved got clear of the fallout area, a general feeling of euphoria would prevail. However, discomfort and anxiety would have followed them. Outnumbering the population of all the rest of the state, they could not be accommodated by existing water, sanitation, housing, or hospital systems. This last in particular would pose a problem since, of the 53,000 hospital beds in the state, the 36,000 in New York City would be in the course of being vacated and abandoned, with many patients needing relocation, just as an enormous new case load was appearing. Except for those who could be taken in by friends, the great majority of these millions of displaced persons would have to be put up in emergency camps. Most would arrive with little or nothing but the clothes on their backs and everything from toothbrushes to typewriters to telephones, would have to be found. With their bank offices closed and no access to paychecks or usual means of livelihood, their every need would have to be supplied from outside. Certainly, if this were the only such instance in the country, there would be no question but that such needs could and would be met; though, even with the utmost efforts, there would be an initial period in which supplies, services, and arrangements would be meager.

It would be a real question whether the needs of such a group could be met adequately and rapidly enough to undercut the basis for outbreaks of antisocial behavior. Their economic prospects in complete disarray, knowing, though not to what degree, that their expectations for healthy, trouble-free years ahead have been impaired, here they are, consigned to a grubby and possibly extended period of shortages and narrowed opportunity. In contrast to the situation following some natural disasters—where it is understood that God chose the victims—it would be realized here that their election to hardship was chargeable to some fault on the part of some government or some society, not necessarily excluding their own; and here they are, confronted with the apparent continuation of a relatively comfortable, well-fed and well-furnished normalcy outside the gates of their emergency camps. Real events of the past—such as when groups of American Indians, Mexican-Americans, persons of Oriental extraction on the West Coast, or even sheep ranchers, held lands or assets coveted by other, larger groups of Americans—are at least as reliable a guide to what to expect as any mystique that Americans are governed in their actions by a solicitous concern for the rights of others. The less reassuring the prospects for early alleviation of discomfort and the longer the period for which evacuation might be expected to be necessary, the more probable that marauding bands would set out to improve things for themselves. Even the mere appearance of well-being might suffice to identify a target, but the farms, homes, and business premises of persons having (or thought to have) a national origin related to that of the enemy in question would, in the Air Force jargon, be prime targets.

Leaving that as it may be (and under very heavy stress it could be very bad indeed), and reverting to the officially favored assumption "that positive, adaptive behavior would prevail over antisocial behavior and that the survivors would support reestablishment of normal cooperative relationships at all levels of community life" (5), it remains to mention that apart from the general effects of a cessation of activity in New York City, and apart from the harrowing prospects faced by the residents, individuals scattered across the country would find their means of support cut back or cut off. With all gainful activity in the city stopped, all payments dependent on current income—whether rents, mortgages, interest, loans or whatever—would do likewise. Any insurance company with a major portion of its assets invested in New York might have to suspend annuity payments. A patrimony of New York–based securities would suddenly become worthless. Disruption on a major scale would result from the closing of New York's large commercial banks and exchanges. Though for outsiders this might be merely a temporary inconvenience since alternative arrangements could soon be put in operation, once this was done New York would probably never regain its hegemony in these matters.

The depth of such peripheral effects, as well as the ultimate severity of the problems confronting the surviving but displaced inhabitants, would depend greatly on the length of time before normal activities could be resumed. A number of factors are involved in this in addition to the natural falloff in radioactivity. These would include the extent to which decontamination measures might be possible or effective, the extent to which mere weathering would reduce the radioactivity, and the standards adopted for acceptability.

With peacetime standards of the sort deemed appropriate in connection with the Reactor Safety Study (3), this might take quite a while. The criterion adopted there is that an area should be evacuated if the residents would be expected to receive an accumulated dose of 10 r during the year following an accident. These standards, of course, are chosen—as are those for the maximum permissible exposure of persons occupationally exposed to radiation (not more than 5 r per year)—with the intent of minimizing the possibility of deleterious effects of radiation on lifespan and heredity. Allowing for the fact that after about 6 months the radioactive decay of fission fragments proceeds considerably more rapidly than the $t^{-1.2}$ law that applies at early times, but not allowing for the effect of weathering, the exposure during the period between 12 months and 24 months at a point that had received 3000 r in the first 96 hours could still be about 80 r, with another 25 r in the year following. Of course, a few good rains would be likely to wash off smooth surfaces such as tin roofs and paved roadways, so that some local areas could be usable long before the natural decay of radioactivity would suggest. At the same time, other patches, such as lawns and gardens or spots where dust might collect—porches, carports, and filters of air-conditioning units—would provide a wide range of radiation levels, including some hot spots; these would have to be examined one at a time, at least if peacetime standards were to be applied.

The National Council on Radiation Protection, which has developed the criteria for use in normal circumstances, has also developed criteria for use in an emergency (4). In that case interest attaches to the largest exposure tolerable without resulting in incapacitation. The NCRP concludes that a brief exposure of 200 r may be regarded as the dividing line between doses that will and will not cause sickness requiring medical care. Since a larger total exposure can be tolerated if it is delivered over a longer period of time, the NCRP has also concluded that no medical care would be necessary for persons receiving a total exposure of 300 r in a period of 4 months, provided that no more than 200 r of this total should be received in any one month, nor more than 150 r in any one week. Even taking full advantage of these standards, a person who wished to spend four months in an area that has received a 96-hour, 1000-r exposure could not enter until after 30 days. Had the 96-hour exposure been 3000 r, he would have to wait until after 110 days. As before, these estimates do not assume any mitigation as a result of weather action, nor do they allow for the possibility that by spending part of his time in a shelter he could enter earlier. They also do not allow for the fact (noted by NCRP) that the survey meters and dosimeters generally available, when properly calibrated and operating correctly, will indicate exposure rates or accumulated exposure only to within about 20 or 25%.

It must be emphasized that these emergency standards guard only against the prospect of being put out of action. They could assist in determining the feasibility of early reentry for some urgently necessary purpose, but even in an emergency they could not provide the basis for a decision to resume normal operations. They allow for a little more exposure than was received by the inhabitants of Rongelap. In addition, it would be expected (6) that a group of persons exposed to 200 r, whether in a short time or over many months, would experience twice as many deaths from cancer over the next 25 years as would a similar group not exposed. (The effect of very small exposures on the cancer death rate is not known; as a matter of prudence, it is usual to assume that the increase in such fatalities is directly proportional to the exposure.) For persons who may subsequently produce children, a similar exposure would be expected to at least double the number of mutations appearing in their offspring, and hence the incidence of those genetically related diseases that are proportional to the mutation rate. Though not all types of abnormalities are directly linked to the mutation rate, it is believed (6) that those that are have together a normal incidence of 1% or more.

At whatever radiation level the area, or some parts of it, might be declared ready for reoccupation, planning and control of the reentry would be required. Transportation, utility, and building services would have to be restored, and some provision made for the protection of still-unoccupied premises. Since normal commercial activity and a general return to the status of gainful employment would build up rather slowly and unevenly, it would remain necessary for some time to supply food and other needs of the returning residents. Again, provided not too many other metropolitan areas in the country were in a similar situation at the same time, all this could certainly be done. Thus, with a great deal of cleanup work, a really well-run organization, perhaps a little help from the weather, and possibly some compromise of the acceptable radiation level, the evacuation of the inhabitants of Long Island might not have to last as long as that of Rongelap (over 3 years; the inhabitants were permitted to return only in July, 1957), nor even much longer than a year, but it certainly could not be much less than a year. Whenever they returned they could still take comfort from the fact that Long Island had not actually been hit.

Had the Bikini explosion occurred on the western border of Iowa, the consequences, though involving many fewer people, would still have been impressive. In this case— given an hour or two's warning and reliable instructions as to where to proceed— prior evacuation of a large fraction of the inhabitants could at least be imagined, provided of course that the family was not scattered, the car was at home and was operable, and so on. The situation of those unable to get away would be very like that already pictured.

A large consideration in such an event would be the effect of heavy fallout on agricultural activity. Iowa, with an area of 56,000 square miles (less than 2% of the United States) provides a tenth of the nation's food supply, has the largest livestock industry, produces 20% of US corn, and ranks among the top states in output of

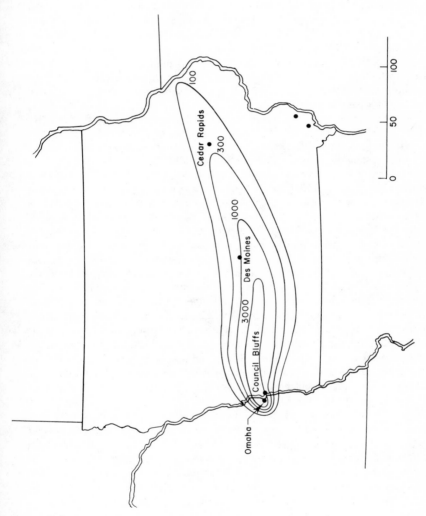

Figure 4 Fallout pattern, Iowa.

several other important crops. In Figure 4 the Bikini fallout pattern is superimposed on Iowa.

The exposure history on the 4-day, 120-r line (Figure 2) is rather close to that chosen by the NCRP as an emergency standard for the purpose of avoiding short-term effects. The area within that contour ($\sim$ 13,000 square miles), where higher exposure levels would apply, is about one-fourth of the area of the state, and in that region agricultural work would necessarily be sharply cut back or suspended. While, in principle, persons outside this line could continue their usual activities without actually becoming ill, the disruption of regular work would obviously extend much farther. The loss of foodstuffs resulting directly from such an interruption would vary with the season, but it would be large if the effect were that crops were not planted or not harvested.

Quite apart from the consequences of a slowdown or stoppage of work, the field crops themselves could be affected by exposure to fallout, either from plant yield being reduced as a result of radiation damage, or from radiological contamination of the products. Clearly, if the plants were already mature at the time the fallout arrived, the amount of their yield would not suffer, though its usefulness might. Under conditions where there is appreciable deposition, direct contamination of the plants is of much greater significance as a source of fission products for animals and man than uptake from the soil (7). Some of this material is absorbed through the leaves, but much of it is retained mechanically. The inflorescences of grain crops, for example, appear to provide an excellent trapping device for deposited particles—indicating a possible inhalation hazard associated with operations such as harvesting, threshing, and milling contaminated crops. The basic question would be whether, in view of its contamination level, the end product would be acceptable in the food chain. In a recent study conducted by the National Academy of Sciences (8) it was noted that with an average of 2.5 Ci/mi^2 of ^{90}Sr some agricultural products would have a contamination level at which control and evaluation would be required. In the present instance the area affected to this extent by early fallout would be at least 15,000 square miles. The actual loss of foodstuffs would depend on the standards deemed appropriate; but in any event for several years the crops from some such area would have to be segregated and monitored so that their disposition could be determined.

Should growing crops be exposed to heavy fallout, there could be some loss in yield, but no generally applicable estimates of the extent of such losses are possible. This is partly because of the great range of sensitivity to radiation among plants of different species; partly because the sensitivity for a given species varies with the stage of development; and partly because, although fallout would expose plants to beta as well as gamma radiation, very few experimental observations of the sensitivity of field crops to β rays, and none on the effects of exposure to the combination of β and γ radiation appear to be available. These might be simply additive, or might be synergistic. A considerable amount of work has been done to determine the effects of gamma radiation alone, and for plants of economic importance a tabulation has been made (9) of the exposure that would reduce the crop yield to 50% of normal (the YD_{50}). The most sensitive group (< 1000 r) includes a variety of pea; wheat and corn appear in the range 2000–4000 r; and among the least sensitive ($>$ 24,000 r) is okra. However, YD_{50} levels less than 1000 r have also been observed for wheat and corn irradiated at particular early stages. At the symposium in 1970 the need for a clearer idea of the effects of β radiation and of combined exposure was recognized as the outstanding requirement before meaningful assessments of the effects of fallout on crop yields could be made. Apparently this requirement is still (1976) outstanding since, starting soon after that symposium, the funding for such study—involving as it does slow-moving, complicated, and rather expensive experiments—was very greatly reduced.

Though not a food crop, coniferous trees are particularly sensitive to exposure to γ radiation. Most deciduous trees can survive exposure to several thousand r, but for almost all coniferous species the LD_{50}, the level at which half are killed, is less than 1000 r and less than 500 for some. With the fallout pattern considered here, essentially all coniferous trees in an area of 4000 square miles would be killed.

The effect on livestock of exposure to fallout is much better known than that on field crops, and since controlled experiments have been possible, it is better known than for humans. The LD_{50} values for exposure to γ radiation for sheep, cattle, and hogs are (9), respectively, 400, 500, and 640 r. A typical barn offers a protection factor of about 2, so that cattle in barns would fare somewhat better—as also, because of self-shielding, would livestock in corrals or feed lots, particularly those with sense enough to stay in the middle of the pack. Though animals in that situation could survive for a while without food, they would have to have water within a very few days. The skin of animals without cover would also be exposed to β radiation from fallout caught in their hair. While this by itself would not be expected to be fatal at the levels where there was hope of surviving the γ exposure, it does contribute to increased mortality rates and the LD_{50} levels are lowered by about 10%.

A more dramatic effect would apply to the ruminants (cattle and sheep) exposed while on pasture or range. From ingesting contaminated forage such animals would undergo β radiation in their gastrointestinal (GI) tract. With the combination of γ radiation and the appropriate relative amount of GI β exposure, the LD_{50} for cattle is lowered to 180 r. For damage-assessment purposes it is assumed that the number of fatalities will be the number exposed to the LD_{50} or more, and that the rest survive; thus in about 20% of the area of the state, cattle left out on pasture would receive a lethal exposure within a few days.

Many of the doomed animals would survive for several weeks and, if slaughtered before they succumbed, would not necessarily be lost to the food supply. Exposure to γ radiation does not affect the food value of the meat, and no instances of bacterial invasion were found in the animals studied. As stated by M. C. Bell (9), "There is no evidence that these animals could not be used for food, especially if food were scarce."

At early times milk products would require the most extensive surveillance to check for possible contamination with iodine: the 20-hour [133]I, and the 8-day [131]I. Ng & Tewes, in a study of the transfer of fallout from forage to milk (9), conclude that an adult who consumed one liter per day of milk from dairy cattle on pasture under fallout conditions about the same as those applying on the 120-r γ contour would be exposed to a dose of a few thousand r of β radiation in his thyroid, where the iodine is concentrated. This would not be expected to have a serious effect on an adult, but it is at least ten times the level that would be expected to show damaging effects in children (who, partly because the size of the gland is smaller, so that a given amount of iodine concentrated there results in a higher localized radiation level, and partly because the body's control of growth processes is handled by this gland, are more susceptible to damage of its functioning), is larger than the exposure believed to have been experienced by those on Rongelap in 1954 and is more than two orders of magnitude larger than the level at which milk was discarded in the United Kingdom after the Windscale reactor accident in 1957. With the Windscale criterion, the area from which milk would have to be impounded would be about 60,000 square miles. Depending on the wind velocity and direction, this region could extend to the Ohio-Pennsylvania border. If, as an emergency measure, it were decided that a ten times less stringent criterion should be used, the area would be only half as large and might extend only to the Illinois-Indiana line. In order actually to meet any particular standard it would be necessary to monitor over a region considerably larger than that in which, on the average, trouble seemed likely. Though the intensity of fallout generally decreases with increasing distance, in detail the pattern can be quite uneven; such things as changes in wind direction, or showers occurring in one locality while not in others, often result in hot spots where the intensity is much greater than suggested by the trend further back. Such a surveillance and control of the milk supply, while requiring considerable effort, would not necessarily result in a large loss of food production. For one thing, if it were available, dairy cattle could be transferred to stored feed. In addition, milk determined to be unsuitable for immediate use could be made into powdered milk or cheese and held until the iodine activity had died away.

Evidently the fallout on Iowa from one Bikini-type weapon would not cause an emergency in the US food supply. Still, the number that would be required to have such a result, should it affect different important agricultural regions around the country, is not really very large, and is in fact very small compared to the number of weapons commonly spoken of in connection with a serious attack.

Presumably agricultural activity would rate a low priority as a target. Though the results could be quite important, they would take effect rather slowly. No doubt the very highest priorities would be awarded to the country's strategic forces and the installations supporting them. There are more than 50 such targets in the 48 contiguous states, reasonably equitably distributed across the country. Some might be deemed remote (as Loring AFB in northern Maine), some are intimately associated with metropolitan centers (Philadelphia naval shipyards), and some are embedded in agricultural areas. For example, a fallout pattern across Iowa of the sort considered could be a purely incidental consequence of a heavy attack using surface bursts on the Headquarters Base of the Strategic Air Command near

Omaha, which is itself the country's largest livestock and meatpacking center, and has a metropolitan area population of over half a million.

To this point, only a surface burst of the Bikini weapon has been considered and it could be felt that this is not particularly relevant to the present situation. For one thing, that 1954 design is obsolete in that ICBM warheads have smaller yields (~ 1 megaton) and would not cause such extensive damage. In addition, since one can obtain a larger radius of blast and thermal damage by choosing an optimum height of burst ($\sim 25\%$ larger for thermal radiation than from a surface burst and $\sim 50\%$ larger for blast at moderate pressure), it is often assumed that an attack on a SAC bomber base, for example, would be made with an air burst, which would lead to little or no close-in fallout.

With respect to the latter point, there is of course nothing to keep an attacker from planning to use air bursts against bomber-base targets, but there is nothing to insure that he will; the choice would be up to him. He might happen to feel that since there might be no aircraft parked at any particular base at any particular moment, a good compromise tactic would be to use a surface burst and render the field and surrounding area unusable for a more extended period. If he were concerned about his delivery accuracy, he might decide that a smaller area of direct blast damage coupled (for 1 megaton) with a 10- by 40-mile patch of 1000-r radiation coverage would be his best choice. Or, since most of the radioactivity from a large-yield air burst rises into the stratosphere and is later distributed fairly uniformly on a worldwide basis while half or more of the radioactivity from a large surface burst is deposited locally (within a few hundred miles) around the target, there could seem to him to be a positive factor in cutting in half the long-range, long-term fallout from his own weapons in his own country. And finally, being in a malevolent frame of mind, he might value the extra damage and disruption his weapons would exact in a surface-burst mode. So far as an attack on a Minuteman silo is concerned, there would be little to recommend an air burst to the attacker. Being a very hard target, a very high pressure (~ 300 psi) is required to inflict damage; for this reason the difference between the optimum range and the range from a surface burst is rather small. A low height of burst is required, and to trigger accurately enough he would have to use a radar (radio altimeter) fuse which certainly, in principle, would be susceptible to jamming. Even without the dividend of greater disruption in the target area, the more straightforward, more reliable, and very nearly as effective contact fuse would seem to be the best choice.

On the matter of yields, though the Soviet Union is said to be capable of delivering up to 25 megatons in a single package, it is supposed that the great majority of their missile warheads are much smaller. However, the following considerations apply. There are nine ICBM installations in the United States: three Titan bases (18 missiles each at locations in Arizona, Arkansas, and Kansas) and six Minuteman fields. Each Minuteman site covers an area of several thousand square miles (being from ~ 60 to ~ 100 miles on a side) and each has 150 or 200 silos. The silos are placed far enough apart that they must be attacked individually (at least with weapons having yields no larger than one or a few megatons). Ascribing an accuracy of at best about a third of a mile to the attacker's missiles (10), even with a yield of a megaton or so any single missile would have an appreciable probability of missing its target silo by a large enough distance to fail to put it out of action. Therefore a reasonably effective attack on the field would require that two warheads be aimed at each silo. Were his accuracy not quite that good, he would favor yields towards the high side of the range. An attack on all the silos of a Minuteman field would, then, entail the explosion of several hundred megatons within the area of the base; if these were surface bursts, the debris clouds would blend and add and produce fallout patterns much more intense and very much more widespread than that from the Bikini weapon.

An attack aimed only at our strategic forces (which could include any or all of the 46 SAC bases, the 1000 Minuteman, and 54 Titan silos, plus possibly the 2 command and control bases, 2 missile submarine support bases, 2 naval shipyards, and 3 surface naval bases located in the United States) is referred to as a limited nuclear attack, to distinguish it from an all-out attack in which urban-industrial centers would also be included, and from which the number of fatalities is commonly estimated as approximately 100 million. Since an all-out attack by us would precipitate an all-out response, it has come to be felt that as long as that was the only option we had, any threat on our part to use our nuclear forces in retaliation for any provocation short of an all-out attack would lack credibility. Because of this, during the past few years (from about 1973) it has been urged that we develop more flexibility, and, in particular, that we acquire the means of conducting a limited nuclear attack as a more credible (and hence more deterrent) response to lesser provocations such as a move against western Europe. This reasoning has undergone at least one iteration, to the effect that if we should be subjected to a limited attack at a time when we had no option available short of an all-out response, we might be self-deterred from making any response at all, whereas if we had the means of a limited attack in return—and this were known to be the case—that would deter anyone from making the limited attack on us in the first place. It is not yet a part of the argument that any adversary is currently in a position to do this, but we should prepare now for the possibility that he might be. (And so on.) As presented to the Senate in mid-1974, a major advantage was said to be that a nuclear exchange confined to limited attacks would involve many fewer fatalities than we have been used to thinking of, or (5, 10): "Compared with the effects of large scale urban/industrial/military attacks within the range of present and projected Soviet capability, the long-term collateral effects on national survival recovery capability of this postulated attack would be relatively small."

Should anyone find this summary presentation less than fully clear and compelling, it may help him to know that some members of the Senate had similar trouble with the full presentation. They were unsure of just how costly such a limited nuclear attack on the United States might actually be; whether there would be only 800,000 fatalities (the number most frequently mentioned in the 1974 DOD presentation), or some higher number (such as 5 or 6 million, also mentioned), or some still larger number (as included in estimates presented by the Arms Control and Disarmament Agency). To clarify some of these points the Committee on Foreign Relations requested the Office of Technology Assessment of the US Congress to arrange for a critical review of the DOD estimates. This was accomplished by mid-1975, as were also additional studies by the DOD.

By these comments it is not intended to attempt to settle—nor even to enter—the ongoing (early 1976) policy debate (11, 12) as to whether US strategic doctrine should be modified to include the capability of conducting limited nuclear warfare or adopting, as it is sometimes called, a "flexible response doctrine;" but merely to indicate the context in which a greatly increased interest has arisen in the effects of a limited nuclear attack on US strategic military targets. The nature and extent of these effects may be viewed in the light of the earlier outline of the possible effects from a single large nuclear detonation.

Since 1974 a number of "attack scenarios," particularly on our ICBM installations, have been considered by the DOD. These have for the most part involved two weapons per silo, with yields ranging between $\sim\frac{1}{2}$ and 3 megatons, and using two airbursts, two surface bursts, or one of each. With two-weapon attacks, the estimates of silo destruction ran (low yield to high) from 40–80%. Naturally, the estimated number of fatalities also covered a considerable range, being higher the larger the yield, and higher in the event of surface bursts, but also depending strongly on the "shelter posture" assumed (whether or not the civilian population was well-prepared and rehearsed and made maximum use of the best shelters available). The new DOD studies resulted in distinctly larger numbers for the fatalities to be expected than had been suggested earlier. In particular, for the scenarios considered (in which airburst attacks on all 46 operational SAC bomber bases were delivered simultaneously with the attack on the ICBMs), the fatality estimates ranged from ~ 3 to ~ 22 million, which means that nothing is really known about them except that the number is likely to be very large. In a letter to the Foreign Relations Committee in July, 1975 (10), the DOD concludes: "...we continue to believe that the 'comprehensive attack' scenario (resulting in 6.7 million fatalities) used in the Secretary of Defense's September 11, 1974 testimony...is the most representative scenario which balances military effectiveness, potential restraint on the part of the Soviets to minimize collateral damage, and the physical uncertainties that could exist at the time of such a postulated attack."

The report of the panel that conducted an analysis for the Office of Technology Assessment is also reproduced in (10). It includes charts (presented by Richard Garwin) of the possible fallout contours from an attack on the ICBM force using two 1-megaton surface bursts per silo, while the other strategic targets were assumed to be attacked by airburst. One of these charts is shown in Figure 5, along with the US target structure. The contours drawn are Garwin's estimates for a 30-day exposure of 1350 r (close to our 96-hr, 900-r level) under the assumption that the wind pattern is of the sort commonly present in the winter. With a March-like wind pattern things would not have been qualitatively very different, except that Chicago could have been included within the contours, while further downwind (where fallout still extends, though at lower levels) both Detroit and Washington DC could receive exposures greater than twice the level experienced by the inhabitants of Rongelap in 1954 (>350 r in 96 hours). Had the attacker not shown such restraint and not followed the DOD scenario, but—perverse as it might seem—used surface bursts against the SAC bomber bases, there would be a need of additional contours of ~1000 mi² each around all the other targets in Figure 5. Before concluding that all one need do to avoid the threat to which the middle of the country appears to be exposed is move to some point on the California coast, one should consider that the scenario could just as well have included heavy attacks on the naval bases at San Diego and Long Beach, and the shipyards at Hunter's Point in San Francisco.

As noted above, large areas downwind from the contours in Figure 5 would be affected. For another ~300 miles, unprotected people would receive lethal exposures of radiation. For ~500 miles (out to the Atlantic coast at the latitudes of Maryland and North Carolina) swathes of land from 50 to 200 miles wide would be subject to such contamination with ^{90}Sr ($\geqq 5$ C/mi²) that by present standards they would have to be withdrawn from agricultural use. Much more severe conditions would

U.S. Target Structure

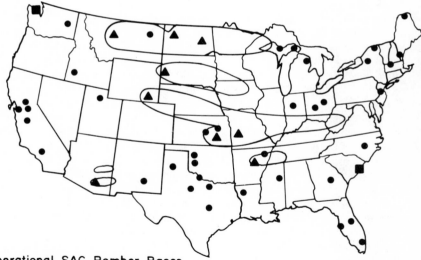

● Operational SAC Bomber Bases

▲ ICBM Fields

■ SSBN Support Bases

Figure 5   US target structure; 96-hr, 900-r fallout contours from attack on US ICBM silos.

be encountered inside the contours. Here, in an area greater than 400,000 square miles (larger, that is, than the combined areas of France, Britain, and West Germany, for example) evacuation prior to the arrival of fallout would be impossible because of the great extent of the regions involved and uncertainty as to the location of safe objectives. Essentially all livestock and the current year's crops would be lost over a very important fraction of the country's agricultural land, with serious question as to how much, if any, of this land could be put back in use in the year or so following. In an area a couple of hundred miles long and 40 or more miles wide downwind from each Minuteman field, the 96-hr exposure could exceed 5000 r. This would probably be enough to kill off all fruit trees and most other deciduous trees. St. Louis, with its metropolitan-area population of 2.5 million, would be well inside this range from Whiteman AFB in western Missouri. To survive under these conditions a person would have to have a larger protection factor and stay in his shelter longer, and, when it came time to leave, be prepared for a longer evacuation run than those considered in connection with Long Island. With the other heavy radiation belts flanking the Whiteman belt both to the north and south, the best bet for a survivor from St. Louis might be to keep going until he got to western Arkansas. Similar considerations would of course apply to many other towns, cities, and localities near the middle of the contours shown.

Even though a large fraction would survive, the tens of millions of people exposed to the effects of this limited nuclear attack would not feel that it was limited nearly enough. Probably only someone engrossed in playing war games with several hundred million expendable pieces at his disposal could fully appreciate the advantages of such a limited attack as compared with other scenarios he might be willing to consider.

2 DISTINCTIVE FEATURES OF NUCLEAR WEAPONS

The importance of nuclear weapons derives from their very high energy density— the fact that it is possible to obtain as much explosive effect from a few kilograms of nuclear fuel as from many thousand tons of chemical explosive. It is not primarily the magnitude of the explosion, although the damage from the first atomic bombs was immense. However, it requires only about 2000 tons of high explosive in bombs one or a few tons in size to produce the same damage by explosion as that from 20,000 tons of TNT in one lump (13); in the early months of 1945 both Germany and Japan had been subjected to numerous raids of about that size or larger (14). Among the targets (with the number of tons of explosive bombs dropped and the date) were Dresden (>2000, February 13–14); Nuremberg (2000, February 20; 1800, February 21); Tokyo (1650, March 9); Essen (4740, March 11); Nagoya (1800, March 11); Dortmund (4900, March 12); Osaka (1730, March 13). It is not primarily the number of fatalities, though that, too, was very large: 70,000 in Hiroshima and 40,000 in Nagasaki. However, approximately 50,000 died from the raids on Hamburg in the last days of July 1943. Twenty-five thousand were killed by the "thousand bomber" raid on Berlin on February 3, 1945 (where, in contrast to Hiroshima and Nagasaki, the inhabitants were able to make extensive use of shelters).

At Dresden, where the raid occurred at a time when the city's normal population of 600,000 was approximately doubled by an uncounted number of refugees, the level of destruction was so severe that the number of fatalities is not really known (15)—the preliminary estimate of the German authorities was 100,000, but later estimates have ranged from 25,000 to 135,000, with the higher number thought (16) to be the more relevant. The March 9 raid on Tokyo, with the possible exception of Dresden, resulted in the largest number (83,800) of civilians killed in any air raid ever. It is not (at least, not in the first atomic bombs) primarily the presence of radiation, though this effect was unprecedented and profoundly insidious. However, radiation did not cause as many as 30% of the fatalities (17) and of the casualties who survived, 70% suffered from injuries caused by blast effects, 65% from serious burns, and only 30% from radiation injury (many, of course, had more than one type of injury). Had there been no radiation at all, the atomic bomb would still have brought on the end of the war with Japan. The staggering effect of the first nuclear weapons was a consequence of their yield-to-weight ratio.

In the massive 1945 raids with conventional explosives, the attacker had to employ from 300 to 1000 or more planes, each carrying several tons of bombs. The bombardment itself required several hours to accomplish and, depending on the state of the defenses, some or many of the attacking planes could be expected to be shot down. The atomic bomb achieved a comparable effect using only a single package dropped from a single plane. By this step the yield-to-weight ratio available in explosive weapons was increased by a factor of more than a thousand, with a corresponding and appalling increase in the ease and cheapness—the casualness, even—with which such destruction could be imposed.

By the same step, not so clearly realized at the time, not only did it become seductively easy for a great power to consider things that previously only it could manage (and that with difficulty), but it was made possible in principle for a wide spectrum of lesser powers to consider them too. With the advent of thermonuclear weapons it became feasible by the late 1950s (18) to produce a million-ton explosion from an object weighing scarcely more than one ton. In addition to enhancing greatly the quality of ease and cheapness, this change brought within reach quantums of destruction that had previously been beyond the means of even a major power.

Before taking up other consequences of this large change in the yield-to-weight ratio, some further distinctive features might be noted. Nuclear weapons require a supply of exotic materials and, if built just for that purpose, plants to produce such materials involve the commitment of large resources (hundreds of millions of dollars). However, once such plants are built and producing, nuclear explosives are cheap. In connection with its program for the peaceful uses of nuclear explosives the Atomic Energy Commission has published (19) a charge of $460,000 for 100 kt in a device specially adapted for use at great depths and hence built under severe constraints and more costly than would be necessary merely to achieve the stated yield. This is 0.25¢ per pound, whereas military explosives will cost more than 25¢ per pound. There is thus a factor of two orders of magnitude or more, by which nuclear explosives are cheaper than chemical—at least in the bulk quantities in which nuclear explosives are usually considered. This is in addition to the ways in which they are operationally and logistically less expensive. On another point, the design of nuclear explosives is evidently technologically complex, as borne out by the continued interest in testing on the part of the major powers. However, this interest by the United States and the USSR is not merely to achieve portable devices with yields of tens of kilotons or a megaton—each was able to do that 15 or 20 years ago—but to improve, adapt, and refine devices for particular applications or more elegant carriers. The mere designing of nuclear weapons has something in common with climbing the Matterhorn: once thought to be impossible, then considered extremely difficult, and now confidently undertaken by parties of informed and prepared tourists. There is some debate as to just what resources of knowledge and capability may be required to design and build an effective nuclear explosive device (whether one versatile and ingenious fellow, a small Manhattan Project, or something in between), but there is no question but that the resources could be assembled by the government of any technologically developed country, or even by very much smaller organizations. Apart from the lack of a motive for doing so, the most serious barrier is that of access to the necessary materials, and that is in the course of being removed for everyone by the worldwide growth of the nuclear power industry.

One direct consequence of the high yield-to-weight ratio of nuclear weapons is that the material of the bomb is, at the moment of explosion, at a temperature higher than any previously observed on earth. Taking an average particle mass of ~ 1 amu (as for completely ionized material), a yield of one megaton per ton would provide about 40 kilovolts of energy per particle. Most of the energy will reside in the radiation field, but for a short time the temperature of the weapon materials will be in the kilovolt range (a few times 10^7 °K). As a black body, this will radiate energy at an exceedingly high rate (20) in the form of kilovolt X rays. In a near vacuum (as at very high altitudes) these X rays can proceed to large distances from the point of the explosion. Should something like a third of the energy of a megaton explosion be radiated away, the X rays would deliver ~ 1000 cal/cm^2 at a radius of a couple of kilometers. This indicates one possible basis for the use of nuclear weapons to attack satellites and missiles at very high altitudes.

In air near sea-level density, kilovolt-like X rays are absorbed in a few tens of centimeters. Thus in a low-altitude explosion a thin layer of air around the weapon is very strongly heated, and it in turn will radiate energy forward. Partly by this means, and partly by the shock wave driven by the weapon materials and the heated air nearby, the energy is spread into a larger and larger mass of air at a progressively lower temperature. As the temperature falls below 10^4 °K, a major part of the spectrum of the photons radiated overlaps the visible range—the band of wavelengths for which air is transparent—and the energy in this band is radiated away from the fireball. By the time the temperature has dropped to ~ 2000 °K, the radiation rate, which varies as the fourth power of the temperature, is so low that this process is no longer important. By then about a third of the energy of the explosion will have been radiated as visible light; by this stage, also, the mass of air heated to ~ 2000 °K or more will be about 1000 tons per kiloton of yield.

In a chemical explosion the temperature of the burned explosive ($\sim 3000\,°K$) is too low for the radiation process just described to be of much significance; little, if any, of the surrounding air is heated to $2000°K$. Thus, the exceedingly high temperatures realized in nuclear explosions account for the distinctive feature that thermal radiation is an important damage mechanism, and that a large mass of air is made hotter than $2000\,°K$. At such temperatures a small percentage of the air is transformed into nitric oxide (NO), and as the air cools to below $2000\,°K$ the rate of the reactions that deplete NO becomes so small that the concentration of NO stabilizes at nearly 1%. By this means about 10^{32} molecules (~ 5000 tons) of NO is produced in nuclear explosions per megaton of yield (21, 22).

The other main distinctive feature of nuclear weapons is the presence of nuclear radiation. As already noted, the prompt radiation, though important at the 20-kt yield level, was by no means predominant as a cause of damage. All three effects were serious at distances of more than a mile. The distance in which the intensity of the prompt radiation is attenuated by a factor of e is 0.2 miles, so that the radius to which a given intensity of prompt radiation extends from an e-times larger explosion is <0.2 miles larger, whereas the range of a given overpressure is increased by a factor $e^{1/3}(\sim 1.4)$, and of a given thermal intensity by a factor $e^{1/2}(\sim 1.65)$. Thus for yields larger than 20 kt, though the range for some particular exposure to prompt radiation will be larger, the increase is much less than that for the blast and thermal effects. For smaller yields the opposite is true, and in the limit a slightly supercritical assembly can provide lethal bursts of radiation without any explosive effects at all. Small tactical nuclear weapons (or "mini-nukes") would be in a low-yield range where prompt radiation could provide the most extensive hazard for exposed personnel. As already seen, from surface bursts of large-yield weapons the effects of the residual radiation could be very severe over much larger distances than any other effect.

Other distinctive consequences of nuclear weapons may be mentioned. In connection with bursts at high altitude, in addition to being able to destroy reentry vehicles, nuclear warheads, and communication or other satellites, the various radiations (neutrons, γ rays, and X rays) can create a cloud of ionized gas that could for a time black out radar or, by disturbing the ionosphere, interrupt long-range radio communication. From bursts at low or medium altitudes, the strong electromagnetic pulse radiated would impose transient currents on electronic circuits, telephone, and power lines, which could put them out of operation if they were not shielded against such effects, and so on. These consequences are transient, and mainly of importance in connection with technical apparatus. In a nuclear-war situation they could add considerably to the complexity of the internecine conflict between weapons systems, but from a civilian point of view they are scarcely of primary importance, since, though they could be used to make it more expensive to accomplish an attack of given severity, they cannot be relied upon to prevent it (if only one in five weapons gets through, prepare to use five times more). The main direct impact of effects of this type on civilians is likely to occur during peacetime, when it may be perceived as necessary to improve or replace weapons systems so as to take advantage of—or counter—such effects.

3   CONSEQUENCES OF LARGE-SCALE NUCLEAR WAR

With respect to a war involving only a few nuclear weapons—whether because the adversaries have only a small stockpile or because the war should be terminated after only a few weapons were used—the consequences (at least the physical consequences in the target areas) can be gauged by reference to the effects observed in Japan, and the worldwide effects, by reference to experience with atmospheric testing. For a tactical nuclear war—a war, that is, in which weapons are used only in or near the battle area (assuming, without any particular confidence, that such a limitation should in fact be possible)—the results might not be encompassed within previous experience. Some thousands of weapons could well be fired and, though they might be said to be small, the radius of destruction of an explosion one eighth the size of Hiroshima (~ 2 kt) would still be half as much as it was there. The setting most often proposed for such an event is the border area between the NATO and the Warsaw Pact powers. Even if we assume that the civilian population could manage to get clear of the battle zone, a large fraction of the targets would involve built-up areas, and the intensity and extent of physical devastation would be more severe than anything yet seen. Wherever surface or near-surface bursts were used, there would be fallout that (in order to avoid exposure greater than 25 r, say) would require the temporary evacuation of some locations as far as 50 to 100 miles downwind. Still, much of the area could be reoccupied in a few weeks or a few months, and only in patches of one to a few square miles around each surface-burst point would reentry at greater than one year result in a lifetime dose exceeding ~ 25 r— and that without benefit of weathering or decontamination. It is not intended to suggest that such effects would be anything but frightful and intolerable—indeed, it is scarcely possible to imagine anything (either physical or political) beside which this scene would seem to be attractive—nevertheless, with respect to total radioactivity and wide-scale effects a full strategic exchange would no doubt be worse.

Before considering this worst case one might ask in what way the possible advent of "mini-nukes" might change the picture. The expression presumably refers to nuclear weapons of quite small yield (possibly ≤ 1 kt) that would result in less collateral damage and could be used in a tidy surgical fashion. Since they are smaller, more would be needed and since they could be used more freely, more would be used. It would be a little like adding rifles to the equipment of a company that previously had only small cannon and bazookas; they would use the rifles, too. If the opponent should not have the same handy gadgets available, he would respond with the more old-fashioned ones he did have (23); the potential for total damage would not necessarily be reduced. What would have changed would be that the dependence on nuclear weapons would increase and the threshold for using them in the first place would be lowered (11), without in any way affecting the high improbability of being able to define, establish, or maintain any line between the use of small weapons at short range, somewhat larger weapons at somewhat larger range, and still larger ones, and so on.

We consider, then, a full strategic exchange—whether induced by the increasing violence of a tactical nuclear war, triggered by miscalculation, or whatever—and suppose it to be between the United States and the Soviet Union, since these are the only contestants who have yet fully qualified for the event. There is a wide spectrum of possible scenarios, none of which seems tolerable, but none of which is inherently impossible. For example, should one consider only a limited nuclear attack or one in which urban/industrial targets were included; an attack with little or no warning, or only after some weeks of growing apprehension; an attack in spring, or accompanied by subzero blizzard conditions in the midwest and New England; an attack including major power plants, on which we have become so heavily dependent, and/or nuclear power plants? In this way some of the more bizarre predictions of opponents of nuclear power could be realized, though the effects might be obscured by the background.

To get free of this predicament we shall merely assume that the scale of a strategic attack on the United States would fall somewhere between the limited attack already identified (~ 2000 Mt of explosions, ~ 7 million fatalities by DOD estimates, and roughly as many nonfatal injuries) and the DOD-posited (24), full-scale attack (6000 Mt—2/3 military, 1/3 civilian, with ~ 100 million fatalities). Possible fallout contours for the smaller attack are shown in Figure 5, and they may not be very much larger for the heavy attack, since, although some types of industrial targets, and possibly dams, might call for low heights of burst involving additional fallout, for urban targets, airbursts at a height of ~ 1 mile (for one megaton) would probably be favored in order to maximize the area exposed to overpressures in the 5–10-psi range. The effects described in Section 1, scaled as necessary, may then be used to picture the consequences in the target areas for any desired scenario.

In 1975 a study was conducted by a committee of the National Academy of Sciences (8) to assess the "Long-Term Worldwide Effects of Multiple Nuclear-Weapons Detonations." Only the physical effects on the environment at locations remote from the target areas (beyond the range of local fallout) were considered, on the assumption that weapons with a total yield of 10^4 Mt (4000-Mt fission) were detonated in the northern hemisphere. The findings are applicable to the situation pictured above, along with a similar attack on the USSR. The main conclusions were (a) with respect to radioactivity: from stratospheric fallout an average of 1 Ci/ km^2 of ^{90}Sr would be deposited in the middle latitudes of the northern hemisphere— this being 20 times the amount observed in the United States as a result of atmospheric tests prior to 1963, which had a total fission yield of ~ 200 Mt. (This average is half the present ERDA standard for use of land for agriculture.) "Hot spots," tens to hundreds of kilometers in extent, in which the contamination level could be 10 or even 30 times larger than the average, could occur at locations remote from the explosions because of variations in precipitation patterns. The southern hemisphere would receive only about a third as much as the northern. Scaling from the conclusion of the United Nations Scientific Committee on the Effects of Atomic Radiation (25) concerning the results of past atmospheric testing, the incremental exposure of the inhabitants of the northern hemisphere over the 30- to 40-year period following the detonation of 4000 Mt of fission would be approximately equal to that provided by natural background. As a result of the long-range fallout, some foods would be contaminated with radionuclides to levels approaching the upper limits of present standards. An increase of 2% over the spontaneous-cancer death rate would be expected in the generation exposed to the fallout; and an increase of 100 to 1000 per million live births (0.2–2%) in the incidence of significant genetic disease among the offspring of the exposed generation (and, at declining rates over succeeding generations, the ultimate integrated number of such cases would be about 5 times that in the first generation). It is possible (though this is still a matter of debate) that, as a result of mutations induced in pathogens by ionizing radiation in the vicinity of the explosions, virulent strains could appear and cause disease epidemics in crops and animals on a global scale. (b) With respect to dust: the amount injected into the stratosphere could be of the same order (10^7–10^8 tons) as that believed to have resulted from the eruption of Krakatoa in 1883. By comparison, this might be expected to lower the average surface temperature by a few tenths of a degree centigrade for a year or so—an effect that would lie within normal climatic variability. (c) With respect to photochemical effects: the amount of NO (10^{36} molecules) injected into the stratosphere would be 5–50 times more than the present natural background of NO in the stratosphere. Such an injection would result in a large initial reduction in the ozone column in the northern hemisphere— possibly by 30–70%—the reduction having a mean life of 2–4 years. A significant change in the amount and distribution of the ozone column would change the pattern of heating of the stratosphere, which in turn would affect air movements in the troposphere, and could lead to significant effects on climate. However, present understanding of climatological phenomena is insufficient to predict these effects—either the amount of change in the average surface temperature, or even the sign. The change would probably lie within global variability (a few tenths of a degree centigrade) but a more dramatic change of a few degrees cannot be excluded. An average cooling by 1°C would, by shortening the growing season, reduce the northern limit for various crops, and eliminate, for example, commercial wheat growing in western Canada. Any reduction in stratospheric ozone would result in an increase in UV radiation reaching the surface (by factors between 2 and 10^2 for a 50% reduction in ozone, depending on the particular wavelength and the latitude considered). A tenfold increase in UV flux would injure or kill some crops, such as peas and onions, and some light-coated animals (e.g. Hereford cattle). For humans the incidence of skin carcinoma and melanoma would be increased at middle latitudes by $\sim 10\%$. Severe sunburn (blistering in possibly 10 minutes) could, for a few years, be experienced in temperate zones, while a greatly increased danger of snow blindness would affect people in northern regions. Toxic effects could result from an increased rate of vitamin D synthesis.

Should the UV and climatological changes in fact lie within the currently expected range, they would cause painful difficulties—but no ultimate catastrophe— in Homo sapiens' attempts to continue to occupy the northern hemisphere. Under these conditions the most disruptive effects would seem likely to be economic and political. The most highly technologically developed countries would be most affected. Their ability to manage or even feed themselves—let alone to provide necessary food, fertilizers, or technological assistance to others—would be disrupted.

Simple-minded notions of "national survival," at least to the extent that this is associated with retaining any of the power and glory we have become used to assuming, would be quite irrelevant. World food production, presently only marginal with respect to supporting the world population, would be seriously impaired. Its restoration would depend on reestablishing the technology base for modern agriculture, and this would no doubt occur. But all that can be said with assurance about the circumstances, location, and political auspices or social arrangements under which this would take place is that they are unlikely to be comfortable extrapolations from the present pattern.

4   IMMANENT EFFECTS OF NUCLEAR WEAPONS

Nuclear weapons have not been used in war since 1945; nevertheless, they have had many and pervasive effects during the past 30 years. No attempt can be made here to discuss these in depth, but a few of the more obvious and troublesome ones can be identified. Though the effects of the appearance and proliferation of nuclear weapons have borne differently on different countries, attention here is limited largely to the effects on the United States, since by means either of action or reaction these have affected all other states, whether nuclear or non-nuclear.

Most obvious, of course, is the end of the traditional invulnerability of the United States. In principle this ended with the advent of nuclear weapons, but it was publicly confirmed when the inevitable loss of the US monopoly was established. It was amplified by the appearance of thermonuclear weapons and compounded by the parallel development of ballistic missiles capable of carrying nuclear warheads. Prior to all this no one could have laid a serious glove on Chicago, for example; by now no place anywhere on the globe is secure. For the United States a tremendous increase in the ability to inflict damage has been accompanied by a corresponding decrease in actual security.

The plants to produce fissile material take a long time to construct and they must be operated over an extended period to accumulate enough material for a significant number of weapons since, for example, it takes a 1000-MW reactor a few days to produce a kilogram of plutonium. The design, testing, and production of long-range carriers, whether planes or missiles, is also a lengthy process. From the fact that a first nuclear attack—or retaliation for an attack—is envisaged in terms of the near-simultaneous launching of many planes or missiles, a nuclear war will be fought (if at all) with forces in being. This is in contrast with past wars in which massive and decisive armament production was mounted only after the outbreak of war. Nuclear weapons, then, have made it necessary for the country to stockpile in peacetime all the armaments it might expect to need in the event of war. Since only weapons-in-being at the outbreak of war will be significant, and since advances in technology keep opening up means of improving the weapons on hand, it has seemed necessary to purchase a new stockpile every few years as older segments of the stockpile are replaced by more capable and costly models. As a result, defense expenditures, from being $<15\%$ ($\sim$\$1.5 billion) of the national budget in 1939 (immediately before World War II), were over 50% ($\sim$ \$40 billion) of the national budget in the years preceding the Vietnam war.

Some credit the fact that there has been no hot war directly between the United States and the Soviet Union during the past 30 years to the sobering influence of the existence of nuclear weapons. Were that indeed the case it would constitute a profound and favorable effect, but the point would seem hard to establish. What is more readily apparent is that for most of this period, and still today, the specter of nuclear war has hung, with paralyzing effect, like an Albatross about the neck of US defense and foreign policy. As Secretary (then Professor) Kissinger observed in 1957 (26), "all-out war has therefore ceased to be a meaningful instrument of policy." For him, this pointed up the need of modifying the traditional American view that the only acceptable outcome of war was the unconditional surrender of the foe. With all-out war no longer a rational option, we would have to make do with limited wars; this, he said, we should recognize and prepare for, since "So long as we consider limited war as an aberration from the 'pure' case of all-out war we will not be ready to grasp its opportunities ..." (27). The "opportunity" referred to was largely that of finding a solution for "this task of posing the maximum *credible* threat" (28). For that he felt a limited nuclear war would be the proper level, in that limited conventional war, though credible, was not sufficiently maximized, whereas it should be possible, he thought, to go as far as limited nuclear war and still have a good chance of avoiding impregnation with the seeds of all-out war. Thus, by nuclear weapons out of Kissinger: the cult of credibility.

A large part of the effort and the agony the United States invested in Vietnam was in the pursuit of this credibility. For example, for the Joint Chiefs in 1962 the importance of the war lay in "the psychological impact that a firm position by the United States will have;" for President Johnson in 1965, to leave Vietnam "would shake the confidence of all these people in the value of an American commitment ...;" for then-Assistant Secretary of Defense John McNaughton in 1966, "The reasons why we went into Vietnam to the present depth are varied; but they are now largely academic. Why we have not withdrawn is, by all odds, one reason: To preserve our reputation as a guarantor ...;" and for President Nixon in 1970, in rejecting for the United States the role of "a pitiful helpless giant" (29). And still in 1975, as already discussed, the nuclear weapons–generated need for credibility had not been met, but required the development of a "flexible response capability"— the means of making (and the concomitant willingness to accept) "limited" nuclear attacks.

In addition to the perceived need of credibility in connection with any threat to use force, the existence of nuclear weapons has had many other notable effects on our thinking and behavior. There is the whole chimerical lexicon developed to describe the intentions we should have (or wished to persuade others we had, or were to suppose that others might have) concerning the use of nuclear weapons: instant massive retaliation, pre-emptive first strike, damage limitation, massive assured destruction, second strike capability, counter force, and limited nuclear warfare. We may yet hear that we should be prepared for anticipatory retaliation. Along with this we have the scenarios: hundreds or thousands of computer studies of the number of fatalities in the event of an attack of this or that intensity on each of so many cities with this or that aiming error (CEP); and which, if it cost so much per kiloton to increase the yield or so much per hundred meters to reduce the CEP, would be the most cost-effective weapons system? Thanks to nuclear weapons we now have answers to more such questions than anyone would care to ask.

Before leaving the matter of the effects that nuclear weapons have had in suppressing, degrading, or at least conditioning official thought, two particular items may be noted. One is the appearance of the expression "national survival" (30). It seems to be assumed that it will be obvious to everyone that no level of devastation and dislocation, and no number of nationals failing to survive, would be too high a price to pay for this objective. The objective itself would seem to consist of the survival of a quorum of some governing board having some residue of power at its disposal. As an objective it may come as strangely as the Ancient Mariner to those celebrating the Bicentennial Year of the Institution of a Government to secure the inalienable Rights of Life, Liberty, and the Pursuit of Happiness. Not all the guests assembled will feel that this is what they came for.

A further indication of the stultifying effects of nuclear weapons on official thought may be found in the response of then-Secretary Schlesinger to questions raised by the Senate Committee on Foreign Relations (31). In September 1974, Dr. F. Ikle, Director of ACDA, had observed that six notable effects of atomic weapons had not been well anticipated. These included the fallout at Bikini, blackout in communications, and the injection of NO into the stratosphere. The DOD was asked if this aspect of unpredictability did not raise questions concerning estimates of the effects of nuclear war. By implication, the Secretary found nothing seriously disturbing in Dr. Ikle's "cauldron of horrors," but did express the hope that "surprise number seven or eight . . . will be such that the implications of using just one, let alone many, nuclear devices would be sufficient to deter the most irrational adversary. Until that time, however, ..." It would seem to be an existence proof: even though fallout doesn't come up to that level, there may be some sort of effect that the DOD would consider intolerable. In the same testimony Secretary Schlesinger, apparently viewing the opposite end of the damage spectrum and referring to an hypothecated single Soviet weapon aimed at the communications facility in Cutler, Maine, felt that ". . . the number of mortalities in the event of such a demonstration strike would be exceedingly low." Too low for what? Too low fully to establish Soviet credibility?

Another sort of peacetime effect of nuclear weapons has been the spontaneous mobilization of many groups devoting serious effort to examining the problems of arms control—the Federation of American Scientists, Pugwash Conferences, and SIPRI among them, as well as the US ACDA. An extensive body of literature has resulted, a notable example (as well as guide to other studies in the field) being the summer 1975 issue of *Daedalus*. Among the thoughtful articles collected there is one (32) discussing the hermetic nature of the process by which new weapons emerge. Partly because of the long lead time from the first general description through assignment of military specifications, detailed design, prototype development and testing—aided by the fact that weapon design patterns and characteristics are secret—a particular new system will have proceeded a good part of the way through the development process before it is considered by Congress for possible acquisition, or even, possibly, before it commands the attention of the staff of the Secretary of Defense. By this time a large organizational momentum will have built up and the Service concerned will have decided that this is what it now most needs. The latest policy of the latest Secretary of Defense may have little to do with the process and politics determining weapons acquisition in any particular year.

Though congenitally well-intentioned, by no means all the theses developed in arms control writing compel full support. Some, from intense concern with nuclear weapons, would seem to imply that conventional war provided an acceptable instrument, in spite of our having managed for eight months during 1972 to shower 100 kt of conventional explosive bombs a month on portions of Vietnam. Some, from assuming that it would be hopelessly unrealistic to campaign for the elimination of nuclear weapons, seek merely to find some legalistic or technical fix to inhibit the growth and elaboration of stockpiles, thus seeming to legitimize the status quo. The probable fact is that constructive thought and tenable policy will continue to elude us as long as nuclear weapons are considered for any wartime role whatever, with the single possible exception of being available to provide a measured repayment in kind (33). Perhaps only when we should seek and earn credibility on such a basis will the Albatross fall off and sink into the sea.

The non-nuclear technology of the twentieth century has already made it sufficiently real and apparent that all the peoples of the earth are traveling on the same spaceship, that everyone's prospects are to an increasing degree affected by the behavior and intentions of the other passengers, and that even those in first class cannot afford to be indifferent to famine, plague, or arson in the hold. At least by the end of World War II it was clear for all who chanced to look that non-nuclear technology had added so much to the means and the range of destructive capability that "all-out war has therefore ceased to be a meaningful instrument of policy." These developments could well have warranted a top billing on the agenda of the United Nations, but to a considerable extent the spotlight was stolen by nuclear weapons, leaving non-nuclear developments relatively free to carry on in the welcome obscurity of the wings.

Almost continuous discussions of nuclear weapons have been conducted in the United Nations, embracing a wide variety of items. Though some progress can be claimed, such as agreements not to establish nuclear arsenals in the Antarctic (1959), in outer space (1967), or on the ocean floor (1971), to some extent these amount to renouncing things no one much wished to do anyway. Almost everyone is on record as devoutly in favor of a complete ban on the testing of nuclear weapons—always provided he could be quite sure that others were also adhering to the ban, and sure, also, that when it went into effect he was at least as far along with his own development as anyone else. Unfortunately, though predictably, the state of understanding of seismic signals and the precision of their recording were (it was said) nowhere near good enough 20 years ago to provide the ineffably high degree of assurance required. In spite of remarkable improvements in seismic detection techniques and interpretation in the interim, a comprehensive ban on testing would still appear remote.

The other major preoccupation of UN discussions of nuclear weapons has been with the Nth country problem, or proliferation, and has resulted in the Non-Proliferation Treaty (NPT) of 1968 (34). Signatories of this treaty who do not possess nuclear weapons undertake not to manufacture, receive, or otherwise acquire such weapons, and to place their relevant nuclear facilities under the safeguards system of the IAEA. Parties possessing nuclear weapons undertake not to transfer such weapons to anyone nor to assist any nonweapon state in manufacturing them, but they do not put their nuclear facilities under the IAEA system. All states undertake to seek a cessation of the nuclear arms race and to pursue nuclear disarmament, and recall their earlier determination to seek to achieve the discontinuance of all test explosions. There is a great asymmetry in the rights and obligations of the two classes of states. Except for the US-USSR agreement to limit weapons tests to yields below 150 kt—which cannot impress many non-nuclear states as a very convincing step towards the cessation of testing—there is no evidence of progress in the intervening years towards nuclear disarmament (the much-celebrated SALT agreements notwithstanding, since they do not address the subject of disarmament).

Recently, great concern has been raised in the United States over the possibility of further proliferation as a result of several proposed sales of nuclear facilities by technologically advanced countries to others, for example, the agreement by West Germany to supply isotope separation equipment to Brazil. This has been given much attention in the hearings on the Senate bill concerning Nuclear Proliferation and Export Policy (35). The measures considered for allaying the present worries do not, however, include the notion of modifying our own insistence on retaining the right to decide to proliferate as we see fit, nor any other measure that would seem to make nonproliferation more palatable. The emphasis instead is on trying to make proliferation by others more difficult to accomplish by adopting a more restrictive export policy and tightening up on information control. Indeed, the thought has been strongly advanced that the United States should take a "strong stand of moral leadership" and seek the cooperation of the Soviet Union in compelling the other nuclear suppliers to agree to the strictest nonproliferation policies. However effective such measures may be in some short range, they would not seem likely to win wholehearted adherence to the NPT on the part of nonweapon states, and until the superpowers should accept more meaningful limitations on their own actions, their appeals and exhortations to the nonweapon states will continue to have a distinctly hollow sound.

In recent years, as the looming development of a worldwide nuclear power industry threatens (unless handled properly) to provide increased opportunities for criminal groups to obtain—by stealth or robbery—possession of weapons-capable materials, nuclear weapons at the subnational level have become a matter of serious concern. This differs from the Nth country problem in at least two ways: the undesired action would be against the local government, rather than conducted by it or on its behalf, so that international sanctions would not be readily applicable; for a change, the interests of the nonweapons and the weapons states would be essentially identical. Since material acquired in one country might be intended for use in any other, there is a clear basis for international recognition of a common interest in having adequate safeguard procedures in effect in all countries. Though such recognition may not be too difficult to achieve, quite sticky problems are likely to arise in attempting to obtain agreement on acceptable measures. For example, the United States may be expected to come up with technologically sophisticated and expensive equipment for the purpose and regard anything less complex as dangerously inadequate, whereas others may consider the US provisions needlessly elaborate and costly. An intriguingly awkward class of problems could emerge from the possible reluctance of many authorities to describe their provisions and procedures in sufficient detail to enable their adequacy (or weaknesses) to be assessed by persons over whom they have no control. Intensive study of means of achieving acceptable safeguards has been taken up in the United States and many other countries, but considerable time will yet be required to establish a worldwide system. A detailed discussion of technical aspects of this problem is given in the 1975 volume of the *Annual Review of Nuclear Science* (36).

There are other instances of nonphysical effects of nuclear weapons, but the final one considered here is that they have clouded the prospects for exploiting nuclear energy—most particularly in the United States. Excited opposition has been directed at nuclear power plants on the grounds of their deleterious effects on health and the environment, while actually, in normal operation, their effluents have a smaller impact on air quality and the health of the public than fossil-fueled plants (37). It has been claimed that they were hazardous to an unacceptable degree, but the critics incline to ignore both the operating record of several hundred reactor-years without harm to any member of the public, and the real deaths and injuries resulting from coal-mine explosions and the spectacular oil fires on the Delaware near Philadelphia, which keep on happening once or twice a year. Great alarm has been raised over the fact that in the event of a really serious accident—one which should release a large amount of radioactivity over a populated area—many fatalities could result; however, the concern is not basically with the potential fatalities, since nothing is said of the fact that the failure of any of a considerable number of dams in the United States would result in very many more fatalities and have a probability of occurrence (in the course of a year) several orders of magnitude larger than that of a serious nuclear power plant accident (38). Evidently (at least in the United States, which first became aware of nuclear energy through its own use of nuclear weapons) what is involved here is that the word *nuclear* is inseparably associated with nuclear war, and aversion to nuclear power is in large measure a fallout from nuclear weapons. The effect is unfortunate, since a visceral antipathy to nuclear power obscures the issues, proliferates error, and diverts attention from the point of real importance, that of the possibility that nuclear weapons might be used. It is also irrelevant: we shall continue to have all the means necessary to engage in nuclear war whether or not we proceed with nuclear power, and other countries will proceed with nuclear power whether we choose to or not. The consequences of any nuclear war, however limited, would be enormously greater than those of even many reactor accidents and, in the judgment of a panel discussing the question "Nuclear War by 1999?" (39), the likelihood of nuclear war by the end of the century is also enormously larger than that of a serious accident with a nuclear reactor.

Opponents of nuclear power have become quite fond of the specter that no matter how much care may be taken in burying radioactive wastes, it is impossible to guarantee that at some point in geologic time some part of them may not again become accessible to the biosphere. Certainly the disposal of radioactive materials deserves (and is receiving) very careful consideration. However, if one could now be sure that in 500 or 5000 years our descendants would find it prudent to re-cover and stabilize the eroding surface of some ancient radioactive burial ground, that would bring the welcome reassurance that we had somehow managed to get through without bringing down the most extreme consequences of an all-out nuclear war.

Literatured Cited

1. Glasstone, S. 1962. *The Effects of Nuclear Weapons.* Washington DC: US DOD & AEC
2. Conard, R. A. et al 1975. *A Twenty-Year Review of Medical Findings in a Marshallese Population.* New York: BNL 50424
3. Reactor Safety Study 1975. WASH-1400 (NUREG-75/014) Washington DC: US NRC
4. National Council on Radiation Protection 1974. *NCRP Rep. No. 42.* Washington DC: GPO
5. 1974. *Briefing on Counterforce Attacks.* Hearing of Committee on Foreign Relations, US Senate, Sept. 1974
6. *Report of the Advisory Committee on the Biological Effects of Ionizing Radiations* 1972. (BEIR Rep.) Washington DC: Natl. Acad. Sci.
7. Garner, R. J. 1972. *Transfer of Radioactive Materials from the Terrestrial Environment to Animals and Man.* Cleveland: CRC
8. 1975. *Long-Term Worldwide Effects of Multiple Nuclear Weapons Detonations.* Washington DC: Natl. Acad. Sci.
9. 1971. *Survival of Food Crops and Livestock in the Event of Nuclear War.* US AEC Symp. Ser. No. 24. Washington DC
10. 1975. *Analysis of Effects of Limited Nuclear Warfare.* Washington DC: Comm. Foreign Relat., US Senate
11. Rathjens, G. W. 1975. *Daedalus.* 104(3): 201–4
12. Drell, S., von Hippel, F. 1976. *Limited Nuclear Warfare—A Viable Option?* Submitted for publication
13. Blackett, P. M. S. 1948. *Fear War and the Bomb,* p. 40. New York: Whittlesey. 244 pp.
14. Craven, W. F., Cate, J. L., eds. 1953. *The Army Air Forces in World War II.* Chicago: Univ. Chicago Press
15. Sallagar, F. M. 1969. *The Road to Total War,* p. 128. New York: Van Nostrand Reinhold. 197 pp.
16. Toland, J. 1966. *The Last 100 Days.* New York: Random House
17. See Ref. 1, p. 551ff
18. York, H. 1970. *Race to Oblivion,* p. 83. New York: Simon & Schuster. 256 pp.
19. ACDA/PAB-253, April 1975
20. See Ref. 1, p. 350ff
21. Foley, H. M., Ruderman, M. A. 1973. *J. Geophys. Res.* 78:441–50
22. Bauer, E., Gilmore, F. R. 1975. IDA P-1076
23. Miettinen, R. K. 1974. Nuclear mini-weapons. In *Nuclear Proliferation Problems,* p. 119. Cambridge, Mass: MIT

Press
24. See Ref. 10, p. 111
25. UNSCEAR 1972. *Ionizing Radiation: Levels and Effects* 1:3. New York: UN Publ. No. E 72, IX·17
26. Kissinger, H. 1957. *Nuclear Weapons and Foreign Policy,* p. 128. New York: Harper. 463 pp.
27. See Ref. 26, p. 145
28. See Ref. 26, p. 191
29. Schell, J. 1975. *The New Yorker.* 51(20): 47 (July 7)
30. See Ref. 10, p. 149
31. See Ref. 10, p. 156
32. Steinbruner, J., Carter, B. 1975. *Daedalus.* 104(3): 131–54
33. Falk, R. A. 1968. *Legal Order in a Violent World,* pp. 425–35. Princeton, NJ: Princeton Univ. Press
34. Text in UN Document A/RES/2373 22; July 1, 1968
35. 1976. US Senate Government Operations Committee. Hearings on S. 1439
36. Taylor, T. B. 1975. *Ann. Rev. Nucl. Sci.* 25:407–21
37. Wilson, R. 1976. *Nucl. News* 19(2): 55–57
38. Okrent, D. 1975. Testimony Comm. on Energy Diminishing Mat., Calif. Legislature, Oct. 29, 1975; Weinberg, A. 1976. *Am. Sci.* 64:16–21
39. Doty, P., Garwin, R., Kistiakowsky, G., Rathjens, G., Schelling, T. Nov. 1975. *Harvard Mag.,* pp. 19–25

Ann. Rev. Nucl. Sci. 1976. 26: 89–149

THE PHYSICS OF e^+e^- COLLISIONS[1]

✻5572

Roy F. Schwitters

Stanford Linear Accelerator Center, Stanford, California 94305

Karl Strauch

Department of Physics, Harvard University, Cambridge, Massachusetts 02138

CONTENTS

[1] Work supported by Energy Research and Development Administration.

1   INTRODUCTION

Within a ten-year period, experiments investigating the products of collisions between positrons and electrons have grown from eclectic curiosities to become major research programs profoundly contributing to our knowledge of elementary particle physics. This metamorphosis has been paced by significant developments in accelerator design that have allowed the promise of e^+e^- physics to be realized in the laboratory. Today, colliding-beam devices and their associated particle detectors are among the most sought-after research tools in high-energy physics.

The importance of e^+e^- collisions to elementary particle physics is derived from the simplicity provided by the e^+e^- initial state. Discrete additive quantum numbers of this state, such as electric charge, baryon number, lepton number, and strangeness are all zero. Thus the total energy of the electron and positron are available to produce any final-state configuration of particles that satisfies energy-momentum conservation and has zero net charge, baryon number, lepton number, strangeness, etc. As is discussed below, the dynamics of the initial-state electron and positron are well described by the theory of quantum electrodynamics, thereby allowing unknown structures and forces of final-state particles to be probed by the known electromagnetic force. At the high energies of interest here, the dominant state formed by the positron and electron has the quantum numbers of a single photon, namely total spin one, negative parity, and negative charge conjugation. Thus e^+e^- collisions provide a unique opportunity to study high-energy systems of particles in a nearly pure quantum state having the quantum numbers of the photon.

In this article we review what has been learned to date from the study of e^+e^- collisions at energies above the threshold for hadron production. In this field, experimental observations are usually on much firmer ground than their theoretical interpretation, so we concentrate almost exclusively on the experiments. After an introduction to nomenclature and kinematics of e^+e^- collisions, we present a brief description of colliding-beam devices and the history of their development. In Section 2 the underlying theoretical principles necessary to the interpretation of experimental results are presented. The e^+e^- system provides important experimental testing grounds for the theory of quantum electrodynamics at high energies and large values of momentum transfer. Experimental tests of QED are discussed in Section 3. In recent years there has been great interest in the study of hadron production in e^+e^- collisions. We report on the current status of this rapidly changing field in Section 4. Other topics, such as photon-photon processes, new particle searches, and weak interaction effects are included where appropriate within these sections.

1.1   Notation and Kinematics

The system of units where $h = c = 1$ is used throughout this article, so that many quantities are expressed in units of energy. The electron rest mass is often ignored in what follows because of the ultrarelativistic nature of electrons in these experiments. The following symbols are frequently used in describing e^+e^- colliding-beam physics:

W = center-of-mass (c.m.) energy of the e^+e^- system.

E = single-beam energy ($= W/2$).

s = square of c.m. energy.

L = luminosity. The conventional units for L are cm^{-2} sec^{-1}.

θ = production angle of any final-state particle with respect to the incident positron direction.

ϕ = azimuthal angle of any final-state particle measured from the plane of the colliding-beam orbit.

R = ratio of the total cross section for hadron production in e^+e^- annihilation to the total cross section for muon pair production.

That colliding-beam systems permit more efficient use of accelerator energy to obtain a high center-of-mass energy is a simple consequence of elementary mechanics that has been taught to generations of students. Consider the collision of two particles, oppositely directed, of total energies E_1 and E_2, and rest masses m_1 and m_2, respectively. The square of the c.m. energy, s, is easily calculated to be

$$s = 2E_1E_2\{1+\sqrt{[1-(m_1/E_1)^2][1-(m_2/E_2)^2]}\}+m_1^2+m_2^2. \qquad 1.1.$$

The most important aspect of Equation 1.1 for high-energy physics is the product E_1E_2; accelerating the "target" particle as well as the "projectile" increases s by essentially the ratio of the target energy to its rest mass, which can be a considerable factor for high-energy beams of light particles such as electrons. In the limit $E_1 = E_2 = E \gg m_1, m_2$, the c.m. energy is simply:

$$W = 2E. \qquad 1.2.$$

However, if a projectile of energy $E \gg m_1, m_2$ strikes a stationary target of mass $m_2 = m$, W is given by:

$$W = \sqrt{2mE}. \qquad 1.3.$$

For instance, Equations 1.2 and 1.3 show that e^+e^- interactions at $W = 10$ GeV can be studied with head-on collisions of two 5-GeV beams or with a 100,000-GeV e^+ beam hitting a stationary target. Obviously, colliding beams provide the only practical means for studying high-energy e^+e^- collisions.

The kinematics of the final state are also extremely simple in colliding beams because the laboratory and c.m. reference systems are the same (or nearly the same if the beams cross at a small angle).

1.2 Colliding-Beam Devices

The advantages in kinematics provided by colliding beams are gained at the expense of rate of interactions. The luminosity L, defined as the interaction rate per unit cross section, is considerably smaller in colliding beams compared to conventional targets because of the extremely low density of particles in a beam. (The interaction rate $\dot{N}$ for a given process is related to the cross section σ for that process by $\dot{N} = \sigma L$.) The task of obtaining sufficiently high luminosities for performing practical experiments is an epic story in itself, performed by a small number of accelerator physicists working with particle physicists. An introduction to the science of colliding-beam devices has been presented by Sands (1); a more advanced discussion is contained in the review article by Pellegrini (2).

All present colliding-beam devices make use of one or two storage rings into which beams of positrons and electrons are introduced. Storage rings confine the beams to periodic orbits within an ultrahigh vacuum so that the beams continue to circulate for periods up to several hours. At several points around the storage ring, the beams intersect and collisions can take place. There are typically two to eight intersection regions in a colliding-beam device, and at these regions the experimental apparatus is mounted. Electromagnetic forces between the particles of a beam and the storage ring structure, the other beam, and particles within the same beam all limit the maximum attainable luminosity. In general, L may be a strong function of energy of a given storage ring and may vary greatly among different colliding-beam devices of comparable energy. For example, L varies as E^4 for $E < 3.5$ GeV at the SPEAR e^+e^- colliding-beam facility. Transverse beam dimensions, typically 1 mm or less, and the spread of energies within the beams [typically $\delta E/E \simeq 0.05\% \times E$ (GeV)] are governed by quantum fluctuations in the synchrotron radiation process.

The birth of experimental high-energy colliding-beam physics occurred at the end of 1963 inside AdA (Anello di Accumulazione) when Bernadini et al (3) observed single bremsstrahlung produced in the collision between 250-MeV positrons and 250-MeV electrons. This pioneering single-ring device (4), designed and built at the Frascati Laboratory in Italy and tested at the Orsay Laboratory in France, achieved its goal of demonstrating that counter-rotating positron and electron beams of sufficient intensity could be built up, stored, and made to collide so that interesting experiments can be performed.

The first two high-energy physics experiments with colliding beams were performed in 1965; they tested the predictions of quantum electrodynamics (QED) for e^-e^- scattering at previously unattained high values of momentum transfer. The Princeton-Stanford group used intersecting double rings (5) located at Stanford initially at $W = 600$ MeV (6), and later on at $W = 1.12$ GeV (7). The Novosibirsk group used a similar though somewhat smaller device, VEPP-1 (8), up to W values of 320 MeV (9). The information obtained and the lessons learned from the building, debugging, and operation of the two double-ring devices (no longer in operation), were most important for the subsequent development of higher-energy colliding-beam systems.

Definite results on hadron production in e^+e^- collisions were first reported at the 1967 Electron-Photon Symposium by the Novosibirsk group (10) and by the Orsay group (11). The Soviet group presented a beautiful excitation curve exhibiting resonance structure for the reaction $e^+e^- \to \pi^+\pi^-$ at values of W corresponding to the ρ mass; these results were obtained with the single-ring system VEPP-2 (12). The French group, also using a single ring called ACO (13), reported the very large cross section at the energy corresponding to the peak of the ρ mass distribution. VEPP-2 is now part of the higher luminosity system VEPP-2M (14). ACO has continued operation.

With the beginning of operation for experiments in 1971 of the single-ring ADONE (15) at the Frascati Laboratory, the available energy region was expanded to $W = 2.5$ GeV (later to 3.1 GeV), and the luminosity increased substantially. The predictions of QED for several reactions have been verified, and hadron production has and is being studied extensively by several groups. A particularly important result was the observation that the total cross section for hadron production was substantially higher than had been expected for values of W well above the resonance region.

During its brief period of operation (1972–1973), the Cambridge Electron Accelerator BYPASS system (16) demonstrated that "low-beta" sections could be used to increase the luminosity and showed that the hadron production cross section exceeded expectations even more at $W = 4$ and 5 GeV than had been observed with ADONE, suggesting the existence of some new hadronic phenomena. Predictions of QED for several reactions were also verified at these two energies.

e^+e^- colliding beam physics entered a new regime of energy, data rates, and detail of experimental information in 1973 with the start of operation of SPEAR and its magnetic detector (17). SPEAR is a single-ring device located at the Stanford Linear Accelerator Center (SLAC); it operates at c.m. energies between 2.5 and 8 GeV with luminosities in the range from 10^{29} to 10^{31} cm^{-2} sec^{-1}. One of its two interaction regions is surrounded by the SLAC/Lawrence Berkeley Laboratory (LBL) magnetic detector, which is able to cover approximately two thirds of 4π solid angle with momentum analysis and particle identifications. Certainly, the zenith of e^+e^- physics to date occurred in November 1974 with the simultaneous announcement of the discovery at SPEAR by the SLAC/LBL group (18) and at the Brookhaven National Laboratory (BNL) by the MIT/BNL group (19) of the very narrow J/ψ resonance, and the subsequent discovery of a second narrow resonance, the ψ', found just ten days after the first at SPEAR (20).

SPEAR was joined in 1974 by DORIS (21) at the German electron-synchrotron laboratory DESY in Hamburg, Germany. DORIS consists of two intersecting storage rings in which the beams collide at an angle of 1.4°. This construction permits future experiments with e^-e^- or e^+e^+ collisions in addition to e^+e^- collisions. DORIS is designed to cover the same energy range as SPEAR with comparable luminosity.

Two new projects are well under way at the end of 1975. The Orsay group is building a novel four-beam device known as DCI (22) that is designed to operate up to $W = 3.6$ GeV with higher luminosity than other systems in this energy region. The Novosibirsk group is working toward the operation of VEPP-4, which should provide e^+e^- collisions up to energies of $W = 14$ GeV (23).

The next generation of colliding-beam devices is planned to commence operation in 1979 to 1980. Two projects, PEP (24) at SLAC and PETRA at DESY (25), are in the final design and initial construction phases. Both are designed to open a new energy regime up to $W \simeq 35$ GeV. A Cornell group (26) has proposed construction of a device to operate up to $W = 18$ GeV.

The operating regions of these e^+e^- colliding beam devices are summarized in Figure 1.

2   THEORETICAL INTRODUCTION

When positrons and electrons collide, they may scatter elastically, annihilate into states containing no e^+e^-, or produce final states containing e^+e^- in addition to other particles. At presently available energies $W \lesssim 10$ GeV, e^+e^- interact primarily through the electromagnetic force and, except for very small and as yet unobserved weak interaction corrections, the e^+e^- system can be described by QED.

2.1   One-Photon Exchange

At lowest order in electromagnetic coupling, e^+e^- annihilation to inelastic final states proceeds through the one-photon [1γ] channel represented by Figure 2. The pioneering work by Tsai (27), Cabibbo & Gatto (28) and a recent paper by Tsai (29) are useful references to the physics of the 1γ channel.

The amplitude for any 1γ process is proportional to

$$A \sim \frac{e^2}{s} j_\mu J^\mu,$$ 2.1.

where

$$j_\mu = \bar{v}\gamma_\mu u$$ 2.2.

is the e^+e^- current. $\bar{v}$ and u are the usual Dirac spinors and γ_μ are Dirac matrices.

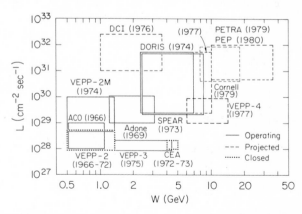

Figure 1 Operating regions of various e^+e^- colliding-beam devices. The indicated luminosities are approximate ranges of peak luminosity and do not represent the actual energy dependence.

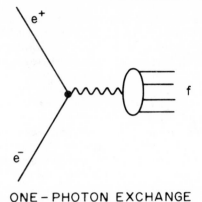

ONE – PHOTON EXCHANGE

Figure 2   Feynman diagram for one-photon exchange processes.

The $1/s$ factor comes from the photon propagator and e is the electron charge. The "physics" of the final state is contained in the current J^μ; the object of many experiments is to measure matrix elements of various J^μ.

j_μ can be computed by standard techniques (30). In the c.m. frame, j_0 is zero (a consequence of current conservation) and the space components of j_μ are

$$\mathbf{j} = \frac{1}{m}\phi^+\left[E\boldsymbol{\sigma} - \frac{\boldsymbol{\sigma}\cdot\mathbf{p}_+}{E+m}\mathbf{p}_+\right]\chi, \qquad\qquad 2.3.$$

where χ and ϕ are two-component Pauli spinors for the e^+e^- spins in their respective rest frames. $\boldsymbol{\sigma}$ are the Pauli spin matrices, m is the electron mass and $\mathbf{p}_+$ is the e^+ 3-momentum. Direct calculation of Equation 2.3 shows that the longitudinal component of $\mathbf{j}$—the component parallel to $\mathbf{p}_+$—is the order of m/E smaller than the transverse component, and therefore negligible in high-energy processes. The transversality of $\mathbf{j}$ is equivalent to the fact that the e^+e^- annihilate only through states of net helicity one.

The beams will naturally polarize in high-energy storage rings due to their interaction with synchrotron radiation (31). A complete discussion of this pheno-menon is given by Baier (32), including calculations of various depolarizing effects. A more pedagogical treatment and an excellent list of references to both the theoretical and experimental literature are contained in the review article by Jackson (33). In the absence of depolarizing effects, the polarization P of each beam builds up in time according to

$$P(t) = P_0[1 - \exp(-t/T)], \qquad\qquad 2.4.$$

$$T \simeq 98 \text{ sec} \times \frac{R_0^3(m)}{E^5\,(\text{GeV})} \times \frac{R_{\text{avg}}}{R_0}, \qquad\qquad 2.5.$$

$$P_0 = \frac{8\sqrt{3}}{15} \approx 0.924, \qquad\qquad 2.6.$$

where R_0 is the bending radius and R_{avg} is the average radius of the storage ring. The positrons (electrons) are polarized parallel (antiparallel) to the guide magnetic field that we choose to lie along the $-y$ axis. Our coordinate system is defined in Figure 3. At low energies P can generally be neglected, but at high energies the characteristic time T is less than typical storage times, and polarization effects become important. With the e^+e^- completely polarized in opposite directions, the only nonvanishing component of j^μ will lie along the mutual polarization direction, in this case the y axis. If this mutual polarization direction were to be aligned with the beam direction, j^μ, and hence 1γ, exchange processes would vanish.

Symmetry principles and gauge invariance restrict the possible forms for the final-state current J^μ. Following Tsai (29), the most general form for single-particle inclusive cross sections in the 1γ channel can be written

$$E_f \frac{d\sigma}{d^3\mathbf{p}_f} = \frac{\alpha^2}{2s^2}\left[(W_1 + W_0) + (W_1 - W_0)(\cos^2\theta + P^2\sin^2\theta\cos 2\phi)\right], \qquad 2.7.$$

where $\mathbf{p}_f$ and E_f are the single-particle momentum and energy. P^2 is the product of the e^+, e^- transverse polarizations, which are assumed to be oppositely directed in the y direction. The structure functions W_0 and W_1 are functions of s, E_f, and the type of particle produced. They are defined by

$$W_0 = \sum_{\substack{\text{all final} \\ \text{states} \\ \text{except } \mathbf{p}_f}} (2\pi)^3\,\delta^4(P_i - P_f)\,|\langle f|J_{z'}|0\rangle|^2, \qquad 2.8.$$

$$W_1 = \sum_{\substack{\text{all final} \\ \text{states} \\ \text{except } \mathbf{p}_f}} (2\pi)^3\,\delta^4(P_i - P_f)\,|\langle f|J_{t'}|0\rangle|^2, \qquad 2.9.$$

where $\langle f|J_{z'}|0\rangle$ is the matrix element of the component of J^μ parallel to $\mathbf{p}_f$ that gives rise to the state f; $\langle f|J_{t'}|0\rangle$ is the matrix element of a component of J^μ perpendicular to $\mathbf{p}_f$; P_i and P_f are the net four-momenta of the initial and final states, respectively; δ is the Dirac delta-function; and W_0 represents the probability that the final state has zero net helicity along $\mathbf{p}_f$, while W_1 gives the probability that the final state has net helicity one along $\mathbf{p}_f$.

Bjorken (34) has argued that at high energies hadron production by 1γ annihilation should exhibit scaling, such that W_0 and W_1 become functions of only one dimensionless quantity x, the ratio of E_f to the beam energy E. If scaling holds, the inclusive cross section can be written

$$\frac{d\sigma}{d\Omega\,dx} = \frac{\alpha^2}{8s}\beta x\{[W_1(x) + W_0(x)] + [W_1(x) - W_0(x)](\cos^2\theta + P^2\sin^2\theta\cos 2\phi)\},$$

$$2.10.$$

where $\beta = p_f/E_f$ is the particle velocity and $x = E_f/E$. The total cross section for a process that exhibits scaling falls with energy as s^{-1}.

Some specific examples of 1γ annihilation processes follow.

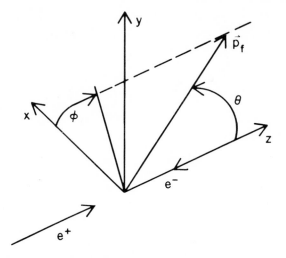

Figure 3 Coordinate system. The incident e^+ direction is the z axis and the guide magnetic field points in the $-y$ direction. Any final-state particle is described by the momentum vector $\mathbf{p}_f$ and polar and azimuthal angles θ and ϕ.

2.1.1 $e^+e^- \rightarrow \mu^+\mu^-$ This is the standard of reference for all 1γ processes. The structure functions for a pair of pointlike spin-1/2 particles can be calculated from QED and are

$$W_0 = 2\delta(E-E_f)m_l^2/E_f,$$
$$W_1 = 2\delta(E-E_f)E_f,$$

2.11.

where m_l is the muon rest mass.

The differential cross section is given by

$$\frac{d\sigma_\mu}{d\Omega} = \frac{\alpha^2}{4s}\beta[(2-\beta^2)+\beta^2(\cos^2\theta+P^2\sin^2\theta\cos 2\phi)].$$

2.12.

As usual, β is the muon velocity.

The total cross section for muon-pair production is

$$\sigma_\mu = \frac{2\pi\alpha^2}{3s}\beta(3-\beta^2).$$

2.13.

At high energies where $\beta \approx 1$, σ_μ is numerically

$$\sigma_\mu \simeq \frac{21.7\,(\text{nb})}{E^2\,(\text{GeV})} = \frac{86.8\,(\text{nb})}{s\,(\text{GeV})^2}.$$

2.14.

2.1.2 $e^+e^- \rightarrow \pi^+\pi^-$ The current for two spinless particles of unknown charge structure contains one unknown function of s, the form factor $F(s)$. The cross section for pion-pair production is

$$\frac{d\sigma}{d\Omega} = \frac{\alpha^2}{8s}|F_\pi(s)|^2\beta^3\sin^2\theta(1-P^2\cos 2\phi).$$ 2.15.

2.1.3 $e^+e^- \to p\bar{p}$ In this case, when a pair of spin-1/2 particles with internal structure are produced, the current will contain two independent form factors. One choice of form factors is to multiply W_0 and W_1 for the μ-pair case (Equation 2.11) by $|G_E(s)|^2$ and $|G_M(s)|^2$ respectively. The cross section for $p\bar{p}$ production can then be written:

$$\frac{d\sigma}{d\Omega} = \frac{\alpha^2}{4s}\beta[|G_M|^2(1+\cos^2\theta+P^2\sin^2\theta\cos 2\phi)$$

$$+(1-\beta^2)|G_E|^2\sin^2\theta(1-P^2\cos 2\phi)].$$ 2.16.

2.2   Other QED Processes of Order α^2

In addition to μ-pair production, two other purely electromagnetic processes of order α^2 occur in e^+e^- interactions. The elastic scattering of e^+e^-, known as Bhabha scattering, may be represented by the Feynman diagrams in Figure 4. In addition to the one-photon exchange contribution discussed above, there is an important t-channel amplitude. The cross section for Bhabha scattering is (29)

$$\frac{d\sigma}{d\Omega} = \frac{\alpha^2}{8s}\frac{1}{(1-\cos\theta)^2}[(3+\cos^2\theta)^2 - P^2\sin^4\theta\cos 2\phi].$$ 2.17.

Small-angle Bhabha scattering is useful for monitoring luminosity because of the relatively large cross section; it is studied at large angles as a test of QED for large spacelike and timelike values of the virtual photon mass.

The second process is e^+e^- annihilation into two photons. Lowest-order diagrams for this process are given in Figure 5 and the cross section is

$$\frac{d\sigma}{d\Omega} = \frac{\alpha^2}{s}\left[\frac{1+\cos^2\theta}{\sin^2\theta} + P^2\cos 2\phi\right].$$ 2.18.

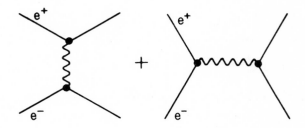

BHABHA SCATTERING

Figure 4   Lowest-order Feynman diagrams for Bhabha scattering.

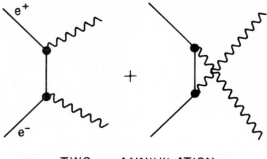

TWO γ ANNIHILATION

Figure 5   Lowest-order Feynman diagrams for e^+e^- annihilation into two photons.

2.3   *Photon-Photon Processes*

It has been pointed out by a number of authors (35–37) that the cross section for the production of systems of invariant mass, which is small compared to W, together with forward-peaked e^+e^-, will compete with one-photon exchange processes at high energies. These reactions, shown schematically in Figure 6, can be thought of as the collision of two nearly real photons emanating from the initial e^+ and e^-, and, unlike 1γ annihilation, they result in states of positive charge conjugation. There is a vast theoretical literature on this subject; it has been reviewed by Terazawa (38).

The essential physics of photon-photon (2γ) processes can be derived from the equivalent photon approximation (38), where the flux of nearly real photons of energy k associated with a high-energy electron is approximately

$$N(k)\,dk \simeq \frac{2\alpha}{\pi} \ln\left(\frac{E}{m}\right)(1 - k/E + k^2/2E^2)\frac{dk}{k},$$ 2.19.

where m is the electron mass.

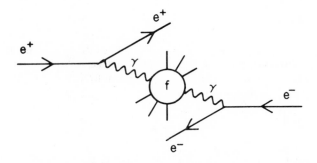

Figure 6   Schematic diagram of photon-photon process leading to the final state e^+e^-f.

The cross section $d\sigma_{ee \to eex}$ for producing the state x through the 2γ process is related to the cross section $d\sigma_{\gamma\gamma \to x}$ for producing x in a collision of two photons by

$$d\sigma_{ee \to eex} = \int dk_1\, dk_2 N(k_1) N(k_2)\, d\sigma_{\gamma\gamma \to x}(k_1, k_2). \qquad 2.20.$$

Even though 2γ processes are higher order in α than 1γ processes, the $[\ln(E/m)]^2$ terms augment the cross section so that at high energies they may exceed 1γ cross sections that are falling as E^{-2}. As opposed to 1γ reactions, 2γ reactions will also occur in $e^- e^-$ or $e^+ e^+$ collisions.

Due to differences in the photon energies, the state x is generally moving relative to the laboratory frame. The spectrum of photons is peaked at very low energies, so it is likely that states of low invariant mass are formed. It is very unlikely to have more than one half of the total $e^+ e^-$ energy appear in the state x. The final-state e^+, e^- are strongly peaked about their initial direction, leading to a characteristic signature for these processes.

2.4   Radiative Corrections

Emission of additional photons, both virtual and real, over those considered in lowest-order results in modifications to lowest-order cross sections and angular distributions. The application of radiative corrections to obtain bare cross sections from experiment is a subtle problem far beyond the scope of this article. The paper by Tsai (39) is a classic in this field. Bonneau & Martin (40) have calculated the lowest-order radiative corrections to one-photon exchange shown in Figure 7.

Briefly, radiative corrections can be categorized into single hard-photon emission, multiple soft-photon emission, and vertex corrections and vacuum polarization. For hard-photon emission the incident beams can be considered as being accompanied by a flux of photons similar to Equation 2.19. The $e^+ e^-$ then annihilate at an energy less than W, where the cross section may be different. The treatment of multiple soft-photon emission dates back to the work of Bloch & Nordsieck (41). In the present context the hard-photon spectrum is usually modified by a factor $(k/E)^t$ to account for multiple soft photons. The effective radiator thickness t is given by

$$t = \frac{2\alpha}{\pi}[\ln(s/m^2) - 1]. \qquad 2.21.$$

Vertex corrections and vacuum polarization effectively modify the flux of the single virtual photons involved in one-photon exchange.

The net radiative correction to 1γ exchange taken from Bonneau & Martin (40) and modified for soft-photon emission is

$$\sigma_{\exp}(s) = \sigma_0(s)\left[1 + \frac{2\alpha}{\pi}\left(\frac{\pi^2}{6.} - \frac{17}{36}\right) + \frac{13}{12}t\right]$$

$$+ t\int_0^E \frac{dk}{k}\left(\frac{k}{E}\right)^t (1 - k/E + k^2/2E^2)[\sigma_0(s - 4Ek) - \sigma_0(s)], \qquad 2.22.$$

where $\sigma_{exp}(s)$ is the experimentally measured cross section and $\sigma_0(s)$ is the one-photon exchange cross section. The first term in Equation 2.22 contains the vertex modification and vacuum polarization corrections; the second term is for external photon emission. Equation 2.22 is displayed here to illustrate the types of radiative corrections involved in e^+e^- collisions. It does not represent a universal formulation to be applied to all cases.

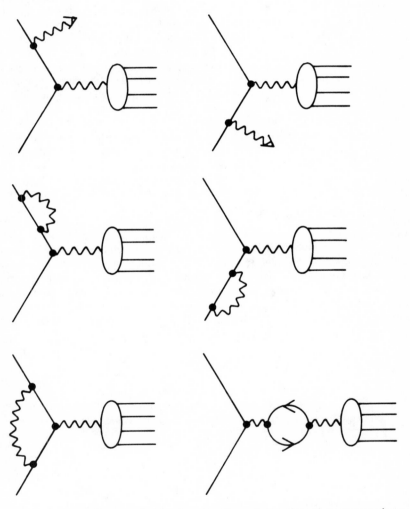

Figure 7   Feynman diagrams showing lowest order radiative corrections to one-photon exchange processes that lead to nonelectrodynamic final states f.

3   ELECTRODYNAMIC FINAL STATES

In this section we review the experimental information on final states of e^+e^- collisions where only leptons and photons are observed. For the most part, the reactions under consideration can be described by the theory of quantum electrodynamics, and therefore these experiments constitute tests of QED at high energies. The relevance of such tests to the theory has been discussed by Brodsky & Drell (42). The basic elements of the theory being tested are the assumptions that both electrons and muons may be treated as pointlike Dirac particles and that the Coulomb field satisfies the $1/r^2$ law. Deviations between experiment and theory would imply one or more of the following: (a) a breakdown of the fundamental assumptions of QED; (b) interference with nonelectromagnetic sources of electrodynamic final states, such as the strong or weak interactions; and (c) the presence of entirely new interactions and particles.

To date, there is no evidence for a failure of the basic assumptions of QED; indeed, it is the most successful theory in physics. There are several examples of hadronic contributions to electrodynamic final states, there is much theoretical anticipation concerning the study of weak interaction effects at the next generation of storage rings (43) but no experimental data, and there is a hint that new electrodynamic particles may be produced in e^+e^- collisions.

3.1   Breakdown Parameters

In experimental tests of QED it is conventional to compare experimental yields with the "predictions of QED" in which the theory, including explicit radiative corrections, is calculated for the particular experimental conditions of angular acceptance, momentum and angle resolution, and kinematic cuts. Deviations from QED are usually parameterized by "breakdown" parameters Λ having dimensions of energy. The scale of distance corresponding to a given Λ is $\hbar c/\Lambda$. There is no unique way to modify QED with breakdown parameters; it has become standard practice (44, 45) to replace the usual photon propagator by

$$\frac{g_{\mu\nu}}{q^2} \to \frac{g_{\mu\nu}}{q^2} \mp \frac{g_{\mu\nu}}{q^2 - \Lambda_\mp^2}, \qquad\qquad 3.1.$$

where $g_{\mu\nu}$ is the metric tensor and q^2 is the invariant mass of the virtual photon. The choice of sign corresponds to "negative" or "positive" metric corrections. Such a modification would change one-photon exchange cross sections by

$$\sigma_{\text{modified}} \simeq \sigma_{1\gamma}\left(1 \pm \frac{2s}{\Lambda_\mp^2}\right). \qquad\qquad 3.2.$$

An alternative parameterization, which leads to modifications similar to Equation 3.2, is to assign form factors to the electron and muon. The charge radius of e or μ would then be proportional to Λ^{-1}.

Based on the work of Kroll (46), reactions involving a lepton propagator, such as $e^+e^- \to \gamma\gamma$ annihilation, are parameterized by

$$\frac{\sigma_{\text{modified}}}{\sigma_{\text{QED}}} = 1 \pm \frac{2q^4}{\Lambda^4_\pm},$$

3.3.

where q^2 is the invariant mass of the virtual lepton.

We wish to stress that breakdown parameters should be used only as a guide in interpreting the data. Only detailed examination of individual experiments will disclose how stringent is the particular test of QED.

3.2 $e^-e^- \rightarrow e^-e^-$

The e^-e^- final state was the first to be studied experimentally with colliding beams at the Stanford storage rings (6, 7) and at Novosibirsk (9). This process, known as Møller scattering, is represented in lowest order by the diagram of Figure 8. The Stanford apparatus consisted of scintillation counters and optical spark chambers covering approximately $\pm 30°$ in ϕ and the polar angle range $35° < \theta < 115°$. There was no magnetic analysis or independent luminosity monitor. The measured angular distribution at $W = 1112$ MeV is presented in Figure 9, along with QED predictions normalized to the same total number of events and corrected for detector acceptance. The data are in excellent agreement with theory and establish lower limits for breakdown parameters in the range 2–3 GeV (95% confidence level).

3.3 $e^+e^- \rightarrow e^+e^-$

Bhabha scattering has been widely studied in e^+e^- colliding-beam experiments. This final state is characterized by a collinear pair of showering particles and it tests the photon propagator for both timelike and spacelike values of momentum transfer. The most readily available theoretical calculation, including detailed radiative corrections, that is suitable for comparison with experiment is given by Berends, Gaemers & Gastmans (47).

The first absolute measurement of Bhabha scattering with colliding beams was performed at ACO (48). A nonmagnetic detector with scintillation counters and

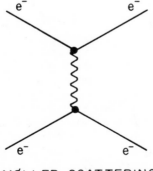

MØLLER SCATTERING

Figure 8   Feynman diagram for Møller scattering.

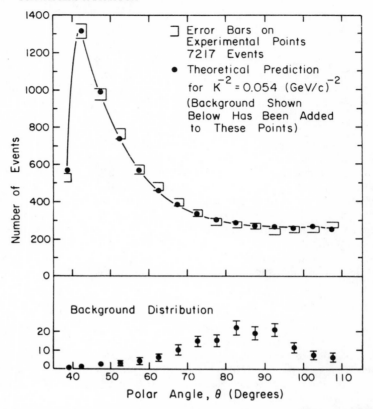

Figure 9   Experimental angular distribution for Møller scattering at $W = 1112$ MeV from (7). The theoretically expected angular distribution, normalized to the total number of events and corrected for detector acceptance, is also shown.

optical spark chambers was used to measure this process at $W = 1020$ MeV. Luminosity was monitored by detecting small-angle gamma rays from double bremsstrahlung ($e^+e^- \to e^+e^-\gamma\gamma$). This experiment also set breakdown parameter limits in the 2-GeV to 3-GeV range.

At ADONE, several groups used optical spark chambers and scintillation counters to detect large-angle Bhabha scattering; luminosity was monitored through small-angle Bhabha scattering. The early experiments (49, 50) verified the validity of QED in absolute normalization and energy dependence to a level of 5% to 10% over the energy range 1.6 GeV $\leq W \leq 2.0$ GeV. The effect of radiative corrections on the collinearity of events was discussed by the "BCF" group (50, 51). The ratio of experimental yields for Bhabha scattering to those theoretically expected from more recent measurements performed by the "boson" group (52, 53) and the "$\mu\pi$" group (54) are plotted in Figure 10 along with results at other energies.

The boson group (53) quoted an average ratio $\bar{R}$ of measured-to-theoretical yield over the energy range 1.4 GeV $\leq W \leq$ 2.4 GeV of

$\bar{R} = 1.05 \pm 0.04$ (± 0.065 systematic).

The $\mu\pi$ group (54) were able to set the limit

$\Lambda_+ > 6$ GeV (95% confidence)

with data over the same energy range. The BCF group (55) described the energy dependence of the Bhabha scattering yield in their detector over the energy range 1.2 GeV $\leq W \leq$ 3.0 GeV by the form:

Yield $= As^n$

and found $A_{\text{exp}}/A_{\text{QED}} = 1.00 \pm 0.02$ and $n = -(0.99 \pm 0.02)$. They also observed acoplanar e^+e^- events that result from radiative corrections (56).

Experiments performed with the nonmagnetic BOLD detector at CEA measured wide-angle Bhabha scattering at $W = 4$ GeV (57) and 5 GeV (58). These measurements, which were normalized to double bremsstrahlung, agreed with QED and

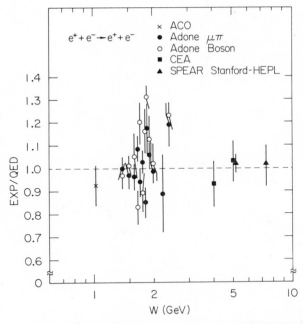

Figure 10 Ratio of experimental yield to that expected from QED for Bhabha scattering. Detector acceptances and normalization procedures differ between experiments. Data points shown are from the following references: ACO (48), Adone $\mu\pi$ (54), Adone Boson (53), CEA (57, 58), and SPEAR (59, 60).

pushed lower limits on breakdown parameters to nearly 10 GeV. These results are summarized in Figure 10. The coplanarity distribution observed in the CEA experiments (58) is in good agreement with theoretical radiative corrections (47).

Two SPEAR groups have reported tests of QED through Bhabha scattering. Using a novel detector consisting of two large NaI crystals, a Stanford-HEPL group (59) measured 90° Bhabha scattering at 5.2 GeV and, more recently (60), have presented preliminary results at 7.4 GeV. These results, plotted in Figure 10, agree with QED and raise the lower limit on breakdown parameters to approximately 20 GeV, an order of magnitude higher than the earliest results discussed above. The SLAC/LBL collaboration (61) measured Bhabha scattering with a magnetic detector at 3.0, 3.8, and 4.8 GeV, and for the first time were able to distinguish the charge of the outgoing particles. The angular distribution obtained in this experiment is presented in Figure 11a and is seen to be in excellent agreement with QED. Their data were fitted to several sets of breakdown parameters; the results are given in Table 1. The lower limits are generally the order of 20 GeV, demonstrating the validity of QED in this process to scales of distance less than 10^{-15} cm.

The beam polarization term in Bhabha scattering (see Equation 2.17) has been

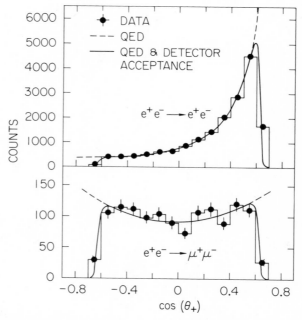

Figure 11 Angular distribution for Bhabha scattering and muon-pair production at $W =$ 4.8 GeV, measured by the SPEAR magnetic detector (61). Predictions of QED normalized to the total number of Bhabha scattering events and corrected for detector acceptance are indicated.

Table 1 Breakdown parameters determined in (61) by fitting to Bhabha scattering and muon-pair production data at three energies, $W = 3.0$, 3.8, and 4.8 GeV[a]

Data used	Model	Fitted parameters (Λ in GeV)	Λ at 95% CL (GeV) positive metric	negative metric
ee only				
	a	$1/\Lambda_S^2 = 0.0008 \pm 0.0022$ $1/\Lambda_T^2 = 0.0013 \pm 0.0031$ correlation coefficient = 0.82	$\Lambda_{S+} > 15$ $\Lambda_{T+} > 13$	$\Lambda_{S-} > 19$ $\Lambda_{T-} > 16$
	b	$1/\Lambda^2 = 0.0007 \pm 0.0022$	$\Lambda_+ > 15$	$\Lambda_- > 19$
$\mu\mu$ and *ee*				
	a	$1/\Lambda_S^2 = 0.0003 \pm 0.0013$ $1/\Lambda_T^2 = 0.0001 \pm 0.0005$ correlation coefficient = 0.23	$\Lambda_{S+} > 21$ $\Lambda_{T+} > 33$	$\Lambda_{S-} > 23$ $\Lambda_{T-} > 36$
	b	$1/\Lambda^2 = 0.0002 \pm 0.0004$	$\Lambda_+ > 35$	$\Lambda_- > 47$
	c	$1/\Lambda_e^2 = 0.0004 \pm 0.0011$ $1/\Lambda_\mu^2 = 0.0014 \pm 0.0021$ correlation coefficient = -0.97	$\Lambda_{e+} > 21$ $\Lambda_{\mu+} > 27$	$\Lambda_{e-} > 19$ $\Lambda_{\mu-} > 16$

[a] The various cases are (*a*) separate breakdown parameters for spacelike (*S*) and timelike (*T*) photons; (*b*) same breakdown parameters for *S*, *T*; and (*c*) separate form factors $ee\gamma$ and $\mu\mu\gamma$ vertices.

verified by a Wisconsin-Pennsylvania group (62) using data from the SLAC/LBL magnetic detector at SPEAR. Their experiment pointed out a sign error in the polarization term that had existed in the theoretical literature until the experiment was performed!

3.4 $e^+e^- \rightarrow \mu^+\mu^-$

The study of muon pair production in e^+e^- collisions is a classic test of QED for one-photon exchange processes and is sensitive to breakdowns in the formulation of the photon propagator for timelike values of momentum transfer or to possible μ-e differences. The complete α^3 cross section has been computed by Berends, Gaemers & Gastmans (63). The lowest-order cross section is given in Equation 2.12. Experiments on μ-pair production can be more demanding than Bhabha scattering measurements because of the lower counting rates and the lack of a characteristic electromagnetic shower to aid in particle identification. Cosmic rays can present serious backgrounds.

Balakin et al (64) measured muon pair production with VEPP-2 at four c.m. energies between 1020 and 1340 MeV. The apparatus included both optical and ferrite-core readout spark chambers, scintillation counters, and water-filled Cerenkov counters. Bhabha scattering events observed in the same appartus served to normalize the data. Their data, presented in Figure 12, agree with QED; the authors set a limit on possible breakdown parameters of $\Lambda_\pm > 3.1$ GeV. These results are not sensitive enough to reflect vacuum polarization effects induced by the ϕ meson; this topic is discussed below.

The μ-π group at ADONE (65) measured μ-pair production between 1.5 GeV and 2.1 GeV c.m. energy. They normalized to wide-angle e^+e^- events; the results, summarized in Figure 12, agree with QED within the estimated systematic errors of $\pm 6.5\%$. An early experiment by the BCF group (66) found that the ratio of $\mu\mu$-to-ee yields agreed with QED within $\pm 8\%$ error. Later results from this group (67) are plotted in Figure 12. They have also observed radiative effects (68) that agree with the theoretical calculations of (63).

At SPEAR, the Stanford-HEPL group has published results at $W = 5.2$ GeV (59) and presented preliminary results at $W = 7.4$ GeV (60). Their data, normalized to small-angle Bhabha scattering, are presented in Figure 12, as are results from the SLAC/LBL group (61), which compared $\mu^+\mu^-$ to e^+e^- yields in their magnetic detector at the three c.m. energies 3.0, 3.8, and 4.8 GeV. The polar angle distribution for muon pairs, obtained in the SLAC/LBL experiment, is shown in Figure 11b and is symmetric about $90°$ as expected for a one-photon exchange process. Various breakdown parameters, summarized in Table 1, were fitted to the SLAC/LBL

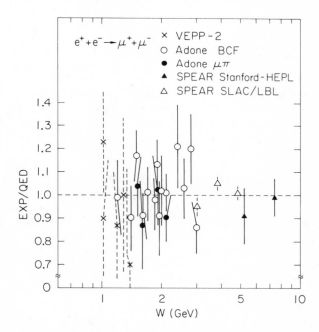

Figure 12 Ratio of experimental yield to that expected from QED for muon-pair production. Detector acceptance and normalization procedures differ between experiments. Data points shown are from VEPP-2 (64), Adone BCF (67), Adone $\mu\pi$ (65), SPEAR Stanford-HEPL (59, 60), and SPEAR SLAC/LBL (61).

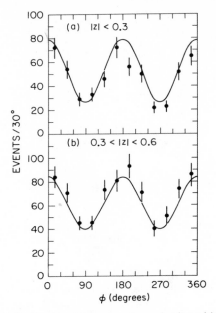

Figure 13 Azimuthal distributions for muon pair production with transversely polarized beams at $W = 7.4$ GeV from (62). Upper data are with the cut $z = |\cos\theta| < 0.3$. Lower data are for $0.3 < |\cos\theta| < 0.6$. The curves are predictions of QED fitted for the observed beam polarization.

data; both SPEAR groups have set lower limits in the vicinity of 30 GeV for possible modifications of the timelike photon propagator.

An azimuthal asymmetry in μ-pair production due to polarized incident beams (see Equation 2.12) has been observed at SPEAR (62) and at VEPP-2M (69). The azimuthal angle distribution from the SPEAR experiment is given in Figure 13, where the strong $\cos 2\phi$ term is evident.

3.5 $e^+e^- \to \gamma\gamma$

e^+e^- annihilation into two photons provides a test of the off-shell lepton propagator for spacelike values of invariant mass. A detailed cross-sectional calculation has been performed by Berends & Gastmans (70); the lowest-order cross section is given by Equation 2.18. Because of the technical problems of photon detection and the small cross section, $e^+e^- \to \gamma\gamma$ is difficult to study experimentally.

In Figure 14 we plot the ratio of experiment to theory, as a function of W for experiments that have measured this reaction (59, 71–74). This ratio has quite different meanings for the various experiments. Balakin et al (71) at VEPP-2 compared 2γ yields to e^+e^- over the same angular range. At ADONE, Bacci et al (72) compared their experimental rate over two regions of polar angle, $20°$–$45°$ and

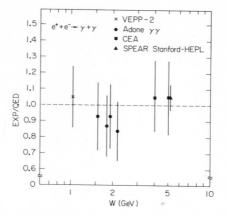

Figure 14   Ratio of experimental yield to that expected from QED for e^+e^- annihilation into two photons. Detector acceptance and normalization procedures differ between experiments. Data points shown are from VEPP-2 (71), Adone $\gamma\gamma$ (72), CEA (73, 74), and SPEAR (59).

70°–110°. The CEA measurements at 4 GeV (73) and 5 GeV (74) observed photons over the range $50° < \theta < 130°$ and normalized to double bremsstrahlung. The Stanford-HEPL group (59) measured gamma rays with their NaI crystals near 90° and normalized to small-angle Bhabha scattering.

Within the rather large errors, all measurements of the reaction $e^+e^- \to \gamma\gamma$ agree with the appropriate QED prediction. The most sensitive test of a possible breakdown of the lepton propagator comes from the SPEAR experiment (59) which sets the limits $\Lambda_+ > 6.2$ GeV, $\Lambda_- > 6.9$ GeV (95% confidence).

3.6 $e^+e^- \to e^+e^-\gamma$

Experiments sensitive to the lepton propagator at timelike values of invariant mass have been reported from ACO (75) and ADONE (76). The process studied gives

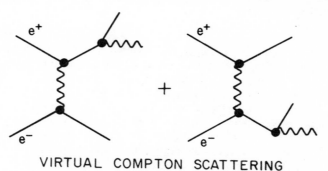

VIRTUAL COMPTON SCATTERING

Figure 15   Feynman diagrams for virtual Compton scattering.

final states of e^+, e^-, and a hard photon at large angles from the charged particles; it is known as virtual Compton scattering and can be represented by the diagram in Figure 15. Both experiments agree with theoretical calculations. Bacci et al (76) present their results in terms of the $e^\pm, \gamma$ invariant mass spectrum, here represented in Figure 16. These data are evidence against any $e\gamma$ excited state in the mass range 0.6 to 2.4 GeV/c^2.

3.7 $e^+e^- \to e^+e^-e^+e^-$, $e^+e^-\mu^+\mu^-$

The first experimental observation of a photon-photon process was reported by Balakin et al (77). They observed approximately 100 events at $W = 1020$ MeV that contained acollinear, coplanar pairs of particles that could not be understood in terms of the usual final states found near the ϕ meson mass. These events were interpreted as coming from the reaction $e^+e^- \to e^+e^-e^+e^-$ and are adequately described by the equivalent photon approximation (38).

Bacci et al (78) observed similar events near $W = 2$ GeV. They demanded a small-angle particle to be tagged in special counters placed near the incident beam line and at least two particles to be detected in their large-angle apparatus. Because of trigger biases, most of their 29 events were interpreted not as coming from the classic photon-photon process, but rather from a similar process where one of the photons is highly virtual. This points out a potentially serious background for experiments intending to exploit the photon-photon process to study nearly real gamma-ray collisions. Parisi (79) has estimated some of these background processes.

Barbiellini et al (80) observed $\mu^+\mu^-$ and e^+e^- production through the photon-photon process. They tagged and momentum-analyzed (using the storage ring magnets) one or both of the forward-going e^+, e^- and detected the other two particles at large angles. Shower criteria were used to separate electrons and muons; kinematic constraints identified the muons. These results are in good agree-

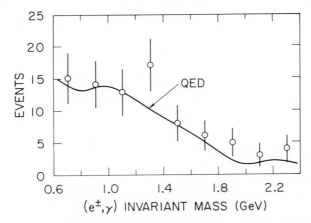

Figure 16   Experimental yield of virtual Compton scattering events from (76) compared to predictions of QED.

ment with the equivalent photon approximation and demonstrate the feasibility of studying $\gamma\gamma$ collisions in e^+e^- colliding beams.

3.8   Hadronic Contributions to e^+e^-, $\mu^+\mu^-$ Final States

There are now several examples (18, 81–88) where the usual formulations of QED fail to describe the cross sections for e^+e^- and $\mu^+\mu^-$ production at certain c.m. energies. These energies correspond to resonance production of various final states through the one-photon channel with the result that interference effects between resonance production and QED can be observed. For example, consider the amplitude for μ-pair production shown schematically in Figure 17. The QED amplitude is real with a value of minus-one unit. The resonance amplitude is described by a Breit-Wigner formula. Unitarity, μ-e universality, and causality demand that the Breit-Wigner amplitude have relative strength $3B_{ee}/\alpha$ (B_{ee} is the resonance branching ratio to lepton pairs) and be superimposed on the QED amplitude as indicated in Figure 17. The resonance amplitude is represented by the circle in Figure 17; passing-through resonance with increasing c.m. energy corresponds to counterclockwise motion of the net amplitude A about the circle. Thus a characteristic interference pattern emerges: below resonance there is destructive interference; then the cross section peaks before returning to the QED value well above the resonant energy. The observed μ-pair cross section σ_{exp} is related to the QED cross section σ_{QED} (Equation 2.13) by

$$\frac{\sigma_{\text{exp}}}{\sigma_{\text{QED}}} = \left| -1 + \frac{3B_{ee}}{\alpha} \frac{\Gamma/2}{M-W-i\Gamma/2} \right|^2, \qquad 3.4.$$

where M, Γ are the mass and full width of the resonance. Of course, Equation 3.4 will be modified by radiative corrections and the effect of energy spread within the beams (89).

The pioneering experimental work on this subject was done by the Orsay group

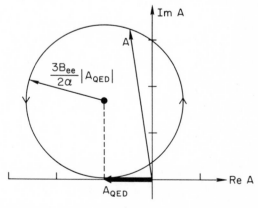

Figure 17   Schematic diagram of the amplitude A for muon-pair production in the vicinity of a vector-meson resonance with electron pair branching ratio B_{ee}.

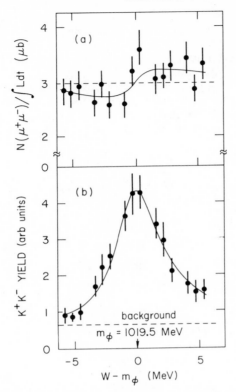

Figure 18 Experimental yield of muon pairs in the vicinity of ϕ meson (indicated by K^+K^- yield of lower figure) from (81).

(81), who studied μ-pair production in the vicinity of the ϕ meson and found evidence for the interference effect. The results from this experiment are presented in Figure 18. The most spectacular demonstrations of hadronic interference effects are the J/ψ (18, 82–86, 88) and ψ' (87) resonances. The SLAC/LBL group has observed the interference term in Equation 3.4 for both ψ particles (86, 87) through the ratio of $\mu^+\mu^-$ to e^+e^- production. Their results are given in Figure 19.

In these experiments we have very clear deviations from the simple predictions of QED which we interpret not as a breakdown of QED, but rather as a method for determining the branching ratios of vector particles to lepton pairs!

e^+e^- annihilation into two photons, which does not proceed via the one-photon channel, shows no interference effects at the J/ψ (88).

3.9 $e^+e^- \rightarrow e\pm\mu\mp, e\pm\mu\mp X$

Processes where an electron and a muon are the only charged particles in the final state of an e^+e^- collision are expected to be extremely rare in all known interactions of e^+e^-. States of $e\mu$ are therefore a unique signal for new interactions or new particle production. One possibility leading to such final states is the production of pairs of particles that subsequently decay to electrons or muons plus undetected neutrals. A heavy charged lepton (90) that decays to either electrons or muons plus neutrinos is an example.

$e\mu$ final states have been searched for without success at ADONE (91–93). Orito et al (93) studied their sensitivity to various types of heavy charged leptons and concluded that no additional charged leptons exist with masses less than 1.15 GeV.

There is now evidence from the SLAC/LBL group reported by Perl et al (94) that the reaction

$$e^+ + e^- \rightarrow e^\pm + \mu^\mp + \geqq 2 \text{ undetected particles}$$

does exist at c.m. energies above 4 GeV. In a sample of two-prong, net charge zero, acoplanar events, where both momenta are greater than 650 MeV/c, they observe an excess of events identified $e^+\mu^-$ or $e^-\mu^+$ with no simultaneous detection of gamma rays. Particle identification relies on Pb-scintillator shower counters and a 20-cm-thick iron hadron absorber. Consequently, there are rather

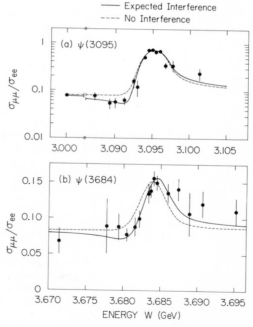

Figure 19   Ratio of muon-pair yields to Bhabha scattering in the vicinity of the $J/\psi(3095)$ and $\psi'(3684)$ resonances showing interference effect expected. The additional t-channel diagram for Bhabha scattering (see Figure 4) gives it a different energy dependence than muon-pair production. Data are from (86) and (87).

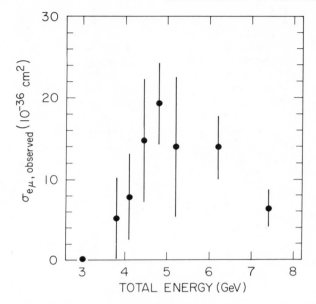

Figure 20   Experimental cross section for anomalous $e\mu$ events reported in (94).

large misidentification probabilities (which are measured with multihadronic final states) for hadrons to simulate electrons and muons. Misidentification results in a 25% contamination of the $e\mu$ signal events by normal processes, but still leaves 64 events for which there is no conventional explanation. The photon-photon process is not expected to contribute significantly; the fact that there is no like-charge signal confirms this. The experimental cross section for $e\mu$ events, given in Figure 20, shows evidence for a threshold near 4 GeV. Kinematic properties of the charged particle lead the authors to state that two or more missing particles are associated with these events.

At present there is no unique hypothesis for the origin of these $e\mu$ events. They certainly indicate the presence of some new physical process occurring above 4 GeV. If this turns out to be another member of the lepton family, it is a major discovery indeed.

4   HADRONIC FINAL STATES

As discussed above, e^+e^- collisions provide a unique tool with which hadrons can be studied in nearly pure quantum states having the quantum numbers of the photon. For purposes of orientation, we present in Figure 21 a summary of the current knowledge of R, the ratio of the total cross section for hadron production by one-photon annihilation to the total μ-pair cross section (Equation 2.13). Data are now available at c.m. energies from approximately 0.5 GeV to nearly 8 GeV.

Many groups at several laboratories have made valuable contributions to the results shown in Figure 21; we examine individual contributions below.

Figure 21 shows that hadron production is comparable to or greater than μ-pair production over the entire range of W studied to date. At certain energies R is punctuated by large peaks that correspond to the direct coupling of vector particles to the virtual annihilation photon. The large cross sections at resonant energies have permitted detailed studies of hadrons produced in decays of these states. These studies lead to determinations of quantum numbers, resonance parameters, and decay-branching ratios for the vector mesons. The very small cross sections away from resonances have limited us to a more qualitative picture of hadron production at these energies. We study these two general regions of cross section in sequence.

4.1   Resonance Production

The resonances observed to date in hadron production by e^+e^- annihilation correspond to vector mesons that are also observed in many other processes. Near a resonance, the cross section σ_f for producing some particular final state f can be described by a Breit-Wigner formula

$$\sigma_f = \frac{12\pi}{s} B_{ee} B_f \frac{M^2\Gamma^2}{(M^2-s)^2+M^2\Gamma^2}. \qquad 4.1.$$

The resonance is assumed to have spin 1. B_{ee} and B_f are branching ratios to electron pairs and the state f, respectively; M and Γ are the mass and full width of

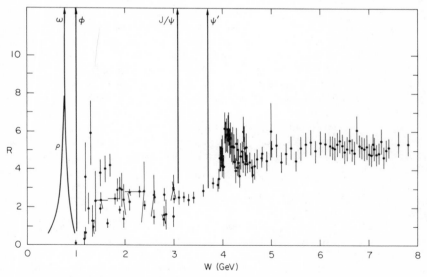

Figure 21 A summary of data for R, the ratio of the cross section for hadron production by e^+e^- annihilation to the muon pair cross section, versus center-of-mass energy W.

the resonance. The observed cross section will be a convolution of the Breit-Wigner shape with the energy spectrum of the incident e^+e^- beams and the effective energy spectrum of the initial state that arises from radiative corrections. The effect of beam energy spread and radiative corrections on resonance line shapes is discussed in (89). In general, radiative corrections give a high-energy "tail" to the observed cross section, the classic example of which is the J/ψ resonance shown in Figure 22. After correction for radiative effects, the value of σ_f at the peak of the resonance can be used to determine the product $B_{ee}B_f$

$$\sigma_f(s = M^2) = \frac{12\pi}{M^2} B_{ee}B_f.$$ 4.2.

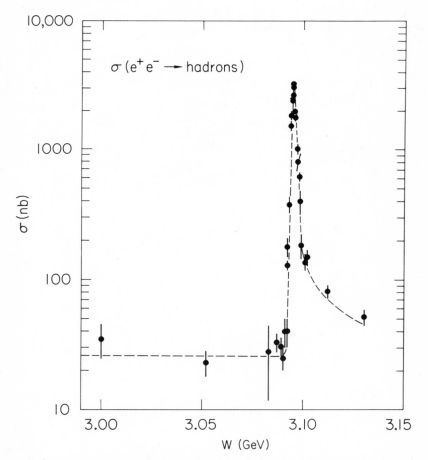

Figure 22   Total hadron production cross section in the vicinity of the J/ψ resonance showing "radiative tail." Data are from (86).

With separate knowledge of B_f, for example, Equation 4.2 gives a determination of B_{ee}. When the full width of a resonance is narrower than the energy spread of the colliding beams, the line shape cannot be resolved, yet the area under the resonance can be measured; this area is related to the resonance parameters by

$$\int \sigma_f \, dW = \frac{6\pi^2}{M^2} B_{ee} B_f \Gamma.$$ 4.3.

The partial width of a resonance decaying to lepton pairs Γ_{ee} ($\Gamma_x \equiv B_x \Gamma$) is proportional to the area under the total production cross section. Γ and B_{ee} can be determined separately from the areas under both the total and elastic resonance cross sections, even in cases where the resonance line shape cannot be resolved.

The coupling of vector mesons to virtual photons is often parameterized by a coupling constant g_V that is related to Γ_{ee} by

$$\frac{g_V^2}{4\pi} = \tfrac{1}{3}\alpha^2 \frac{M_V}{\Gamma_{Vee}},$$ 4.4.

where V refers to the particular vector meson and M_V is its mass.

4.1.1 THE ρ, ω, ϕ RESONANCES   The first hadronic state to be studied in e^+e^- annihilation was $\pi^+\pi^-$ in the vicinity of the ρ meson (95–97). The most recent results from the Novosibirsk group (98) and Orsay group (99) are shown in Figure 23. In this figure, the square of the pion form factor $|F_\pi|^2$, defined by Equation 2.15, is plotted.

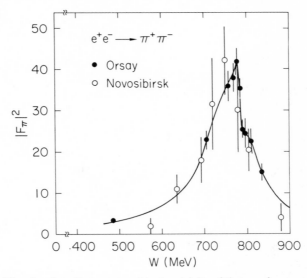

Figure 23   The pion form factor squared in the vicinity of the ρ and ω resonances. Data are from Orsay (99, 102) and Novosibirsk (98). The curve is a fit to the Gounaris-Sakurai formula (100), including ρ-ω interference.

Table 2 Parameters of the vector mesons ρ, ω, ϕ, J/ψ, and ψ' as measured in e^+e^- colliding-beam experiments

V	ρ (99)[a]	ω (104)[a]	ϕ (Ref.)	J/ψ (86)[a]	ψ' (87)[a]
m (MeV)	772.3 ± 5.9	no precision value reported	1019.4 ± 0.3 (111)	3095 ± 4	3684 ± 5
Γ (MeV)	135.8 ± 15.1	9.1 ± 0.8	4.09 ± 0.29 (109) 4.67 ± 0.42 (107) 3.81 ± 0.37 (108)	0.069 ± 0.015	0.228 ± 0.056
Γ_{ee} (keV)	5.8 ± 0.5	0.76 ± 0.08	1.41 ± 0.12 (109) 1.31 ± 0.12 (107) 1.27 ± 0.11 (110)	4.8 ± 0.6	2.1 ± 0.3
$B_{ee} = \Gamma_{ee}/\Gamma$	$(4.2 \pm 0.4) \times 10^{-5}$	$(0.83 \pm 0.10) \times 10^{-4}$	$(3.45 \pm 0.27) \times 10^{-4}$ (109) $(2.81 \pm 0.25) \times 10^{-4}$ (107) $(3.3 \pm 0.3) \times 10^{-4}$ (110)	0.069 ± 0.009	0.0093 ± 0.0016
$g_V^2/4\pi$	2.38 ± 0.18	18.4 ± 1.8	12.9 ± 1.1 (109) 13.8 ± 1.3 (107) 14.3 ± 1.2 (110)	11.4 ± 1.4	31.1 ± 4.5

[a] Numbers in parentheses are references.

The data do not follow a simple Breit-Wigner curve. There are two reasons for this. First, in studying the analytic properties of F_π, Gounaris & Sakurai (100) showed that a more complicated formulation than an s-wave Breit-Wigner amplitude is appropriate for the ρ because of its large width and the fact that the $\pi^+\pi^-$ are in a relative P wave. Second, the ω meson, to be discussed below, has a G parity–violating decay mode to $\pi^+\pi^-$ that interferes with the ρ amplitude and gives the sharp high-energy edge to $|F_\pi|^2$. The Orsay group (99) have fitted both the Novosibirsk and Orsay data to the Gounaris-Sakurai form factor F_{GS}, plus an interference term from ω decay

$$|F_\pi|^2 = \left| F_{GS} + Ae^{i\phi} \frac{M_\omega^2}{M_\omega^2 - s - iM_\omega\Gamma_\omega} \right|^2. \qquad 4.5.$$

A represents the strength of the ω coupling to e^+e^- and its subsequent G parity–violating decay, ϕ is the phase of the $\rho\omega$ interference, and M_ω, Γ_ω are the mass and full decay width of the ω meson. Resonance parameters describing the ρ and ω mesons determined from the data of Figure 23 are summarized in Table 2. ϕ was determined to be $88.3 \pm 15.8°$ (99), so the coefficient of the ω contribution to the $\pi^+\pi^-$ amplitude is almost purely imaginary. If we consider the G parity–violating transition as the ω coupling to a virtual ρ, which then decays to $\pi^+\pi^-$, then the $\omega\rho$ coupling is nearly real and the ρ propagator evaluated near M_ρ^2 gives the G parity–violating amplitude its large imaginary part (101). There is a recent point, also plotted in Figure 23, that was measured with the new Orsay magnetic detector at the c.m. energy $W = 484$ MeV (102). This point agrees with the Gounaris-Sakurai formula.

The $\pi^+\pi^-\pi^0$ decay mode of the ω meson has been studied with two different experimental setups at Orsay (103, 104). The later apparatus (105), shown schematically in Figure 24, subtended a large solid angle ($0.6 \times 4\pi$ sr) with cylindrical

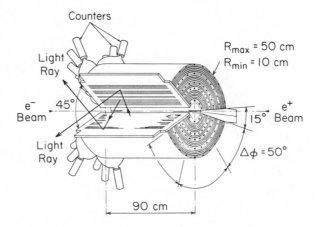

Figure 24   Schematic diagram of the Orsay detector described in (105).

spark chambers and scintillation counters. The inner spark chambers were used to track charged particles and the outer chambers were interspersed with Pb sheets to identify electron- and photon-induced showers. This apparatus was used in many of the Orsay experiments and has been one of the most successful detectors in e^+e^- physics. The cross section for $\pi^+\pi^-\pi^0$ production in the ω region is shown in Figure 25; resonance parameters derived from these data (104) are given in Table 2.

The ϕ meson has been studied extensively in e^+e^- annihilation. Excitation curves for the final states $K_S^0 K_L^0$ (106–109) and K^+K^- (106, 107, 110) have been measured at Novosibirsk and Orsay. Resonance parameters determined from these experiments are given in Table 2. The most precise determination of the ϕ decay width comes from the experiment of Bizot et al (109). In a very interesting experiment performed at Novosibirsk, Bukin et al (111) were able to accurately calibrate the VEPP-2M energy by measuring the electron spin precession frequency of particles stored in the machine. This calibration yields a very precise determination of the ϕ mass.

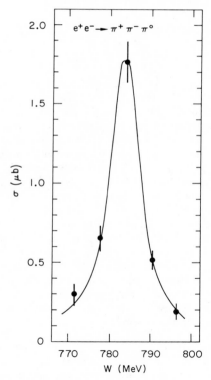

Figure 25   Cross section for the reaction $e^+e^- \to \pi^+\pi^-\pi^0$ in the vicinity of the ω meson from (104).

Table 3 Branching ratios for radiative decays of ω and ϕ determined from e^+e^- colliding-beam experiments

Decay mode	Ratio (%)	Ref.
$\dfrac{\Gamma(\omega \to \pi^0\gamma)}{\Gamma(\omega \to \pi^+\pi^-\pi^0)}$	10.9 ± 2.5	(112)
$\dfrac{\Gamma(\omega \to \eta\gamma)}{\Gamma(\omega \to \pi^0\gamma)}$	<27 (90% CL)	(112)
$\dfrac{\Gamma(\omega \to \pi^0\pi^0\gamma)}{\Gamma(\omega \to \pi^0\gamma)}$	<15 (90% CL)	(112)
$\dfrac{\Gamma(\phi \to \eta\gamma)}{\Gamma(\phi \to \text{all})}$	1.5 ± 0.4	(102)
$\dfrac{\Gamma(\phi \to \pi^0\gamma)}{\Gamma(\phi \to \text{all})}$	0.14 ± 0.05	(102)
$\dfrac{\Gamma(\phi \to \pi^+\pi^-\gamma)}{\Gamma(\phi \to \text{all})}$	<0.7 (90% CL)	(110)

Radiative decays of the ω and ϕ mesons leading to $\pi^0\gamma$ or $\eta\gamma$ states were studied by Benaksas et al (112). They detected events with three coplanar γ rays. By measuring the angles between the γ rays, the energies of all three γ rays can be calculated, assuming there were no missing particles in the event. The three energies can then be plotted on the triangular Dalitz plot shown in Figure 26a. Events where two γ rays have a definite invariant mass are confined to three straight lines on the Dalitz plot. If the γ energies are ordered, only the triangular sector of the Dalitz plot shown in Figure 26b will be populated. An example of one of these Dalitz plots for 3 γ events at the ϕ mass (112) is given in Figure 27. The $\pi^0\gamma$ and $\eta\gamma$ bands are broadened by experimental resolution, yet they clearly show an accumulation of events. Various branching ratios for radiative decays of ω and ϕ are given in Table 3. The ϕ results are from some very recent work performed by the Orsay group (102).

In a remarkable tour de force reported in (102), the Orsay group of Parrour et al have evidence for an interference between the ω and ϕ in the $\pi^+\pi^-\pi^0$ channel. In spite of their narrow widths and wide spacing, the ω and ϕ seem to interfere as if the phase were simply given by the ω propagator evaluated near the ϕ mass, but with coupling constants of opposite sign (112a).

4.1.2 THE NARROW ψ RESONANCES In the brief period of time since their discovery (18–20), the J/ψ and ψ' resonances have received an extraordinary amount of attention from both an experimental and theoretical point of view. It is much too soon to present a complete review of the properties of these states; we cover only a small fraction of the experimental work performed to date.

Two pieces of experimental apparatus have made major contributions to our knowledge of the new particles. The SLAC/LBL magnetic detector of SPEAR (61), shown schematically in Figure 28, consists of a large solenoid magnet with magnetostrictive spark chambers for tracking charged particles and measuring their momenta. Time-of-flight counters, shower counters, and range chambers are used for particle identification. The full momentum-analysis and particle-identification capabilities of this device extend over 0.65 of 4π sr solid angle. Two charged particles are necessary to trigger the device. The double-arm spectrometer DASP (113) at DESY consists of two symmetric, high-resolution, magnetic spectrometers, each covering approximately 0.45 sr with excellent particle identification capability. Surrounding 0.7 of 4π sr solid angle is a nonmagnetic inner detector, consisting of scintillation counters and proportional chambers, which is well suited for tracking

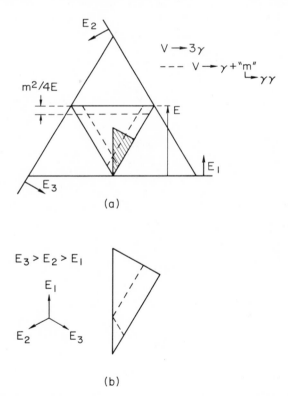

Figure 26 (a) Dalitz plot for 3γ final states of e^+e^- annihilation. Events populate the interior triangle. Dashed lines represent regions populated by events where two of the gamma rays are decay products of a state of mass *m*. (b) Same as *a*, except only the populated region where the gamma rays are ordered by energy is shown. This corresponds to the shaded region of *a*.

charged particles and γ rays. A schematic drawing of the DASP detector is given in Figure 29.

The SLAC/LBL collaboration measured the total hadronic, e^+e^-, and $\mu^+\mu^-$ cross sections at the J/ψ (86) and ψ' (87) and from these derived the resonance parameters given in Table 2. From the angular distributions of e^+e^- and $\mu^+\mu^-$ final states and the interference with QED in the $\mu^+\mu^-$ channel, noted in Section 3, the SLAC/LBL group conclusively demonstrated that the new resonances are vector states.

The most remarkable properties of the new particles are their exceedingly narrow decay widths. Even though they are over three times as massive as the ρ, ω, and ϕ mesons, their decay widths are two to three orders of magnitude smaller. A detailed theoretical understanding of the decay rates of the new particles is one of the challenging problems in elementary particle physics today. Below, we briefly discuss some of the theoretical ideas concerning the new particles.

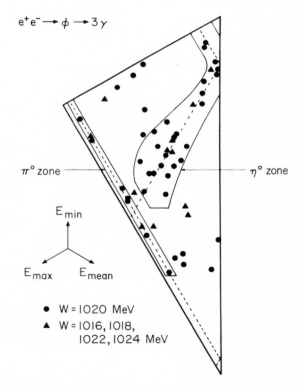

Figure 27   Dalitz plot for 3γ final states in the vicinity of the ϕ meson from (112). The π^0 and η^0 bands are indicated.

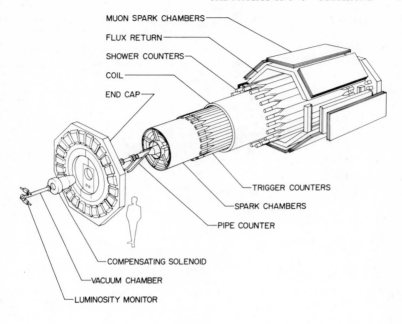

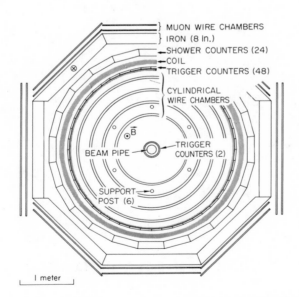

Figure 28   Schematic diagrams of the SLAC/LBL SPEAR magnetic detector.

There are two general decay modes for these resonances: direct decays and second-order electromagnetic decays. These are shown schematically in Figure 30. Direct decays display the intrinsic properties of the resonance, while second-order electromagnetic decays display the same characteristics as one-photon exchange

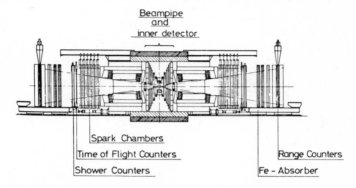

Figure 29   Schematic diagram of the DASP double arm spectrometer at DESY.

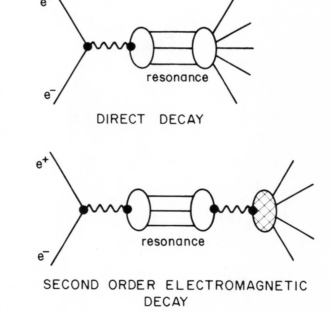

Figure 30   Schematic diagram showing the difference between "direct" and "second-order electromagnetic" decays of a resonance in e^+e^- annihilation.

processes at nearby nonresonant energies. The μ-pair final state is an excellent monitor of second-order electromagnetic processes, so that comparison of particular decay rates to the μ-pair rate, both on and off resonance, will differentiate between the two modes.

Jean-Marie et al (114) compared ratios of multipion final states to μ-pair production for data taken at the J/ψ energy and $W = 3.0$ GeV. Both sets of data were taken with the SLAC/LBL magnetic detector. The results, given in Figure 31, show states with an even number of pions to be second-order electromagnetic decays, while states with an odd number of pions come from direct decays of the J/ψ. Thus G parity is a good quantum number in J/ψ decays to pion states with the value -1. In the same paper Jean-Marie et al studied isotopic ratios for the states $\rho\pi$ and showed that the isotopic spin of the J/ψ is zero. Corroborating evidence for this assignment comes from observations of decays to $\Lambda\bar{\Lambda}$ (115), $p\bar{p}$ (115, 116), and the stringent upper limits placed on $\pi^+\pi^-$ and K^+K^- final states by the DASP group (116). Many hadronic decays of the J/ψ have been identified by the SLAC/LBL group; a summary of these is given in (115).

The ψ' decays predominantly to states containing a J/ψ (117, 117a). The presence of J/ψ among decay products of the ψ' is most easily established through the $\mu^+\mu^-$ decay mode of the J/ψ. The invariant mass spectrum for μ-pair candidates in

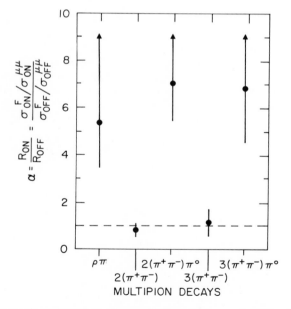

Figure 31   Comparison of multipion decays of J/ψ resonance to an adjacent nonresonance energy to find direct decay modes of J/ψ. ON and OFF refer to data taken on the resonance and just below the resonant energy, respectively. Values of α greater than one indicate direct decays of the J/ψ. Data taken from (114).

the SLAC/LBL magnetic detector (117), shown in Figure 32, has two peaks; one corresponds to μ-pair production at the ψ' mass, the other corresponds to J/ψ decays. At present, the following cascade decays have been identified:

$$\psi' \rightarrow J/\psi + \pi^+\pi^- \qquad (113, 117), \qquad\qquad\qquad 4.6a.$$

$$\psi' \rightarrow J/\psi + \pi^0\pi^0 \qquad (113, 179), \qquad\qquad\qquad 4.6b.$$

$$\psi' \rightarrow J/\psi + \gamma\gamma \qquad (118, 119), \qquad\qquad\qquad 4.6c.$$

$$\psi' \rightarrow J/\psi + \eta \qquad (113, 120). \qquad\qquad\qquad 4.6d.$$

Branching ratios for these decays are given in Table 4.

These important cascade decays point out the close relationship between the J/ψ and ψ'. The relative rates for the dipion decays and the η decay mode indicate that the isotopic spin of the ψ' is the same as J/ψ, namely zero.

4.1.3   THE NEW PARTICLE SPECTROSCOPY

In a study of reaction 4.6c, the DASP collaboration (118) reported evidence that the γ-ray cascade proceeds through an intermediate state in the sequence:

$$\psi' \rightarrow \gamma + P_c \qquad\qquad\qquad\qquad\qquad\qquad 4.7.$$
$$\quad\ \ \hookrightarrow \gamma + J/\psi,$$

where the P_c had a unique mass either near 3500 or 3300 MeV. The SLAC/LBL group (119) also observed this decay scheme by detecting the $\mu^+\mu^-$ decay of the J/ψ and one or both γ rays. In seven of their events, one of the γ rays converts to an e^+e^- pair in the beam vacuum chamber so that its energy can be measured accurately from the e^+e^- momenta. From this measurement and more recent DASP results (113), the observed width of the P_c state is compatible with the experimental resolution of approximately ± 10 MeV. In principle, the present ambiguity in the mass of the P_c can be resolved by observing Doppler broadening of the second γ ray; as yet, this has not been done.

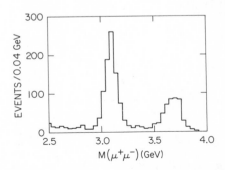

Figure 32 Invariant-mass spectrum of muon-pair candidates in decays of the ψ' from (117). The peak near 3.7 GeV corresponds to muon-pair production; the peak near 3.1 GeV indicates the presence of J/ψ in decays of the ψ'.

Table 4   Branching ratios for cascade decays of $\psi' \to J/\psi$

Decay mode	Ratio (%)	Ref.
$\dfrac{\Gamma(\psi' \to J/\psi + \text{anything})}{\Gamma(\psi' \to \text{all})}$	57 ± 8	(117)
$\dfrac{\Gamma(\psi' \to J/\psi + \pi^+\pi^-)}{\Gamma(\psi' \to \text{all})}$	31 ± 4 36 ± 6	(117) (113)
$\dfrac{\Gamma(\psi' \to J/\psi + \pi^0\pi^0)}{\Gamma(\psi' \to \text{all})}$	18 ± 6	(113)
$\dfrac{\Gamma(\psi' \to J/\psi + \eta)}{\Gamma(\psi' \to \text{all})}$	4.3 ± 0.8 3.7 ± 1.5	(120) (113)
$\dfrac{\Gamma(\psi' \to J/\psi + \gamma\gamma)}{\Gamma(\psi' \to \text{all})}$	3.6 ± 0.7 4 ± 2	(119) (113)
$\dfrac{\Gamma(\psi' \to J/\psi + \text{neutrals})}{\Gamma(\psi' \to J/\psi + \text{anything})}$	44 ± 3	(117)
$\dfrac{\Gamma(\psi' \to J/\psi + \pi^0\pi^0)}{\Gamma(\psi' \to J/\psi + \pi^+\pi^-)}$	64 ± 15	(179)

Evidence for radiative decays of the ψ' to hadronic states has been reported by the SLAC/LBL group (121). They observed charged hadrons and inferred the presence of single γ rays from the energy-momentum balance of the charged particles. These results indicate the decay scheme:

$$\psi' \to \gamma + \chi$$
$$\hookrightarrow 4\pi^\pm, 6\pi^\pm, \pi^\pm K^\pm, \pi^+\pi^-, K^+K^-. \qquad 4.8.$$

At least two different χ states are present, a narrow state at 3410 ± 10 MeV and a broader state at 3530 ± 20 MeV. The state at 3410 MeV decays to $\pi^+\pi^-$ and K^+K^-, therefore its spin and parity must be even. It is tempting to associate the P_c with the $\chi(3530)$, but the broader width of the $\chi(3530)$ implies either that they are not the same, or that there is more than one state near 3500 MeV in mass.

Finally, two DESY groups (113, 122) have examined coplanar 3γ-ray final states of the J/ψ to look for radiative decays to new states that subsequently decay to 2γ rays. Using a Dalitz plot analysis technique similar to that discussed above in connection with ω and ϕ radiative decays, they have evidence for a new state near 2800 MeV that has been designated $X(2800)$.

We have summarized the current state of knowledge of the new particle spectroscopy in the energy-level diagram of Figure 33. The higher-mass solution was assigned to the P_c. It should be noted that the information on many of these new states is very preliminary. All of the events observed to date at the P_c, χ, and X states number only a few hundred, compared to the hundreds of thousands of J/ψ

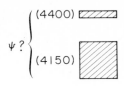

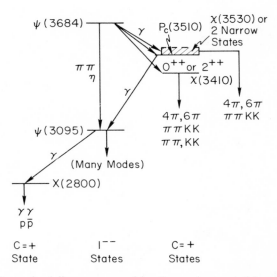

Figure 33   Energy-level diagram summarizing the new heavy-particle spectroscopy.

and ψ' decays that have been studied. The existence of a new heavy-particle spectroscopy is well established; the job of determining the masses, quantum numbers, and decay rates of the new particles lies before us.

4.1.4 OTHER RESONANCES IN e^+e^- ANNIHILATION The first resonance beyond the ρ, ω, and ϕ to be studied in e^+e^- colliding beams was the $\rho'(1600)$. It was first observed in e^+e^- collisions as a broad enhancement in the $2\pi^+\pi^-$ final state at ADONE (123). There is a complete discussion in (124) of the apparatus and analysis methods used by the $\mu\pi$ group that has performed much of the work on the $\rho'(1600)$ discussed here. This state has also been observed in photoproduction (125). The most recent data (126) indicate that the mass of the $\rho'(1600)$ is (1550 ± 60) MeV and its total decay width is (360 ± 100) MeV. Ceradini et al (127) studied 23 events from the ρ' energy region with four charged prongs in a nonmagnetic detector and concluded that most of the decays proceeded through the quasi–two-body final state $\rho^0\varepsilon^0$. Given the small sample of events, the broad ε^0 width used in the analysis, and the small available phase space, the notion of an ε^0 in decays of the $\rho'(1600)$ seems dubious. However, ρ^0 production does seem to be important at this energy. Bernardini et al (128) and Alles-Borelli et al (129) showed that the 4π enhancement at 1.6 GeV may simply be a consequence of the opening of four-pion phase space coupled with a general decrease in production amplitude that could give a peak near 1.6 GeV.

Conversi et al (126) speculate on the existence of a resonance at 1250 MeV in the $\pi^+\pi^-\pi^0\pi^0$ final state with limited statistics. The analysis depended on a calculation of the decay $\rho'(1600) \to \pi^+\pi^-\pi^0\pi^0$ that was subtracted from the measured yields. Statistically marginal departures of the measured pion form factor (129, 130) from the Gounaris-Sakurai formula (100) is also taken as evidence for a $\rho'(1250)$. More experimental information is required before a $\rho'(1250)$ is established.

The SLAC/LBL group (131) reported a broad structure in R at 4.15 GeV. The area under this peak is comparable to the narrow ψ resonances. Preliminary data from a detailed scan of the 4-GeV region (132) shown in Figure 34 confirmed the structure at 4.1 GeV and indicated that there may be another, narrower resonance near 4.4 GeV and that the 4.1-GeV peak may contain a substructure.

The 4-GeV region in e^+e^- annihilation appears to contain a vast richness of structure that we are only beginning to resolve at present. Note also that this is the same energy as the apparent threshold for μe events discussed in Section 3.

4.1.5 SEARCHES FOR NEW RESONANCES e^+e^- colliding-beam devices are ideally suited for high-resolution scanning for new vector states. After the discovery of the J/ψ, several groups (20, 132–136) performed scans where short data runs were taken at fine c.m. energy intervals. Typical c.m. energy spacings were of order 2 MeV. By far the most successful of these scans was the first, which discovered the ψ' (20). Since then, no additional narrow resonances have been uncovered. The c.m. energy ranges scanned to date are 770 to 1340 MeV (133); 1910 to 2545 MeV and 2970 to 3090 MeV (134, 135); and 3.2 to 7.7 GeV (132, 136). The sensitivity of these

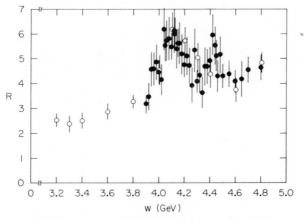

Figure 34   *R* vs *W* in the 4-GeV region. Open points are from (131); closed points are preliminary measurements reported in (132). These data indicate there are at least two and possibly more resonances in the 4-GeV region.

searches varies greatly, but narrow ($\Gamma \lesssim 10$ MeV) resonances decaying to lepton pairs at partial widths of order 20% or larger of the J/ψ value should have been observed. Undoubtedly, more sensitive searches will continue to be made until new resonances are found or the experimentalists become exhausted.

4.2 Nonresonant Hadron Production

4.2.1 TWO-BODY FINAL STATES

Measurements of hadronic pair production at non-resonant energies in e^+e^- annihilation provide information on electromagnetic form factors of the hadrons for timelike values of momentum transfer. In general, these measurements are difficult to perform because of the very small cross sections involved and the large competing reactions such as Bhabha scattering and μ-pair production, which also give two final-state particles.

The pion form factor has been studied at energies above the ρ mass in experiments at Novosibirsk (130) and Frascati (137, 138). The Novosibirsk group used threshold water Cerenkov counters to separate π and K mesons. The Frascati $\mu\pi$ group (137) used sufficient absorber so that at $W = 1.25$ and 1.52 GeV, only pions could be detected; at higher energies no separation of π and K mesons was

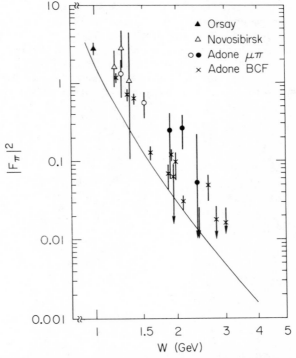

Figure 35 $|F_\pi|^2$ vs W above the ρ mass region. Data are from Orsay (99), Novosibirsk (130), Adone $\mu\pi$ (137) (full circles represent upper limits), and Adone BCF (138).

possible and the corresponding results represent upper limits to the pion form factor. For $W \leqq 1.40$ GeV the BCF group (138–140) detected only pions; between 1.50 and 1.70 GeV, π and K mesons were separated by a range method; above $W = 1.85$ GeV no such separation was possible and a theoretical ratio of $(\pi^+\pi^-)$ and (K^+K^-) pairs based on SU(3) was applied to the data to calculate the pion form factor. These results are shown in Figure 35, where we also plot an extrapolation of the Gounaris-Sakurai formula (100) for these energies. Particularly between 1.0 and 1.5 GeV, the data are consistently higher than the theoretical extrapolation. As noted previously, this has been taken as evidence for the existence of additional resonances.

The information on the K^+ form factor obtained from experiments of the Novosibirsk group (130) and the BCF group (139, 140) suggests that the kaon and pion form factors are comparable within the rather large errors.

The $p\bar{p}$ final state has been studied by the Naples group (141) at ADONE. Their result for the $p\bar{p}$ cross section is $\sigma(e^+e^- \rightarrow p\bar{p}) = (0.91 \pm 0.22)$ nb at the energy $W = 2.1$ GeV. The $p\bar{p}$ were identified through ionization in a nonmagnetic detector. It is impossible to separately extract the form factors G_E and G_M (see Equation 2.16) for this measurement. Assuming the magnitudes of the two to be equal, they find

$$|G_M| = |G_E| = 0.27 \pm 0.04.$$

4.2.2 TOTAL HADRONIC CROSS SECTION   The preliminary results on multihadron production from groups working at ADONE were presented at the Kiev Conference (142), causing great excitement because of the "large" total hadronic cross section σ_T. Since then we have come to realize the fundamental role of σ_T in the study of hadron structure. In studying the Bjorken scaling (34) properties of inclusive proton production by e^+e^- annihilation and its connection to deep inelastic electron scattering, Drell, Levy & Yan (143) predicted a $1/s$ energy behavior for σ_T, using parton model arguments. The parton model (144) is a very convenient framework in which to organize the experimental data, and, as we shall see, provides a simple intuitive picture of many of the phenomena actually observed. In this picture hadrons are built out of constituents, the partons, which have pointlike couplings to the electromagnetic current. For example, spin-$\frac{1}{2}$ partons would couple to the virtual-annihilation photon in the same way as muons, except for differences in net electric charge. As indicated schematically in Figure 36, hadron production is viewed as

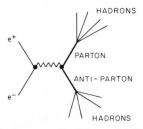

Figure 36   Schematic diagram of hadron production in the parton model.

the production of pairs of parton-antipartons that subsequently decay to hadrons. At sufficiently high energies, where hadron masses can be neglected, the total hadronic cross section is equal to the sum of the individual parton cross sections that are proportional to σ_μ, the ratio σ_T/σ_μ, denoted R, therefore depends only on the sum of squares of parton charges (145, 146):

$$R = \sum_{J=\frac{1}{2}} Q_i^2 + \tfrac{1}{4} \sum_{J=0} Q_i^2. \qquad 4.9.$$

The indicated summations are over spin-$\frac{1}{2}$ and spin-0 partons, respectively. The Q_i are parton charges in units of e. R is a convenient experimental quantity as well, because experimenters usually normalize their hadronic yields to the yield from some electrodynamic process that would normally be proportional to σ_μ. If R is a constant over some range of energies, then hadron production is said to exhibit "scaling" over that energy range. The actual value of R gives information about the number and properties of partons.

The important factors limiting the precision to which R can be measured are (a) systematic errors in hadronic event detection efficiency, (b) systematic errors in luminosity monitoring, and (c) statistical errors of the event sample. In virtually all measurements of R, the detection efficiency is estimated by an "unfold" procedure. The number of events N_q observed in some configuration q (e.g. the number of observed charged prongs) is related to the number of events $\tilde{N}_p$ produced in the configuration p by

$$N_q = \sum_p \varepsilon_{qp} \tilde{N}_p, \qquad 4.10.$$

where ε_{qp} is the appropriate detection efficiency for observing an event in configuration q when it was produced in configuration p. The ε_{qp} are usually computed by Monte Carlo techniques. Known properties of the apparatus and a model of representative final states are necessary ingredients in the calculation. Parameters of the model are varied to obtain agreement with observed quantities such as angle, multiplicity, and momenta spectra. Such calculations are discussed in reference 147. Equation 4.10 is then inverted, usually by maximum likelihood methods, subject to the constraint $\tilde{N}_p \geq 0$. The average detection efficiency $\bar{\varepsilon}$ is defined by

$$\bar{\varepsilon} = \frac{\sum N_q}{\sum \tilde{N}_p}, \qquad 4.11.$$

and the total cross section is computed from:

$$\sigma_t = \frac{\sum N_q}{\bar{\varepsilon} \int L \, dt}, \qquad 4.12.$$

where $\int L \, dt$ is the time-integrated luminosity. Usually, no corrections are made for all-neutral final states; an all-π^0 state is excluded by charge conjugation invariance. The advantage of the unfold procedure is that maximum use is made of experimental data; model dependence does not directly enter into the calculation of $\bar{\varepsilon}$, but enters in an average way through the ε_{qp}. Generally, σ_t and the mean charged-particle multiplicity $\langle n_{ch} \rangle$ depend only weakly on a particular choice of model. However, individual partial cross sections usually are not well determined by these methods.

Several experimental groups at Frascati have measured σ_t. The first group to publish multiparticle results was the boson group (148, 149); they concluded that σ_t was greater than 30 nb for the c.m. energy range 1.4–2.4 GeV. A final report of this experiment is contained in (53). The $\gamma\gamma$ group (150, 151) measured σ_t between 1.4 and 3.0 GeV and presented positive arguments that these multiparticle events were hadronic. A detailed description of the apparatus and data analysis methods used by the $\mu\pi$ group can be found in (124). Their most recent cross section results were reported by Ceradini et al (152). The BCF group calibrated their apparatus in beams of hadrons and leptons and verified the hadronic nature of the multiparticle events. Their most recent results on σ_t are given in (153). In this paper two different methods were used to compute $\bar{\varepsilon}$ and they did not agree at all energies.

The most recent of the ADONE results on R are plotted in Figures 21 and 37, along with lower-energy data from Novosibirsk (154) and Orsay (155). In spite of the large errors and disagreements between various groups, it is clear that multihadron production is significant at energies above $W = 1.2$ GeV. By far the most serious problems with these pioneering measurements of R come from the relatively small

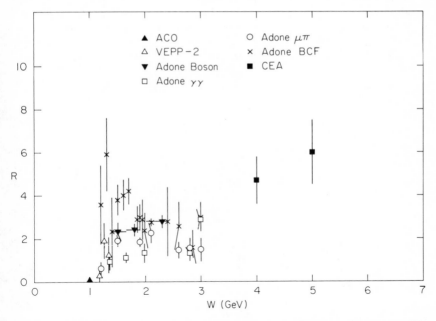

Figure 37 Results on R vs W from ACO (155), Novosibirsk (154), Adone Boson (53), Adone $\gamma\gamma$ (151), Adone $\mu\pi$ (124, 152), Adone BCF (153), and CEA (156, 157).

solid angle of the detectors. Typically, the early Frascati detectors covered only about 20% of 4π sr solid angle, required at least two charged particles to trigger, and had interaction-region volumes that were nearly as long as the detectors. Thus correction factors as large as 10 to 50 were applied to the experimental cross sections to obtain R. The model dependence and other systematic errors in $\bar{\varepsilon}$ cannot be reliably controlled with such small acceptance. These detectors were simply not prepared for the new world of multihadron production that they discovered.

The next round of σ_t information came from the nonmagnetic BOLD detector at CEA (156, 157). The larger solid angle of this device was much better suited for determining R, but low counting rates limited measurements to the two energies $W = 4$ and 5 GeV. Nevertheless, these results, plotted in Figure 37, were a forerunner of the "new physics," shortly to be discovered.

The SLAC/LBL group (131) published measurements of R at several c.m. energies in the range from 2.4 to 5.0 GeV. In Figures 21 and 38, the published SPEAR results and recently presented values for R up to $W = 7.8$ GeV (132) are shown. The large solid-angle and momentum-analysis capability of the SPEAR magnetic detector have substantially reduced the correction factors needed to determine R. The SLAC/LBL group estimate the systematic uncertainty in the absolute value of R to be $\pm 10\%$. A further $\pm 15\%$ smooth variation could occur between lowest and highest energies covered.

Aside from the narrow resonances J/ψ and ψ', the most conspicuous features of our present knowledge of R at energies above 1 GeV are the two regions where R is roughly independent of c.m. energy and the transition region between these. As shown in Figures 21 and 38, between 2.4 and 3.5 GeV the data are consistent with the value $R = 2.5$. The data between 1 and 2.4 GeV could also have this same value for R, but the large experimental uncertainties prevent us from drawing any

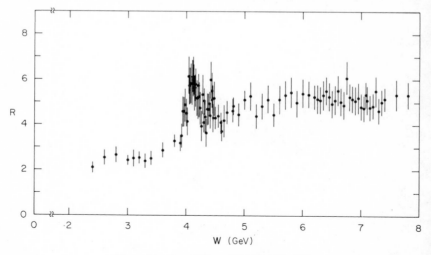

Figure 38   R vs *W* from SPEAR (131, 132).

conclusions except that there probably exists at least one broad resonance here, the $\rho'(1600)$. Above 5.0 GeV R is again nearly constant at a level approximately twice that of the lower-energy region. Between 3.5 and 5.0 GeV there is a complicated transition region with several possible resonances.

From the constancy of R we see that hadron production exhibits scaling in two different regions of c.m. energy. In the context of the parton model, it follows from Equation 4.9 that new partons must be coming into play in the transition region of R to effect an increase in R. The lower-energy scaling region, where R is approximately 2.5, is compatible with parton models where the partons are nine Gell-Mann/Zweig quarks (158, 159) arranged in three "flavors" and three "colors." This quark arrangement, so very successful in explaining the spectroscopy of "old" hadrons, predicts $R = 2$. The increase in R in going to the higher-energy scaling region indicates that new processes are adding 2 to 3 units of R to the "old physics" of the lower-energy scaling region. The fact that this rise in R occurs at energies just above the masses of the J/ψ and ψ' suggests that the "new physics" represented by the increase of R is related to the new particles. The rich structure observed in the transition region also hints at such a relationship; it may reflect the breaking of the selection rules responsible for the very long lifetimes of the new particles. The anomalous $e\mu$ events (94) may also be related to an increase in R near $W = 4$ GeV if the parent particles have significant decays to hadrons (160).

The notions of "new" and "old" physics are made explicit in the theoretical models that attempt to explain the properties of the new particles (161). The charmed-quark model (162–164) proposes that the J/ψ and ψ' are bound states of a fourth type of quark and its antiquark, which possess a new quantum number called charm. Above the threshold for production of particles having this new quantum number (charmed particles), charmed-quark models predict that R should approach $3\frac{1}{3}$; below threshold, R is equal to 2. The additional $1\frac{1}{3}$ units of R come from charmed particle production. Another feature of the charm models (163, 164) is the atomlike spectroscopy of the bound charmed quark-antiquark system. In addition to the J/ψ and ψ', a host of intermediate states having various quantum numbers and broad states above the ψ' were predicted. Indeed, these predictions motivated much of the experimental effort that ultimately led to the new particle spectroscopy summarized in Figure 32. While the data on the new particle spectroscopy (113, 165) and R disagree in detail with present theoretical calculations, the charmed-quark hypothesis provides an impressive framework in which the energy dependence of R and the new particle spectroscopy can be understood in qualitative terms. A search for charmed particles was carried out in the "new" physics region at $W = 4.8$ GeV by the SLAC/LBL group (166) with no success. As pointed out by Einhorn & Quigg (167) and others, the null results of the SPEAR experiment are unpleasant but not catastrophic for the charm model.

In another class of models, which follow from the quark model of Han & Nambu (168), this general picture of isolated new particle states connected to an increase in R recurs. In these models the new physics of so-called colored states adds 2 units of R to the low-energy value, which is also equal to 2.

4.2.3   PHOTON-PHOTON PROCESSES   An early criticism of experiments measuring R was that the photon-photon process could be responsible for a substantial fraction of the large multihadron cross section. Most of the groups that have published the results summarized in Figure 21 discussed possible contamination from photon-photon processes. Experiments (53, 151, 156) calculated possible contaminations and concluded that they were negligible. Experiments described in (131, 152, 157) used small-angle tagging counters to detect the forward electrons in coincidence with multihadron events; they experimentally verified that the contamination of R by photon-photon processes can be neglected at present energies.

Hadronic final states produced through photon-photon processes have been observed, albeit with very low rates. Orito et al (169) observed two events consistent with the reaction

$$e^+e^- \to e^+e^-\pi^+\pi^-. \qquad\qquad 4.13.$$

Paoluzi et al (170) reported evidence for the production of acoplanar pairs of pions

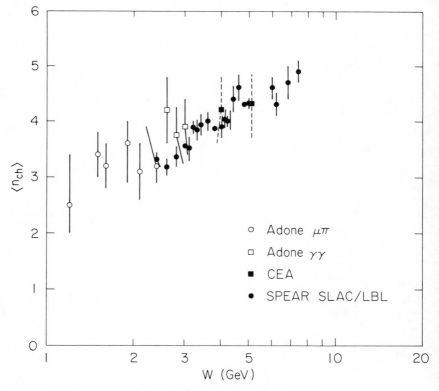

Figure 39   Mean charged multiplicity vs W for hadron production in e^+e^- annihilation. Data are from Adone $\mu\pi$ (171), Adone $\gamma\gamma$ (172), CEA (156, 157), and SPEAR (131, 132).

that would indicate the presence of additional neutral particles in the final state. Again, two events were detected. Both experiments used the same apparatus and demanded that the small-angle e^+ and e^- be tagged and momentum-analyzed.

4.2.4 AVERAGE PROPERTIES OF MULTIHADRON FINAL STATES The mean charged-particle multiplicity $\langle n_{ch} \rangle$ can be computed from the unfold procedure used to determine R. The data from experiments at ADONE (124, 152, 171, 172), CEA (156, 157), and SPEAR (131, 132) are presented in Figure 39. The energy dependence of these data is consistent with the logarithmic growth

$$\langle n_{ch} \rangle = a + b \ln s, \qquad\qquad 4.14.$$

where (165) $a = 1.93$, $b = 0.75$, and s is in units of GeV^2. Of course, some other slow growth with energy cannot be ruled out. This behavior is reminiscent of the multiplicity growth in many other hadronic processes at comparable energies (173). There is no evidence for abrupt changes in $\langle n_{ch} \rangle$ in the transition region of R, although the experimental uncertainties are rather large and could obscure important effects. Bjorken & Brodsky (174) have discussed the energy dependence of $\langle n_{ch} \rangle$ in terms of either parton or statistical models of final states. The observed slow growth of $\langle n_{ch} \rangle$ favors the parton picture.

The mean fraction of c.m. energy appearing in charged particles was measured with the SLAC/LBL magnetic detector (132) and is shown in Figure 40. In the analysis, all particles were assumed to be pions and corrections were applied for expected geometrical losses of particles. An important feature of Figure 40 is the

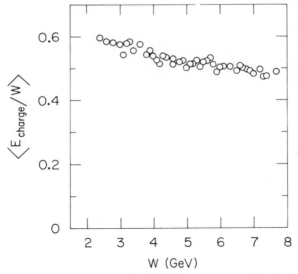

Figure 40 Average fraction of c.m. energy appearing in charged hadrons from the SPEAR experiments (132).

fact that the charged-energy fraction is less than the naive expectation of $\frac{2}{3}$ at all energies. That it is equal to 0.6 at low energies is probably a consequence of the production of heavy particles such as nucleons, kaons, and etas, in addition to pions. Why it should fall with increasing c.m. energy is not understood.

At present there is little experimental information available on the relative abundance of different types of hadrons in multihadron events. The most detailed information has been obtained by the DASP group (113) on the mass of the J/ψ. Preliminary nonresonant data were presented in (172) by the SLAC/LBL group. The Princeton Pavia Maryland group (175) published results of a single-arm spectrometer measurement at nonresonant energies. In the broadest terms, these measurements have shown that pions dominate the population of final-state particles, while roughly 10% of charged particles are kaons and 5% or so are protons.

4.2.5 INCLUSIVE MOMENTUM SPECTRA   The scaling of R immediately suggests (143) that single-particle momentum spectra may show Bjorken scaling (34). The SLAC/ LBL group (132) has measured inclusive momentum spectra at several c.m. energies. Their results for the quantity $s\,d\sigma/dx$ at three c.m. energies are plotted versus x in Figure 41. In this case, the scaling variable x is computed from the particle momentum rather than total energy because no hadron identification was available.

The spectra for all three energies shown in Figure 41 rise sharply at small values of x, peak at relatively low x, then fall with increasing x. Above $x = 0.4$, the spectra are equal within experimental error and thus are consistent with Bjorken scaling. To study scaling more critically, the SLAC/LBL results for $s\,d\sigma/dx$ are plotted versus W for several x intervals in Figure 42. Bjorken scaling implies

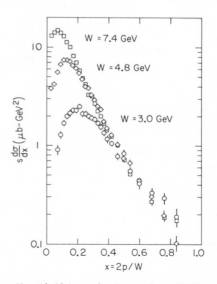

Figure 41 $s(d\sigma/dx)$ vs x for three values of W from (132).

(see Equation 2.10) that $s\,d\sigma/dx$ should not change with W at fixed values of x. For the lowest x interval near $x = 0.1$, scaling is badly broken. By $x = 0.2$, the data are roughly independent of W for W greater than 4 GeV. For $x \geq 0.4$, the data exhibit Bjorken scaling to the 20% level over the entire energy range studied.

The scaling observed for x values greater than 0.4 is quite remarkable in light of the doubling of R over this same energy range, and suggests that the "new" physics is confined to relatively small values of x. However, the interplay between the changes in R, the mean charged-particle energy, and inclusive momentum spectra as a function of W have made it impossible to isolate two components of the data relating to "old" and "new" physics.

4.2.6 ANGULAR DISTRIBUTIONS   The most general single-particle inclusive angular distribution for hadrons produced through one-photon exchange has been given in Equation 2.7. The coefficient of the $\cos^2 \theta$ term α is defined by:

$$\alpha = \frac{W_1 - W_0}{W_1 + W_0} \qquad\qquad 4.15.$$

where W_1 and W_0 are the structure functions defined in Section 2. α is bounded between -1 and 1. Experiments measuring only polar angle distributions have been

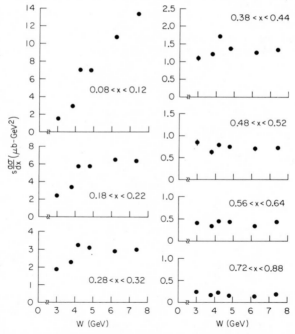

Figure 42 $s(d\sigma/dx)$ vs W for various regions of x from (132). Bjorken scaling implies $s(d\sigma/dx)$ should be independent of W for fixed values of x.

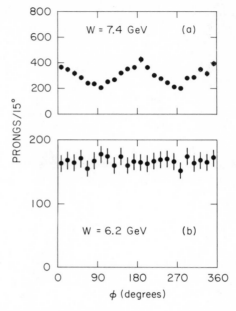

Figure 43   Azimuthal distributions for hadrons with $x > 0.3$ at two values of W from (177). At $W = 7.4$ GeV the e^+e^- beams were polarized, while at $W = 6.2$ GeV they were not.

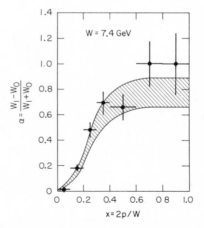

Figure 44 α vs x at $W = 7.4$ GeV determined from the θ and ϕ dependence of inclusive hadron production in the case of polarized e^+e^- beams (177). The shaded region is the prediction of the "jet" model of (178) where $\alpha_{jet} = 0.78 \pm 0.12$.

reported by the BCF group at ADONE (176) and the SLAC/LBL group (172). The
ADONE group covered the c.m. energy range from 1.2 to 3.0 GeV; the SPEAR
experiment covered 3.0 to 4.8 GeV. In both cases the observed angular distributions
were essentially isotropic, but α was poorly determined because of the relatively
small range of $\cos^2 \theta$ covered in these experiments. The SLAC/LBL group (177)
have measured α as a function of the scaling variable x at $W = 7.4$ GeV, where the
beam polarization is significant. In this experiment α was determined from the
azimuthal, as well as the polar angle, distribution of hadrons where complete
coverage was possible. Two examples of their azimuthal distributions are given in
Figure 43. At $W = 6.2$ GeV there is no azimuthal dependence because the beams
are depolarized due to a spin-precession resonance (32). At $W = 7.4$ GeV the
polarization was large and the strong $\cos 2\phi$ term evident in Figure 43 could be used
to determine α. The values for α as a function of x are given in Figure 44. It is seen
that particles with low x are produced isotropically, while hadrons of large x are

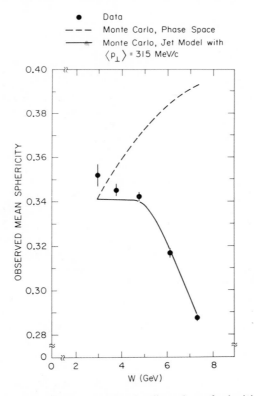

Figure 45   Mean sphericity vs W from (178). Smaller values of sphericity indicate events
are more jetlike. The predictions of two models of final states are shown.

produced predominantly through a coupling to the virtual photon that is transverse to their direction of motion.

4.2.7 JET STRUCTURE   An important prediction of the parton model (144) is that jets of hadrons moving in opposite directions should be produced in e^+e^- collisions. The schematic drawing of hadron production by partons given in Figure 36 provides an intuitive picture of jets—a pair of partons are produced and they subsequently radiate hadrons along their common direction of motion with limited transverse momentum relative to that axis. An operational definition of jets is any multiparticle correlation that leads to some preferred direction in a system of hadrons about which the transverse momentum is limited.

The SLAC/LBL group (178) has searched for jet structure in multihadron events by finding that axis in each event that minimizes the sum of squares of charged particle momenta perpendicular to it. For each event, a parameter called the sphericity, S, was computed. S is a measure of the jetlike character of events and is defined by:

$$S = \frac{3 \sum_i p_{\perp i}^2}{2 \sum_i p_i^2} .$$

4.16.

where summations run over all charged particles observed in the event, p_i is the particle momentum, and $p_{\perp i}$ is the momentum perpendicular to the jet axis. S is bounded between 0 and 1; events with small S are jetlike, events with large S are not.

The mean sphericity reported by (178) is plotted versus c.m. energy in Figure 45; it decreases substantially with increasing c.m. energy, suggesting the presence of jets in the high-energy data. Two models of final states, an invariant-phase space model, and a jet model that modified phase space to give limited transverse momentum about a jet axis, were compared with the data. The mean transverse momentum of particles in the jet model was 315 MeV/c at all c.m. energies. In Figure 45 the jet model is seen to agree with the data at all c.m. energies, while the invariant-phase space model fails to describe the data. The authors argue that the observed small values of S are not simple kinematic reflections, but represent multiparticle correlations that cannot be explained by energy-momentum conservation alone.

The complete (θ and ϕ) angular distribution for the jet axis at $W = 7.4$ GeV displays the usual one-photon exchange form with $\alpha = 0.78 \pm 0.12$, where α is defined by Equation 4.15. This is very close to the μ-pair distribution where $\alpha = 1$, and strongly implies that if partons are responsible for jets, then the parton spin is $\frac{1}{2}$.

The jet model of (178) is able to reproduce well the inclusive angular distribution of hadrons at $W = 7.4$ GeV. The predictions of this model are given by the shaded area in Figure 44. The simple jet picture provided by the parton model with spin-$\frac{1}{2}$ partons gives a remarkably accurate description of the sphericity and angular distributions of hadrons produced in e^+e^- annihilations at high energy.

5   CONCLUSIONS

In this article we have reviewed experimental results from the study of e^+e^- interactions at high energies. This field is in a period of exceptionally rapid growth at present, so that we have only been able to outline current areas of research.

In many examples we have seen the importance of the simple initial state provided by the e^+e^- system. Throughout the studies of quantum electrodynamics and hadron production, the one-photon exchange channel has played a dominant role. Its associated simple angular distribution has been observed in electrodynamic processes as well as in hadron production. Radiative corrections are necessary for a quantitative understanding of all processes, but have not introduced any essential difficulties to understanding experimental results.

The photon-photon process has played a minor role to date, but has been observed in electrodynamic and hadronic final states and promises to be more important in the future.

Final states involving only electrodynamic particles are well described by the theory of quantum electrodynamics when calculable hadronic vacuum polarization corrections are included. The only exception to this rule is the sample of anomalous $e\mu$ events discovered at SPEAR. A great deal more experimental information is required in order to establish the origin of these very significant events.

Predictions of QED have been experimentally verified to the 10% to 20% level in many reactions over a broad range of energies. Lower limits on possible breakdown parameters are in the vicinity of 30 GeV. Equivalently, these experiments have checked the validity of QED down to scales of distance less than 10^{-15} cm. Hadronic corrections to QED have been observed in the vicinity of the narrow resonances ϕ, J/ψ, and ψ'.

e^+e^- experiments have played a major role in the discovery and determination of properties of the vector mesons. The spectrum of vector mesons now includes the $\rho, \omega, \phi, J/\psi, \psi', \rho'(1600)$, and several broad states with masses near 4 GeV. Because of their extraordinary properties, the J/ψ and ψ' appear to represent new hadronic degrees of freedom not present in ordinary hadrons. A new heavy-particle spectroscopy related to the narrow resonances is presently being explored. e^+e^- colliding-beam experiments are uniquely responsible for our present depth of understanding of the new particles.

The total hadronic cross section is scaling in two different energy regions. For c.m. energies between 2.4 and 3.5 GeV, R is approximately 2.5, while between 5.0 and 7.8 GeV, R is roughly 5. A complicated transition region connects these scaling regions. The mean charged-particle multiplicity slowly increases with energy, while the mean fraction of c.m. energy appearing in final-state charged hadrons decreases as the c.m. energy increases. Single-particle inclusive momentum spectra exhibit Bjorken scaling for $x > 0.4$ over the energy range $3.0 \text{ GeV} \leq W \leq 7.4 \text{ GeV}$. There is jet structure in multihadronic final states at high energies. The average transverse momentum of hadrons with respect to the jet axis is approximately

315 MeV/c. Jets couple to the virtual photons mainly throu$_{\zeta}$, $_{1}$ a transverse coupling similar to muon pairs.

Qualitative features of hadron production by e^+e^- colliding beams are remarkably well described by the quark-parton model. The constancy of R, scaling of inclusive momentum spectra, and jets are natural consequences of the parton model. The azimuthal dependence of hadron production with polarized beams is strong evidence for the spin-$\frac{1}{2}$ nature of the partons.

Specific quark models are able to describe qualitative features of the data, but fail quantitatively. The new resonances J/ψ and ψ' and their associated heavy-particle spectroscopy look very much like bound states of a new charmed quark, but details of the spectrum and decay rates disagree with current estimates. A direct connection between the "new" physics represented by the increase in R between 3.5 and 5.0 GeV and the new particles has not yet been established.

Much experimental data is still needed in order to understand hadron production in e^+e^- collisions. Specifically, the c.m. energy region 1.0 to 2.5 GeV needs to be explored with new detectors more suited to studying multihadron final states. The approach to scaling and spectrum of vector mesons in this energy range is of crucial importance to our understanding of the structure of the "old" physics. Likewise, detailed information on R in the 4-GeV transition region and knowledge of the spectrum of states, quantum numbers, and decay rates for the new particle spectroscopy are essential for understanding the "new" physics.

With the new storage rings now being built, a vast, uncharted sea of e^+e^- physics awaits us. Our guides, the theory of QED and the parton ideas, will again be tested. We can expect to encounter new physics involving the weak interactions. From past experience we have every reason to believe that the next generation of experiments will uncover new phenomena as rich and important to our understanding of fundamental processes as those we have outlined here.

Literature Cited

1. Sands, M. 1971. *Proc. Int. Sch. Phys.* "Enrico Fermi," Course 46, pp. 257–411. New York/London: Academic
2. Pellegrini, C. 1972. *Ann. Rev. Nucl. Sci.* 22:1
3. Bernardini, C. et al 1964. *Nuovo Cimento* 34:1473
4. Bernardini, C., Corazza, G. F., Ghigo, G., Touscheck, B. 1960. *Nuovo Cimento* 18:1293
5. O'Neill, G. K. 1959. *Proc. Int. Conf. High Energy Accel. Instrum., Geneva*, p. 125
6. Barber, W. C., Gittelman, B., O'Neill, G. K., Richter, B. 1966. *Phys. Rev. Lett.* 15:1127
7. Barber, W. C., Gittelman, B., O'Neill, G. K., Richter, B. 1971. *Phys. Rev. D* 3:2796
8. Budker, G. I. et al 1965. *Proc. Int. Conf. High Energy Accel., 5th, Frascati*, p. 390
9. Budker, G. I. et al 1965. *At. Energ.* 19: 497
10. Auslander, V. L. et al 1967. *Proc. Int. Symp. Electron Photon Interactions High Energies, Stanford*, pp. 385, 601
11. Augustin, J.-E. et al 1967. See Ref. 10, pp. 385, 600
12. Auslander, V. L. et al 1965. See Ref. 8, p. 394
13. Orsay Storage Ring Group 1965. See Ref. 8, p. 271
14. Budker, G. I. et al 1972. *Sov. Conf. Charged Part. Accel.* Moscow, 1:318
15. Amman, F. et al 1963. *Proc. Int. Conf. High Energy Accel., Dubna*, p. 249
16. Hofmann, A. et al 1967. *Proc. Int. Conf. High Energy Accel., 6th, Cambridge*, p. 112
17. SPEAR Storage Ring Group 1971. *Proc. Int. Conf. High Energy Accel., 7th,*

Geneva, p. 145
18. Augustin, J.-E. et al 1974. *Phys. Rev. Lett.* 33 : 1406
19. Aubert, J. J. et al 1974. *Phys. Rev. Lett.* 33 : 1404
20. Abrams, G. S. et al 1974. *Phys. Rev. Lett.* 33 : 1453
21. DESY Storage Ring Group 1974. *Proc. Int. Conf. High Energy Accel., 9th, Stanford,* p. 43
22. Marin, P. 1974. See Ref. 21, p. 49
23. Sidorov, V. A. 1975. Private communication
24. LBL-SLAC PEP Study Group 1974. LBL Rep. No. 2688, SLAC Rep. No. SLAC-171
25. PETRA Storage Ring Group 1974. DESY Proposal
26. Cornell Storage Ring Group 1975. Cornell Rep. CLNS-301
27. Tsai, Y. S. 1960. *Phys. Rev.* 120 : 269
28. Cabibbo, N., Gatto, R. 1961. *Phys. Rev.* 124 : 1577
29. Tsai, Y. S. 1976. *Phys. Rev. D* 12 : 3533
30. Bjorken, J. D., Drell, S. D. 1964. *Relativistic Quantum Mechanics.* New York : McGraw-Hill
31. Sokolov, A. A., Ternov, I. M. 1964. *Sov. Phys. Dokl.* 8 : 1203
32. Baier, V. N. 1971. See Ref. 1, pp. 1–49
33. Jackson, J. D. 1976. *Rev. Mod. Phys.* 48 : 417
34. Bjorken, J. D. 1969. *Phys. Rev.* 179 : 1547
35. Brodsky, S. J., Kinoshita, T., Terazawa, H. 1970. *Phys. Rev. Lett.* 25 : 972
36. Arteaga-Romero, N., Jaccarini, A., Kessler, P., Parisi, J. 1971. *Phys. Rev. D* 3 : 1569
37. Baier, V. N., Fadin, V. S. 1971. *Nuovo Cimento Lett.* 1 : 481
38. Terazawa, H. 1973. *Rev. Mod. Phys.* 45 : 615
39. Tsai, Y. S. 1960. *Phys. Rev.* 120 : 269
40. Bonneau, G., Martin, F. 1971. *Nucl. Phys. B* 27 : 381
41. Bloch, F., Nordsieck, A. 1937. *Phys. Rev.* 52 : 54
42. Brodsky, S. J., Drell, S. D. 1970. *Ann. Rev. Nucl. Sci.* 20 : 147
43. Godine, J., Hankey, A. 1972. *Phys. Rev. D* 6 : 3301; Cung, V. K., Mann, A. K., Paschos, E. A. 1972. *Phys. Lett. B* 41 : 355
44. Drell, S. D. 1958. *Ann. Phys.* 4 : 75
45. Lee, T. D., Wick, G. C. 1969. *Nucl. Phys. B* 9 : 209
46. Kroll, N. M. 1966. *Nuovo Cimento A* 45 : 65
47. Berends, F. A., Gaemers, K. J. F., Gastmans, R. 1974. *Nucl. Phys. B* 68 : 541
48. Augustin, J.-E. et al 1970. *Phys. Lett. B*
31 : 673
49. Bartoli, B. et al 1970. *Nuovo Cimento A* 70 : 603
50. Alles-Borelli, V. et al 1972. *Nuovo Cimento A* 7 : 345
51. Alles-Borelli, V. et al 1971. *Phys. Lett. B* 36 : 149
52. Bartoli, B. et al 1971. *Phys. Lett. B* 36 : 593
53. Bartoli, B. et al 1972. *Phys. Rev. D* 6 : 2374
54. Borgia, B. et al 1971. *Phys. Lett. B* 35 : 340
55. Bernardini, M. et al 1973. *Phys. Lett. B* 45 : 510
56. Bernardini, M. et al 1973. *Phys. Lett. B* 45 : 169
57. Madaras, R. et al 1973. *Phys. Rev. Lett.* 30 : 507
58. Newman, H. et al 1974. *Phys. Rev. Lett.* 32 : 483
59. Beron, B. L. et al 1974. *Phys. Rev. Lett.* 33 : 663
60. Hofstadter, R. 1975. *Proc. 1975 Int. Symp. Lepton Photon Interactions High Energies, Stanford,* p. 869
61. Augustin, J.-E. et al 1975. *Phys. Rev. Lett.* 34 : 233
62. Learned, J. G., Resvanis, L. K., Spencer, C. M. 1975. *Phys. Rev. Lett.* 35 : 1688
63. Berends, F. A., Gaemers, K. J. F., Gastmans, R. 1972. *Nucl. Phys. B* 57 : 381; 75 : 546
64. Balakin, V. E. et al 1971. *Phys. Lett. B* 37 : 435
65. Borgia, B. et al 1972. *Nuovo Cimento Lett.* 3 : 115
66. Alles-Borelli, V. et al 1972. *Nuovo Cimento A* 7 : 330
67. Alles-Borelli, V. et al 1975. *Phys. Lett. B* 59 : 201
68. Bollini, D. et al 1975. *Nuovo Cimento Lett.* 13 : 380
69. Kurdadze, L. M. et al 1975. Novosibirsk Prepr. 75-66
70. Berends, F. A., Gastmans, R. 1973. *Nucl. Phys. B* 61 : 414
71. Balakin, V. E. et al 1971. *Phys. Lett. B* 34 : 99
72. Bacci, C. et al 1971. *Nuovo Cimento Lett.* 2 : 73
73. Hanson, G. et al 1973. *Nuovo Cimento Lett.* 7 : 587
74. Law, M. E. et al 1974. *Nuovo Cimento Lett.* 11 : 5
75. Cosme, G. et al 1973. *Nuovo Cimento Lett.* 8 : 509
76. Bacci, C. et al 1973. *Phys. Lett. B* 44 : 530
77. Balakin, V. E., Bukin, A. D., Pakhtusova, E. V., Sidorov, V. A., Khabakhpashev,

A. G. 1971. *Phys. Lett. B* 34:663
78. Bacci, C. et al 1972. *Nuovo Cimento Lett.* 3:709
79. Parisi, J. 1974. *J. Phys. Paris Colloq. C2* 35:51
80. Barbiellini, G. et al 1974. *Phys. Rev. Lett.* 32:385
81. Augustin, J.-.E. et al 1973. *Phys. Rev. Lett.* 30:462
82. Barbiellini, G. et al 1974. *Nuovo Cimento Lett.* 11:718
83. Baldini–Celio, R. et al 1974. *Nuovo Cimento Lett.* 11:711
84. Braunschweig, W. et al 1974. *Phys. Lett. B* 53:393
85. Braunschweig, W. et al 1975. *Phys. Lett. B* 56:491
86. Boyarski, A. M. et al 1975. *Phys. Rev. Lett.* 34:1357
87. Lüth, V. et al 1975. *Phys. Rev. Lett.* 35:1124
88. Ford, R. L. et al 1975. *Phys. Rev. Lett.* 34:604
89. Jackson, J. D., Scharre, D. L. 1975. *Nucl. Instrum. Methods* 128:13
90. Tsai, Y. S. 1971. *Phys. Rev. D* 4:2821
91. Alles-Borelli, V. et al 1970. *Nuovo Cimento Lett.* 4:1151
92. Alles-Borelli, V. et al 1970. *Nuovo Cimento Lett.* 4:1156
93. Orito, S. et al 1974. *Phys. Lett. B* 48:165
94. Perl, M. L. et al 1975. *Phys. Rev. Lett.* 35:1489
95. Auslander, V. L. et al 1967. *Phys. Lett. B* 25:433
96. Augustin, J.-E. et al 1968. *Phys. Rev. Lett.* 20:126
97. Augustin, J.-E. et al 1969. *Phys. Lett. B* 28:508
98. Auslander, V. L. et al 1969. *Sov. J. Nucl. Phys.* 9:114
99. Benaksas, D. et al 1972. *Phys. Lett. B* 39:289
100. Gounaris, G. J., Sakurai, J. J. 1968. *Phys. Rev. Lett.* 21:244
101. Gourdin, M., Stodolsky, L., Renard, F. M. 1969. *Phys. Lett. B* 30:347
102. Bemporad, C. 1975. See Ref. 60, p. 113
103. Augustin, J.-E. et al 1969. *Phys. Lett. B* 28:513
104. Benaksas, D. et al 1972. *Phys. Lett. B* 42:507
105. Cosme, G., Jean-Marie, B., Jullian, S., Lefrançois, J. 1972. *Nucl. Instrum. Methods* 99:599
106. Augustin, J.-E. et al 1969. *Phys. Lett. B* 28:517
107. Balakin, V. E. et al 1971. *Phys. Lett. B* 34:328
108. Cosme, G. et al 1974. *Phys. Lett. B* 48:159
109. Bizot, J. C. et al 1970. *Phys. Lett. B* 32:416
110. Cosme, G. et al 1974. *Phys. Lett. B* 48:155
111. Bukin, A. D. et al 1975. Novosibirsk Prepr. 75-64
112. Benaksas, D. et al 1972. *Phys. Lett. B* 42:511
112a. Renard, F. M. 1974. *Nucl. Phys. B* 82:1
113. Wiik, B. H. 1975. See Ref. 60, p. 69
114. Jean-Marie, B. et al 1976. *Phys. Rev. Lett.* 36:291
115. Abrams, G. S. 1975. See Ref. 60, p. 25
116. Braunschweig, W. et al 1975. *Phys. Lett. B* 57:297
117. Abrams, G. S. et al 1975. *Phys. Rev. Lett.* 34:1181
117a. Hilger, E. et al 1975. *Phys. Rev. Lett.* 35:625
118. Braunschweig, W. et al 1975. *Phys. Lett. B* 57:407
119. Tanenbaum, W. et al 1975. *Phys. Rev. Lett.* 35:1323
120. Tanenbaum, W. et al 1976. *Phys. Rev. Lett.* 36:402
121. Feldman, G. J. et al 1975. *Phys. Rev. Lett.* 35:821
122. Heintze, J. 1975. See Ref. 60, p. 97
123. Barbarino, G. et al 1972. *Nuovo Cimento Lett.* 3:689
124. Grilli, M. et al 1973. *Nuovo Cimento A* 13:593
125. Bingham, H. H. et al 1972. *Phys. Lett. B* 41:635
126. Conversi, M. et al 1974. *Phys. Lett. B* 52:493
127. Ceradini, F. et al 1973. *Phys. Lett. B* 43:341
128. Bernardini, M. et al 1974. *Phys. Lett. B* 53:384
129. Alles-Borelli, V. et al 1975. *Nuovo Cimento A* 30:136
130. Balakin, V. E. et al 1972. *Phys. Lett. B* 41:205
131. Augustin, J.-E. et al 1975. *Phys. Rev. Lett.* 34:764
132. Schwitters, R. F. 1975. See Ref. 60, p. 5
133. Aulchenko, V. M. et al 1975. Novosibirsk Prepr. 75-65
134. Bacci, C. et al 1975. *Phys. Lett. B* 58:481
135. Esposito, B. et al 1975. *Phys. Lett. B* 58:478
136. Boyarski, A. M. et al 1975. *Phys. Rev. Lett.* 34:762
137. Barbiellini, G. et al 1973. *Nuovo Cimento Lett.* 6:557
138. Bollini, D. et al 1975. *Nuovo Cimento Lett.* 14:418
139. Bernardini, M. et al 1973. *Phys. Lett. B* 44:393

140. Bernardini, M. et al 1973. *Phys. Lett. B* 46:261
141. Castellano, M. et al 1973. *Nuovo Cimento A* 14:1
142. Wilson, R. 1972. *Proc. Int. Conf. High Energy Phys., 15th,* Kiev, p. 219
143. Drell, S. D., Levy, D., Yan, T. M. 1969. *Phys. Rev.* 187:2159
144. Feynman, R. P. 1972. *Photon-Hadron Interactions.* Reading, Mass: Benjamin
145. Cabibbo, N., Parisi, G., Testa, M. 1970. *Nuovo Cimento Lett.* 4:35
146. Ferrara, S., Greco, M., Grillo, A. F. 1970. *Nuovo Cimento Lett.* 4:1
147. Feldman, G. J., Perl, M. L. 1975. *Phys. Rep. C* 19:235
148. Bartoli, B. et al 1970. *Nuovo Cimento A* 70:615
149. Bartoli, B. et al 1971. *Phys. Lett. B* 36:598
150. Bacci, C. et al 1972. *Phys. Lett. B* 38:551
151. Bacci, C. et al 1973. *Phys. Lett. B* 44:533
152. Ceradini, F. et al 1973. *Phys. Lett. B* 47:80
153. Bernardini, M. et al 1974. *Phys. Lett. B* 51:200
154. Kurdadze, L. M., Onuchin, A. P., Serednyakov, S. I., Sidorov, V. A., Eidelman, S. I. 1972. *Phys. Lett. B* 42:515
155. Cosme, G. et al 1972. *Phys. Lett. B* 40:685
156. Litke, A. et al 1973. *Phys. Rev. Lett.* 30:1189
157. Tarnopolsky, G. et al 1974. *Phys. Rev. Lett.* 32:432
158. Fritzsch, H., Gell-Mann, M. 1972. *Proc. Int. Conf. High Energy Phys., 16th,* Batavia, Ill.
159. Bjorken, J. D. 1973. *Proc. Int. Symp. Electron Photon Interactions High Energies, 6th,* Bonn, p. 25
160. Gilman, F. J. 1975. See Ref. 10, p. 131
161. Harari, H. 1975. *Proc. 1975 Int. Symp. Lepton Photon Interactions High Energies,* Stanford, p. 317
162. Gaillard, M. K., Lee, B. W., Rosner, J. L. 1975. *Rev. Mod. Phys.* 47:277
163. Appelquist, T., De Rújula, A., Politzer, H. D., Glashow, S. L. 1975. *Phys. Rev. Lett.* 34:365
164. Eichten, E. et al 1975. *Phys. Rev. Lett.* 34:369
165. Feldman, G. J. 1975. See Ref. 60, p. 39
166. Boyarski, A. M. et al 1975. *Phys. Rev. Lett.* 35:196
167. Einhorn, M. B., Quigg, C. 1975. *Phys. Rev. Lett.* 35:1114
168. Han, M. Y., Nambu, Y. 1963. *Phys. Rev. B* 139:1006
169. Orito, S., Ferrer, M. L., Paoluzi, L., Santonico, R. 1974. *Phys. Lett. B* 48:380
170. Paoluzi, L. et al 1974. *Nuovo Cimento Lett.* 10:435
171. Ceradini, F. et al 1972. *Phys. Lett. B* 42:501
172. Richter, B. 1974. *Proc. Int. Conf. High Energy Phys., 17th, London,* pp. IV-37
173. Whitmore, J. 1974. *Phys. Rep. C* 10:273
174. Bjorken, J. D., Brodsky, S. J. 1970. *Phys. Rev. D* 1:1416
175. Atwood, T. L. et al 1975. *Phys. Rev. Lett.* 35:704
176. Bernardini, M. et al 1975. *Nuovo Cimento A* 26:163
177. Schwitters, R. F. et al 1975. *Phys. Rev. Lett.* 35:1320
178. Hanson, G. et al 1975. *Phys. Rev. Lett.* 35:1609
179. Hilger, E. et al 1975. *Phys. Rev. Lett.* 35:625

Ann. Rev. Nucl. Sci. 1976. 26:151–98

THE FERMI NATIONAL ACCELERATOR LABORATORY

✳5573

James R. Sanford

Fermi National Accelerator Laboratory,[1] Batavia, Illinois 60510

CONTENTS

[1] Operated by Universities Research Association, Inc. under contract with the US Energy Research and Development Administration.

1   INTRODUCTION

The Fermi National Accelerator Laboratory was designed and constructed by a number of dedicated individuals to provide frontier facilities in the United States for the study of particle physics. Today the 400-GeV[2] proton accelerator and extensive research facilities are operated routinely for research in high-energy particle and accelerator physics. These research facilities at Fermilab are used by a large number of scientists from the United States and many foreign countries. Both experimental and theoretical physicists are studying the properties of subnuclear particles at high energies and their interactions with matter and energy.

This article gives a general overview of Fermilab, how it came into being, what it is today, and some of the hopes for the future. After a general introduction concerning the history of the project, the site in Illinois, and the people who built the accelerator, the article goes on to technical aspects of the project. The accelerators culminating in the 400-GeV proton synchrotron are described. This accelerator provides a high-intensity proton beam that can be extracted and made available for experimental use. The four experimental areas are described, as are the properties of the secondary beams and detector facilities located within each area. A general description of the extent of the experimental program is provided and information about the process followed in selecting and initiating experiments is covered. Some of the recent experimental accomplishments are reviewed and a brief description of the plans for higher energy capabilities at Fermilab is provided.

Before describing the Laboratory and its research facilities, a brief discussion of nomenclature and terms is appropriate. The primary parameters describing any accelerator are the energy and intensity of the accelerated particles. The energy is usually expressed in electron volts (eV), a unit that describes the energy acquired by a particle of charge e accelerated in an electric field created by a voltage difference V. For example, a proton accelerated from a $+1,000,000$-V terminal to ground acquires 10^6 eV or 1 MeV of energy. The unit for 10^9 eV is 1 GeV. In conventional physics units the energy of a 400-GeV proton is 6.4×10^{-8} J.

The number of protons accelerated to the desired energy during one cycle or pulse of the accelerator is called the beam flux. It is usually expressed in number of protons/pulse. When the repetition rate of the accelerator is taken into account, the average intensity of particles can be expressed in number of protons/sec. Literally speaking, a beam of 10^{13} protons/sec represents a current of 1.6 μA. A high-intensity beam of energetic particles contains a large amount of power. For example, a 10^{13}/sec beam of 400-GeV protons contains 640 kW of beam power. Obviously, the acceleration and handling of such a beam must be done with great care. That is only one of the many challenges that faced the physicists who set out to build what today is known as Fermilab.

---

[2] 500 GeV have recently been reached.

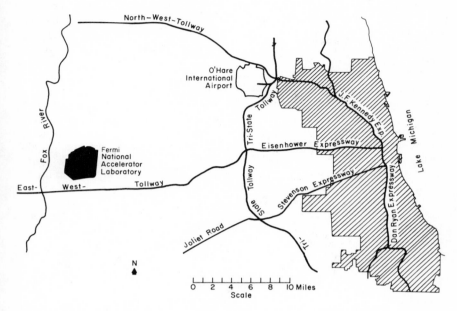

Figure 1   Fermilab is located about 35 miles west of Lake Michigan in Batavia, Illinois.

2   THE BUILDING OF FERMILAB

The Fermi National Accelerator Laboratory is located 35 miles west of Chicago on a flat plain near the Fox River in Illinois (Figure 1). On 6800 corn-filled acres a small group of physicists, engineers, technicians, and construction workers began building a new laboratory in 1967. The project had its beginning years before in discussions held by a number of physicists concerning the next accelerator project in the United States. The Alternating Gradient Synchrotron (AGS) at Brookhaven National Laboratory, a 30-GeV proton synchrotron, began operation in 1961 and provided a wide variety of particles for experimentation (1). Some physicists recognized that a yet-higher-energy proton synchrotron would be necessary in order to look at the finer grain structure within the nucleons.

Realizing that technical evaluations and recommendations should be made about the future course of high-energy physics, the Atomic Energy Commission (AEC) in 1963 convened a panel under Professor Norman Ramsey of Harvard University to study the future direction of the national program. One feature of the panel's report, the first recommendation in fact, was that a 200-GeV proton synchrotron be built as rapidly as possible (2).

Physicists at the Lawrence Radiation Laboratory in Berkeley, California, had been looking at some preliminary designs for an accelerator capable of producing a beam of 200-GeV protons. They then made detailed evaluations and submitted a design study to the AEC in 1965 (3). Special studies with broad participation were held in Berkeley in the summers of 1965 and 1966 to explore the experimental use of such an accelerator (4). Those studies indicated that the interest in such a facility was very high throughout the United States.

The AEC, recognizing the broad national interest in such a project, determined that site proposals for the location of the laboratory should be invited. More than 100 sites were proposed and a committee of the National Academy of Sciences narrowed the choice to six locations (5). On December 16, 1966, the AEC announced that the site at Weston, Illinois had been selected. While the State of Illinois went about purchasing the land as a gift to the Federal Government, the AEC drew up a contract with Universities Research Association, Inc. (URA) for the building and operating of the new laboratory. URA had been organized during this time in order to be the contractor with the AEC for the building and operation of the new laboratory (6). This organization originally consisted of 34 universities and today numbers 52. The President of URA is Professor N. Ramsey of Harvard.

Professor R. R. Wilson of Cornell was appointed the Director of the new laboratory in the spring of 1967. He was soon joined by Professor Edwin L. Goldwasser of the University of Illinois, the Deputy Director. They began to assemble a staff of physicists to make a detailed design for the facility and to initiate construction. The first of what were to become annual users' meetings was held at Argonne National Laboratory in April, 1967 (7). Following a number of talks and a visit to the site, the participants undertook the formation of the Users Organization. Through the years that organization has helped in communications between the Laboratory and the experimenters and has set an example for discussions and informed criticism.

The project that the staff set out to build was to include the highest-energy accelerator and the most extensive research facilities possible within the funds that the AEC intended to provide. The project, as originally conceived at Berkeley, was expected to cost in excess of $300 million and to require 8 years to construct (8). After consulting with the Bureau of the Budget, the AEC determined that such a sum was excessive and ordered a reduced scope for the project, namely to a lower beam intensity and less extensive experimental facilities. This led eventually to a project authorized for $250 million.

In order to respond to these constraints an austere design, but one that had within it possibilities for growth, was adopted. The small staff and a number of invited visitors came together in Oak Brook, Illinois, on June 15, 1967 to begin work. The terrain of the site was examined and the optimal dimensions and position of the main accelerator and experimental areas were determined. The location of the accelerator was chosen to provide optimal use and to insure that future expansion or new projects could be accommodated on the remainder of the site. Throughout this period and during construction the laboratory staff was assisted by DUSAF,[3] the architectural and engineering group selected to design and supervise the construction of the structures, roads, and utility systems for the Laboratory. Members of DUSAF worked very closely with Laboratory employees to design an economical facility.

The design concepts were incorporated into a Design Report issued in January 1968 and preceded in October by a specific project proposal to the AEC (9). Initially, only partial authorization was granted by Congress and annual appropriations did not come as rapidly as had been expected. Consequently, the construction proceeded in many phases and segments. The linear accelerator and booster construction began, but there was money for only one sixth of the Main Ring. In the meantime the technical components were designed, prototyped, and specified for industrial production. Eventually the remainder of the Main Ring was placed under contract and the design of the experimental areas specified. While beam acceleration was being attempted, the long-awaited building of the Central Laboratory began. Eventually the Central Laboratory was occupied and the completion of experimental areas brought the major construction activity to a close (Figure 2). The phased method of construction added to the tension and excitement at the Laboratory (10).

Summer studies involving experienced physicists played a major role in the design of the research facilities at the Laboratory. Beginning at Berkeley in 1965 and 1966 and continuing at Aspen in 1968 and 1969, participants in these discussions proposed and explored possible configurations of experimental areas, secondary beams, and detectors for use in the new energy regime (11). By exploiting the concept of splitting one extracted proton beam, it was possible to consider separate and distinct experimental areas. The plans developed at the studies revolved around research facilities for neutrinos. Beams for neutrinos and muons became the basis of Area 1, located directly in front of the extraction point and later appropriately named the Neutrino Area. Area 2 ideas concentrated on the use of common secondary beams; today this is known as the Meson Area. Area 3, lying to the other side of the extracted beam line, is now called the Proton Area.

Considerable attention was focused on the need for a large bubble chamber at NAL. Such a device had been included in the early plans (12), but did not remain when the project was pared down to $250 million. By the summer study in 1970 the decision was made to build a smaller chamber largely out of equipment funds and construction of a chamber with a length of 15 feet and a volume of 30,000 liters was under way. At the study controversies raged over where to locate the chamber in the Neutrino Area. No early commitments were made to other major research facilities for the new laboratory, but it was recognized that large arrays of electronic detectors would grow in time, particularly for counter neutrino and muon physics.

---

[3] DUSAF was a joint venture of four architectural and engineering firms to help build the Laboratory. The firms represented were Daniel, Mann, Johnson & Mendenhall; The Office of Max O. Urbahn; Seelye Stevenson Value & Knecht, Inc.; and George A. Fuller Company.

Figure 2 An aerial view of the Fermilab site with the main accelerator in the foreground and the experimental areas to the left beyond the Injector and Central Laboratory.

By 1970 the scope of the research facilities was tentatively specified (13), and physicists were invited to submit proposals for experiments. This marked the end of the general and broad deliberations on research facilities, and now the self-interest of different groups came forward. The summer study that year was very specific and argumentative (14). That study was followed by the consideration of the initial proposals. To consider these proposals a Program Advisory Committee (PAC) was formed, and the first summer meeting of the PAC was held in Aspen, Colorado. This established the practice of a week-long annual review of the research program of the Laboratory. This tradition continues today and provides an in-depth examination of approved experiments as well as consideration of new proposals.

After completing the initial design work at Oak Brook, the Laboratory staff moved in the fall of 1968 to a former housing development located in the Village of Weston, which was within the boundaries of the new site. Using houses that were already there, rearranged somewhat for more efficient use, the work of the Laboratory progressed (15). The ion source, Pre-Accelerator, and first tank of the Linac were developed in a building hastily erected at the edge of the Village. Prototype units of the magnets and radiofrequency systems were soon under way, while development of modern power supply modules was initiated in other buildings.

Ground was broken for the first accelerator enclosure on December 1, 1968. Work progressed rapidly on the Linac and Booster structures throughout 1969 and 1970 (16). Since funds did not exist for building the Central Laboratory where the Control Room was to be located, a structure called the Cross Gallery containing the main control room was built, tying together the galleries in the "Footprint" area. As soon as the last magnet was put into place within the main ring in April 1971, attempts were made to accelerate a beam. For months on end the beam would make a few turns only to be lost without appreciable acceleration. Obstructions were found within the vacuum chamber and extensive searches were undertaken. Then the magnets began to fail. This was apparently due to the fact that when the tunnel was completed in the winter of 1970–1971 frozen earth was used to cover the concrete tunnel sections. This left the walls and tunnel environment very cold during spring and summer of 1971, and the humid Illinois air condensed on everything, including the magnets. Water got into small cracks in the magnet insulation, causing many magnets to fail when energized. A year of hectic work went on while the tunnel was dried out, and reworked, improved magnets installed. This delay postponed the start of the accelerator and confounded the announced plans of the Director to have a 200-GeV beam in July 1971, a year ahead of the date in the Design Report.

Most of the staff were still located in the Village when the first beam of 200-GeV protons was achieved on March 1, 1972. A year later some hearty souls began to move into the unfinished Central Laboratory. Now more than 1300 individuals work at Fermilab occupying space in the Central Laboratory, Village buildings, and structures in the experimental areas. In addition, some 200 contract personnel such as guards, maintenance workers, riggers, and electricians augment the Laboratory staff. Several hundred visiting physicists, students, and technicians who use the facilities are normally in attendance. The mix of people leads to an exciting and stimulating environment within the architecturally striking Central Laboratory, which supports the intellectual life of the Laboratory (Figure 3).

The Director and his staff were determined to build the Laboratory within the authorized $250 million. In fact, their goal was to exceed the objectives set forth in the Design Report and restore potentially lost capabilities. In almost all aspects the performance of the Laboratory has exceeded their expectations even in the face of reduced project funding. The entire facility, including the 400-GeV accelerator, 4 experimental areas, numerous research facilities, and the Central Laboratory was provided for $243.5 million. This was accomplished through the intense work and sacrifices of every person at the Laboratory during the difficult early years.

3   DESCRIPTION OF THE ACCELERATORS

The Main Ring of the 400-GeV accelerator presented a formidable challenge. The construction of an accelerator this large had never before been attempted. Over 1000 magnets would have to be produced and installed within the more than

6 km of underground tunnel. While a large part of the Fermilab staff concentrated on this accelerator, others were working on the lower-energy accelerators that formed the injector to the Main-Ring accelerator.

3.1 *Injector*

The protons that eventually are accelerated to 400 GeV begin as hydrogen gas in an ion source. The gas emerges from a bottle into the intense but localized electric

Figure 3 A view of the Central Laboratory taken across the reflecting lake. Sixteen floors are available for staff and visitors use.

field within the ion source. Here the electrons are stripped from the hydrogen molecules and the protons stream from the source. The entire assembly, including gas bottle, ion source, power supplies, and instrumentation is located in the dome of the Pre-Accelerator (Figure 4).

The Pre-Accelerator standing at the entrance to the Linac is in fact an accelerator in its own right. The large dome containing the pulsed ion source is maintained at a constant high voltage by a Cockcroft-Walton power supply providing a dome

Figure 4 The Pre-Accelerator: the ion source is located in the high-voltage dome. The protons stream into the Linac through the column to the left.

voltage of 750,000 V. The pulse of positively charged protons that stream out of the source is pushed away from the dome by the electric field created by the +750,000 V on the dome. The protons are accelerated in the resulting electric field and gain an energy of 0.75 MeV after going about 30 cm down the high-voltage column. Traveling in a vacuum through the carefully shaped column, the beam consists of a current of protons measuring 200 mA. This is a large current compared to the number of protons eventually accelerated, but there are losses during all stages of acceleration. In order to match the beam with the time varying electric field of the Linac, the stream of protons is bunched into bursts of beam with the frequency of the Linac.

The Linac (18) consists of a series of accelerating cavities extending a distance of 175 m. Over that distance a burst of protons is accelerated from 0.75 MeV to 200 MeV. The Linac consists of 9 copper tanks each 1 m in diameter and about 16 m long tuned to resonate at 201 MHz. Arrayed within the tanks are drift tubes containing focusing quadrupoles (Figure 5). The entire tank is evacuated, and the beam goes down the centerline through the 2-cm aperture in the quadrupoles.

The Linac, similar in design to several others in the United States, accelerates protons by the application of time-varying electric fields. When the timing is adjusted correctly, the burst of protons from the Pre-Accelerator is attracted toward the

Figure 5   An interior view of the drift tubes within an accelerating cavity of the Linac.

first of 295 drift tubes. By the time the protons enter the drift tube, the polarity of the electric field has reversed. Within the drift tube the protons are shielded from the electric field and are not affected by the reversed and now decelerating electric field. As the protons emerge from the drift tube the electric field again has the proper polarity and the burst of protons is further pushed towards the next drift tube. This process is repeated all along the linear accelerator so that a stream of protons measuring up to 160 mA is available at 200 MeV for injection into the Booster accelerator.

Acceleration within the Linac is made possible by the intense electric fields generated as standing waves within the cavities. The resonant cavities are excited by nine very large power supplies, each containing a superpower triode. A peak radio-frequency power of 37 MW is provided during each pulse. The 9 cavities are pulsed 15 times a second and use an average rf power of 110 kW. During each pulse the electric field builds up within the Linac cavity and provides a stream of protons extending for about 10 μsec at the high-energy end of the accelerator.

The basic pulsing frequency of 15 Hz for the Linac is determined by the needs of the Booster accelerator. As the third accelerator in the process, it accelerates a beam of protons from 0.2 to 8.0 GeV every 1/15 of a second. The Booster is an alternating gradient synchrotron (19), in which each magnet has focusing as well as bending fields. A major difference of this compared with the AGS at Brookhaven is the rapid cycling capability of the Booster. A resonant network involving the magnets and power supplies allows the accelerator to operate steadily at 15 Hz.

The Booster consists of 96 10-foot long magnets arranged in a circle with a radius of 75 m. The magnets bend the 0.2-GeV protons that emerge from the Linac into a circular orbit. The protons coast around within the evacuated gap in the magnets until acceleration begins with the energizing of electric fields in 18 distributed rf cavities. As the energy of the protons increases, the magnetic field changes from the value of 0.5 kG at injection to 6.7 kG at 8.0 GeV. During this time the circle of protons passes through the accelerating cavities nearly 16,000 times. On the average the protons gain 0.5 MeV of energy per turn and achieve full energy after 33 μsec. During acceleration the velocity of the protons changes from $\beta = 0.57$ to $\beta = 0.99$. Consequently, the beam moves more rapidly around the ring and the frequency of the rf must change from 30 to 53 MHz. After the beam is transferred to the Main Ring accelerator the process is repeated.

Near the minimum of the magnetic-field cycle it takes 2.8 μsec for the injected 0.2-GeV protons to circle the Booster. Since the Linac pulse is about 10 μsec long, several turns of beam can be stored in the Booster. It is hoped that eventually 4 turns of protons will fit within the Booster so that 4×10^{12} protons per Booster cycle can be accelerated to 8 GeV. This is one way of achieving the intensity goal of the Main Ring accelerator.

The Booster is an interesting accelerator in many respects; among the foremost of these are its rapid cycling characteristics. This put many constraints upon its design, starting with the magnets. The laminated iron magnets are optimized for fast pulsing. In order to avoid eddy currents no metallic vacuum chamber could be used in the magnet gap; the vacuum necessary for the beam is created by encapsulating the entire magnet within an outer, airtight skin of stainless steel. The magnet leads emerge through ceramic insulators into air and are connected to capacitors and chokes located in a space provided in the supporting truss. In order to achieve the rapid cycle time, each pair of magnets with an inductance of 20 mH forms a resonant circuit at 15 Hz frequency together with the capacitors and chokes. The circuit is driven by four power supplies located in the Booster gallery above the accelerator enclosure. In this manner the magnetic field oscillates between 488 G where injection from the Linac takes place to 6700 G at 8 GeV.

The Pre-Accelerator, Linac, and Booster can all be considered as the Injector to the Main Ring accelerator. Table 1 reviews the properties of the Injector. The data in the table use typical parameter values for the already achieved Booster flux of 2.5×10^{12} protons/Booster pulse (20). This corresponds to an intensity of 4×10^{13} protons/sec, making the Booster the most intense high-energy proton synchrotron.

3.2 Main Ring Accelerator

The Main Ring Accelerator is located in a tunnel 6 m below ground. The components that produce the final stage of acceleration, bringing the proton beam to an energy of 400 GeV (21), are located in a circular enclosure with a radius of 1 km. The bending magnets that produce the circular orbit are the dominant elements occupying over 70% of the available space (Figure 6). Other magnets such as quadrupoles and magnetic correction elements hold the protons together in the vacuum pipe. The magnets of the Main Ring are divided into 6 circular sectors

Table 1 Injector properties

Pre-Accelerator		
Peak energy	0.750 MeV	
Beam current	200 mA	
Emittance	50π mm-mrad	
Linac		
Peak energy	200 MeV	
Beam current	100 mA	
Pulse length	10 μsec	
Emittance	10π mm-mrad	
Booster		
Peak energy	8 GeV	
Repetition rate	15 Hz	
Beam current	250 mA	
Particle flux	2.5×10^{12} protons/pulse	
Emittance	1π mm-mrad	

divided by 50-m long field-free straight sections. One straight section contains the rf cavities that accelerate the protons, while another is used for injection and extraction of the protons.

The heart of a synchrotron is the magnet (22). At Fermilab there are more than 1000 magnets in the Main Ring. There are 774 20-ft long bending magnets or dipoles. In order to produce an economical design the magnet aperture is sized to match the beam properties. Half of the dipoles have an aperture 5-in wide × 1.5-in high and the other half are 4 × 2 in. The dipoles are all connected in series and the magnetic fields are set at 0.36 kG during injection at 8 GeV. When the beam has been accelerated to 400 GeV, the field is 18 kG ; a magnetic field produced by an excitation current of about 5000 A.

There are 192 regular quadrupole magnets in the Main Ring, each 7 ft long. In addition, there are 48 4-ft long quadrupoles combined with 7-ft quadrupoles to adjust properties of the beam at the straight sections. Each type of focusing and defocusing quadrupoles is connected in series to separate circuits of power supplies. Separate power supplies for the bending magnets are also located in the 24 service buildings around the Main Ring and complete the third circuit. The dipole and quadrupole magnets are arranged around the ring in a definite lattice consisting of cells of magnets and straight sections. The usual cell consists of 8 bending magnets

Figure 6   The main-ring accelerator tunnel contains magnets that bend the proton beam into a circle with a circumference of 6.2 km.

(B), and 2 quadrupoles (Q). These 10 magnets are arranged in the order QBBBBQBBBB. This pattern repeats itself 84 times except when interrupted by a long or short straight section. In addition, there are many corrective magnets and other pulsed devices located in the Main Ring.

The straight sections are used for many purposes. There are six 50-m long straight sections around the ring. One, labeled A-0, is used for the injection of protons from the Booster and ejection of the 400-GeV beam from the Main Ring. At A-0 the 8-GeV protons from the Booster are steered into position and injected vertically into the magnet aperture. Later in the cycle other elements are used to extract the circulating proton beam horizontally from the accelerator at the same straight section.

Another long straight section, C-0, houses a modest experimental area (23). Here the first experiments at the Laboratory were done using the proton beam as it circulated within the Main Ring. Initially using rotating filament targets and later jets of hydrogen gas, Russian and American experimenters were able to study the systematic behavior of *p-p* and *p-d* elastic scattering as a funtion of proton energy. A room for a recoil spectrometer that will be used for a variety of new experiments has recently been added to this area.

The rf-accelerating cavities are located at F-0 (Figure 7) where 18 cavities are

Figure 7   The radio frequency accelerating cavities are located within the main ring at the F-0 straight section.

lined up one after another. These cavities provide a burst of energy to the protons on every traversal (24). About 2.8 MeV of energy can be imparted to the protons each time they go through the cavities at F-0 without voltage breakdown. At the fully relativistic velocity of the protons, the beam goes around the circumference of the Main Ring nearly 50,000 times per second. This results in a rate of energy increase of about 125 GeV/sec. This rate of rise determines the cycle rate of the Main Accelerator. Nearly 3 sec are needed to accelerate the beam from 8 to 400 GeV; then the beam is used for 1 sec and the magnets return to their injection field ready to receive protons from the Booster. The cycle repeats every 10 sec, providing full-energy protons for 10% of the time.

The accelerating cavities operate at the same frequency as the Booster cavities at the end of the Booster cycle. Since the protons at 8 GeV are traveling at nearly the speed of light, the change of frequency during acceleration is quite small. The cavities operate near 53 MHz. This means that the protons are forced into bunches separated by 19 nsec. Considering the 6.3-km long ring, one finds that 1100 bunches are distributed around its circumference, each spaced from its neighboring bunch by 5.6 m.

There are a number of distinguishing features of the Main Ring accelerator. One is that the separated function design allows for high-peak magnetic fields in the bending magnets. This has made it feasible for the magnets to be excited to fields equivalent to 500 GeV. Another innovation is the use of solid-state rectifiers within the power supplies rather than the customary motor-flywheel-generator sets for providing the slowly pulsed magnet current from the usual 60-Hz alternating current electrical power (25). The power supply system is controlled by firing circuits whose programmed activation determines the shape of the voltage cycle applied to the magnets. This technology lends itself to computer control such that great flexibility exists in the pulsing of the Main Ring accelerator. At Fermilab the power supplies in every service building are connected directly to the 13.8-kV electrical power lines distributed around the Main Ring. These lines are in turn connected to a 345-kV/13.8 kV transformer in the substation appropriately built for pulsed service and located in the master substation.

Major innovations exist with respect to the magnet cooling system. Early in the project it was decided that a water-treatment plant was unnecessary. This meant that the water-cooling systems for magnets that require low conductivity water (LCW) had to be self-contained. Through the use of heat exchangers, the LCW

Table 2   Main-Ring properties

Accelerator radius	1 km
Energy	400 GeV
Peak proton flux	2.0×10^{13} protons/pulse
Rate of energy gain	125 GeV/sec
Flattop duration	1 sec
Cycle period at 400 GeV	10 sec
Beam emittance	0.1π mm-mrad

that circulates through the magnet is cooled by ordinary site water drawn from a canal formed around the main ring. The water is pumped into the service building and through a spray back into the canal. In this manner the LCW is cooled and the magnets can operate efficiently.

The principal characteristics of the Main Ring accelerator are displayed in Table 2 (26).

4   ACCELERATOR OPERATION AND BEAM EXTRACTION

4.1   *Accelerator Operation*

Proper operation of the accelerators requires a high level of coordination. Operators in the Main Control Room monitor and control all the accelerators, support systems, beam transfer from the Booster, and the extraction processes. Assisted by beam sensors, equipment controllers, and computers, the operators set up the pulsing of the accelerators and continually trim and tune all devices for high efficiency. The operation and control of the accelerator systems is possible by implementing recent computer techniques (27). Thousands of individual devices are controlled by minicomputers (MAC 16's) located in remote areas. They monitor and set the digital parameters required for each device under supervision of a large computer. This computer (Xerox 530) is located in the Cross Gallery, where it spans the communications between the minicomputers and the consoles in the Control Room. The accelerator operators sit at the consoles and make use of interactive displays. They are able to connect their dial to any controllable device and see the effect of changes upon the performance of the accelerator. The computer also produces colored displays that help in the analysis of accelerator and system performance (28).

A closely related activity includes the use of closed-circuit television throughout the Laboratory. Commercially available sets originally carried passive information such as pictures of dials, oscilloscope traces, etc to the Control Room. The service was later extended to the experimental areas, and there are now 7 systems each with the standard 12 VHF channels plus mid-band channels. Recent developments made it possible to display alphanumeric information on the TV screens by incorporating teletype keyboards in the system. Among the most frequently seen displays are the oscilloscope traces representing the energy and intensity of each accelerator pulse. This is often overlaid with schedule information including announcements of current events.

Two traces representative of a 300-GeV pulse in the Main Ring are shown in Figure 8. The upper trace represents the energy increase of the beam and the lower trace the intensity. In the upper trace the energy is held steady at 8 GeV for nearly 1 sec while 13 pulses of protons from the Booster are injected into the Main Ring. Since the circumference of the Booster is about $\frac{1}{13}$ that of the Main Ring, 13 pulses of the Booster provide a segment-to-segment line of protons arrayed around the circumference of the larger machine. The lower trace shows the increase of intensity as the 13 Booster pulses are injected into the Main Ring. Since each Booster beam contains 2×10^{12} protons, the 13 pulses provide up to 2.5×10^{13} protons for acceleration in the Main Ring. Unfortunately, not all of the beam is retained and at present only 2×10^{13} protons survive to the full energy.

After injection is complete the phase of the rf field is adjusted so that acceleration begins, and the magnetic field of the main-ring magnets increases to keep pace with the increasing beam energy. The top trace shows how the energy increases as the protons go through the accelerating cavities. This trace also is representative of the increasing magnetic field as the circulating protons accelerate to 300 GeV. Before and after the beam reaches full energy, the protons are used in different modes as illustrated on the intensity trace. Some are used on targets at C-0 in the Internal Target Area. Although the amount of beam used is in reality very small, the loss is

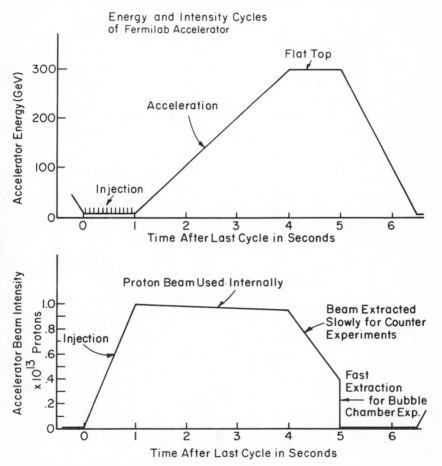

Figure 8 Curves representing the energy and intensity of the accelerated protons during a nominal 300-GeV cycle.

exaggerated in the figure. After the beam reaches full energy, it is extracted from the accelerator in several ways. Shown here is slow extraction where about half of the beam is peeled out of the Main Ring during 1 sec followed by a fast spill where the remaining protons are very rapidly removed.

After each cycle the magnetic field decreases, putting energy back into the electrical system through the reversible power supplies, and the field returns to its injection value. After a pause the accelerator is ready to repeat the process. The frequency of pulsing of the Main Accelerator is largely determined by the electrical power consumed. At 400 GeV, 50 MW of average power is needed for a repetition frequency of 6 pulses per minute. Adding the 25 MW of electrical power needed by the site and experimental areas, more than 40,000 MW-hr of energy are used per month in operating the facilities at Fermilab. As is described later, this large and expensive need for electrical power has given added emphasis to applying superconductivity to the construction of accelerator and beam line magnets.

4.2   Beam Extraction

Making it possible to use the accelerated protons conveniently is one of the triumphs of the art of building accelerators. As mentioned, a very small percentage of protons is targeted within the accelerator itself. Because of limitations imposed by radioactivity, it is not feasible to target more than about 10^{10} of the protons internally. The main use of the protons is in the external experimental areas, where they are studied directly or are targeted to produce other particles, which in turn are formed into secondary or even tertiary particle beams. This requires that the circulating proton beam be removed from the accelerator (29).

The most efficient manner of extracting the beam is by removing it during one turn; this provides a beam spill 21 μsec in duration. For this purpose pulsed magnets are energized that drive the beam from its normal position in the vacuum chamber. The displaced beam eventually goes through a series of stronger magnets. In this manner the beam is deflected out of the main-ring guide field so that at the long A-0 straight section it will travel tangent to the ring and be free for transport to the external experimental areas.

For most purposes the circulating proton beam should be extracted as slowly as possible. In this manner a steady stream of protons is provided to electronic detectors so that experimenters can take data more efficiently. Slow extraction of the beam is somewhat complicated because it requires peeling the beam out of the accelerator in a uniform manner without high losses. This mode of extraction is achieved by changing the focusing within the accelerator such that the circulating beam is held at constant energy but expands in size. Extraction begins by gently changing the tune of the accelerator. Normally the Main Ring is operating at a betatron frequency of 19.45. This means that there are 19.45 oscillations of a typical proton about the equilibrium orbit during one turn about the Main Ring. To maintain a stable beam, destructive resonances that occur when the betatron frequency is equal to 19.50, 19.33, and other values must be avoided. Just the opposite is the case in order to extract the proton beam. Here the magnetic field in the quadrupoles in the accelerator are changed in strength such that the betatron

Table 3 Modes of extracting the proton beam

Mode	Duration	Method
Fast	21 μsec	single-turn
Coherent fast	200 μsec	resonant
Multiple pings	1 msec/ping	resonant
Slow	1 sec	resonant

value is brought near to a resonance point. This causes the oscillation of the protons about the equilibrium orbit to grow steadily larger within the accelerator vacuum chamber. As a result, some of the protons cross the wires of the first extraction device, an electrostatic deflector. Most of the protons make it across the thin wires into the region of electric field that deflects them slightly. They are bent into the next device downstream, a septum magnet, and eventually out of the accelerator at the long straight section A-0. Less than 2% of the protons strike the wires and are lost in the process. This means that an extraction efficiency of approximately 98% has been achieved. The duration of the slow spill is determined by the length of time that the accelerator is held at fixed energy with beam emerging into the extraction channel. Typically, a 1-sec spill repeated every 10 sec characterizes the present-day 400-GeV operation of the accelerator. This provides a beam for 10% of the time and is called a 10% duty factor so far as counter experiments are concerned.

There are a number of other possible beam spills. One that is quite important for use in the experimental program periodically enhances the slow spill and produces several spikes of beam during the 1-sec long slow extraction. These spikes are used for bubble-chamber experiments where the chamber is rapidly cycling and able to take a new burst of particles every quarter of a second or so. Table 3 shows some of the most frequently used modes of extracted beam and describes their basic characteristics.

4.3   Switchyard

Once the beam is free of the Main Ring accelerator and directed toward the experimental areas it comes into a vast complex of underground tunnels called the Switchyard (30). In this area the now-external proton beam is manipulated and shaped in preparation for being split among the three external experimental areas. Proceeding from the extraction point the first devices that the beam encounters are those that shave off some of the beam and deflect it to the right for the Proton Area. The beam deflectors for the Meson Area are found further downstream. The remaining beam then goes straight on to the Neutrino Area. The Switchyard deflecting elements are similar to those used within the Main Accelerator for beam extraction. They consist of electrostatic septa (Figure 9) followed by magnetic septa called Lambertson magnets after their inventor.

In order to split off some fraction of the beam efficiently, the beam of protons is made into an elliptical configuration. Some of the beam goes within the positive electric-field region of the electrostatic splitter, while the remaining portion of beam is in the negative-field region. In this manner the electric fields affect portions of the beam differently and impart small deflections. One deflected segment of the beam next encounters a magnetic field and an additional deflection is provided, while the other is restored to the center line as if unaffected. By these processes the net bend of

Figure 9 Interior view of the electrostatic wire septum used in the Switchyard to split the external proton beam.

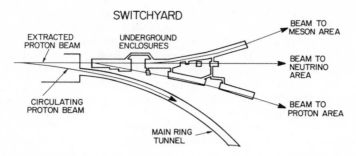

Figure 10 Expanded view of the Switchyard area where the extracted proton beam is split to the three external experimental areas.

one beam becomes large enough to miss the return yoke of a downstream conventional magnet. When the deflected beam enters the gap of the separate magnet the beam is bent into the appropriate experimental area and the undeflected beam goes on to be targeted elsewhere.

To give some feeling for the vastness of the Switchyard Area, there are about 2 km of tunnel in which bending magnets are located (Figure 10). To bend the beam into the Proton Area 40 magnets are used, and nearly 60 are used to create the 10-degree left bend to the Meson Area. The Neutrino Area, since it lies tangential to the A-0 sector, requires no lateral deflection, although there is a 2.5-m change of elevation. Four magnets are needed to bring the beam up and four to level it off and aim it into the targeting region of the Neutrino Area. The Switchyard and the extraction processes are both controlled from the Main Control Room, where operators must make frequent changes in beam properties in order to achieve the desired splitting ratios of protons among the three areas as well as to control beam spot sizes.

Before talking about the formation of secondary beams, it is important to make some general comments about the use of the protons and the targeting of them. Within an external experimental area protons are used to produce other particles such as π mesons, neutrinos, muons, etc, which in turn are used as initiators of particle interactions. At the heart of each area is a target hall when the protons strike a target, producing many other particles. The target area and downstream beam-forming elements must be heavily surrounded with steel, concrete, and finally earth to filter out unwanted hadrons and muons (31).

The protons are focused as a thin pencil of particles into the targeting region. Typical beam spot sizes are 1 to 2 mm with as many as 10^{13} protons. Such a beam contains an enormous amount of power. For example, a 400-GeV beam of 10^{13} protons/sec contains over 640 kW of beam energy and must be handled and targeted safely. The target is typically a rod of aluminium or tungsten 10 to 30 cm long and about 2 mm in diameter. Usually several targets are available for use and are located on a movable frame in the target box so that the best target can be selected for the purpose (Figure 11). Only a fraction of the beam energy ends up in the target and the rest of the protons go into a massive aluminum beam dump that is cooled by a sealed system of circulating water. At Fermilab the target and beam dump are usually contained within a target box. This is a large steel and concrete bunker within which railroad tracks have been installed. Rail cars containing the target, dump, and special trim magnets needed to direct the beam properly into the target are located on rail cars. When the train of cars is in the target box the protons can be safely stopped because the massive target box is also surrounded by a large mound of earth. In this manner people can work in the nearby service building or downstream experimental area without encountering unnecessary radiation. If a problem arises, the target train can be moved from the box into a special shielded hall where the elements on the train can be serviced or repaired. In extreme cases the beam elements are removed to a Target Service Building, where special remote manipulators are available so that radiation exposure to individuals is greatly reduced (32).

5   THE EXTERNAL EXPERIMENTAL AREAS

There are three research areas external to the accelerator and an internal area at C-0. As seen in Figures 12 and 13, the longest external one extends for nearly 3 km from the proton-beam extraction point in the transfer hall of the Main Ring. There are common features to all the external areas such as proton beam transport

Figure 11   An assemblage of targets located on a train. A beam detector is seen at the right.

Figure 12 An aerial photograph of the experimental areas looking back towards the main accelerator.

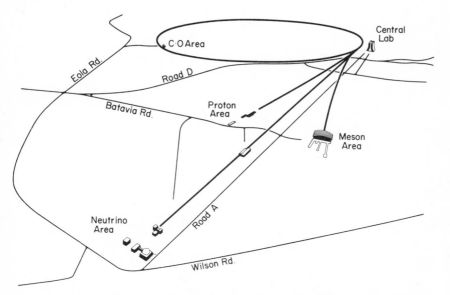

Figure 13 A matching perspective drawing to Figure 12 identifying the experimental areas.

elements, targets, beam dumps, collimators, and secondary beam formation equipment, but each area has its own special characteristics and purpose. It was recognized early in the project that a range of experimental resources would be needed to match the broad interests of the prospective users (33).

In the Meson Area, which is designed more conventionally, a series of secondary beams are produced from one target struck by the proton beam. Here six secondary beams span out into a common experimental hall where researchers are able to install apparatus and do experiments. The Neutrino Area contains specialized beams. Here the secondary particles are handled so that the mesons decay into neutrinos and muons. Some of the muons are bent aside and used, while all other particles but the neutrinos are absorbed before reaching the detectors at the end of the area. The third experimental area, the Proton Area, has been initially used for the study of direct proton-induced reactions. As this field of research reaches completion the Proton Area is slowly being modified to provide secondary beams with special characteristics. Figure 14 shows the beams and general facilities available in the experimental areas (34).

Before describing the individual experimental areas, some general comments are in order concerning the production and formation of secondary beams. When high-energy protons strike a target such as those shown in Figure 11, an enormous number of secondary particles of lower energy emerge from the target. As is characteristic of most hadron-production processes, the secondary particles are concentrated in the forward direction constrained by a transverse momentum of about 300 MeV/c. This can be represented by $p_\perp = p\theta \approx 300$ MeV/c. For example, the major flux of 300-GeV/c particles are found between 0 and 0.001 rad, while lower-energy particles extend to larger angles. These secondary particles can be

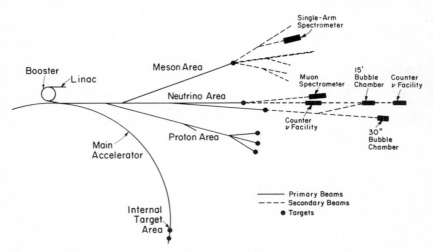

Figure 14 An expanded drawing showing the target locations, secondary beams, and detector facilities at Fermilab.

gathered together, provided suitable quadrupoles and dipole magnets are positioned near the target. Unfortunately, because of the need for a beam dump for unused protons, collimators, and adequate shielding, it is not possible for the active secondary beam elements to get closer to the target than 4 m in the usual arrangements. When a number of beams must collect particles from the same target, this requires that the individual magnets be displaced even farther away. Typically, solid angles for acceptance of produced particles of about 2 μsr result from practical designs. After sign selection and momentum determination using dipole magnets, the beam of secondary particles is transported into an experimental zone where detector facilities or individual experimental apparatus are located.

Figure 15 shows the spectrum of secondary particles in a beam produced from 200-GeV protons (35). As can be seen, the flux of particles is quite large, up to half the momentum of the incident proton. Beyond that, the number of higher-energy particles falls off rapidly. This illustrates the importance of having a high intensity of incident protons in order to boost the flux of secondary particles. In addition, the flux of secondaries in the higher momentum range can be dramatically increased when higher-energy protons are targeted. In almost all beams there is a need for the enrichment of one type of particle compared to another. This process of enrichment is one of the major distinguishing features among secondary beams in the three experimental areas. In addition, beam particles are individually identified wherever possible so that the incident particle in a subsequent interaction can be determined.

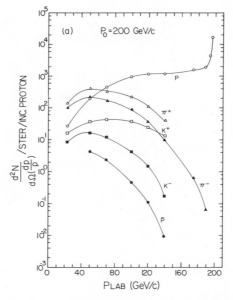

Figure 15 Flux of secondary particle produced by 200-GeV protons in the M1 beam of the Meson Area at an angle of 3.5 mr.

5.1   Meson Area

Originating from one targeting station in the Meson Area are six secondary beams. After traveling 425 m these secondary beams are shaped and spread apart so that they are all directed into a common experimental hall. This area, pictured in Figure 16, shows how five of the secondary beams come into the hall and the sixth, a below-ground beam, is available in a small experimental pit downstream from the building. The Meson Area provides this wide variety of conventional secondary beams for research using general-purpose experimental equipment (36). By the use of branches in the charged beams and multiple experimental locations along beam lines, it is possible to have nearly eight experiments set up at one time. Several of the experimental groups are installing or testing their apparatus while four or five others are taking data. Table 4 shows the major characteristics of the beams in the Meson Area.

M1 is a secondary beam with a maximum momentum of 280 GeV/c. This beam is equipped with a beam-particle identifier designed at BNL that can separate π's from K's up to the full momentum of the beam. Equipped with an experimental enclosure that extends downstream from the building, this beam with its two branches provides space for four experimental setups. These apparatus locations are reserved for routine use by experimenters who come for six months or so at a time. M2 was designed as a diffracted proton beam. As such it accepts elastically scattered protons produced at a very small angle from the primary target and transports them into the experimental hall. It has been possible to use up to 10^9 protons/pulse for experiments in the neutral hyperon facility located at the end of the M2 beam. The downstream beam enclosure covers experiments that use the tertiarily produced Λ°

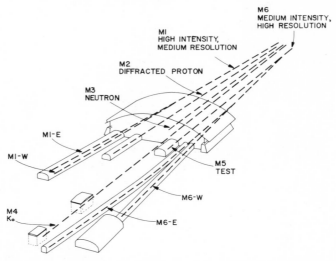

Figure 16   Perspective drawing of the beams and Detector Building in the Meson Area.

Table 4 Meson area beams

Beam	Production angle (mrad)	Maximum momentum (GeV/c)	Solid angle (μsr)	Momentum acceptance ($\Delta p/p$, in %)	Approximate flux per 10^{13} protons at 400 GeV
M1—medium resolution	3.5	280	2.0	$\pm 0.1 \rightarrow 1.5$	$10^7 \pi^-$ at 200 GeV/c
M2—diffracted protons	1.0	400	0.2	$\pm 0.1 \rightarrow 1.4$	$10^9 p$ at 400 GeV/c
M3—neutrons	1.0	—	$\sim 10^{-4}$		$10^7 n$'s/cm^2
M4—K°'s	7.0	—	variable		$10^6 K^\circ$s/cm^2
M5—test beam	20.0	50	5.0	$\pm 0.05 \rightarrow 0.5$	$10^5 \pi^-$ at 50 GeV/c
M6—high resolution	3.0	200	1.3	$\pm 0.01 \rightarrow 1.0$	$10^7 \pi^-$ at 150 GeV/c

particles. Ahead of the hyperon facility an experiment can be set that uses the protons parasitically. Frequently the beam has been changed to negative particles so that a moderate flux of π^- mesons up to 300 GeV/c can be provided. The π^-'s have been used for charge exchange and inclusive photon-production experiments.

The M3 beam is a neutral beam rich in neutrons. It is produced at near 1 mrad and contains high-energy neutrons produced from the target. The neutron spectrum peaks at ~ 300 GeV and usually contains about 10^7 neutrons/pulse. This beam has been used for neutron dissociation, neutron scattering, and particle-search experiments. M4, located below ground, is another neutral beam at 7 mrad. At this angle the ratio of neutral kaons to neutrons is larger than at 1 mrad; there are nearly 10^6 K°'s per pulse in that beam. The M5 beam, although produced at the largest angle (~ 20 mrad) from the target, is bent back into the experimental hall, and used primarily as a test beam. It is limited in momentum to 50 GeV/c and has a low flux of particles. As a test beam it is eagerly used by many experimental groups in order to get their equipment in operating order. M6, a high-resolution secondary beam with an upper limit of 200 GeV/c, is primarily intended to be used by the semi-fixed detector facilities located in each branch of the beam. One is the single-arm spectrometer that analyzes the high-energy forward-going particle from nuclear interactions. Space is available around the hydrogen target for recoil particle detectors. The other branch of the M6 contains a multiparticle spectrometer built around a superconducting analysis magnet. This detector, when equipped with Cerenkov counters and photon detectors for electronically detecting the scattered particles, will become a powerful research device in the Meson Area.

5.2   Neutrino Area

The Neutrino Area is the longest experimental area on site (Figure 17). It is composed of a decay region measuring 400 m followed by 800 m of earth shielding (37). The π and K mesons produced from the target are formed into a nearly parallel beam by special magnetic elements on the target train. One form of focusing results from passing the mesons through a pulsed, shaped current funnel called a "horn." Another incorporates conventional quadrupoles and dipoles in the target region. In both cases a fraction of the mesons in the transmitted beam decay into neutrinos and muons within the long evacuated decay pipe. A small fraction of the muons are deflected out of the decay pipe and formed into a beam transmitted to the Muon Laboratory. The unused protons, mesons, and unwanted muons are all absorbed in a beam dump and in the earth mounded up at the end of the decay pipe. Emerging at the end of the earth shield are the neutrinos to be used in three experimental facilities at the end of the Neutrino Area. The centermost facility is the 15-ft cryogenic bubble chamber (38). This vessel contains nearly 30,000 liters of liquid that when pulsed can be made sensitive. Interactions on nuclei in the working liquid can be photographed and studied at a later time. The chamber is usually filled with liquid hydrogen, but fillings of liquid neon and deuterium are also available. Surrounding a portion of the downstream area of the chamber is the External Muon Identifier built by physicists from the University of Hawaii and Lawrence Berkeley Laboratory. This device is used to identify muons produced in the most common neutrino interactions. Upstream and downstream from the chamber are experimental halls intended for neutrino experiments using electronic detectors. In each hall large toroidal magnets are used in the identification and measurement of the momentum of outgoing muons from the interactions. Ahead of the muon magnets are target calorimeters where the neutrino interacts. The location and energy of the other produced particles are measured in the calorimeter. Both buildings contain large installations where several hundred tons of material are available for detecting the neutrino reactions. Even with this mass of material in the usual flux of 10^8 neutrinos/ m^2, only a few interactions per pulse are produced and a small fraction of these are used in the analysis of neutrino interactions at the highest energies.

Table 5 shows the properties of the beams in the Neutrino Area. Several methods of producing the neutrinos are indicated in the table under the different N0 designations. N1 is the muon beam with a maximum momentum of 300 GeV/c that makes use of a quadrupole-focusing system. The strongly interacting particles accompanying the muons in this beam are readily removed by filtering the particles through a 12-m long polyethylene filter near the end of the beam line. When this filter is removed

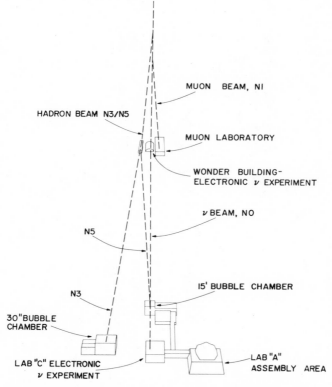

Figure 17   Perspective drawing of the beams and research buildings in the Neutrino Area.

Table 5 Neutrino area beams

Beam	Production angle (mrad)	Maximum momentum (GeV/c)	Solid angle (μsr)	Momentum acceptance ($\Delta p/p$)	Approximate flux per 10^{13} protons at 400 GeV
N0-D—narrow-band sign-selected ν's	0	300	10	$< \pm 25\%$	2×10^8 ν/m^2
N0-H—broad-band horn ν's	0	500	2800	5–500 GeV	10^{10} ν/m^2
N0-T—broad-band quadrupole ν's	0.5	500	4–16	20–300 GeV	7×10^8 ν/m^2
N1—muons	0	300	4–16	$\pm 2\%$	10^6 μ^+ at 150 GeV/c
N3—30-in hadron beam	0–1	500	0.3	$\pm 0.07\% \rightarrow \pm 1.2\%$	few/pulse
N5—15-ft hadron beam	0–1	500	0.3	$\pm 0.02\% \rightarrow 0.6\%$	few/pulse

the beam can be converted to a low-intensity pion beam used for hadron research in the Muon Laboratory. The Muon Laboratory building has space for two experimental setups. One is built around the magnet formerly used as part of the University of Chicago cyclotron. Here a group of physicists from Harvard University, Oxford University, the University of Chicago, and the University of Illinois have developed a large facility using detectors before and after the magnet to measure deep inelastic muon scattering. The muon beam continues farther downstream into the experimental hall where up to 30 m of space is available for general experimentation.

The N3 and N5 beams are low-intensity but usually enriched beams of secondary particles for use in the bubble chambers. N5 is directed to the 15-ft chamber and can be used in conjunction with the neutrino beam when the chamber is pulsed twice per accelerator cycle. The N3 beam that branches off from it is directed towards the 30-in bubble chamber. The 30-in chamber, built by the Midwestern Universities Research Association and used at Argonne National Laboratory for a number of years, was brought to Fermilab in 1971. At Fermilab it has taken over two million pictures and is one of the major devices used in exploring multi-hundred GeV physics at Fermilab. It is capable of taking up to six photographs per accelerator cycle. The chamber is equipped with a downstream counter system in order to help in the identification and measurement of particles produced from the nuclear collisions. Both the N3 and the N5 beams are equipped with particle detectors that locate and identify the type of particle incident on the bubble chambers. This additional information has extended the usefulness of the chambers into the high-energy region.

5.3   Proton Area

The Proton Area is really three small areas gathered together (39). Ahead of the Area the proton beam is split again into three segments so that the three separate experimental lines can be used as once (Figure 18). The protons emerge into experimental regions that are located 4 m below the ground surface. Surrounded by earth, a high-intensity proton beam can be used safely. Up to 10^{13} protons/pulse have been targeted in this region with a tolerable radiation level above ground. The experimental areas are called Proton West, Center, and East.

Proton West currently makes use of low-mass targets for transmission experiments. For these experiments a small number of protons interact and studies such as elastic scattering at very large angles can be undertaken. Proton Center is used for dilepton experiments in which proton-nucleus interactions produce electrons and muons on each side of the target; the particles are studied in coincidence. Proton East, although initially used for studies of single particles produced at large transverse momenta, now is the source of two secondary beams. P1 is a wide-band photon beam found by sweeping aside the charged particles. The neutrons in the resulting beam of neutral particles are attenuated through absorption and scattering in a 30-m long liquid deuterium filter. The remaining photons emerge into a downstream experimental hall. P2 is an electron-tagged photon beam where electrons are produced by γ rays resulting from the decay of π^0 mesons. The electrons are

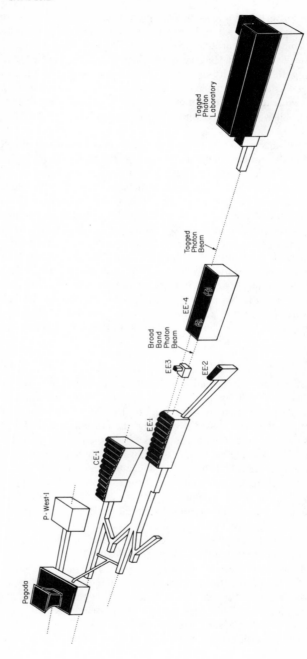

Figure 18 Perspective drawing of the proton beam branches and experimental regions in the Proton Area.

formed into a momentum-selected beam and transported to a radiator where photons are produced. The energy of the degraded electron is measured, thus tagging the energy of the produced photon. At the end of the beam an experimental area has been built for photon total cross section and photon inelastic scattering experiments.

In addition to these beams, two new beam facilities are planned for the Proton Area. Downstream from Photon West a high-energy, high-intensity secondary beam (P3) is under construction. Beyond the previously described transmission target the protons will be targeted once again to produce a very high-intensity (10^{10} π's/pulse) secondary beam. This beam has been designed to use superconducting magnets so that it will have an ultimate momentum capability of 1000 GeV/c. Finally, in Proton Center it is hoped to build a charged hyperon facility. This will make efficient use of the protons after the experiments in Proton Center. Table 6 shows the properties of the existing and planned beams for the Proton Area.

6   HIGH-ENERGY PHYSICS RESEARCH PROGRAM

Nearly 500 experiments have been proposed to use the Fermilab research facilities, and half of them have been approved. Experiments proposed to Fermilab are described in writing and are studied by the laboratory staff prior to their consideration by the Program Advisory Committee (PAC). This committee, appointed by the Director, reviews all of the proposed experiments and makes recommendations concerning their merit to the Director, who, makes the final decisions. Detailed information about the procedures to be followed in doing research at the Laboratory are contained in the booklet "Procedures for Experimenters" (40). In addition, information on the status of all proposed experiments and the overall extent of the research program is summarized annually in publications available from the Directors Office and used by the PAC in their general review of the experimental program (41).

6.1   *Extent of the Research Program*

Figure 19 shows the status of proposals over the past five years. One can see the progress with the research program including the completion of data-taking for a number of approved experiments. As shown, a large number of experiments have completed their taking of data, while about 30 experiments are in the process of taking data or tuning up their apparatus at a given time. The information in this table is extracted from the "Situation Report" that concerns all experiments approved and pending at the Laboratory. The Situation Report, published quarterly in NALREP, the monthly publication of Fermilab, contains information about all the approved experiments by categories that best describe their circumstance at publication time. This document provides the first step in the long-range planning process that goes on at the Laboratory. Periodically, more detailed plans for specific experiments for the coming months are issued by the Directors Office. These show how the research facilities are to be used and what experiments are to be located in the secondary beams of the experimental areas. These plans are discussed weekly and

Table 6 Proton Area beams

Beam	Produced angle (mrad)	Maximum momentum (GeV/c)	Solid angle (μsr)	Momentum acceptance ($\Delta p/p$, in %)	Approximate flux per 10^{13} protons at 400 GeV
P1—wide-band photons	0	—	40×10^3	—	10^5
P2—electrons/tagged photons	0–2	300	1.2	± 2.3	7×10^7 electrons at 100 GeV/c
P3—high energy high intensity (under construction)	0–7	1000	30	$\pm\frac{1}{2} \to 5$	$2.5 \times 10^{10} \, \pi^-$ at 400 GeV when 1000-GeV protons are available

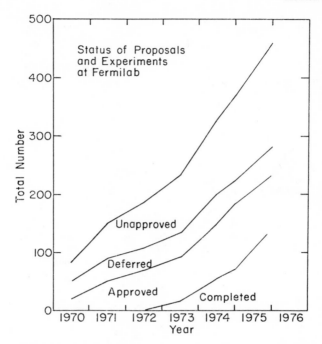

Figure 19   Status of proposals and experiments at Fermilab since 1970. The labels refer to the number of proposals or experiments in each category.

a very detailed day-by-day schedule is issued. This weekly schedule coordinates the activities of all the research areas and provides specific information to the accelerator and area operators so that optimum utilization can be realized.

A quantitative examination of the research program at a given time shows the extent of the program. Table 7 provides a snapshot of the overall status of the research program at the beginning of 1976. The first category contains information about electronic experiments and shows clearly the work completed and that which remains to be done. These experiments usually require dedicated use of the particle beams and represent the largest commitments in the research program. Bubble-chamber experiments are collected into the next category, where the overall number of pictures accomplished and pending has been indicated. The third category includes miscellaneous experiments that can usually be accommodated without major impact. The latter experiments do not add substantially to the extent of the program. This category includes experiments such as emulsion exposures, nuclear chemistry irradiations, and tests of experimental apparatus.

To give some idea about the ability of the Laboratory to accomplish its objectives, information about the performance of the accelerator and research facility is needed.

Table 7   Extent of the research program (as of January 1, 1976)

Electronic experiments

Complete and in progress	60 at 64,350 beam-hr
Set up now to 1 year	40 at 21,260 beam-hr
Unscheduled	16 at 9050 beam-hr
Pending proposals	28 at 19,925 beam-hr

Bubble-chamber experiments

Complete and in progress	33 at 2876 K pictures
Set up now to 1 year	8 at 1600 K pictures
Unscheduled	6 at 950 K pictures
Pending proposals	34 at 11,385 K pictures

Other experiments

Complete and in progress	67 experiments
Set up now to 1 year	1 experiment
Unscheduled	2 experiments
Pending proposals	2 experiments

Total number approved:	234
Total number pending:	64

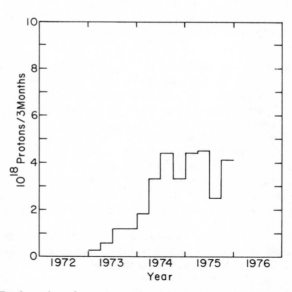

Figure 20   Total number of protons accelerated per 3-month period since experimental use of Fermilab facilities began in 1972.

Figure 20 shows the number of accelerated protons per quarter for the last few years and demonstrates the increased intensity capability of the accelerator systems as well as the improved reliability. Recent performance has been somewhat reduced over that achieved in the early part of 1975, largely because of complications arising from changing the basic accelerator operating energy from 300 to 400 GeV and the reduced repetition rate at the higher energy. Many of the energy and intensity accomplishments are noted in Figure 21, where progress since 1972 is recorded.

The accelerator and research areas are scheduled to operate continuously except during times of required maintenance. The goal is to achieve more than 100 hours of time for high-energy physics every week. This goal is exceeded many weeks, but there are times when accelerator or equipment failures limit the usefulness of the beam. Accurate records of utilization are maintained and reported monthly and yearly in NALREP. The amount of beam time used for each electronic experiment is kept along with the number of bubble-chamber pictures taken. During each month, the hours or pictures accumulated for all experiments are summed together. Since many experiments operate simultaneously, the number of beam hours for all experiments operating during the month greatly exceeds the hours of accelerator operation provided for high-energy physics research. In fact, the ratio of the number of beam hours to operating hours gives a measure of the multiple, simultaneous use of the beam facilities. This ratio is usually about 7, which means that on the average, during every hour of operation about 7 experiments collect data. This achievement results from the ability of the Switchyard to supply a beam to several targets simultaneously and the independent performance of each area. The overall capacity of the research facilities is described in Table 8.

An average over the recent experience shows that nearly 2700 beam hours are available per month for electronic experiments, and about 100,000 bubble-chamber

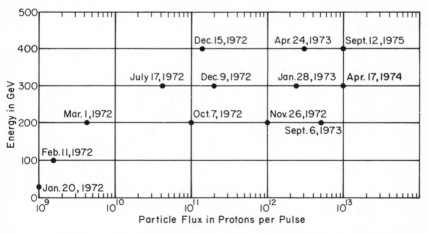

Figure 21 Milestone dates associated with the attainment of the energy and intensity goals at Fermilab.

Table 8 Beam capacity at Fermilab

Number of experimental areas		4
Primary target locations		7
Number of secondary beams		11
Meson Area	5	
Neutrino Area	3	
Proton Area	3	
Number of beams and branches		15
Minimum number of locations for experiments		24
Typical number of experiments taking data or testing at one time		12

pictures are taken per month. When the extent of the research program as displayed in Table 7 is considered in these terms, one can see that about 18 months of work remains to be done on the approved experiments. When the pending proposals are considered, an almost endless program of work lies ahead.

6.2   Research Accomplishments

It is beyond the scope of this paper to describe in detail the extensive research results from the early years of operation. These have been reported widely in the physics journals and conference proceedings. Nevertheless, it is tempting to mention briefly some of the highlights of the research at Fermilab. After the 200-GeV proton beam first became available in 1972, experiments were undertaken within the Main Ring, followed soon after by initiating research in the external experimental areas.

In the Neutrino Area, two experiments have measured the cross section for neutrino interactions. They have observed the linear rise of the cross section unsuppressed by the production of an intermediate vector boson (42). As well as studying the general characteristics of neutrino scattering, experiments followed up the observation of neutral-current phenomena where the absence of a muon in the interaction was examined. Most recently, interactions yielding two muons have been discovered (43), and now both counter and bubble-chamber experiments are providing a number of unusual events. While this paper was in preparation, several neutrino events were reported in the 15-ft chamber showing the presence of a K° in conjunction with a positron. Assuming that an invisible neutrino is also present, one is tempted to view this as the decay of some new particle, perhaps a "charmed" meson (44).

Muon experiments in the Neutrino Area have studied the scaling of the cross section for muon scattering from low to high energies. The data in the muon experiment depart from the predicted value based upon the theory of electron scattering. Perhaps that indicates a difference between the muon and electron or an absence of scaling. Meanwhile, other experiments are examining deep inelastic muon scattering. Bubble-chamber exposures to hadrons have shown the rising cross sections and high multiplicity of particles produced and have been an unbiased recorder of new phenomena.

In the Meson Area searches for real quarks and magnetic monopoles were an important part of entering into a new energy regime. Although none was found, the searching contributed to an early excitement and a fast start to research. Quantitative experiments in the Meson Area began by looking at the total cross section for K mesons and antiprotons on hydrogen. Rising cross sections of the K-p interactions were observed for the first time (45). A series of elastic scattering experiments have taken data and have studied πp elastic scattering from momentum transfers of 0.002 to 3 $(GeV/c)^2$. Now, inelastic scattering experiments have begun in the Meson Area, some looking at dissociation of beam or target particles and others for jets of particles indicative of subnuclear structure. Currently, many Meson Area experiments are searching for high-mass states related to ψ/J phenomena or for evidence of new particles.

The Proton Area was the last area to come into existence. There the initial studies have concentrated on the production of particles at high transverse momenta, thus examining the smallest granularity of the nucleons (46). Recently, the production of e^+e^- pairs has yielded unusual events including evidence for a new mass peak at about 6 GeV/c^2 (47). Other experiments have made use of the wide-band photon beam for the production of ψ's and study of the production process (48).

As can be felt by anyone working at Fermilab in early 1976, it seems that new results are announced daily. Excitement is quite high, and seminars are crowded with speakers as well as members of the audience. Theorists contribute strongly to the understanding of the experiments, and their seminars reveal many striking results. To see the richness of possibilities for new results, an overview of recent research accomplishments is given in Table 9 and a *Physics Report* article published in 1975 (49). In that article all the approved experiments are described in general terms according to their primary objectives. The results of Fermilab experiment are being reported with increasing frequency in the physics literature. Important reviews of the research were included in the 1974 London Conference (50) while the current accomplishments will be reported at Tbilisi, USSR, in July 1976.

7   SUPPORT SERVICES AT THE LABORATORY

With a site of nearly 7000 acres containing a complicated physical plant, the Site Services personnel have a major responsibility in maintaining roads, utilities, and emergency services in a safe and continuous manner. This is often a major challenge since the weather in the Chicago area sometimes turns quite severe. The dominant utility service begins at the master substation where four transformers are located, each connected to the 345,000-V electrical power line coming onto the Laboratory. Two of these transformers are used for pulsed power to the Main Ring, another one for power to the experimental areas and a fourth for electrical service to buildings and the rest of the Laboratory. The average power consumption for the entire site is nearly 75 MW. This would increase rapidly if the Main Accelerator were pulsed more frequently. An early estimate of the eventual power needed for rapid pulsing at 400 GeV was 160 MW. Unfortunately, the increased cost of electrical power has not made such rapid pulsing feasible, and the full potential of the accelerator has not been realized.

Table 9 Major research program accomplishments at Fermilab

A. Hadron interactions

Charged hadron and neutron total cross sections at 23–280 GeV/c
Charged hadron elastic scattering at 50–200 GeV/c
Direct production of e's, μ's from proton interactions
Measurement of neutral hyperon production and total cross sections
K° regeneration from carbon and CH_2 at 30–120 GeV/c
π^--p charge exchange at 20–200 GeV/c
γ and π^0 production from π-p and p-p interactions
Neutron elastic scattering at 100–350 GeV/c
Proton-nucleon inelastic scattering at 50–400 GeV/c
Total of 2500 K pictures in 30 in with H_2, D_2, and Ne
Total of 200 K pictures in 15 ft with H_2 and H_2/Ne

B. Electromagnetic interactions

Test of scaling of μ interactions at 56, 150 GeV/c
Inelastic μ scattering from hydrogen at 100, 150 GeV/c
Photoproduction from heavy targets at 25–330 GeV/c
Measurement of pion form factor at 100 GeV/c

C. Weak interactions

Study of ν interactions in electronic detectors
Investigation of ν interactions without μ's and production of di-muons
Measurement of ν and $\bar{\nu}$ total cross sections at 50–200 GeV/c
Total of 350 K pictures of ν and $\bar{\nu}$ interactions in 15 ft

D. Particle searches

Searches for quarks and magnetic monopoles
Photoproduction of heavy particles decaying to hadrons and leptons
Missing-mass search with proton and pions
Neutron production of heavy particles decaying to hadrons and μ's
Dilepton and hadron pair production from proton interactions at 400 GeV/c
Production of μ pairs by pions and protons at 150 and 230 GeV/c
Study of muon polarization from proton interactions at 400 GeV/c

With respect to other services, the Laboratory maintains houses in the Village and some apartments in converted farmhouses. The farmhouses that existed on the site for nearly 100 years have been moved together so that they will be preserved for use for years to come. This housing is eagerly sought by visiting experimenters who work at Fermilab so that they can be close to their experiments during the intense time of data taking. The Village also includes recreational facilities and a Users Center for staff and visitors.

Considerable technical support is available at the Laboratory, centered largely around the building of experimental apparatus to be used for particle physics research. Here the Laboratory plays a role in assuring itself that safe and dependable equipment is provided. Examples of apparatus designed and operated at the Laboratory are hydrogen targets, magnets for secondary beams, analysis magnets used in the experiments, on-line computers, high-speed electronics, and general-purpose detectors. Much of the development of electronics and detectors has been to provide instrumentation for the beam lines, although some advice and development capability is made available to the experimental groups.

The Laboratory maintains more than 20 small computers for the collection and sampling of research data. These computers are located near the apparatus and the fast electronics associated with the experiments. Some are connected to the central computers in the Central Laboratory, while others are connected by a "bicycle link." In the Central Laboratory are two CDC 6600 computers sharing a common memory. The size of the computing system at Fermilab is barely adequate to monitor the collection of data and to do some preliminary analysis (51). Since the capacity of the computers is very limited, most experimenters are obliged to take their data to their home institutions for full analysis. It is hoped in the future that a large computing system will be located at the Laboratory to make the analysis of the physics results more efficient and responsive.

A large laboratory requires extensive administrative services. Most of these were developed during the construction stage of the laboratory but continue today as the experimental areas come to full strength and the experimental program expands. Personnel in the business office often assist experimenters in acquiring the apparatus needed for their research activities. For example, much of the fast electronics used in the counter experiments is available commercially, but must be specified and acquired rapidly. Frequently contracts must be let for the construction of the specialized gear used in the research program.

A discussion about Fermilab cannot be complete without review of the cost of constructing the Laboratory and the current funding requirements. After the project was partially authorized in 1967, funds were made available by the AEC on a yearly basis. The initial money was used to start prototype work on the different components. Later, the construction was initiated and the last monies were used for the completion of the experimental facilities. A summary of the major costs of construction is presented in Table 10. Here one sees how $243.5 million were used out of the original $250 million authorization. With this money all of the major goals of the Laboratory were achieved or exceeded except for the attainment of the peak proton intensity. In many areas, such as maximum beam energy and scope of research facilities, the objectives were considerably exceeded.

An important component in the financial picture was the availability of equipment funds. This money was used to provide the beam lines and most of the experimental apparatus used in the research areas. As such, these monies formed a vital part of the overall funding of the Laboratory. The amount of equipment money received through the time of the project funds is indicated in the same table. Today the largest part of the laboratory support comes from yearly operating funds. In fiscal year (FY)

Table 10 Fermilab costs

Project construction costs (in millions of dollars)	
Site and utilities	30.6
Accelerator structures (including Switchyard and Target Areas)	30.4
Accelerator components (including Switchyard and Target Areas)	118.1
Experimental area structures	32.5
Other laboratory buildings	31.9
Total	243.5
Equipment costs through FY 1975 (in millions of dollars)	
Secondary beams	18.2
Apparatus for experiments	16.3
Bubble-chamber and detector facilities	12.7
General equipment	16.2
Total	63.4
Yearly operating costs (in millions of dollars)	
FY 1968–1971 (average)	5.1
FY 1972	12.8
FY 1973	19.2
FY 1974	29.0
FY 1975	35.8
FY 1976	42.7
FY 1977 (estimate)	49.7

1976 these totaled $42.7 million and were accompanied by $1.3 million of facility maintenace and improvement funds. It is expected that substantial new funds will come into the Laboratory in support of the Energy Doubler/Saver Project and projects yet to be proposed.

8   PRESENT AND FUTURE PROJECTS AT FERMILAB

In a laboratory as large as Fermilab there are many research and development activities under way at any time. Some are directed toward an improvement of the high-energy physics research capability while others attempt to push the frontiers of accelerator technology. At present, major efforts are under way to increase the intensity of protons available from the Main Ring. This involves improvements at every stage of acceleration. The ion source will be changed and a shorter column prepared to improve the beam going into the Linac. As the Linac intensity goes up, the pulse length shortens and different methods of injecting into the Booster accelerator are called for. Efforts are now under way to fill the Booster by injecting two short pulses of high-intensity protons from the Linac, each filling half the

Booster instead of multi-turn injection filling the radial aperture. There needs to be about a factor of two more proton flux from the Booster on every pulse in order to come within range of accomplishing the intensity goals in the Main Ring.

In the Main Ring accelerator the emphasis is on improved transmission. Under low-intensity conditions the transmission has been as high as 90%, but with the more intense beam the transmission falls off rapidly. Correction magnets have significantly broadened the Main Ring acceptance. That, coupled with improved rf acceleration power and beam dampers, is thought to be enough to maintain the higher beam intensities. Meanwhile the Switchyard and experimental areas are being retooled in order to handle the 5×10^{13}-MeV beam when it is available. It is essential that hardened beam splitters be used in the Switchyard under these conditions.

Periodic advances are made on the energy front with the goal of achieving, at least for special experiments, 500-GeV protons from the accelerator. The magnets in the Main Ring have been excited to the equivalent of 480 GeV, but no beam accelerated at that time. With a beam the accelerator has operated briefly at 450 GeV. The major obstacle to be overcome is the voltage fluctuation on the main electrical lines bringing power to the Laboratory site. Without compensation the fluctuations during 400-GeV pulsing are nearly 2%. This is more than is considered acceptable and compensating capacitors are to be installed. These will smooth out the voltage fluctuations and make 500 GeV a possible objective. It may not be feasible to have sustained operation at 500 GeV because of the high electrical power requirements of the main ring magnets. That gives rise to one of the major driving forces behind research on superconductivity at the Laboratory.

Superconductivity developments at Fermilab have concentrated on magnets for accelerator applications and for beam line use. The beam line need is to upgrade the energy capabilities of secondary beams. When the Meson Area was first constructed, it was designed as a 200-GeV area and all of the secondary beams were limited to 200 GeV. It has been possible to insert additional conventional magnets to raise the upper limits. Nevertheless, superconducting magnets will be needed to get the entire area working at 500 GeV. In the Neutrino Area some of the magnets will be changed to superconducting to reduce electrical power consumption. The Proton Area will make extensive use of such magnets since it is the one area readily prepared for 1000-GeV physics. The high-energy secondary beam being designed (P3) assumes the use of high-field superconducting magnets. This not only makes a secondary beam of 1000 GeV possible, it also offers the possibility of a high-purity electron beam. Here the electrons can be separated from the accompanying π^- mesons because they will radiate and lose energy in the very strong magnetic fields. The resulting change in electron energy and momentum allows one to separate the two species of particles.

Of course, the main objective behind the superconductivity program is to build a higher-energy accelerator. At Fermilab this takes the form of the Energy Doubler/Saver Project (52). Here the objective is to use the same main ring tunnel but to add an additional set of magnets that are twice as strong so that the proton beam can be raised to 1000 GeV. These magnets would accept protons at 200 GeV where the beam would be transferred from the Main Accelerator and accelerate it up to 1000 GeV. At this energy the magnetic field would be about 45 kG. This cycle would take place in less than 60 sec, and the Main Accelerator could be used for other purposes in the intervening time. At 1000 GeV the circulating proton beam would be used within the Main Ring at the Internal Target Area or extracted into the Neutrino and Proton experimental areas.

Because the beam is quite small at 200 GeV, the aperture of the superconducting magnet does not have to be large. At present a somewhat less than 3-in magnet bore is being developed for the Energy Doubler magnets. This aperture is adequate to allow for the beam size growth during slow extraction. The coils of the super-conducting magnet are formed around the vacuum tube. Niobium-titanium wire is formed into two layer coils and held in place by banding. Super insulation and a vacuum jacket completes the cryostat around which the iron is placed. The outside dimensions of the 20-ft long magnet are 13 by 9 in. More than a thousand magnets must be built and installed within the main ring during pro-grammed interruptions to the ongoing research program. The installation is a major technical challenge to the entire development.

These magnets and the Energy Doubler accelerator can also be used in another mode that has many interesting features. This mode is called the Energy Saver. For example, the injected proton beam would be accelerated to 400 GeV and held within the superconducting accelerator for quite a long time. During this time the beam would be slowly extracted, increasing the duty factor of the accelerator very considerably, while greatly reducing the electrical energy consumption compared to the present facility. The need for a longer duty factor of the proton beam at 400 GeV exists in the present experimental program where the beam is available only 10% of the time. With the Energy Saver a beam approaching an 80% duty factor would be feasible. It appears that the superconducting Energy Doubler/Saver will offer major new opportunities for the research program at Fermilab.

A longer-range project at the Laboratory is to construct proton-proton storage rings (53); 1000-GeV protons in one ring will collide with 1000-GeV protons in the other, giving all their energy to the interaction. The energy release is equivalent to that of a 2,000,000-GeV conventional proton accelerator. The intersecting storage rings would be built in a remote location on the Laboratory site where the 1000-GeV protons from the Energy Doubler could be directed to one ring and then to another. After each ring is filled with circulating currents of 10 A, taking about an hour, the injection process would stop and the circulating beam would be allowed to interact for a period of several hours until depleted, after which the filling is repeated. The name of this project is POPAE (54), standing for Protons On Protons And Electrons, indicating that an eventual electron ring may also be constructed. The plans for POPAE are very preliminary and are being studied jointly by physicists from Fermilab and Argonne National Laboratory. Another suggestion for higher energy capability concerns the possibility of a several thousand-GeV accelerator on the Fermilab site. The accelerator enclosure would circle the site with the extracted proton beam bent within the ring for experimentation, perhaps extending once again the usefulness of the existing experimental facilities.

All of these ideas have been considered by the community of high-energy physicists and are incorporated in the third recommendation of the 1975 "Woods Hole" panel report (55):

> ... that funds be provided to support an accelerator development program at FERMILAB directed toward the long-term goal of fixed target and/or colliding beam systems in the region of 1000 GeV or above, and we recognize the Energy Doubler/Saver as a proper step to this end.

This recommendation recognizes what has been one of the guiding policies at the Laboratory—the pursuit of increasingly higher energies.

9 SUMMARY

In a ten-year period from 1967 to 1976 a new laboratory was built in Illinois for the study of high-energy particle physics. The Fermi National Accelerator Laboratory staff designed and built the first accelerator to exceed 100 GeV, which is now operating at 400 GeV.[2] The entire facility was built in an economical and efficient manner incorporating advances in technology and methods of construction at every step of the way. Prior to groundbreaking in December 1968 and well beyond the achievement of a 200-GeV beam in March 1972, the Laboratory staff worked long and hard to provide a national research facility.

The Main Accelerator on the Laboratory site is a 400-GeV proton synchrotron located in a below ground tunnel. Here more than 1000 magnets bend the beam in a circular orbit with a circumference of 6.5 km. The beam is injected into this accelerator from an 8-GeV rapid cycling synchrotron that in turn is supplied with protons from a 200-MeV Linac. The rapid cycling booster provides 13 pulses in less than 1 sec to fill the Main Ring and the circulating beam is accelerated to 400 GeV in 3 sec. At 400 GeV the beam is slowly extracted from the accelerator for electronic experiments and rapidly extracted for bubble-chamber exposures.

A remarkable feature of the accelerator facility is the switchyard complex located below ground near the extraction point. Here the proton beam is divided and sent to the three external experimental areas. The largest fraction of the beam is used in the Neutrino Area for neutrino experiments in the 15-ft bubble chamber or large electronic counter arrays. A muon beam with modest intensity is also located here. The Meson Area contains a variety of secondary beams, each one providing particles for individual experiments. In the Proton Area the protons are used directly or to produce specialized secondary beams containing photons and electrons.

There are ambitious plans at the Laboratory to upgrade the energy capability of the Main Accelerator. By the installation of superconducting magnets within the existing main-ring tunnel, it is expected that the resulting new accelerator will provide 1000-GeV protons to be used in the Internal Target Area and in at least two of the external experimental areas. This new accelerator will open new physics possibilities while saving electrical energy. The energy-saving mode of this device will substantially reduce the yearly expenditure of funds for electrical power and in turn save natural resources.

At present the 400-GeV accelerator is being used vigorously by many high-energy physicists. A broad program is under way, using the protons directly or to produce numerous secondary beams. Experiments with hadron beams as probes have been examining the cross sections, elastic scattering, and simple inelastic processes in experiments located in the Meson Area. The Neutrino Area has concentrated on deep inelastic neutrino and muon scattering experiments including particle searches. Exposures have been taken in both the 15-ft and 30-in bubble chambers to study the behavior of hadron interactions at 100, 200, 300, and 400 GeV. The Proton Area has concentrated on particle production at large momentum transfers and particle search experiments. The experiments in all three areas will be greatly improved as higher-energy and higher-intensity beams are available in the future.

In closing, tribute should be given to the dedicated individuals who worked in the building of Fermilab. Starting with one and eventually involving thousands of individuals, it has been possible to build four accelerators and four experimental areas including numerous research facilities. Perhaps more than can be seen in the physical plant, the opportunity has existed to create an interesting scientific environment. The development of a park-like site, an appreciation of exciting architecture, and the courage to adopt daring approaches to doing physics has made the Fermi National Accelerator Laboratory a successful project of which staff members, visitors, and citizens can be proud.

Literature Cited[4]

1. Courant, E. D. 1968. *Ann. Rev. Nucl. Sci.* 18:435–64
2. Joint Committee on Atomic Energy, US Congress. 1965. *High Energy Phys. Progr.: Rep. Natl. Policy Background Inform.*, pp. 85–121. Washington DC: GPO. 176 pp.
3. 200-GeV Accelerator Design Study and Summary. 1965. *Lawrence Berkeley Lab. Tech. Rep. UCRL 16000*, Vols. 1–3: 446 pp., 244 pp., 25 pp.
4. 200-GeV Accelerator: Studies on Experimental Use 1964–1965. 1966. *Lawrence Berkeley Lab. Tech. Rep. UCRL-16830*

[4] Note: Much of the published literature concerning the Fermi National Accelerator Laboratory is found in the proceedings of national and international conferences and physics journals, although descriptions of specific devices are often found in the technical notes and reports of Fermilab. Using a variation of the approach adopted by Courant (1), recent national and international accelerator conferences are designated as follows:

Conf. 67: Proc. Int. Conf. High Energy Accel., 6th, Cambridge, Mass., 1967, ed. R. A. Mack. Cambridge, Mass.: Cambridge Electron Accel. 620 pp.
Conf. 69a: US Natl. Particle Accel. Conf. Washington DC, 1969. IEEE Trans. Nucl. Sci. NS-16, Vol. 3. 1136 pp.
Conf. 69b: Int. Conf. High Energy Accel., 7th, Yerevan, USSR, 1969, Vols. 1 & 2. Yerevan, USSR: Am. Acad. Sci. Sci. 694 pp., 751 pp.
Conf. 70: Proton Linear Accel. Conf., 7th, Batavia, Ill., 1970. Batavia, Ill: Natl. Accel. Lab. 1160 pp.
Conf. 71a: US Natl. Particle Accel. Conf., Chicago, Ill., 1971. IEEE Trans. Nucl. Sci. NS-18, Vol. 3. 1161 pp.
Conf. 71b: Int. Conf. High Energy Accel., 8th, Geneva, Switzerland, 1971. Geneva: CERN. 614 pp.
Conf. 72: Proc. Proton Linear Accel. Conf., 8th, Los Alamos, NM, 1972. Springfield, Va: Natl. Tech. Inform. Serv. 436 pp.
Conf. 73: US Natl. Particle Accel. Conf., San Francisco, 1973. IEEE Trans. Nucl. Sci. NS-20, Vol. 3. 1080 pp.
Conf. 74: Proc. Int. Conf. High Energy Accel., 9th, Stanford, Calif., 1974. Springfield, Va: Natl. Tech. Inform. Serv. 769 pp.
Conf. 75: US Natl. Particle Accel. Conf., Washington DC, 1975. IEEE Trans. Nucl. Sci. NS-22, 3:906–1982

(*UCID-10184*), Vols. 1 & 2. 429 pp., 312 pp.; 200-GeV Accelerator: Studies on Experimental Use 1966. 1967. *Lawrence Berkeley Lab. Tech. Rep. UCRL16830.* 336 pp.

5. National Academy of Sciences. 1966. *Rep. Site Evaluation Comm.* Washington DC: Natl. Acad. Sci. 78 pp.
6. Livingston, M. S. 1968. *NAL Tech. Rep. NAL-12.* Batavia, Ill: Fermilab. 32 pp.; Ramsey, N. F. 1975. *History of the Fermilab Accelerator and URA.* Presented at Ann. Meet. Fermilab Users Organ., Batavia, Ill.
7. Research at 200 GeV. 1967. *Univ. Res. Assoc. Tech. Rep. URA-1,* Washington DC. 48 pp.
8. 200-GeV Accelerator: Preliminary Project Report. 1966. *Lawrence Berkeley Lab. Tech. Rep. UCRL-16606.* 59 pp; see Ref. 3, Summary, pp. 4–5
9. *Natl. Accel. Lab. Des. Rep.* 1968. Batavia, Ill: Univ. Res. Assoc. 232 pp. 2nd ed.
10. Livingston, M. S. 1969. *Natl. Accel. Lab. Tech. Rep. NAL-26,* Batavia, Ill. 41 pp.
11. *1968 Summer Study.* Batavia, Ill: Natl. Accel. Lab. Vols. 1–3. 340 pp, 342 pp, 344 pp; *1969 ,Summer Study.* Batavia, Ill: Natl. Accel. Lab. Vols. 1–4. 470 pp, 316 pp, 342 pp, 324 pp.
12. 25-Foot Cryogenic Bubble Chamber Design Report. 1969. *Brookhaven Natl. Lab. Tech. Rep. 14160.* 180 pp.
13. Read, A. L., Sanford, J. R. 1970. *Design of Experimental Areas and Facilities at the National Accelerator Laboratory.* Batavia, Ill: Natl. Accel. Lab. 20 pp.
14. *1970 Summer Study.* Batavia, Ill: Natl. Accel. Lab. 426 pp.
15. Wilson, R. R. 1968. *CERN Cour.* 8:156–58; Goldwasser, E. L. 1969: *Bull. At. Sci.* 25(8):7–10; Cole, F. T., Larsen, C. W. 1970. *Nucl. News* 13:38–44; Wilson, R. R. 1971. *Science* 171:362–64
16. Wilson, R. R. 1967. *Conf. 67:* 210–20; Collins, T. L. 1969. *Conf. 69a,* pp. 12–14; Wilson, R. R. 1969. *Conf. 69b* 1: 103–5; Cole, F. T. 1971. *Part. Accel.* 2:1–10
17. Wilson, R. R. 1974. *Sci. Am.* 230(2):72–83
18. Young, D. 1970. *Conf. 70* 1:15–27; Curtis, C. D., Gray, E. R., Livdahl, P., Owen, C. W., Shea, M. F., Young, D. E. 1972. *Conf. 72,* pp. 17–24
19. Billinge, R., Snowden, S. C., van Steenbergen, A. 1969. *Conf. 69a,* pp. 969–74; Hubbard, E., ed. 1973. *Natl. Accel. Lab. Tech. Rep. TM-405,* Batavia, Ill. 119 pp.
20. Livdahl,.P. 1975. *Conf. 75,* pp. 929–35; Reardon, P. J. 1974. *Conf. 74,* pp. 17–18;

Young, D. E. 1973. *Conf. 73,* pp. 191–97
21. Malamud, E. 1971. *Conf. 71a,* pp. 948–52; Wilson, R. R. 1971. *Conf. 71b,* pp. 3–13
22. Hinterberger, H., Satti, J., Schmidt, C., Sheldon, R., Yamada, R. 1971. *Conf. 71a,* pp. 853–56; Hinterberger, H., Pruss, S., Satti, J., Schivell, J. 1971. *Conf. 71a,* pp. 857–59
23. Jovanovic, D. June 1975. *NALREP,* pp. 1–8
24. Griffin, J. E., Kerns, Q. A. 1971. *Conf. 71a,* pp. 241–43
25. Cassel, R. 1973. *Conf. 73,* pp. 355–59; Cassel, R., Pfeffer, H. 1971. *Conf. 71a,* pp. 860–63
26. Livdahl, P. 1975. *Conf. 75,* pp. 929–35
27. Daniels, R. E., Goodwin, R. W., Storm, M. R. 1973. *Conf. 73,* pp. 505–9
28. Shea, M. F. 1975. *Conf. 75,* pp. 1458–65
29. Edwards, H. T. 1974. *Conf. 74,* pp. 447–50; Andrews, R. A., Edwards, H. T., Fisk, H. E., Hornstra, F., McCarthy, J. D., Pfeffer, H., Rode, C. H., Walton, J. 1974. *Conf. 74,* pp. 456–61; Walton, J., Andrews, R. A., Edwards, H. T., Palmer, M. 1975. *Conf. 75,* pp. 1091–93
30. Oleksiuk, L. W., Andrews, R. A., Bleser, E. J., Rode, C. H. 1973. *Conf. 73,* pp. 428–33
31. Theriot, D., Lee, K., Awschalom, M. 1971. *Conf. 71a,* pp. 750–52; Kang, Y. W., Roberts, A., Theriot, D. 1971. *Conf. 71a,* pp. 753–54
32. Lindberg, J., Grimson, J., Simon, J., Theriot, D. 1975. *Proc. Conf. Remote Syst. Technol., 23rd,* pp. 24–50. Hinsdale, Ill.: Am. Nucl. Soc.
33. See Ref. 9, pp. 14/1–14/15; Bleser, E. J., Collins, T. L., Maschke, A., Moll, D., Nezrick, F. A., Read, A. L., Shoemaker, F. C. 1969. *NAL Tech. Rep. TM-175,* Batavia, Ill. 24 pp.
34. Toohig, T. E. 1974. *Design Description of the External Experimental Areas at the Fermi National Accelerator Laboratory.* Batavia, Ill: Fermilab. 46 pp.
35. Baker, W. F., Carroll, A. S., Chiang, I.-H., Eartly, D. P., Fackler, O., Giacomelli, G., Koehler, P. F. M., Kycia, T. F., Li, K. K., Mazur, P. O., Mockett, P. M., Pretzl, K. P., Pruss, S. M., Rahm, D. C., Rubinstein, R., Wehmann, A. A. 1974. *Phys. Lett. B* 51:303–5
36. Orr, J. R., Read, A. L. 1971. *Meson Lab.: Prelim. Des. Rep.* Batavia, Ill: Natl. Accel. Lab. 75 pp.; Koehler, P. Aug. 1975. *NALREP,* pp. 1–8
37. See Ref. 34, pp. 19–31; Huson, F. R. Apr. 1975. *NALREP,* pp. 1–11
38. Fowler, W. Oct. 1973. *NALREP,* pp.

1–12; *Safety Rep., 15-Foot Bubble Chamber,* July 1972. Batavia, Ill: Natl. Accel. Lab. Vols. 1–3. 320 pp, 312 pp, 280 pp.

39. Peoples, J. Jr. Nov. 1974. *NALREP,* pp. 1–17

40. Sanford, J. R., ed. 1976. *Procedures for Experimenters.* Batavia, Ill: Fermilab. 60 pp.

41. Greene, A. F., Sanford, J. R. 1975. *Fermilab Research Program Workbook.* Batavia, Ill: Fermilab. 80 pp; Greene, A. F. Feb. 1976. *NALREP,* pp. 1–14

42. Barish, B. C., Bartlett, J. F., Buchholz, D., Humphrey, T., Merritt, F. S., Sciulli, F. J., Stutte, L., Shields, D., Suter, H. 1975. *Phys. Rev. Lett.* 35:1316–20; Benvenuti, A., Cline, D., Ford, W. T., Imlay, R., Ling, T. Y., Mann, A. K., Messing, F., Piccioni, R. L., Pilcher, J., Reeder, D., Rubbia, C., Stefanski, R., Sulak, L. 1974. *Phys. Rev. Lett.* 32:125–28

43. Benvenuti, A., Cline, D., Ford, W. T., Imlay, R., Ling, T. Y., Mann, A. K., Orr, R., Reeder, D. D., Stefanski, R., Sulak, L., Wanderer, P. 1975. *Phys. Rev. Lett.* 35:1199–1202

44. von Krogh, J., Fry, W., Camerini, U., Cline, D., Loveless, R. J., Mapp, J., March, R., Reeder, D., Barbaro-Galtieri, A., Bosetti, P., Lynch, G., Marriner, J., Solmitz, F., Stevenson, M. L., Haidt, D., Harigel, G., Wachsmuth, H. 1976. *Phys. Rev. Lett.* 36:710

45. Carroll, A. S., Chiang, I.-H., Kycia, T. F., Li, K. K., Mazur, P. O., Mockett, P., Rahm, D. C., Rubinstein, R., Baker, W. F., Eartly, D. P., Giacomelli, G., Koehler, P. F. M., Pretzl, K. P., Wehmann, A. A., Cool, R. L., Fackler, O. 1974. *Phys. Rev. Lett.* 33:928–31, 932–35

46. Cronin, J. W., Frisch, H. J., Shochet, M. J., Boymond, J. P., Mermod, R., Piroué, P. A., Sumner, R. L. 1974. *Phys. Rev. D* 10:3093–94

47. Hom, D. C., Lederman, L. M., Paar, H. P., Snyder, H. D., Weiss, J. M., Yoh, J. K., Appel, J. A., Brown, B. C., Brown, C. N., Innes, W. R., Yamanouchi, T., Kaplan, D. M. 1976. *Phys. Rev. Lett.* 36:1236

48. Knapp, B., Lee, W., Leung, P., Smith, S. D., Wijangco, A., Knauer, J., Yount, D., Nease, D., Bronstein, J., Coleman, R., Cormell, L., Gladding, G., Gormley, M., Messner, R., O'Halloran, T., Sarracino, J., Wattenberg, A., Wheeler, D., Binkley, M., Orr, R., Peoples, J., Read, L. 1975. *Phys. Rev. Lett.* 34:1040–43, 1044–47

49. Giacomelli, G., Greene, A. F., Sanford, J. R. 1975. *Phys. Rep. C* 19:171–232

50. *Proc. Int. Conf. High Energy Phys., 17th, London, 1974,* ed. J. R. Smith. Chilton, Didcot: Rutherford Lab., Sci. Res. Counc. 900 pp.

51. Brenner, A. E. Sep. 1975. *NALREP,* pp. 1–12

52. *The Energy Doubler/Saver Design Proposal.* 1975. Batavia, Ill: Fermilab. 63 pp.; Edwards, D. A., Fowler, W. B., Reardon, P. J., Richied, D. E., Strauss, B. P., Sutter, D. F. 1974. *Conf. 74,* pp. 184–90

53. Teng, L. C. 1968. *Proton-Proton Colliding Beam Storage Rings for the National Accelerator Laboratory.* Design Study 1968. Batavia, Ill: Natl. Accel. Lab. 112 pp.

54. *1973 Summer Study.* 1973. Batavia, Ill: Natl. Accel. Lab. Vols. 1, 2. 300 pp, 314 pp; Collins, T. L., Edwards, D. A., Ingebretsen, J., Teng, L. C. 1975. *Fermilab Tech. Rep. TM-547.* Batavia, Ill. 98 pp.; Collins, T. L., Edwards, D. A., Ruggiero, A. G., Teng, L. C. 1975. *Fermilab Tech. Rep. TM-600.* Batavia, Ill. 33 pp.

55. Energy Research and Development Administration. 1975. *Rep. 1975 Subpanel New Facil. High Energy Phys. Advis. Panel.* Washington DC: ERDA. 25 pp.

Ann. Rev. Nucl. Sci. 1976. 26: 199–238

THE PARTON MODEL[1] ×5574

Tung-Mow Yan[2]

Laboratory of Nuclear Studies, Cornell University, Ithaca, New York 14853

CONTENTS

1   INTRODUCTION

Since its inception in 1968 Feynman's parton model (1, 2) has attracted a great
deal of attention both theoretically and experimentally. Many theoretical ideas and
experiments have been proposed to test the model [for a summary see (3)]. Data
on various reactions are now available for comparison with the theoretical predictions. In this paper we review the theoretical development of the model and
present a brief comparison with the available data. We find that the success of
the parton model is quite impressive. We restrict ourselves to only those applications
that require minimum extension of the parton idea, and thus the most fundamental
aspects of the model.

We begin with a short review of kinematics for the electron (and muon), and
neutrino scattering, and the recent experimental status of Bjorken scaling (4). We
then develop the formulation of the parton model for electron and neutrino scattering, and discuss the basic ideas and assumptions of the model. The model is then
applied to the inclusive one-hadron spectrum in the final states of electron scattering,
electron-positron annihilation into hadrons, and massive lepton pair production

---

[1] Supported by the National Science Foundation.
[2] Alfred P. Sloan Foundation Fellow.

in proton-proton collisions. It is then followed by a generalization of the simple parton model to incorporate the renormalization group approach of scale invariance at short distances. Finally, we mention a few outstanding problems in the quark parton model.

Our intention of making our presentation intuitive and simple prevents us from discussing a more rigorous and perhaps more precise formulation of the parton model—the covariant parton model (5), and the formal approach of light cone algebra [see the reviews of (6)]. Other important topics omitted in our review include fixed poles [see the reviews of (7)], inelastic Compton scattering (2, 8), shadowing in electron-nucleus scattering [see (9) for a review], and large transverse momentum processes in purely hadronic processes [see the reviews of (10)], etc.

We are aware of alternative interpretations of the deep inelastic processes discussed in this paper.[3] In particular, the concept of generalized vector dominance is certainly useful in the diffraction region of electron-nucleon scattering.[4] The parton model, especially if the quarks are the partons, has the advantage that it provides a unified and intuitive description of these processes and makes very specific predictions. Therefore it could be more readily confronted with experiments. It is hoped that we succeed in conveying the message that it is possible to understand on an intuitive and simple basis the problems of short-distance properties in particle physics: the inner structure of hadrons as probed most effectively by the electromagnetic and weak currents.

2   KINEMATICS AND BJORKEN SCALING

In the one-photon exchange approximation the inclusive electron scattering cross section from a unpolarized target is given by (13):

$$\frac{d\sigma}{d\Omega \, dE'} = \left(\frac{d\sigma}{d\Omega}\right)_{\text{Mott}} \left[W_2(Q^2, \nu) + 2W_1(Q^2, \nu) \tan^2 \frac{\theta}{2} \right], \qquad 2.1.$$

$$\left(\frac{d\sigma}{d\Omega}\right)_{\text{Mott}} = \frac{\alpha^2 \cos^2 \frac{\theta}{2}}{4E^2 \sin^4 \frac{\theta}{2}}, \qquad 2.2.$$

$$Q^2 = -q^2 = 4EE' \sin^2 \frac{\theta}{2}, \qquad 2.3a.$$

$$\nu = E - E' = \frac{q \cdot P}{M}, \qquad 2.3b.$$

$$W^2 = (P+q)^2, \qquad 2.3c.$$

where E, E', and θ are the initial energy, the final energy, and the scattering angle of the electron in the laboratory system; P_μ is the four-momentum of the target

[3] See, for example, the space-time approach of suri & Yennie (11).
[4] For a detailed discussion see the forthcoming review by Bauer et al (12).

whose mass is denoted by M. In Equation 2.1 the electron mass has been set to zero. Throughout this paper, this approximation is made. The structure functions W_1 and W_2 are related to the total absorption cross sections of a transverse virtual photon σ_T and that of a longitudinal virtual photon σ_L introduced by Hand (14) by

$$W_1 = \frac{K}{4\pi^2 \alpha} \sigma_T, \qquad\qquad 2.4a.$$

$$W_2 = \frac{K}{4\pi^2 \alpha} \frac{Q^2}{Q^2 + v^2}(\sigma_T + \sigma_L), \qquad\qquad 2.4b.$$

$$K \equiv \frac{2Mv - Q^2}{2M}. \qquad\qquad 2.4c.$$

One also introduces the ratio

$$R = \sigma_L/\sigma_T. \qquad\qquad 2.5.$$

Bjorken's scaling variable is defined as

$$\omega = 2Mv/Q^2 = 1/x. \qquad\qquad 2.6.$$

Phenomenologically, one also introduces a new variable

$$\omega' \equiv \omega + M^2/Q^2 = \frac{1}{x'}. \qquad\qquad 2.7.$$

The structure functions W_1 and W_2 can be defined in terms of the electromagnetic current-current correlation function

$$W_{\mu\nu} = 4\pi^2 \frac{E_p}{M} \int d^4x \exp{(iq \cdot x)} \langle P | J_\mu(x) J_\nu(0) | P \rangle$$

$$= -\left(g_{\mu\nu} - \frac{q_\mu q_\nu}{q^2}\right) W_1(Q^2, v) + \frac{1}{M^2}\left(P_\mu - \frac{P \cdot q}{q^2} q_\mu\right)\left(P_\nu - \frac{P \cdot q}{q^2} q_\nu\right) W_2(Q^2, v),$$
$$\qquad\qquad 2.8.$$

where a spin average is understood.

The kinematics for neutrino and antineutrino scattering are similar to that for electron scattering. We define, for neutrino scattering,

$$W_{\mu\nu}^\nu = 4\pi^2 \frac{E_p}{M} \int d^4x \exp{(iq \cdot x)} \langle P | J_\mu^{c\dagger}(x) J_\nu^c(0) | P \rangle$$

$$= -g_{\mu\nu} W_1^\nu + \frac{1}{M^2} P_\mu P_\nu W_2^\nu + i\varepsilon_{\mu\nu\lambda\kappa} \frac{P^\lambda q^\kappa}{2M^2} W_3^\nu + \dots, \qquad\qquad 2.9.$$

where J_μ^c is the Cabibbo hadronic weak current; similarly, we define $W_{1,2,3}^{\bar{\nu}}$ for antineutrino scattering with the replacement $J_\mu^c \leftrightarrow J_\mu^{c\dagger}$. The dots in Equation 2.9 denote additional terms proportional to q_μ or q_ν. The inclusive cross sections for neutrino and antineutrino scattering are given by

$$\frac{d^2\sigma^{\nu,\bar{\nu}}}{dE'\,d\Omega} = \frac{G^2}{2\pi^2}(E')^2\left[W_2^{\nu,\bar{\nu}}\cos^2\frac{1}{2}\theta + 2W_1^{\nu,\bar{\nu}}\sin^2\frac{\theta}{2} \pm \left(\frac{E+E'}{M}\right)\sin^2\frac{\theta}{2}\,W_3^{\nu,\bar{\nu}}\right].\qquad 2.10.$$

In 1965, based on Gell-Mann's current algebra (15) for charge densities and the assumption of an unsubtracted dispersion relation, Adler (16) derived a sum rule for neutrino scattering,

$$\int d\nu\,[W_2^{\bar{\nu}p}(Q^2,\nu) - W_2^{\nu p}(Q^2,\nu)] = 1 \qquad (\cos\theta_c = 1). \qquad 2.11.$$

From this sum rule Bjorken (17) in 1966 obtained an inequality for inelastic electron-nucleon scattering,

$$\int d\nu\,[W_2^{ep}(Q^2,\nu) + W_2^{en}(Q^2,\nu)] \geqq \tfrac{1}{2}. \qquad 2.12.$$

This follows from

$$W_2^{\bar{\nu}p} - W_2^{\nu p} \leqq W_2^{\nu n} + W_2^{\nu p} \leqq 2(W_2^{en} + W_2^{ep}). \qquad 2.13.$$

The most interesting aspect of Equations 2.11 and 2.13 is the absence of Q^2 dependence on the right-hand side even at very large values of Q^2. It means that the·total neutrino, antineutrino, or electron scattering cross section should be as big as pointlike. In 1968 Bjorken (4) suggested that this can be easily understood if the integrand has the scaling behavior

$$\nu W_2(Q^2,\nu) \xrightarrow[\substack{Q^2,\nu\to\infty \\ Q^2/\nu\ \text{fixed}}]{} \underset{Bj}{\text{Lim}}\ \nu W_2(Q^2,\nu) = F_2\left(\frac{2M\nu}{Q^2}\right). \qquad 2.14.$$

Then as $Q^2 \to \infty$,

$$\int d\nu\,W_2(Q^2,\nu) = \int\frac{d\nu}{\nu}\,\nu W_2(Q^2,\nu) = \int\frac{d\omega}{\omega}\,F_2(\omega), \qquad 2.15.$$

which is independent of Q^2. This is Bjorken's famous prediction of scaling. The scaling behavior, Equation 2.14, is now known as Bjorken scaling.

The pioneering work on electron scattering at Stanford Linear Accelerator Center (SLAC) [summary given in (18)] has been extended to much higher energy at Fermilab, using a muon beam (19). The kinematic domain explored so far covers the incident lepton (electron or muon) energy up to 150 GeV and invariant momentum transfer Q^2 up to about 40 $(\text{GeV}/c)^2$.

Ideally, one obtains the structure functions W_1 and W_2 from measurements of differential cross sections and tests for the Bjorken scaling, Equation 2.14. In practice, there are various difficulties of testing scaling in the region where the kinematical variables Q^2 and ν are not truly asymptotic. For example, apparent nonscaling behavior can be at least partly removed by choice of a new scaling variable. As the range and precision of the data increase, it becomes more and more important to have a better theoretical understanding of how to handle the effects of finite Q^2 (20).

To approximately a 20% level, all experimental data (19, 21) have shown that Bjorken's predictions have withstood the strain of increasing ν by almost an order of magnitude. A look at the more detailed behavior of the data reveals some systematic trend of deviation from exact scaling.

The SLAC data for the proton target collected in the last six years are shown in Figure 1. The main features of the SLAC data follow (21).

1. νW_2^p and $2MW_1^p$ do not scale in x nor in x'. They both break the scaling in the same manner. However, the observed scaling violation is not large. For instance, for $0.5 < x' < 0.7$ about a 1% decrease in the structure functions brings about an increase of 1 $(\text{GeV}/c)^2$ in Q^2. Data for νW_2^p at $x = 0.1$ show a rise with increasing Q^2. However, these data at low x correspond to low values of Q^2.
2. Neutron structure functions break scaling in a different manner than the proton's.
3. R_p is consistent with a constant $R_p = 0.14 \pm 0.06$ and it is also consistent with a behavior such as $1/Q^2$ or $1/(\ln Q^2)$. (R_p is the ratio σ_L/σ_T for the proton).
4. $R_d = R_p$ within experimental errors. (R_d is the ratio σ_L/σ_T for the deuteron.)

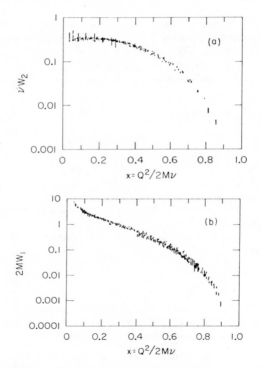

Figure 1   Values of νW_2 and $2MW_1$ for the proton with $Q^2 > 1$ GeV^2 and $W^2 > 4$ GeV^2. $R = \sigma_L/\sigma_T$ is assumed to be zero.

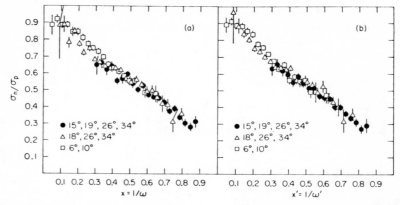

Figure 2 Ratio of inelastic electron-neutron to electron-proton cross sections versus x ar d x' (84).

5. The ratio of the neutron to proton cross sections is shown in Figure 2. The neutron cross section is extracted from electron deuterium scattering. The effects of deuterium become increasingly important as $x \to 1$. The behavior of the ratio σ_n/σ_p near $x = 1$ is important in the quark parton model.

6. The data for $2MW_1$ can be made to scale much better if one uses the ad hoc variable

$$x_s = \frac{1}{\omega_s} = \left(\omega + \frac{1.5\ \text{GeV}^2}{Q^2}\right)^{-1}.$$ 2.16.

Near $x_s = 1$, if W_1 is fitted with the form

$$W_1 = \sum_{i=3}^{i=7} \beta_i (1 - x_s)^i,$$ 2.17.

it is found that β_3 is small and consistent with zero, and β_4 is large and positive. However, the absence of the cubic term depends on the choice of variable. It will be present if the variable x instead of x_s is used. Previously, a good fit (22) to νW_2 was obtained by

$$W_2 = \sum_{i=n}^{i=m} \alpha_i (1 - x')^i,$$ 2.18.

with $n = 3$, $m \geq 7$. The presence or absence of the cubic term in the structure functions is relevant in discussing the connection between the structure functions near $x = 1$ and the asymptotic electromagnetic elastic form factor of the nucleon.

There have been two muon experiments at Fermilab (19). In the experiment of the Cornell–Michigan State–Berkeley–La Jolla collaboration (23) the Bjorken scaling is tested directly by a ratio method. The experiment is designed such that the ratio for data taken at two different incident energies

$$r(\omega, q^2) = \frac{E\left(\dfrac{d^2\sigma}{dx\,dy}\right)_{E=\lambda E_0}}{E\left(\dfrac{d^2\sigma}{dx\,dy}\right)_{E=E_0}}$$ 2.19.

will remain constant for all values of ω and q^2 if Bjorken scaling holds. The ratio r for $E_0 = 150$ GeV and $E = \lambda E_0 = 56$ GeV is shown in Figure 3. The scaling appears to be violated.

The same collaboration also compared their measured cross sections with a scale-invariant Monte Carlo calculation, using the νW_2 function measured at SLAC as input with the assumption $R = 0.18$. Ratios of observed-to-simulated event rates are shown in Figure 4. These plots indicate a distinctive feature that the deviation from Bjorken scaling as a function of q^2 has a negative slope at values of ω less than 5; this slope becomes positive at values of ω greater than 5.

The second experiment from Fermilab is on muon scattering from a liquid hydrogen or deuterium target by a Chicago-Harvard-Illinois-Oxford collaboration (19). Values of νW_2 were extracted with the assumption of $R = \sigma_L/\sigma_T = 0.18$ and also with $R = 0$. The result is not sensitive to the values of R assumed. Within errors the data agree with the data from SLAC. It establishes Bjorken scaling

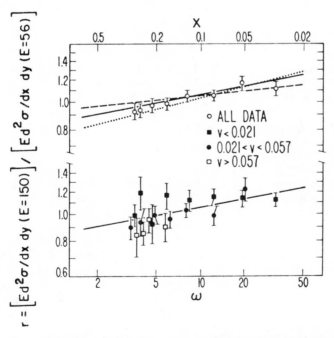

Figure 3   The result for $r(\omega, q^2)$. Solid lines are power-law fits to the data. The effect of changing the spectrometer calibration by 1% is indicated by the dashed line.

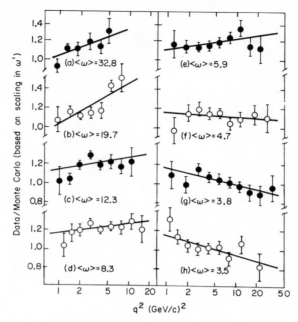

Figure 4   Ratio of observed to "Monte Carloed" event rates vs q^2.

within the errors. The data show a falloff of νW_2 at large ω. It is probably due to the fact that the values of Q^2 in the region of large ω are rather small, and νW_2 is constrained to vanish at $Q^2 = 0$.

3   PARTON MODEL FOR DEEP INELASTIC LEPTON SCATTERING

A classical example of studying the structure of matter by means of the electromagnetic interaction is Rutherford's experiment. In this experiment a beam of energetic α particles is incident upon a target. The α particles are deflected by the electrostatic force between the α particles and the charged constituents of the atoms in the target. From the angular distribution of the scattered α particles it is possible to establish the nuclear model of the atom—that the atom consists of a concentrated charge, the nucleus, with electrons orbiting around it. Inelastic electron scattering is a Rutherford-like experiment on a nucleon rather than on an atom. The basic difference between the two is that in the Rutherford scattering the nucleus retains its identity, so the reaction is elastic. In the inelastic electron scattering the nucleon is usually "broken up;" namely, other particles are produced. So in addition to the scattering angle of the incident electron, there is a new variable associated with the energy loss of the electron. They are related to the relativistic kinematic variables q^2 and ν defined in the last section.

In a nonrelativistic description of an elastic Coulomb scattering between an electron and a composite system, the differential cross section measures the Fourier transform of the spatial charge density distribution inside the system. We may ask what information the inelastic electron scattering tells us about the structure of the nucleon.

Before we attempt to answer this question, we must recognize a fundamental difference between a nucleus and a nucleon. The dynamics of the constituent nucleons inside a nucleus is essentially nonrelativistic. The identity of the constituent nucleons is clearly maintained. On the other hand, the constituents inside a nucleon, if they exist, have never been identified, and their dynamics are presumably relativistic. Nevertheless, it is assumed that the physical hadron states can be expanded as a linear combination of a Fock space basis. The particles in this basis are called partons by Feynman (1).

In a relativistic theory the electromagnetic current not only can scatter a charged constituent but can also create a pair of charged quanta. The scattering reveals the inner structure of the hadron, but the creation act mainly reflects the property of the virtual photon. The relative importance of the two processes depends on, among other things, the frame of reference one chooses. It is not a Lorentz-invariant procedure to decompose an invariant Feynman amplitude into scattering and creation parts according to the time-ordering. For these reasons the answer to the posed question depends on which aspect one emphasizes. The parton model we discuss emphasizes the scattering aspect. It makes use of a special class of coordinate systems—the so-called infinite momentum frames—to satisfy two requirements:

1. The dissociation of the virtual photon into hadronic states is minimized. This can be achieved by choosing a frame in which the energy of the (spacelike) virtual photon is as small as possible. Intuitively, the dissociation would be forbidden if the lifetime for the dominant intermediate states is short. The lifetime of a particular state with mass M_n is estimated by the uncertainty principle

$$\tau_\gamma \sim \frac{1}{E_n - q^0} = \frac{1}{(q_0^2 + Q^2 + M_n^2)^{1/2} - q_0}$$
$$\cong \frac{1}{(Q^2 + M_n^2)^{1/2}}$$

3.1.

if $q_0^2 \ll Q^2$. In the deep inelastic region $Q^2 \to \infty$ (then $v \to \infty$ automatically since $2Mv - Q^2 \gtrless 0$), $\tau_\gamma \to 1/Q \to 0$.

2. Impulse approximation can be applied to the scattering from the individual constituent. This requires that the lifetime of the important intermediate states in the decomposition of the physical nucleon state be long enough so that they behave as free during the scattering act. For this we choose an infinite momentum frame for the proton. Then the lifetime of an intermediate state with mass M_n is

$$\tau \sim \frac{1}{(M_n^2 + \mathbf{P}^2)^{1/2} - (\mathbf{P}^2 + M^2)^{1/2}} \xrightarrow[P \to \infty]{} \frac{2P}{M_n^2 - M^2}$$

3.2.

Thus it grows with P. This is the well-known relativistic time-dilation effect. It is implicitly assumed that the values of M_n for the important intermediate states are bounded.

When these two conditions are fulfilled, one only has to consider the incoherent elastic scattering of the "free constituents" from the electron. This is the parton model. This discussion shows why the parton model must be formulated in a special class of infinite momentum frames. We give here two examples of such frames that have frequently been used.

1. The infinite momentum center-of-mass system of the electron and nucleon. Let $\mathbf{P}$, the momentum of the proton, be in the third axis, then

$$q^0 = \frac{2M\nu - Q^2}{4P}, \qquad q^3 = \frac{-2M\nu - Q^2}{4P}, \qquad |\mathbf{q}_\perp| = (Q^2)^{1/2} + 0\left(\frac{1}{P^2}\right). \qquad 3.3.$$

The limit $P \to \infty$ is taken before $Q^2, \nu \to \infty$. This frame has been used in the earlier studies (2, 24, 47).

2. Breit frame of the virtual photon and the parton with which it interacts. In this frame we have

$$P^0 = P + \frac{M^2}{2P}, \qquad P^3 = P, \qquad \mathbf{P}_\perp = 0,$$

$$q^0 = 0, \qquad q^{3\prime} = -2Px, \qquad \mathbf{q}_\perp = 0, \qquad 3.4.$$

where $x = Q^2/2M\nu$. Notice that the two limits $P \to \infty$ and the Bjorken limit are taken simultaneously. This frame is extensively used by Feynman (3).

In one of these frames the cross section for inelastic electron scattering is given by the product of the probability of finding a charged parton in a particular configuration and the elastic cross section of the parton-electron scattering. The ratio $x = Q^2/2M\nu$ determines the longitudinal momentum fraction of the scattered parton. Since the cross section of a pointlike particle is known, the longitudinal momentum distribution of charged partons is determined by the deep inelastic electron-nucleon scattering. We show that deep inelastic electron scattering also measures the spins of the charged partons.

The probability of finding a parton in a particular configuration, of course, is determined by the wave function for the nucleon. In infinite momentum frames the wave function and the amplitudes have a limiting form if they are expressed in terms of transverse momenta measured on an absolute scale, but the longitudinal momenta are measured in fractions.

There are two qualifications for the validity of the above statement. First, wee partons need special attention. Wee partons are those with x so small that xP is of the order of 1 GeV. Wee partons obviously cannot be regarded as moving with infinite momentum, but the results of deep inelastic electron scattering do not depend on the details of the wee partons. Let $\psi(x_1, x_2, \ldots, \mathbf{k}_{\perp_1}, \mathbf{k}_{\perp_2}, \ldots)$ be the wave function of the nucleon in a sector with partons labeled by longitudinal momentum fractions

x_i and transverse momenta $\mathbf{k}_{\perp_i}$. For the validity of the parton model it suffices to require the weaker condition that

$$f(x_1) = \int |\psi(x_1, x_2, \ldots, \mathbf{k}_{\perp_1}, \mathbf{k}_{\perp_2}, \ldots)|^2 \, dx_2 \ldots d^2\mathbf{k}_{2\perp} \ldots \qquad 3.5.$$

be finite for all values of x_1 not in the wee region as $P \to \infty$. Second, it was realized in the very beginning that the limiting behavior, Equation 3.5. is not valid in renormalizable field theories, at least in perturbation expansion, unless a transverse momentum cutoff is present. This means that we have to introduce by hand a transverse momentum cutoff. This procedure was adopted in the early studies (24). The existence of a transverse momentum cutoff is suggested by phenomenological considerations. Transverse momenta of final products in high-energy hadron collisions are always limited. It is hard to imagine this situation if the constituents themselves do not possess such a property. This was emphasized explicitly in (24). Furthermore, electromagnetic form factors of hadrons show that the charge distribution inside a hadron is extended. Thus the average momentum of the constituents should be limited by the size of the hadron according to the uncertainty principle.

The parton model result for the structure function $W_{\mu\nu}$, Equation 2.8, can be summarized by (24)

$$\lim_{Bj} W_{\mu\nu} = 4\pi^2 \frac{E_p}{M} \int d^4x \exp(iqx) \langle UP | j_\mu(x) j_\nu(0) | UP \rangle, \qquad 3.6a.$$

where $j_\mu(x)$ is the electromagnetic current in terms of the free fields representing the partons, and $|UP\rangle$ is the expansion of the physical nucleon state in terms of the parton basis in an infinite momentum frame. The scattering in the Breit frame mentioned above is shown in Figure 5. Let us introduce the notations

$f_i(x) \, dx =$ number of partons of type i with longitudinal momentum fraction between x and $x + dx$,

$e_i =$ electric charge of parton of type i in units of electron charge, and

$\bar{f}_i(x) \, dx$ is the corresponding quantity for antipartons of type i.

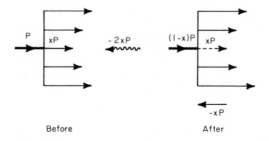

Before   After

Figure 5 Parton configurations before and immediately after the collision in the Breit frame.

Then the structure function $W_{\mu\nu}$ of Equation 3.6a is given by

$$W_{\mu\nu} = \frac{1}{2M}\int_0^1 \frac{d\xi}{\xi}\sum_i e_i^2[f_i(\xi)+\bar{f}_i(\xi)]w_{\mu\nu}^{(i)}(p,q), \qquad \text{3.6b.}$$

where $w_{\mu\nu}^{(i)}(p,q)$ is the analog of $W_{\mu\nu}$ for the elastic scattering of parton i with four-momentum p_μ, and $p^3 = \xi P$. Therefore, $w_{\mu\nu}^{(i)}(p,q)$ is proportional to $\delta[(p+q)^2 - m^2 - \Delta]$, where Δ measures the distance away from the mass shell of the scattered parton. The assumptions of the parton model discussed above imply that Δ is finite, much smaller than Q^2 but it can vary with ξ. The tensor structure of $w_{\mu\nu}^{(i)}$ depends on the spin of the parton. In the Bjorken limit it is a good approximation to set $p^\mu = \xi P^\mu$ and to neglect m^2 and Δ in comparison with Q^2 and $M\nu$.

For spin-0 partons we have

$$\begin{aligned} w_{\mu\nu}^{(i)} &= (2p_\mu+q_\mu)(2p_\nu+q_\nu)\delta(Q^2-2p\cdot q) \\ &= 4\xi^2 P_\mu P_\nu \delta(Q^2-2M\nu\xi). \end{aligned} \qquad \text{3.7.}$$

For spin-$\frac{1}{2}$ partons we have

$$\begin{aligned} w_{\mu\nu}^{(i)} &= \tfrac{1}{2}\mathrm{Tr}\{(\not{p}+m)\gamma_\mu(\not{p}+\not{q}+m)\gamma_\nu\}\,\delta(Q^2-2p\cdot q) \\ &= (4\xi^2 P_\mu P_\nu - 2\xi M\nu g_{\mu\nu})\delta(Q^2-2M\nu\xi). \end{aligned} \qquad \text{3.8.}$$

In Equations 3.7 and 3.8 we dropped terms proportional to q_μ or q_ν, since they are not necessary in identifying W_1 and W_2. Thus for spin-0 and spin-$\frac{1}{2}$ partons we get

$$F_2(x) = \operatorname*{Lim}_{Bj}\nu W_2(Q^2,\nu) = x\sum_i e_i^2[f_i(x)+\bar{f}_i(x)], \qquad \text{3.9.}$$

and for spin-0 partons

$$\operatorname*{Lim}_{Bj} W_1 = 0, \qquad \text{3.10.}$$

while for spin-$\frac{1}{2}$ partons we have

$$xF_1(x) = \operatorname*{Lim}_{Bj} 2MW_1(Q^2,\nu)x = F_2(x). \qquad \text{3.11.}$$

The relative importance of W_1 and W_2 is measured in terms of the ratio

$$R = \frac{\sigma_L}{\sigma_T} = \frac{\left(1+\dfrac{\nu^2}{Q^2}\right)W_2 - W_1}{W_1}. \qquad \text{3.12.}$$

In the Bjorken limit R reduces to

$$R = \frac{F_2(x)-xF_1(x)}{xF_1(x)}, \qquad \text{3.13.}$$

which would vanish if all the charged partons are spin-$\frac{1}{2}$ particles (25). Experimentally R is small, as discussed in the last section.

The smallness of R has led to the proposal that all charged partons are spin-$\frac{1}{2}$

particles. A natural candidate for the partons is the Gell-Mann–Zweig (26) fractional charged quark. That $R \neq 0$ is then attributed to the fact that Q^2 and v are not truly asymptotic. According to this hypothesis, R should approach zero as Q^2 and v approach infinity. As mentioned in the last section, presently available data for R_p cannot distinguish a constant behavior or falling with increasing Q^2.

We now discuss some consequences of the parton model in deep inelastic electron scattering. In the parton model one finds a relation between the asymptotic electro-magnetic elastic form factor and the structure function $F_2(x)$ near the threshold $x = 1$ (27–29). The basic ingredient is the boundness of the transverse momenta of the constituents. The following derivation is due to Feynman (3), which is essentially the same as in (27) but uses a different coordinate system. Consider the elastic electron-proton scattering at large Q^2 in the coordinate system

$$q^\mu = (q^0 = 0, q_\parallel = -Q, \mathbf{q}_\perp = 0), \qquad\qquad 3.14a.$$

$$P^\mu = (E, P, 0, 0), \qquad\qquad 3.14b.$$

$$P'^\mu = (E, -P, 0, 0), \qquad\qquad 3.14c.$$

$$Q^2 = 4P^2. \qquad\qquad 3.14d.$$

For large Q (hence large P), the initial state of the proton is described as a sum of various components for different configurations. Let us suppose that there is one component with more than one (say three) non-wee partons plus some wee partons. The configurations before and after the collision are shown in Figure 6. Clearly, the final state does not resemble at all a proton with momentum $-P$, since it contains backward-moving partons relative to $-P$. Therefore such a configuration cannot contribute to the elastic scattering. To scatter elastically, it must have a configuration with only one parton that carries almost all the momentum, i.e. $x \simeq 1$, and all the rest are wee partons, with $x \sim 1/P \simeq 1/Q$. Now the momentum of the single hard parton is reversed by the incident photon, with little energy change, and the final configuration is possible for a proton with momentum $-P$ since wee partons do not have a definite sense of direction. Therefore the elastic scattering amplitude (the electromagnetic form factor) is proportional to the probability that the proton has one and only one hard parton. This is precisely determined by $F_2(x)$ near $x = 1$. Suppose that as $x \to 1$

$$F_2(x) = (1-x)^p; \qquad\qquad 3.15.$$

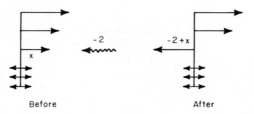

Before   After

Figure 6   Parton configurations before and after an elastic scattering.

then the elastic form factor $G(Q^2)$ at large Q^2 behaves as

$$G(Q^2) \propto \int_{1-\varepsilon/Q}^{1} (1-x)^p \, dx = \left(\frac{\varepsilon}{Q^2}\right)^{1/2(p+1)}.$$ 3.16.

The limits of integration are specified by the conditions that wee partons can have momenta from 0 up to the order of 1 GeV. So the hard parton can have its x from $1-(\varepsilon/Q)$ to 1. Experimentally

$$p \simeq 3$$ 3.17.

from elastic form-factor data. This implies

$$F_2(x) \propto (1-x)^3 \qquad x \simeq 1.$$ 3.18.

Data on νW_2 are not in disagreement with this prediction (22). On the other hand, the recent data on W_1 show that (from Equation 2.17)

$$F_1(x) \propto (1-x_s)^4,$$ 3.19.

where $1/x_s = \omega + 1.5 \; \text{GeV}^2/Q^2$. The comparison between data and the prediction (Equations 3.15 and 3.16) is complicated by several factors. Although the variables ω, ω', and ω_s all approach to the same asymptotic value as $Q^2 \to \infty$, they do differ for finite values of Q^2. The difference is most important for $\omega \simeq 1$. Furthermore, the original theoretical consideration assumes the validity of Bjorken scaling and a power-law falloff for the asymptotic elastic form factor. The violation of Bjorken scaling of the structure functions may have now been observed experimentally. Theoretically, in the asymptotically free gauge theory (to be discussed later), which predicts approximate Bjorken scaling with logarithmic violation, the connection between Equations 3.15 and 3.16 has to be modified. We discuss these matters later.

Insight about the other end $x \to 0$, the so-called wee region, has been gained by Feynman's study of hadron-hadron collisions in the parton model (1, 3). Consider the collision in the center-of-mass system of the two incident hadrons. Their wee partons do not move in a definite direction, and therefore they can be exchanged between the two oppositely moving hadrons. Feynman argues that this is the basic mechanism responsible for the strong interactions between energetic hadrons. It is found that a constant cross section within logarithms requires the wee parton distribution to be

$$n(x) \, dx = c \, \frac{dx}{x}$$ 3.20.

for the number of partons in the small-x region. Equation 3.20 predicts that $F_2(x)$ should approach a constant as $x \to 0$. This is consistent with available data within experimental errors (Figure 1). It is generally assumed that the wee partons as a whole do not carry quantum numbers other than energy and momentum. Thus we expect $F_2(x)$ for the proton and the neutron to be equal as $x \to 0$. It appears to be the case experimentally (Figure 2).

Let us return to the suggestion that the spin-$\frac{1}{2}$ partons are the quarks of Gell-

Mann & Zweig (26). This question is reinforced by the observation that if the partons carry one or zero units of charge, the fraction of the total longitiduinal momentum deposited in the charged partons is given by the integral

$$\int_0^1 dx\, F_2(x) \cong 0.18,$$ 3.21.

which is unreasonably low. Since $F_2(x)$ is weighted by the charge-squared of the partons, (Equation 3.9), the fraction will be higher if the partons carry fractional charges as the Gell-Mann–Zweig quarks do. Many experimentally testable predictions follow from the identification of the charged partons with quarks. So far these predictions have not been contradicted by experiments as long as they do not require actual detection of quarks themselves.

The following results are derived to show how experiments could lead to further identification of the character of partons.

We use the fractionally charged quarks as an example. The main motivation is that it has been studied extensively in low-energy hadron spectroscopy with great success,[5] and by introducing color vector gluons one can construct a gauge field theory that is asymptotically free at short distances (31). The latter may explain the Bjorken scaling observed in deep inelastic lepton scattering.

We assume that quarks possess two different quantum numbers called color and flavor. There will always be three colors that refer to a perfect SU(3) symmetry and all observed hadrons are assumed to be color singlets. Color (32) was introduced originally to resolve the well-known difficulty of quark statistics. Since then it has found support in the PCAC calculation of the decay $\pi^0 \to 2\gamma$ (32, 33) and as mentioned above, it has played an important role in understanding the possible origin of Bjorken scaling. Flavor refers to the ordinary quantum numbers such as isospin and strangeness. Since the discovery of new particles, there are indications that additional quantum number(s) like charm (34) may be needed. For illustration we assume there are four flavors and each flavor comes with three colors. The color quantum number is suppressed in general. The four flavors are denoted by u, d, s, and c. Two of these, u and d, form an isospin doublet. The other two are isospin singlets. To describe the quark distribution in a proton we need eight functions. Let $u(x)\,dx$ denote the number of u quarks with longitudinal momentum fraction between x and $x+dx$, $\bar{u}(x)\,dx$ the corresponding quantity for anti-u quarks. Similarly, we define $d(x)$, $\bar{d}(x)$; $s(x)$, $\bar{s}(x)$; and $c(x)$, $\bar{c}(x)$ for other quark species.

The property of the proton is characterized by zero strangeness and charm, two net u quarks and one d quark:

$$\int_0^1 dx\, [u(x) - \bar{u}(x)] = 2,$$ 3.22a.

$$\int_0^1 dx\, [d(x) - \bar{d}(x)] = 1,$$ 3.22b.

[5] For a review on the quark model see (30).

$$\int_0^1 dx \, [s(x) - \bar{s}(x)] = 0, \qquad\qquad\qquad\qquad 3.22c.$$

$$\int_0^1 dx \, [c(x) - \bar{c}(x)] = 0. \qquad\qquad\qquad\qquad 3.22d.$$

If we call

$$f_p(x) = \text{Lim} \, 2MW_1 \qquad\qquad\qquad\qquad 3.23.$$

for the proton, then

$$f_p(x) = \tfrac{4}{9}(u + \bar{u}) + \tfrac{1}{9}(d + \bar{d}) + \tfrac{1}{9}(s + \bar{s}) + \tfrac{4}{9}(c + \bar{c}). \qquad 3.24.$$

For a neutron target the corresponding function $f_n(x)$ can be obtained from Equation 3.24 by an isospin rotation. Thus

$$f_n(x) = \tfrac{1}{9}(u + \bar{u}) + \tfrac{4}{9}(d + \bar{d}) + \tfrac{1}{9}(s + \bar{s}) + \tfrac{4}{9}(c + \bar{c}), \qquad 3.25.$$

where u, $\bar{u}$, etc, refer to the quark and antiquark distributions inside a proton. Since

$$\tfrac{1}{9}(u + \bar{u} + d + \bar{d} + s + \bar{s} + c + \bar{c}) \leqq f_p(x), f_n(x) \leqq \tfrac{4}{9}(u + \bar{u} + d + \bar{d} + s + \bar{s} + c + \bar{c}),$$

$$3.26.$$

we notice immediately that

$$\frac{1}{4} \leqq \frac{f_n(x)}{f_p(x)} \leqq 4. \qquad\qquad\qquad\qquad 3.27.$$

In the region $x \to 0$, the ratio is nearly unity experimentally. The same data show that as $x \to 1$ the ratio is less than unity, and is probably greater than $\tfrac{1}{4}$ (Figure 2).

We can also study the total momentum carried by each kind of quarks. Let us define

$$U = \int_0^1 dx \, x(u + \bar{u}), \qquad\qquad\qquad\qquad 3.28a.$$

$$D = \int_0^1 dx \, x(d + \bar{d}), \qquad\qquad\qquad\qquad 3.28b.$$

$$S = \int_0^1 dx \, x(s + \bar{s}), \qquad\qquad\qquad\qquad 3.28c.$$

$$C = \int_0^1 dx \, x(c + \bar{c}). \qquad\qquad\qquad\qquad 3.28d.$$

These quantities, U, D, S, and C, are the total fraction of longitudinal momentum carried by u and $\bar{u}$ quarks, etc. From data we have

$$\int_0^1 dx \, x f_p(x) = 0.18, \qquad\qquad\qquad\qquad 3.29a.$$

$$\int_0^1 dx\, x f_n(x) = 0.12.$$ 3.29b.

Therefore

$$\tfrac{4}{9}U + \tfrac{1}{9}D + \tfrac{1}{9}(S+4C) = 0.18,$$ 3.30a.

$$\tfrac{1}{9}U + \tfrac{4}{9}D + \tfrac{1}{9}(S+4C) = 0.12.$$ 3.30b.

If we assume that there are no neutral partons besides quarks, all the momentum of the proton must reside on the quarks. Consequently

$$U + D + S + C = 1.$$ 3.31.

We find

$$D = 0.03 - C \le 0.03,$$ 3.32a.

$$U = 0.21 - C \le 0.21,$$ 3.32b.

$$C = S - 0.76 \ge 0.$$ 3.32c.

This result is clearly unreasonable—that more than $\tfrac{3}{4}$ of the momentum of the nonstrange proton is to be found in the strange quarks. This suggests most strongly that there must be some neutral partons. Could these be the color vector gluons in an asymptotically free gauge field theory?

Additional information about the partons can be obtained by measurement of the polarization asymmetry in the deep inelastic scattering of polarized electrons on a polarized target. In the quark parton model the structure function $g_1(x, Q^2)$ that describes this asymmetry measures the difference of the distribution of quarks with spins parallel and antiparallel to the nucleon's spin. Bjorken (35) shows that it satisfies the sum rule,

$$\int_0^1 dx\, [g_{1p}(x, Q^2) - g_{1n}(x, Q^2)] = \frac{1}{6}\left|\frac{g_A}{g_V}\right|,$$ 3.33.

where g_A and g_V are the axial vector and vector coupling constant of the nucleon.

Neutrino and antineutrino scattering on nucleon targets provide more information on the quark parton distribution functions. Let

$$\underset{Bj}{\text{Lim}}\; \nu W_2^{\nu N} = f_2^{\nu N},$$ 3.34a.

$$\underset{Bj}{\text{Lim}}\; 2M W_1^{\nu N} = f_1^{\nu N},$$ 3.34b.

$$\underset{Bj}{\text{Lim}}\; \nu W_3 = f_3^{\nu N}.$$ 3.34c.

For spin-$\tfrac{1}{2}$ quarks,

$$f_2^{\nu N} = x f_1^{\nu N}.$$ 3.35.

In a quark model with charm the Cabibbo weak current is given by

$$J_\mu^c = \bar{u}\gamma_\mu(1-\gamma_5)(d \cos\theta + s \sin\theta) + \bar{c}\gamma_\mu(1-\gamma_5)(-d \sin\theta + s \cos\theta).$$ 3.36.

The quantum number charm (34) was originally introduced to eliminate the first-order $\Delta s = 2$ weak interaction effects. It has now been exploited to accommodate the new resonance and the rising behavior in the ratio of the total cross section of e^+e^- annihilation into hadrons to that into μ-pairs near the total center-of-mass energy about 3.9 GeV [see (36) for a review]. If the rise is interpreted as a threshold effect, the quark additivity will set the charm quark mass at between 1.5 and 2 GeV. We assume that scaling works below or above the charm threshold. We discuss the two regions separately.

Below the charm threshold, the current that involves charm quark is inoperative. We find

$$f_1^{\nu p} = 2(\bar{u} + d \cos^2 \theta + s \sin^2 \theta),$$ 3.37a.

$$f_1^{\bar{\nu} p} = 2(u + \bar{d} \cos^2 \theta + \bar{s} \sin^2 \theta),$$ 3.37b.

$$f_3^{\nu p} = 2(\bar{u} - d \cos^2 \theta - s \sin^2 \theta),$$ 3.37c.

$$f_3^{\bar{\nu} p} = -2(u - \bar{d} \cos^2 \theta - \bar{s} \sin^2 \theta).$$ 3.37d.

Corresponding quantities for a neutron target can be obtained from the above by the interchange $u \leftrightarrow d$, $\bar{u} \leftrightarrow \bar{d}$. Above the charm threshold, all quarks contribute. The above equations are modified. Again we find for a proton target

$$f_1^{\nu p} = 2(\bar{u} + d + s + \bar{c}),$$ 3.38a.

$$f_1^{\bar{\nu} p} = 2(u + \bar{d} + \bar{s} + c),$$ 3.38b.

$$f_3^{\nu p} = 2(\bar{u} - d - s + \bar{c}),$$ 3.38c.

$$f_3^{\bar{\nu} p} = -2(u - \bar{d} - \bar{s} + c).$$ 3.38d.

Many relations and sum rules come from these equations. We give a few examples:

Adler's sum rule (16),

$$\int_0^1 dx \, (f_1^{\bar{\nu} p} - f_1^{\nu p}) = 2(2 - \cos^2 \theta)$$ 3.39a.

$$= 2;$$ 3.39b.

Gross & Llewellyn Smith sum rule (37),

$$\int_0^1 dx \, (f_3^{\nu p} + f_3^{\bar{\nu} p}) = 2(-2 - \cos^2 \theta)$$ 3.40a.

$$= 2(-2 - 1) = -6.$$ 3.40b.

The two results in Equations 3.39 and 3.40 are for below and above the charm threshold, respectively.

If one assumes that strange quarks and charm quarks make negligible contributions to electron and neutrino scattering, then one finds the approximate sum rule ($\cos \theta_c \simeq 1$)

$$\frac{1}{2} \int_0^1 dx \, (f_2^{\nu N} + f_2^{\bar{\nu} N}) = \frac{18}{5} \int_0^1 dx \, f_2^{eN},$$ 3.41.

where

$$f_2^{eN} = \tfrac{1}{2}(F_2^{ep} + F_2^{en}).$$ 3.42.

Agreement of this sum rule with experiment is remarkable (38).

There are simpler predictions in the quark parton model without the need for separation of the structure functions. We observe that the differential cross sections for neutrino and antineutrino scattering in the scaling limit can be written as

$$\frac{d\sigma^{\nu N}}{dx\,dy} = \frac{G^2}{\pi} ME[q_{\nu N}(x) + \bar{q}_{\nu N}(x)(1-y)^2],$$ 3.43a.

$$\frac{d\sigma^{\bar{\nu} N}}{dx\,dy} = \frac{G^2}{\pi} ME[\bar{q}_{\bar{\nu} N}(x) + q_{\bar{\nu} N}(x)(1-y)^2],$$ 3.43b.

where, neglecting charm,

$$q_{\nu p}(x) = 2x(d\cos^2\theta + s\sin^2\theta),$$ 3.44a.

$$\bar{q}_{\nu p}(x) = 2x\bar{u},$$ 3.44b.

$$q_{\nu p}(x) = 2xu,$$ 3.44c.

$$\bar{q}_{\nu p}(x) = 2x(\bar{d}\cos^2\theta + \bar{s}\sin^2\theta).$$ 3.44d.

Corresponding quantities for a neutron target can be obtained from those for a proton by the interchange $u \leftrightarrow d$, $u \leftrightarrow d$. Under the assumption that the antiquark contributions in a nucleon target are negligible and the approximation $\theta_c \simeq 0$, the averaged cross sections take the simple form

$$\frac{1}{2}\left(\frac{d\sigma^{\nu p}}{dx\,dy} + \frac{d\sigma^{\nu n}}{dx\,dy}\right) = \frac{G^2}{\pi} ME\,q(x),$$ 3.45a.

$$\frac{1}{2}\left(\frac{d\sigma^{\bar{\nu} p}}{dx\,dy} + \frac{d\sigma^{\bar{\nu} n}}{dx\,dy}\right) = \frac{G^2}{\pi} ME\,q(x)(1-y)^2.$$ 3.45b.

Two predictions immediately follow:

$$\sigma_{tot}^{\nu,\bar{\nu}} = c^{\nu,\bar{\nu}}E,$$ 3.46a.

$$\frac{\sigma^{\nu p} + \sigma^{\nu n}}{\sigma^{\bar{\nu} p} + \sigma^{\bar{\nu} n}} = \frac{1}{3}.$$ 3.46b.

The data (38) on the total cross sections are consistent with a linear rise with energies. The best fits are, for $2 < E < 14$ GeV,

$$\sigma_\nu = 0.74 \times 10^{-38}E \text{ (GeV) cm}^2/\text{nucleon}$$ 3.47a.

$$\sigma_{\bar{\nu}} = 0.28 \times 10^{-38}E \text{ (GeV) cm}^2/\text{nucleon}.$$ 3.47b.

The mean value for the ratio of cross sections in the same energy range is also close to $\tfrac{1}{3}$ (38). It is

$$\frac{\sigma^{\nu p} + \sigma^{\nu n}}{\sigma^{\bar{\nu} p} + \sigma^{\bar{\nu} n}} = 0.37 \pm 0.02.$$ 3.48.

However, the Harvard-Pennsylvania-Wisconsin-FNAL collaboration has reported recently the observation of anomalies in antineutrino y-distribution (39). For $E_{\bar{\nu}} > 30$ GeV and $x < 0.1$, the following features are observed: (a) the y distribution appears to be flat rather than $(1-y)^2$ as predicted above and (b) near $y \simeq 0 (\sigma^{\bar{\nu}}/\sigma^{\nu}) = \frac{1}{3}$. Feature a has not been confirmed by the Caltech experiment (40). It would be difficult to explain this effect by introducing a large antiquark contribution, since it will contradict the CERN results (Equation 3.47) at lower energies. The effect of feature b depends on the normalization procedure. If confirmed, it implies a large violation of charge symmetry, since charge symmetry and positivity lead to the constraint

$$(1-y)^{-2} \geq \frac{\sigma^{\bar{\nu}}(Q^2, \nu, y)}{\sigma^{\nu}(Q^2, \nu, y)} \geq (1-y)^2 \qquad\qquad 3.49.$$

for any admixture of V and A if strangeness-changing weak current is neglected $(\theta_c \simeq 0)$.

The real significance of these effects is not clear at the present, both theoretically and experimentally. These effects may also be related to the so-called dimuon events observed in neutrino scattering, where two muons have been detected in the final states (39, 40). The HPWF analysis (39) concludes that these dimuon events can only be explained in terms of production of new particles with new quantum numbers.

4   APPLICATIONS OF PARTON MODEL

In this section we apply the parton model to several processes:[6] (a) inclusive single-hadron cross section in deep inelastic electron scattering; (b) inclusive single-hadron spectrum and the total cross section in electron-positron annihilation; and (c) massive lepton pair production in hadron-hadron collisions.

We show that these processes can be discussed very naturally within the framework of the parton model. Yet except in the total cross section of e^+e^- annihilation, they are not accessible in the formal approach of light cone algebra. In parallel to the parton distributions in a hadron, the distribution functions of hadrons in a parton are introduced to discuss the detection of hadrons in the final states. We do not attempt to answer the important and serious question of why quarks do not come out if they are indeed the partons. Comments on this question will be made later, however.

4.1   *Inclusive Single-Hadron Spectrum in Deep Inelastic Electron Scattering*

The differential cross section for the inclusive process.

$e + p \rightarrow e' + h + $ anything

---

[6] See (3) and (41) for discussions of rapidity distributions, multiplicities of final states, and other related topics not treated in this section.

is given by (42),

$$\frac{d^4\sigma}{dQ^2\, dv\, d\kappa_1\, dv_1} = \frac{4\pi\alpha^2}{(Q^2)^2}\left(\frac{E'}{E}\right)\left[\mathscr{W}_2\cos^2\frac{\theta}{2} + 2\mathscr{W}_1\sin^2\frac{\theta}{2}\right],$$ 4.1.

where Q^2, v, E, E', and θ are the familiar variables defined in Section 2. We have introduced new variables in terms of the four-momentum P_1 and mass M_1 of the detected hadron

$$v_1 = P_1 \cdot q/M_1,$$ 4.2a.

$$\kappa_1 = P \cdot P_1/M.$$ 4.2b.

The two structure functions $\mathscr{W}_{1,2}$ are defined by

$$\mathscr{W}_{\mu\nu} = 4\pi^2 \frac{E_P}{M}\sum_n \int d^3P_1\, \delta\left(\kappa_1 - \frac{P\cdot P_1}{M}\right)\delta\left(v_1 - \frac{P_1\cdot q}{M_1}\right)\langle P|J_\nu(0)|P_1 n\rangle$$

$$\times \langle nP_1|J_\nu(0)|P\rangle(2\pi)^4\delta^4(q + P - P_1 - P_n)$$ 4.3.

$$= -\left(g_{\mu\nu} - \frac{q_\mu q_\nu}{q^2}\right)\mathscr{W}_1(q^2, v, \kappa_1, v_1) + \frac{1}{M^2}\left(P_\mu - \frac{P\cdot q}{q^2}q_\mu\right)$$

$$\times \left(P_\nu - \frac{P\cdot q}{q^2}q_\nu\right)\mathscr{W}_2(q^2, v, \kappa_1, v_1),$$

where J_μ is the hadronic electromagnetic current and a spin average is understood. We have made the immediate simplification of integrating over the azimuthal angle of the detected hadron. If we describe the deep inelastic electron scattering in the Breit frame as depicted in Figure 5, it is clear that we must distinguish whether the detected hadron originates from the group of partons going in the same direction as the initial proton (target fragmentation) or from the struck parton moving in the opposite direction (current fragmentation).

In the current fragmentation region it is obvious from Figure 5 that the probability of detecting a particular hadron is proportional to the product of the distribution of a parton in the proton and the distribution of the hadron inside the struck parton. The precise connection is (42–44)

$$\frac{d^4\sigma}{dx\, dy\, dz\, dp_\perp^2} = \left(\frac{d\sigma}{d\Omega}\right)_{\text{Mott}}(2\pi)\left(\frac{M}{E}\right)\frac{1}{1-y}\left\{x\sum_i c_i^2[f_i(x) + \bar{f}_i(x)]D_h^i(z, p_\perp)\right.$$

$$\left. + \frac{x}{2}\sum_i e_i^2[f_i'(x) + \bar{f}_i'(x)]D_h^i(x, p_\perp)\frac{y^2}{1-y}\right\}.$$ 4.4.

where $y = v/E$ and

$$f_i' = \bar{f}_i' = 0$$ 4.5.

for spin-0 partons and

$$xf_i' = f_i, \qquad x\bar{f}_i' = \bar{f}_i$$ 4.6.

for spin-$\frac{1}{2}$ partons. The function $D_h^i(z, p_\perp)$ is the probability of finding the hadron

h with longitudinal momentum fraction z and transverse momentum $|p_\perp|$ in a parton i. The fraction z can be related to the invariant kinematical variables by (42)

$$z = \kappa_1/\nu = P \cdot P_1/P \cdot q. \qquad 4.7.$$

Equation 4.4 predicts a generalized Bjorken scaling for the inclusive one-particle cross section. If one further assumes that quarks are the partons, then many relations can be derived from Equation 4.4 (3).

We learn from target fragmentation little more than that (a) the detected hadron moves with finite momentum in the laboratory frame—it is one of the fragments left behind after the impact of the virtual photon; and (b) as in the previous case $2M\nu W_1/\nu^2 W_2 = \omega$ for spin-$\frac{1}{2}$ partons and $M\nu W_1 = 0$ for spin-0 partons.

There are only limited data on the inclusive single-hadron cross section. Pipkin's group from Harvard has made measurements at the Cornell synchrotron that thoroughly explore the forward region over a range of $0.6 < Q^2 < 3.9$ GeV^2, and missing mass W between 2.2 and 3.1 GeV. Measurements were made on both H_2 and D_2 (45) for both π^+ and π^-. According to Equation 4.4, at a fixed value of z the cross section should be a function of x. The distributions have a discernible dependence on W but apparently no variation with Q^2. That is to say, the invariant structure function for π's does not seem to scale with ω![7] On the other hand, the SLAC data (46) indicate that the z-scaling works well.

4.2   Electron-Positron Annihilation

In the center-of-mass frame, the differential cross section for

$$e^+ + e^- \to h + \text{anything}$$

is given by (47),

$$\frac{d^2\sigma_h}{dE\,d\cos\theta} = \frac{4\pi\alpha^2}{(q^2)^2} \frac{M^2\nu}{(q^2)^{1/2}}(1 - q^2/\nu^2)^{1/2}\left[2\overline{W}_1(q^2, \nu) + \frac{2M\nu}{q^2}(1 - q^2/\nu^2) \right.$$
$$\left. \times \frac{\nu\overline{W}_2(q^2, \nu)}{2M} \sin^2\theta \right], \qquad 4.8.$$

where E is the energy of the detected hadron; θ is the angle of the hadron momentum $\mathbf{P}$ with respect to the axis defined by the incident colliding e^+ and e^- beams; and ν is defined as

$$\nu = P \cdot q/M. \qquad 4.9.$$

The structure functions $\overline{W}_{1,2}$ are defined by

$$\overline{W}_{\mu\nu} = 4\pi^2 \frac{E_P}{M} \sum_n \langle 0|J_\mu(0)|P_n\rangle\langle nP|J_\nu(0)|0\rangle (2\pi)^4 \delta^4(q - P - P_n)$$
$$= -\left(g_{\mu\nu} - \frac{q_\mu q_\nu}{q^2}\right)\overline{W}_1(q^2, \nu) + \frac{1}{M^2}\left(P_\mu - \frac{P \cdot q}{q^2}q_\mu\right)\left(P_\nu - \frac{P \cdot q}{q^2}q_\nu\right)\overline{W}_2(q^2, \nu).$$
$$4.10.$$

[7] Though this piece of data does not seem to support the parton prediction, it is likely that Q^2 values are still too low to see current fragmentation.

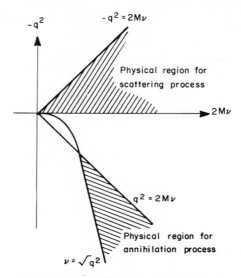

Figure 7 Physical regions in the $(-q^2, 2M\nu)$ plane for inelastic scattering from a proton and e^+e^- annihilation to a proton plus anything.

A spin average over the detected hadron is understood. In Figure 7 is shown the physical region in the $(-q^2, 2M\nu)$ plane corresponding to inelastic scattering from a proton and to e^+e^- annihilation to a proton plus anything.

In the deep inelastic region, electron-positron annihilation into hadrons in the parton model is described by the creation of a parton-antiparton pair and their subsequent decay into final hadrons (Figure 8). In this limit the cross section (Equation 4.8) becomes the scaling form (43, 44, 47)

$$\frac{d\sigma_h}{dz\, d\cos\theta} = \frac{4\pi\alpha^2}{3q^2}\left[\frac{3}{8}(1+\cos^2\theta)\sum_{\substack{i\\ \text{spin-}\frac{1}{2}}} e_i^2 D_h^i(z) + \frac{3}{16}\sin^2\theta\sum_{\substack{i\\ \text{spin-}0}} e_i^2 D_h^{(i)}(z)\right],\qquad 4.11.$$

where

$$z \equiv \frac{2M\nu}{q^2} = \frac{2P^0}{\sqrt{q^2}}\qquad\qquad\qquad 4.12.$$

and the first (second) sum refers to the hadron originating from spin-$\frac{1}{2}$ (spin-0) partons; and $D_h^i(z)$ here is related to $D_h^i(z, p_\perp)$ in Equation 4.4 by

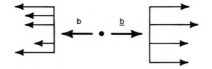

Figure 8 e^+e^- annihilation in the parton model.

$$D_h^i(z) = \int dp_\perp^2 \, D_h^i(z, p_\perp),$$ 4.13.

and it is the probability distribution of a hadron h with a longitudinal momentum fraction z in a parton i. Notice that the angular distribution of the detected hadron is characterized by the spin of its parent parton.

Now by momentum conservation, we have

$$\sum_h \int dz \, z D_h^i(z) = 1,$$ 4.14.

which is a statement that the sum of hadron momenta must be equal to the parton momentum. The relation between inclusive and total cross section

$$\sigma_h = \sum_h \int dz \, d\cos\theta \, z \frac{d\sigma_h}{dz \, d\cos\theta}$$ 4.15.

implies (48)

$$\sigma_h = \frac{4\pi\alpha^2}{3q^2}\left(\sum_{\substack{i \\ \text{spin-}\frac{1}{2}}} e_i^2 + \frac{1}{4}\sum_{\substack{i \\ \text{spin-0}}} e_i^2\right).$$ 4.16.

Recall that the total cross section for e^+e^- annihilation into a muon pair at high energy is

$$\sigma_\mu = \frac{4\pi\alpha^2}{3q^2}.$$ 4.17.

Thus Equation 4.16 implies a constant ratio for σ_h/σ_μ:

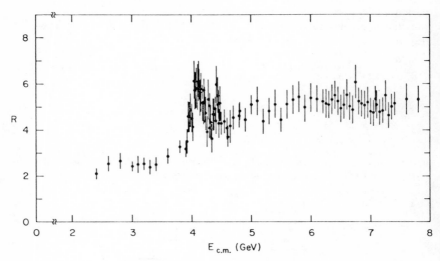

Figure 9 $R_{e^+e^-}$ vs $E_{\text{c.m.}} = (q^2)^{1/2}$.

$$R_{e^+e^-} = \sigma_h/\sigma_\mu = \sum_{\substack{i \\ \text{spin-}\frac{1}{2}}} e_i^2 + \frac{1}{4} \sum_{\substack{i \\ \text{spin-0}}} e_i^2.$$ 4.18.

If quarks are the partons we find

$$R_{e^+e^-} = 2 \qquad \text{(3 triplets)},$$ 4.19a.

$$R_{e^+e^-} = \tfrac{10}{3} \qquad \text{(4 triplets)}.$$ 4.19b.

We now summarize and compare with data the predictions of the parton model in electron-positron annihilation into hadrons at asymptotic energy (49).

1. The data for $R_{e^+e^-}$ from SPEAR are shown in Figure 9. It exhibits three distinct regions over the SPEAR energies. Below $(q^2)^{1/2} = 3.5$ GeV, $R_{e^+e^-}$ is approximately constant at a value near 2.5. Above $(q^2)^{1/2} = 5$ GeV, $R_{e^+e^-}$ is again nearly constant with a value of about 5. Between these two scaling regions there is a complicated transition with several possible broad resonances. This rising transition region for $R_{e^+e^-}$ suggests strongly that there exists a threshold for a new quantum number at this energy range. The discrepancy between the predicted value of 2 and the observed value of 2.5 for $(s)^{1/2} < 3.5$ GeV could be partly or totally attributed to the nonasymptotic nature of the present experiments. If we interpret the rise in $R_{e^+e^-}$ between 3.5–5 GeV as due to the charm production threshold, $R_{e^+e^-}$ should be $\tfrac{10}{3}$ when it flattens out. The observed value of 5 seems to be too large to attribute all the discrepancy to the nonasymptotic corrections. Now, there is some evidence that a heavy lepton pair may have been produced around 4 GeV (50). If confirmed, it will remove most of the discrepancy between theory and experiment.

2. The data for the inclusive single-particle spectrum exhibits the scaling behavior predicted by Equation 4.13 for $z > 0.4$ in the entire energy region 3.0 Gev $< (q^2)^{1/2} < 7.4$ GeV (Figure 10). Above the 4-GeV region, scaling obtains for $z > 0.2$.

3. At $(q^2)^{1/2} = 7.4$ GeV the inclusive angular distribution of hadrons was measured. Particles with low z are produced isotropically, while particles with $z > 0.2$ are produced with an angular distribution

$$\frac{d\sigma}{d\cos\theta} \propto 1 + \alpha\cos^2\theta,$$ 4.20.

$$\alpha = 0.78 \pm 0.12.$$ 4.21.

The observed value (Equation 4.21) shows that partons are predominantly spin-$\frac{1}{2}$ particles. This is consistent with the observation of a small σ_L/σ_T in the deep inelastic electron scattering.

4. In the parton model the picture of e^+e^- annihilation into hadrons is that the virtual photon first creates a pair of quark and antiquark. The quark and antiquark subsequently decay into hadrons with limited transverse momenta with respect to the axis defined by the momentum of the quark and that of the antiquark. Consequently, one expects that for a given event the final hadrons should exhibit

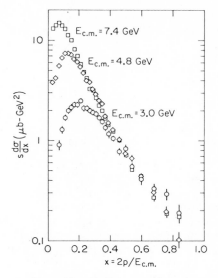

Figure 10 $s(d\sigma/dx)$ vs $x = 2p/E_{c.m.}$.

jet structure. There is strong experimental evidence for jet structure in hadron production by e^+e^- annihilation.

5. It has been conjectured (47) that if we define the scaling functions

$$\underset{Bj}{\text{Lim}}\,(-2M)\bar{W}_1(q^2, v) = \bar{F}_1\!\left(\frac{q^2}{2Mv}\right),\tag{4.22a}$$

$$\underset{Bj}{\text{Lim}}\,v\bar{W}_2(q^2, v) = \bar{F}_2\!\left(\frac{q^2}{2Mv}\right),\tag{4.22b.}$$

then $\bar{F}_1$ and $\bar{F}_2$ are the continuations of the corresponding functions $F_1(x)$ and $F_2(x)$ from $x < 1$ to $x > 1$. This conjecture implies that $\bar{F}_2$ near $x \sim 1$ can be predicted from the measurement of F_2 near $x \sim 1$. An analysis of field theoretical models shows that this is not true in general because of "double discontinuity" terms (51). Gribov & Lipatov (52) have obtained a somewhat general relation

$$\bar{\omega}F_1(\bar{\omega}) = F_1(1/\bar{\omega} = \omega), \qquad \bar{\omega}^3\bar{F}_2(\bar{\omega}) = F_2(1/\bar{\omega} = \omega), \qquad \left(\bar{\omega} \equiv \frac{2P\cdot q}{q^2}\right).\tag{4.23.}$$

If we assume quarks are the partons, one finds that

$$\frac{\bar{\omega}}{\sigma_{\mu\mu}}\frac{d\sigma}{d\bar{\omega}} = \bar{\omega}^3\bar{F}_2(\bar{\omega}),\tag{4.24.}$$

where the right-hand side is also equal to $F_2(1/\bar{\omega} = \omega)$ according to Gribov & Lipatov (52).

Gilman (53) has made the comparison of $(\bar{\omega}\, d\sigma/d\bar{\omega})/\sigma_{\mu\mu}$ for $e^+e^- \to p+$ anything with $F_2(\omega = 1/\bar{\omega})$ for $ep \to e+$ anything. The agreement is good within the uncertainties.

4.3   Massive Lepton Pair Production in Hadron Collisions

Another interesting process that can be analyzed by the parton model method is the massive μ-pair production in hadronic collisions (54),

$$h_1 + h_2 \to \mu^+\mu^- + \text{anything}.$$

In the following we implicitly assume that both h_1 and h_2 are protons. The analysis, of course, goes through with little change for other hadrons.[8] A similar analysis also applies to production of the hypothetical intermediate bosons $W^\pm$. The general expression for the cross section is

$$\frac{d\sigma}{dQ^2} = \frac{4\pi\alpha^2}{3Q^2}\left(1 - \frac{4m^2}{Q^2}\right)^{1/2}\left(1 + \frac{2m^2}{Q^2}\right)$$

$$\times \{[s-(M_1+M_2)^2][s-(M_1-M_2)^2]\}^{-1/2}W(Q^2, s), \qquad 4.25.$$

where m is the muon mass, P_1, M_1 and P_2, M_2 are the 4-momenta and masses of the initial hadrons, and

$$s = (P_1 + P_2)^2. \qquad 4.26.$$

The spin-averaged invariant function $W(Q^2, s)$ is defined by

$$W(Q^2, s) = -16\pi^2 E_1 E_2 \int d^4q\, \delta(q^2 - Q^2) \int d^4x\, \exp(-iq\cdot x)$$

$$\times \langle P_1 P_2 | J_\mu(x) J^\mu(0) | P_1 P_2 \rangle. \qquad 4.27.$$

When the invariant-mass–squared Q^2 of the μ-pair is large, the μ-pairs are produced by the annihilation of a parton from one hadron and an antiparton from another (54) (Figure 11). Thus the cross section is determined by the product of the parton and antiparton distribution functions in their respective parent hadron.

Let x_1 and x_2 be the longitudinal momentum fraction of the parton and antiparton, repectively. Energy and momentum conservation give the relation

$$x_1 x_2 = \frac{Q^2}{s} \equiv \tau. \qquad 4.28.$$

The differential cross section (Equation 4.25) now assumes the simple form in the scaling limit [54, 56–58]

$$\frac{d\sigma}{dQ^2} = \frac{4\pi\alpha^2}{3Q^2}\frac{\tau}{Q^2}\int_0^1 dx_1 \int_0^1 dx_2\, \delta(x_1 x_2 - \tau)\frac{1}{N}\sum_i e_i^2[f_i(x_1)\bar{f_i}'(x_2) + \bar{f_i}(x_1)f_i'(x_2)],$$

$$4.29a.$$

[8] Application of the parton-antiparton annihilation model to photoproduction of muon pairs is discussed by Jaffe (55).

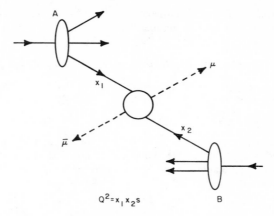

Figure 11   Massive μ-pair production in parton model.

where N is the number of color and f_i and $\bar{f}_i$ (f_i' and $\bar{f}_i'$) refer to the parton and antiparton distribution, respectively, in the first (second) hadron that appears in the structure functions of electron scattering, Equation 3.6. If, in addition to the invariant mass, the longitudinal momentum of the μ-pair in the center-of-mass (c.m.) system is also measured, the differential cross section becomes (59)

$$\frac{d\sigma}{dQ^2\,dp_\parallel} = \frac{4\pi\alpha^2}{3Q^2}\frac{\tau}{Q^2}\int_0^1 dx_1 \int_0^1 dx_2\,\delta(x_1 x_2 - \tau)\delta\!\left(P_\parallel - \frac{(s)^{1/2}}{2}\right)(x_1 - x_2)$$

$$\times \frac{1}{N}\sum_i e_i^2\left[f_i(x_1)\bar{f}_i'(x_2) + \bar{f}_i(x_1)f_i'(x_2)\right].$$

4.29b.

If both initial hadrons are protons and quarks are the partons, we can express $f_i, \bar{f}_i$, etc in terms of the quark distributions inside the proton introduced in the last section.

We would like to make a few remarks on our results.

1. One may be concerned that the parton distributions are first disturbed by the strong interactions of the initial hadrons before they make the μ-pair. Feynman (3) argued that since hadron-hadron interactions are due to exchange of wee partons, the distributions for hard partons will not be altered. We must therefore exclude the wee region and keep Q^2/s finite. A more formal study has been made in the covariant parton model (60, 61). It has been found that Equation 4.29 is the correct result.

2. In principle, the virtual photon that creates the μ-pair can be produced by bremsstrahlung (62) as well as by annihilation. However, kinematics of large Q^2 forces the parton that emits the photon to be far off the mass shell. These diagrams are small by the usual assumption in the parton model that partons are all close to the mass shell.[9]

[9] For a comparison of the parton-antiparton annihilation model to other approaches see Wilson (6).

3. In the quark model the proton is made up by three quarks. Neutrino scattering also suggests that there are very few antiquarks present inside the proton, at least when x is not small. Therefore the μ-pair production from proton-proton collision is difficult in the annihilation picture. It has been suggested that the cross section should be bigger in πp collisions since a pion is made up by a quark and an antiquark. Palmer et al (63) have made a theoretical comparison of the quark parton model without color with three different experiments for $\bar{l}l$ continuum production. (a) The old BNL-Columbia experiment (64) at 29.5 GeV/c, with dilepton laboratory longitudinal momentum cutoff at $p_l \geqq 12.5$ GeV/c. The mass of the muon pair varies from 1 to 5 GeV. There is a shoulder in the data starting at μ-pair mass of 3 GeV. It has never been understood until the discovery of ψ/J. In this comparison the ψ/J

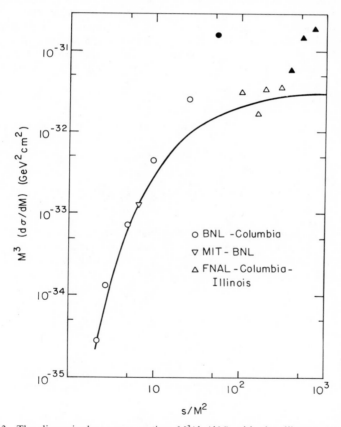

Figure 12   The dimensionless cross section $M^3(d\sigma/dM)$ with the dilepton momentum integrated over the undetected momentum range according to the model used in (63). The black points have $M < 1.5$ GeV. The solid curve is the prediction based on the parameterization of (67).

is subtracted by a Breit-Wigner fit. (b) The MIT-BNL experiment (65) at 30 GeV/c, with dilepton c.m. longitudinal momentum $p_l^* = 0$. The dilepton mass is at 3 GeV (below the J peak). (c) The FNAL-Columbia-Illinois experiment (66) at 320 GeV/c with dilepton laboratory longitudinal momentum cutoff at $p_l \geqq 60$ GeV/c. The dilepton mass varies from 1 to 2.8 GeV.

Notice that these three experiments do not overlap in the variable $\tau = Q^2/s$. These authors used the quark and antiquark distributions obtained by Pakvasa, Parashar & Tuan (67). This particular choice maximizes the lepton-pair cross sections and gives the best agreement with data in the small s/Q^2 region. The result of their analysis is shown in Figure 12.

If one excludes experimental data points for lepton pair masses less than 1.5 GeV, where the contributions from other sources may dominate, the agreement between experiment and theory appears remarkable. However, it must be remembered that if color is present then the theoretical curves are lowered by a factor of three and a significant discrepancy may exist. In any case, a smooth curve can be drawn through the BNL and FNAL points, suggesting that the data are consistent with scaling.[10]

Data have also been reported on e^+e^- pair production in p-Be collisions at 400 GeV. The pair mass region studied is between 2.5 and 20 GeV (68). Duong-van (69) has demonstrated that it is possible to fit the continuum of e^+e^- pairs both in the J/ψ region as well as the region $Q \geqq 4.5$ GeV, if a flat x distribution near $x = 0$ is used $[x \equiv (p_+ + p_-)_{||}/\frac{1}{2}(s)^{1/2}$ in the c.m. system] and a linear dependence on the atomic number A is assumed.

5   QUANTUM FIELD THEORY AND SCALE-INVARIANT PARTON MODEL

The parton model discussed so far assumes that as Q^2 becomes sufficiently large, the virtual photon will see the pointlike and structureless constituents of the hadrons, the partons. In field theory this means that the strength of interactions weakens rapidly as the momentum passes a characteristic mass scale. These are the super-renormalizable field theories characterized by coupling constants with dimension of masses. On the other hand, the coupling constants in renormalizable field theories are dimensionless. There is no time and length scale beyond which interactions become weak. In these theories the constituents of the hadrons, if they exist, will never be precisely structureless and pointlike. A parton seen by a virtual photon at certain values of Q^2 will reveal further substructure for a photon of higher Q^2 (70). Kogut & Susskind (71) have extended the intuitive parton model to describe such scale-invariant interactions. Their approach is based on the Wilson-Kadanoff theory of scaling phenomena (72). Following Wilson (70, 72), it is assumed that matter organizes itself into clusters. For example, molecules are made up of atoms; atoms are made up of nuclei; and nuclei are made up of nucleons, etc.

---

[10] A fairly model-independent comparison of theory and data can be done for the FNAL-Columbia-Illinois experiment. They agree within the uncertainty about experimental normalization (66).

The time and length scale of ordinary hadrons is given by

$$R_0 = 10^{-13} \text{ cm.}$$ 5.1.

Denote the hadrons as $N = 0$ clusters. We assume that these clusters can be described as composites of $N = 1$ clusters that can in turn be described as composites of $N = 2$ clusters, etc. To simplify the discussion we assume that the length scale is reduced by a constant factor Λ as N increases by 1. Thus

$$R_N = \Lambda^{-N} R_0.$$ 5.2.

To resolve the clusters of type N the external probe must have a wavelength comparable to or less than the length scale R_N. If the external probe is a virtual photon, then

$$Q \simeq \frac{1}{\lambda} \gtrsim \frac{1}{R_N}$$ 5.3.

In this section we discuss this intuitive approach to quantum field theory. We find that many of the formal properties established in the renormalization group can be understood in very simple terms. The importance of the renormalization group approach to strong interactions has been emphasized by Wilson (73).

5.1 Coupling Constant Renormalization[11]

It is well known that in field theories coupling constant requires renormalization. In quantum electrodynamics the relation between the renormalized coupling constant e_R and the unrenormalized coupling constant e_0 is

$$e_R = e_0(1 + ce_0^2 \ln \Lambda/m),$$ 5.4.

where Λ is a cutoff introduced to regulate divergences at short distances. Only the second-order term is included in Equation 5.4. This is one of the simplest examples of the relation between the quantities associated with different levels of clusters. Here e_0 is the charge of the (pointlike) bare electrons and e_R is the charge of the (dressed) physical electrons. In the description in terms of discrete scales, we expect a relation to exist between quantities associated with adjacent levels of clusters. In particular, for the coupling constant we write

$$g_{N+1} = f_N(g_N).$$ 5.5.

where g_N is the coupling constant for type-N clusters. The main idea of the renormalization group is the recognition that as N becomes large,

$$f_N \underset{N \to \infty}{\to} f,$$ 5.6.

i.e. the relation between the adjacent levels of description is universal. Equation 5.5 becomes

$$g_{N+1} = f(g_N) \qquad (N \gg 1).$$ 5.7.

[11] Material in this subsection is based on K. G. Wilson's unpublished lectures.

In perturbation expansion,

$$f(g) = g[1 + ag^2 + bg^4 + \ldots].$$ 5.8.

Suppose that the g^4 term can be neglected. In the large-N limit Equation 5.7 gives

$$dg_N/dN = ag_N^3,$$ 5.9.

which can be solved to obtain

$$g_N^2 = \frac{g_0^2}{1 + 2aNg_0^2}.$$ 5.10.

The behavior of g_N as $N \to \infty$ depends critically on the sign of a. We discuss the two cases separately.

In the case $a > 0$ as $N \to \infty$ we obtain

$$g_N^2 = \frac{1}{2a}\frac{1}{N} \qquad (N \gg 1).$$ 5.11.

Notice that the asymptotic value for g_N is independent of the reference value g_0. Furthermore, g_N becomes smaller and smaller as N gets larger and larger. That is, the coupling constant for smaller clusters is smaller. This weakening of coupling strength at short distances is known as asymptotic freedom (31). The only known renormalizable field theory in four dimensions to exhibit this property is the Yang-Mills non-Abelian gauge theories. The decrease of g_N with N is not sufficiently fast to make the field theory truly free at short distances. This is more obvious if we write Equation 5.11 as

$$g^2(Q^2) = \frac{1}{a}\frac{1}{\ln Q^2},$$ 5.12.

where we have made use of Equations 5.2 and 5.3. The approach of $g(Q^2)$ to zero is only logarithmic with Q^2. Bjorken scaling observed in deep inelastic processes may be a manifestation of the asymptotic freedom of the underlying dynamics. This suggests that strong interactions are governed by a non-Abelian gauge field theory.

In the case $a < 0$ (as in QED, where polarization of the vacuum makes a necessarily negative), the result, Equation 5.10, breaks down when

$$2|a|Ng_0^2 \simeq 1.$$ 5.13.

Thus we must keep all the higher-order terms. The differential equation we have to study takes the general form

$$\frac{dg_N}{dN} = \beta(g_N).$$ 5.14.

This gives

$$N = \int_{g_0}^{g_N} \frac{dg}{\beta(g)}.$$ 5.15.

This is the famous Callan-Symanzik equation (74, 75) for the coupling constant. Since the left-hand side of Equation 5.15 diverges as $N \to \infty$, the integral on the right-hand side must also diverge. Apart from the obvious possibility that $g_N \to \infty$ as $N \to \infty$, it suggests the interesting possibility of a fixed point such that

$$g_N \to g_\infty < \infty,$$ 5.16a.

$$\beta(g_\infty) = 0.$$ 5.16b.

Existence of such a fixed point would lead to important and interesting consequences, some of which are discussed below. However, it is not known at the present time whether any of the familiar field theries ($\lambda\phi^4$, QED, etc) possesses such a property.

5.2   Deep Inelastic Electron Scattering

To simplify the discussion we ignore all quantum numbers except energy and momentum. Casher, Kogut & Susskind (76) have shown how to include them in a more detailed study. Equations 5.2 and 5.3 imply a relation between the size of type-N clusters and the wavelength $\lambda = 1/Q$ of the virtual photon that resolves these clusters:

$$N = \ln Q/\ln \Lambda.$$ 5.17.

The parton distributions in the naive parton model are now replaced by the distributions of the just-resolved clusters. At each value of Q^2 the structure function $F(\xi, N)$ then measures the longitudinal momentum distribution of the clusters of type $N \sim \ln Q^2$. As Q^2 increases until $(N+1)$-type clusters are resolved, the clusters of type $N+1$ are now the partons. One can write an integral equation for $F(\xi, N+1)$ in terms of $F(\xi, N)$ by introducing a function $[f_{N+1,N}(x)]/x$ that gives the probability per unit x to find a cluster of type $N+1$ and longitudinal momentum fraction x in a cluster of type N. Then the distribution of clusters of type $N+1$ having a longitudinal momentum fraction satisfies the equation

$$\frac{F(\eta, N+1)}{\eta} = \int_\eta^1 \frac{f_{N+1,N}(\eta/\xi)}{(\eta/\xi)} \frac{F(\xi, N)}{\xi} \frac{d\xi}{\xi}.$$ 5.18.

To solve this equation let us introduce the moments

$$M_\alpha(N+1) = \int_0^1 \eta^\alpha F(\eta, N+1) \frac{d\eta}{\eta}.$$ 5.19.

Substituting Equation 5.18 into Equation 5.19 and interchanging the order of integrations, and we get

$$M_\alpha(N+1) = m_\alpha(N+1, N) M_\alpha(N),$$ 5.20.

where

$$m_\alpha(N+1, N) = \int_0^1 d\eta\, \eta^\alpha \frac{f_{N+1,N}(\eta)}{\eta}.$$ 5.21.

Equation 5.21 is the main result of Kogut & Susskind's work.

Depending on the behavior of g_N for large N, we have to distinguish two cases in discussing Equation 5.21. Consider first the case of asymptotic freedom. As $N \to \infty$ the coupling constant g_N becomes small. Thus $m_\alpha(N+1, N)$ approaches unity, the free field value. Its deviation from unity is proportional to g_N^2. So

$$m_\alpha(N+1, N) = 1 - c_\alpha g_N^2 \qquad 5.22.$$

and

$$M_\alpha(N+1) = (1 - c_\alpha g_N^2) M_\alpha(N). \qquad 5.23.$$

This converts into a differential equation for large N:

$$\frac{dM_\alpha(N)}{dN} = -c_\alpha g_N^2 M_\alpha(N)$$

$$= -\frac{d_\alpha}{N} M_\alpha(N), \qquad 5.24.$$

where use has been made of Equation 5.11 and d_α is defined as

$$d_\alpha \equiv -\frac{c_\alpha}{2a}. \qquad 5.25.$$

Consequently

$$M_\alpha(N) = \left(\frac{1}{N}\right)^{d_\alpha} M_\alpha(N = 0) \qquad 5.26.$$

or

$$M_\alpha(Q^2) = \int_0^1 \frac{d\xi}{\xi} \xi^\alpha F(\xi, Q^2) = M_\alpha (\ln Q^2)^{-d_\alpha}, \qquad 5.27.$$

where

$$M_\alpha \equiv M_\alpha(N = 0). \qquad 5.28.$$

Thus the moments of the structure function are power-behaved in $\ln Q^2$ with α-dependent powers.[12] These results have been obtained by formal means of Wilson's operator product expansion (77) and renormalization-group equations in non-Abelian gauge theories (78). The dimensions d_α are calculable in perturbation theory.

In the case of a fixed point, as $N \to \infty$ the coupling constant g_N tends to a limit independent of N. So $m_\alpha(N+1, N)$ also approaches an N-independent limit:

$$m_\alpha(N+1, N) \underset{N \to \infty}{\to} m_\alpha. \qquad 5.29.$$

Then

$$M_\alpha(N+1) = m_\alpha M_\alpha(N) \qquad 5.30.$$

has the solution

$$M_\alpha(N) = m_\alpha^N M_\alpha(N = 0) \qquad 5.31.$$

[12] A general technique to derive the moment sum rules from operator expansion (77) is given in (79).

or

$$\int_0^1 \frac{d\xi}{\xi} \, \xi^z F(\xi, Q^2) = (Q^2)^{-d_z} M_z,$$
5.32.

where

$$d_\alpha = -\frac{\ln m_\alpha}{\ln \Lambda^2}.$$
5.33.

In this case the moments of the structure function are power-behaved in Q^2 rather than in $\ln Q^2$, as in the previous case. In contrast with the previous case, the anomalous dimensions d_z are not calculable in perturbation.

5.3   Elastic Form Factors

In the simple parton model described in Section 3 a relation has been derived between the elastic form factor and the deep inelastic structure function. We have argued that only if the struck parton in the elastic scattering carries almost all the longitudinal momentum of the hadron can the resulting parton distribution form an outgoing hadron. This condition requires that the longitudinal fraction of the struck parton satisfies

$$1 - \frac{\varepsilon}{Q} \le x \le 1.$$
5.34.

In the scale-invariant parton model partons are the clusters just resolved by the virtual photon of momentum Q^2. The relation applies to the range of Q^2 when only the type of $N = 1$ clusters are resolved. Let us denote by $G_0(Q^2)$ the elastic form factor appropriate for these values of Q^2. Then

$$G_0(Q^2) \sim \int_{1-\varepsilon/Q}^1 F_2(x, N = 0) \, dx.$$
5.35.

Now let us increase the momentum Q until it lies in the second scaling region. The photon is now absorbed by a cluster of type $N = 2$ that is within a particular cluster of type $N = 1$. Since the $N = 1$ cluster is now a composite system, we must apply the previous argument again so that the $N = 1$ cluster will not be broken as well as the hadron. So for Q^2 in this region we have

$$G_1(Q^2) = G_0(Q^2) g_1(Q^2).$$
5.36.

As Q^2 increases further,

$$G_N(Q^2) = G_0(Q^2) g_1(Q^2) \dots g_N(Q^2),$$
5.37.

where g_i can be regarded as the probability that an ith cluster will not break into smaller clusters. Thus the connection between the elastic form factor and the deep inelastic structure function is more complicated in the scale-invariant parton model. To proceed further, additional assumptions are needed. Kogut & Susskind (71) found that in asymptotically free gauge theories, the form factor behaves as[13]

---

[13] A similar result has also been obtained by Gross & Treiman (80).

$$G(Q^2) \to G_0(Q^2)(\varepsilon/Q)^{\bar{g}^2[\ln(\ln Q^2)-1]}. \qquad\qquad 5.38.$$

where $\bar{g}$ is defined by $g_N^2 \to \bar{g}^2/N$. In this model the elastic form factor falls off faster than any power of Q^2. Their result for theories with a fixed point is

$$G(Q^2) \to G_0(Q^2)(\varepsilon/Q)^{-\ln z}, \qquad\qquad 5.39.$$

where $z < 1$ so $-\ln z > 0$.

5.4   Comparison with Experiments

The data presented in Section 2 show that Bjorken scaling works quite well within accuracy of 20–30%. Nevertheless, in finer details the data also show a systematic deviation from the simple prediction. Renormalizable field theories discussed in this section do predict definite patterns of scaling violation. An interesting question is whether the theoretical predictions are compatible with experiments. This question has been studied in detail by W. K. Tung (81) for field theories both with a fixed point and with asymptotic freedom.

Parisi (82) has proposed a procedure to reconstruct the structure functions from the moment sum rules (Equations 5.27 or 5.32). If we assume that for this purpose we can neglect the nonleading contributions on the right-hand side of these sum rules, then these equations (5.27 and 5.32) can be written, respectively, as

$$M_n(Q^2) = \int_0^1 d\xi\, \xi^{n-1} F(\xi, Q^2) = M_n(Q_0^2)\left[\frac{\ln(Q^2/\mu^2)}{\ln(Q_0^2/\mu^2)}\right]^{-d_n} \qquad 5.40.$$

$$M_n(Q^2) = \int_0^1 d\xi\, \xi^{n-1} F(\xi, Q^2) = M_n(Q_0^2)\left[\frac{Q^2}{Q_0^2}\right]^{-d_n} \qquad 5.40b.$$

The structure function $F(x, Q^2)$ can be recovered from the Mellin transform

$$F(x, Q^2) = \frac{1}{2\pi i}\int_{\tau-i\infty}^{\tau+i\infty} dn\, x^{-n} M_n(Q^2), \qquad\qquad 5.41.$$

provided d_n and $F(x, Q_0^2)$ are given.

The anomalous dimension d_n in an asymptotically free gauge theory is completely determined apart from an overall constant factor that depends on the symmetry group (78). In a fixed point theory the d_n's are not known. Tung uses a simple model for d_n that satisfies general constraints imposed by positivity requirement (83). The SLAC data (22, 84) at $Q_0^2 = 10$ (GeV)^2 are then used to calculate $M_n(Q_0^2)$. The scale μ in the asumptotically free case is chosen to be 1 GeV. Tung's conclusions follow.

1. At the present energies available it is impossible to discriminate between one theory and the other. Both asymptotic freedom and fixed point give very similar results.

2. For $0.25 < x < 1$, $F_2(x, Q^2)$ is a decreasing function of Q^2. The rate of decrease in Q^2 is greatest for $x = 0.4$ and tapers off at both ends.

3. In the vicinity of $x \simeq 0.2$, $F_2(x, Q^2)$ has little or no dependence on Q^2.

4. For $0 < x < 0.15$, $F_2(x, Q^2)$ is an increasing function of Q^2. The rate of increase grows monotonically as x approaches zero.

These predictions agree rather well with the μ-N inelastic scattering data from FNAL where the data (Figure 4) have been compared with calculated extrapolations from SLAC data assuming strict scaling.

6 CONCLUSIONS

We have seen that the parton model in its simplest form or as modified by Kogut & Susskind is capable of providing an intuitive and unified description to the complicated processes of deep inelastic electron, muon, and neutrino, as well as antineutrino scattering, e^+e^- annihilation; and perhaps massive muon-pair production in hadron collisions. The parton-model interpretation of these processes suggests that the experimental data can provide direct information on the substructure of hadrons. The identification of charged partons with fractionally charged quarks has led to numerous predictions that all seem to be consistent with experimental data available. Reinforced by the success of the nonrelativistic quark model in predicting the low-lying hadron spectroscopy, we are tempted to believe in quarks being the fundamental constituents of hadrons. However, many difficult problems remain to be solved before one can construct a dynamical theory of strong interactions based on quarks. In this final section of the review, we would like to discuss several of these problems.

So far there is no conclusive evidence that fractionally charged quarks are the fundamental constituents of the hadron (85). In fact, the color quarks of Gell-Mann & Zweig (without charm) (26) and the Han-Nambu's triplet model (86) with integer charge are indistinguishable on the phenomenological level. This follows from Lipkin's observation (87) that as long as the hadrons are singlets in the "Han-Nambu color quantum number" and if the color excitations are not yet accessible experimentally, there is no difference between the two models.

There are certain theoretical arguments that favor Gell-Mann & Zweig's quarks over Han-Nambu's. The former can be readily incorporated into an elegant relativistic quantum field theory with the introduction of color gauge fields. These color gluons are singlets in ordinary SU(3) and do not participate in the electron and neutrino scattering. Their presence does not invalidate the sum rules derived in previous sections. Yet these gluons may help account for the large fraction of momentum not carried by the charged quarks. More importantly, it exhibits asymptotic freedom. It is therefore unique in understanding the Bjorken scaling.

Han-Nambu's color is a quantum number in the full symmetry group. To construct a non-Abelian gauge theory, the vector gauge fields introduced participate in electromagnetic and weak interactions. For instance, the longitudinal cross section in deep inelastic scattering is not expected to be small; the total cross section for e^+e^- annihilation is not simply given by Equation 4.18. Moreover, the electromagnetic and weak currents are not singlets with respect to the gauge transformations. Strong, electromagnetic, and weak interactions should be treated as a whole to maintain gauge invariance and unitarity. This should be done at least in principle at sufficiently high energies.

The most serious problem in the quark parton model is how the quarks can sustain the deep inelastic impacts without breakups into pieces of isolated quarks. The absence of isolated free quarks suggests a strong binding between them. Yet Bjorken scaling and pointlike cross sections indicate that quarks, if they exist, are free most of the time when they are inside the hadrons.

It is not difficult to devise a phenomenological potential description that incorporates both aspects. The potential between quarks must grow without bound with distance to prevent them from escape.[14] The potential must become weak at short distances so that the quarks are temporarily free during the sudden impact of a virtual photon. Among the many models of quark confinement,[15] the linear potential model[16] seems to enjoy special attention. It has been suggested by studies in lattice gauge theories (90, 91). Its application to the charmonium model (93) for J/ψ has led to surprisingly successful predictions for the spectrum and their radiative transitions among the ψ particles (94–103) [for a review see (36)].

How a quark-confining potential such as the linear potential emerges from a fundamental field theory remains one of the most important problems in particle physics. If, indeed, non-Abelian gauge theory provides the correct description for Bjorken scaling and other scaling phenomena, we would like to know what this very same theory predicts for the spectroscopy of low-lying hadrons and their low-energy properties. A correct theory should at least explain hadron dynamics at both ends, high-energy and low-energy.

The rise in $R_{e^+e^-}$ near the $(s)^{1/2} = 3.7$ GeV and the discovery of narrow resonances J/ψ and ψ' in the vicinity strongly suggest the onset of a new threshold. If we continue to employ the quark description of hadron structure, we are led to conclude that extra-heavy quarks exist. This is welcome, since a new quantum number, charm, is long anticipated from theoretical considerations. Most recently it is reported that new and narrow resonances may exist at even higher masses. Are they bound states of even heavier quark-antiquark pairs? Are there any manifestations of purely gluonic degrees of freedom? Is it too naive to imagine that there are only a finite number of constituents for hadrons? Perhaps as we explore the shorter and shorter distances or higher and higher energies new degrees of freedom will reveal themselves.[17]

---

[14] One can visualize such a potential as a string connecting the quarks. Consider a quark-antiquark pair. When the $q\bar{q}$ separation is sufficiently large, it may become energetically more favorable to break the string by creating a $q\bar{q}$ pair in the middle from the vacuum. The original widely separated q-$\bar{q}$ pair can now combine with the $q\bar{q}$ pair created to form two pairs of $q\bar{q}$ systems with much smaller separations. That is, the original $q\bar{q}$ pair has materialized as two physical mesons.

[15] A few examples are the MIT bag (88), the SLAC bag (89), and the lattice gauge theory (90, 91).

[16] The relevance of a linear potential for quark confinement first appears in Tryon's work (92).

[17] For a review on models with more quarks or leptons see (104) and (105).

Literature Cited

1. Feynman, R. P. 1969. *Phys. Rev. Lett.* 23: 1415; Feynman, R. P. 1969. *High Energy Collisions,* New York: Gordon & Breach
2. Bjorken, J. D., Paschos, E. A. 1969. *Phys. Rev.* 185: 1975
3. Feynman, R. P. 1972. *Photon-Hadron Interactions,* New York: Benjamin
4. Bjorken, J. D. 1969. *Phys. Rev.* 179: 1547
5. Landshoff, P. V., Polkinghorne, J. C. 1972. *Phys. Rep. C* 5: 1
6. Wilson, K. G. 1971. *Proc. 1971 Int. Symp. Electron Photon Interactions,* Cornell Univ., Ithaca, NY; Frishman, Y. 1972. *Proc. Int. Conf. High Energy Phys., 14th,* Batavia, Ill.
7. Harari, H. *Proc. 1971 Int. Symp. Electron Photon Interactions,* Cornell Univ., Ithaca, NY; see also Wilson, K. G., Ref. 6
8. Bjorken, J. D., Paschos, E. A. 1970. *Phys. Rev. D* 1: 1450; Brodsky, S. J., Roy, P. 1971. *Phys. Rev. D* 3: 2914
9. Silverman, A. 1975. *Proc. 1975 Int. Symp. Lepton Photon Interactions,* Stanford, Calif.
10. Landshoff, P. V. 1974. *Proc. Int. Conf. High Energy Phys., 17th.* Didcot, Engl: Rutherford Lab.; Sivers, D., Brodsky, S. J., Blankenbecler, R. 1976. *Phys. Rep. C* 23: 1; DiLella, L. 1975. See Ref. 9
11. suri, A., Yennie, D. R. 1972. *Ann. Phys.* 72: 243
12. Bauer, T., Pipkin, F., Spital, R., Yennie, D. R. *Rev. Mod. Phys.* Submitted for publication
13. Drell, S. D., Walecka, J. D. 1964. *Ann. Phys.* 28: 18
14. Hand, L. 1964. *Phys. Rev.* 129: 1834
15. Gell-Mann, M. 1962. *Phys. Rev.* 125: 1067
16. Adler, S. L. 1966. *Phys. Rev.* 143: 1144
17. Bjorken, J. D. 1966. *Phys. Rev. Lett.* 16: 408
18. Kendall, H. 1971. See Ref. 6
19. Mo, L. 1975. See Ref. 9
20. Georgi, H., Politzer, H. D. 1976. *Phys. Rev. Lett.* 36: 1281; Witten, E. 1976. *Nucl. Phys, B* 104: 445
21. Taylor, R. 1975. See Ref. 9
22. Miller, G. et al 1972. *Phys. Rev. D* 5: 528
23. Watanabe, Y. et al 1975. *Phys. Rev. Lett.* 35: 898; Chang, C. et al 1975. *Phys. Rev. Lett.* 35: 901
24. Drell, S. D., Levy, D. J., Yan, T. M. 1969. *Phys. Rev. Lett.* 22: 744; Drell, S. D., Levy, D. J., Yan, T. M. 1969. *Phys. Rev.* 187: 2159; Drell, S. D., Levy, D. J., Yan, T. M.

1970. *Phys. Rev. D* 1: 1035
25. Callan, C. G., Gross, D. 1969. *Phys. Rev. Lett.* 22: 156
26. Gell-Mann, M. 1964. *Phys. Lett.* 8: 214; Zweig, G. 1964. *CERN Rep.* TH401, TH412
27. Drell, S. D., Yan, T. M. 1970. *Phys. Rev. Lett.* 24: 181
28. West, G. 1970. *Phys. Rev. Lett.* 24: 1206
29. Bloom, E., Gilman, F. 1970. *Phys. Rev. Lett.* 25: 1140; Bjorken, J. D., Kogut, J. 1973. *Phys. Rev. D* 8: 1341
30. Lipkin, H. J. 1973. *Phys. Rep. C* 8(3): 173
31. Politzer, H. D. 1973. *Phys. Rev. Lett.* 30: 1346; Politzer, H. D., Gross, D., Wilczek, F. 1973. *Phys. Rev.* 30: 1343; t'Hooft, G. Unpublished
32. Greenberg, O. W., Nelson, C. A. 1968. *Phys. Rev. Lett.* 20: 604; Greenberg, O. W., Nelson, C. A. 1969. *Phys. Rev.* 179: 1354; Bardeen, W. A., Fritzsch, H., Gell-Mann, M. 1973. *Scale and Conformed Symmetry in Hadron Physics,* ed. R. Gatto. New York: Wiley
33. Crewther, R. J. 1972. *Phys. Rev. Lett.* 28: 1421
34. Bjorken, J. D., Glashow, S. L. 1964. *Phys. Lett.* 11: 225; Glashow, S. L., Iliopoulos, J., Maiani, L. 1970. *Phys. Rev. D* 3: 1285
35. Bjorken, J. D. 1967. *Phys. Rev.* 163: 1767
36. Gilman, F., Harari, H. 1975. See Ref. 9
37. Gross, D. J., Llewellyn Smith, C. H. 1969. *Nucl. Phys. B* 14: 337
38. Cundy, D. C. 1974. See Ref. 10; Cundy, D. C., Perkins, D. H. 1975. See Ref. 9
39. Rubbia, C. 1975. See Ref. 9
40. Barish, B. C. 1975. See Ref. 9
41. Bjorken, J. D. 1971. See Ref. 6
42. Drell, S. D., Yan, T. M. 1970. *Phys. Rev. Lett.* 24: 855
43. Berman, S., Bjorken, J. D., Kogut, J. 1971. *Phys. Rev. D* 4: 3388
44. Colglazier, E. W., Ravndal, F. 1973. *Phys. Rev. D* 7: 1537; Gronau, M., Ravndal, F., Zarmi, Y. 1973. *Nucl. Phys. B* 51: 611; Dakin, J. T., Feldman, G. J. 1973. *Phys. Rev. D* 8: 2862
45. Hanson, K. 1975. See Ref. 9
46. Bunnell, K. et al 1976. *Phys. Rev. Lett.* 36: 772
47. Drell, S. D., Levy, D. J., Yan, T. M. 1970. *Phys. Rev. D* 1: 1617
48. Cabibbo, N., Parisi, G., Testa, M. 1970. *Lett. Nuovo Cimento* 4: 35
49. Schwitters, R. F. 1975. See Ref. 9
50. Feldman, G. J. 1975. See Ref. 9
51. Landshoff, P. V., Polkinghorne, J. C. 1973. *Nucl. Phys. B* 53: 473; Dahmen,

H. D., Steiner, F. 1973. *Phys. Lett. B* 43:217; Gatto, R., Vendramin, I. 1973. Unpublished
52. Gribov, V. N., Lipatov, L. N. 1972. *Sov. J. Nucl. Phys.* 15:675; Gatto, R., Menotti, P., Vendramin, I. 1972. *Lett. Nuovo Cimento* 5:754; Gatto, R., Menotti, P., Vendramin, I. 1973. *Phys. Rev. D* 7:2524
53. Gilman, F. 1974. See Ref. 10
54. Drell, S. D., Yan, T. M. 1970. *Phys. Rev. Lett.* 25:316; Drell, S. D., Yan, T. M. 1971. *Ann. Phys.* 66:578
55. Jaffe, R. L. 1971. *Phys. Rev. D* 4:1507
56. Kuti, J., Weisskopf, V. F. 1971. *Phys. Rev. D* 4:3418
57. Landshoff, P. V., Polkinghorne, J. C. 1971. *Nucl. Phys. B* 33:221; 1972. 36:643
58. Altarelli, G. et al 1975. *Nucl. Phys. B* 92:413
59. Farrar, G. 1974. *Nucl. Phys. B* 77:429
60. Cardy, J. L., Winbow, G. A. 1974. *Phys. Lett. B* 52:95
61. DeTar, C., Ellis, S., Landshoff, P. V. 1975. *Nucl. Phys. B* 87:176
62. Berman, S., Levy, D. J., Neff, T. 1969. *Phys. Rev. Lett.* 23:1363
63. Palmer, R. B. et al 1975. Brookhaven prepr. To be published
64. Christenson, J. H. et al 1970. *Phys. Rev. Lett.* 25:1523
65. Aubert, J. J. et al 1974. *Phys. Rev. Lett.* 33:1404
66. O'Halloran, T. 1975. See Ref. 9
67. Pakvasa, S., Parashar, D., Tuan, S. F. 1974. *Phys. Rev. D* 10:2124; 1975. 11:214
68. Hom, D. C. et al 1976. *Phys. Rev. Lett.* 36:1236
69. Duong-van, M. Unpublished
70. Wilson, K. G. 1971. *Phys. Rev. Lett.* 27:690
71. Kogut, J., Susskind, L. 1974. *Phys. Rev. D* 9:697, 3391
72. Wilson, K. G., Kogut, J. 1974. *Phys. Rep. C* 12(2):75; Wilson, K. G. 1975. *Rev. Mod. Phys.* 47:773
73. Wilson, K. G. 1971. *Phys. Rev. D* 3:1818
74. Gell-Mann, M., Low, F. E. 1954. *Phys. Rev.* 95:1300
75. Callan, C. G. 1970. *Phys. Rev. D* 2:1541; Symanzik, K. 1970. *Commun. Math. Phys.* 18:227
76. Casher, A., Kogut, J., Susskind, L. 1974. *Phys. Rev. D* 9:706
77. Wilson, K. G. 1969. *Phys. Rev.* 179:1499
78. Gross, D., Wilczek, F. 1973. *Phys. Rev. D* 8:3633; 1974. 9:980; Georgi, H., Politzer, H. D. 1974. *Phys. Rev. D* 9:416
79. Christ, N., Hasslacher, B., Mueller, A. 1972. *Phys. Rev. D* 6:3543
80. Gross, D., Treiman, S. 1974. *Phys. Rev. Lett.* 32:1145
81. Tung, W. K. 1975. *Phys. Rev. Lett.* 35:490
82. Parisi, G. 1973. *Phys. Lett. B* 43:107; Parisi, G. 1974. *Phys. Lett. B* 50:367
83. Nachtman, O. 1973. *Nucl. Phys. B* 63:237
84. Bodek, A. et al 1974. *Phys. Lett. B* 52:259
85. Llewellyn-Smith, C. H. 1975. See Ref. 9
86. Han, M. Y., Nambu, Y. 1965. *Phys. Rev. B* 139:1006
87. Lipkin, H. J. 1972. *Phys. Rev. Lett.* 28:63
88. Chodos, A. et al 1974. *Phys. Rev. D* 9:347; 10:2599
89. Bardeen, W. A. et al 1975. *Phys. Rev. D* 11:1094
90. Wilson, K. G. 1974. *Phys. Rev. D* 10:2445; Wilson, K. G. 1975. Erice Lect. Notes. To be published
91. Kogut, J., Susskind, L. 1975. *Phys. Rev. D* 11:395
92. Tryon, E. P. 1972. *Phys. Rev. Lett.* 28:1605
93. Appelquist, T., Politzer, H. D. 1975. *Phys. Rev. Lett.* 34:43
94. Appelquist, T. et al 1975. *Phys. Rev. Lett.* 34:365
95. Eichten, E. et al 1975. *Phys. Rev. Lett.* 34:369
96. Eichten, E. et al 1976. *Phys. Rev. Lett.* 36:500
97. Callan C. G. et al 1975. *Phys. Rev. Lett.* 34:52
98. Harrington, B., Park, S. Y., Yildiz, A. 1975. *Phys. Rev. Lett.* 34:168, 706
99. Borenstein, J., Shankar, R. 1975. *Phys. Rev. Lett.* 34:619
100. Kogut, J., Susskind, L. 1975. *Phys. Rev. Lett.* 34:767; Kogut, J., Susskind, L. 1975. *Phys. Rev. D* 12:2821
101. Barbieri, R. et al 1975. *CERN Rep.* TH-2026, TH-2036
102. Pumplin, J., Repko, W., Sato, A. 1975. *Phys. Rev. Lett.* 35:1538
103. Schnitzer, H. J. 1975. *Phys. Rev. Lett.* 35:1540
104. Barnett, M. 1976. *Phys. Rev. D* 13:671
105. Lee, B. W. 1975. See Ref. 9

Ann. Rev. Nucl. Sci. 1976. 26: 239–317
Copyright © 1976 by Annual Reviews Inc. All rights reserved

THE VARIABLE MOMENT ✻5575
OF INERTIA (VMI) MODEL AND
THEORIES OF NUCLEAR
COLLECTIVE MOTION

Gertrude Scharff-Goldhaber and Carl B. Dover
Brookhaven National Laboratory,[1] Upton, New York 11973

Alan L. Goodman[2]
Carnegie-Mellon University, Pittsburgh, Pennsylvania 15213

CONTENTS

[1] Work supported by the US Energy Research and Development Administration.
[2] Work supported by the National Science Foundation. Present address: Tulane University,
New Orleans, Louisiana 70118.

239

Foreword

From the discovery of the atomic nucleus in 1911 arose a remarkably successful (though even now incomplete) theory of atomic structure, in which the nucleus was at first treated as a point. With the discovery of the neutron in 1932 began a slow evolution of a class of models for nuclear structure, now known as "microscopic theories." In such models the nucleus is treated as a collection of A bodies, Z protons, and $N = A - Z$ neutrons interacting with each other through instantaneous two-body potentials. That these are only approximate models, rather than fundamental theories, becomes clearer with time, as the importance of relativistic effects, meson degrees of freedom, and internal nucleon structure is recognized.

Even without these difficulties, the A-body problem is so formidable that "realistic" calculations for $A > 3$ are simplified versions of a fully microscopic theory, in which one truncates the number of degrees of freedom and often makes use of pheno-menologically determined parameters. It cannot be denied that much has been learned, especially about some specific aspects of nuclear structure and the response of the nucleus to external probes, but it is obvious that nuclear theory has not achieved a completeness approaching that of atomic physics and quantum electrodynamics. Therefore it is valuable to study in a phenomenological way nuclear processes that may correspond to excitation of only one or a few degrees of freedom. Understanding of these degrees of freedom and the reasons for their dominance permits development of our intuition about a unique physical object, the quantum droplet.

For such studies the natural sequence might be to examine first the systematics of ground-state properties, and then go on to look for patterns in spectra of low-lying excited states, which could be related to the most "macroscopic" aspects of the nuclear droplet. The recognition of patterns in the ground-state bands of nuclei with even Z and N was slow, and only recently has been quantified in terms of simple and surprisingly accurate empirical rules. These rules indicate that nuclei with widely differing ground-state structure respond in similar ways when they acquire a small amount of energy—they rotate. However, the effect of the rotational motion seems to be different for different ground-state structure; highly deformed, stable nuclei simply rotate, so that the moment of inertia is almost independent of angular velocity. Unstable and less deformed nuclei show a tendency to become more deformed as they rotate faster.

The studies leading to these conclusions depend heavily on information about both excitation energies and electric quadrupole moments. The interplay between the results of these empirical studies and the formulation of various nuclear models is briefly sketched in the historical introduction presented in Section 1. In Section 2 the phenomenological variable moment of inertia (VMI) model and its ramifications are discussed. Section 3 deals with the relation of moments of inertia to transition quadrupole moments and presents two simple models, and Section 4 discusses the nature of transitions from the "VMI phase." Sections 5–8 give a comprehensive survey of theoretical approaches which, while still unable to derive quantitatively the VMI rule, have overcome various limitations arising from previous model assumptions. At the end, we present our conclusions and outline areas requiring further research.

1   Historical Introduction

1.1   ATOMIC HYPERFINE STRUCTURE AND NUCLEAR QUADRUPOLE MOMENTS   When Wolfgang Pauli (1) interpreted Nagaoka's discovery of hyperfine structure in the spectra of Hg atoms as being caused by the angular momentum or spin of the atomic nucleus, he pointed out that the electric quadrupole moments of Hg nuclei appear to be considerably smaller than one might expect ($Q \ll Zed^2$; $d = A^{1/3}$ 10^{-13} cm), implying a high degree of spherical symmetry. He expressed the hope that the study of atomic hyperfine structure might eventually provide important insight into the detailed structure of the nucleus, which he referred to as "das Kerngebäude." Almost twelve years passed until this challenge was taken up by Casimir (2), who evolved a theory of the effect of the nuclear electric quadrupole moment on hyperfine structure. From the measurements by Schüler & Schmidt (3),

Casimir was able to deduce very large quadrupole moments for three nuclei in the rare earth region, namely ^{151}Eu (1.5b), ^{153}Eu (3.2b), and ^{175}Lu ($\sim$ 6.3b).

1.2 PACKING FRACTION, THE CHARGED LIQUID DROP, AND EARLY CLUES TO NUCLEAR SHELL STRUCTURE A very important clue to nuclear structure had meanwhile evolved from a quite different experimental approach. In the late 1920s Aston's pioneering mass spectroscopic results had become extensive enough to allow the conclusion that the packing fraction, which is a measure of the mean binding energy per nucleon, has an almost constant value except for the lightest nuclei. This suggested that forces between nucleons are qualitatively similar to the forces between identical atoms, which Heitler & London (4) had been able to ascribe to the short-range quantum mechanical exchange force between electrons. On this basis Gamow (5) proposed a quite remarkable liquid drop model of the nucleus, consisting mainly of α particles, and held together by surface tension.

Deviations from constancy of the packing fraction for certain mass and charge values, on the other hand, led Stefan Mayer to point out "that the packing fraction seems to show a *periodic* behavior which is related to the well-known periods in the natural classification of the elements" (6). This was clearly an invitation to invent the nuclear shell model!

With the discovery of the neutron and the introduction of the concepts of isotopic spin and charge independence of the strong nucleon-nucleon forces, it became possible to define nuclear models with some precision. Weizsäcker (7) proposed an approximate semiempirical formula for the nuclear ground-state binding energy, which is presented here in the form stated by Bohr & Mottelson (8):

$$B = b_{vol} A - b_{surf} A^{2/3} - \frac{1}{2} b_{sym} \frac{(N-Z)^2}{A} - \frac{3}{5} \frac{Z^2 e^2}{5 R_c},$$

1.1.

where the first term denotes the volume energy. b_{vol} gives the binding energy per nucleon for infinite nuclear matter (9).[3] The second term corresponds to the surface energy E_s, the third term to the symmetry energy, and the fourth term to the Coulomb energy E_c. Here

$b_{vol} \approx 16$ MeV, $b_{sym} \approx 50$ MeV,

$b_{surf} \approx 17$ MeV, $R_c = \gamma_0 A^{1/3} \approx 1.24 A^{1/3}$ fm.

Simultaneously, early versions of the shell model were based on the regularities discovered in relative and absolute isotopic abundances, which gave further evidence on "magic numbers," and the systematics of ground-state spins and magnetic moments of light nuclei. Difficulties were encountered, however, in deducing the larger magic numbers from a plausible coupling scheme (11).

1.3 SEARCH FOR ROTATIONAL BANDS Thibaud (12) in 1930 had suggested that rotations of nuclei may reveal themselves in the fine structure of α spectra. Teller &

---

[3] Unfortunately, recent variational methods for the theoretical calculation of this energy have suggested new uncertainties about the reliability of microscopic theory in its current form (10).

Wheeler (13), following this suggestion, carried out a search for rotational bands in alpha- and beta-ray spectra, in the hope of being able to determine nuclear moments of inertia. Their search was unsuccessful, partly because it concentrated mainly on the spectrum of ^{208}Pb, now known to be a doubly magic nucleus, and partly because the nuclear spectroscopy of the second nucleus investigated, ^{238}U, was still quite incomplete. It is worth mentioning, however, that the authors of this paper expected a spin sequence $J = 1, 2, 3, 4, \ldots$, even parity, as in a polar molecule. They predicted energy levels of 3.3, 10, 20, ... keV according to the equation

$$E_J = \frac{\hbar^2}{2} \frac{J(J+1)}{\mathscr{I}},$$ 1.2.

with $\mathscr{I}$ being the rigid moment of inertia of a sphere with uniform mass distribution and $r_0 = 2 \times 10^{-13}$ cm.

1.4 SLOW NEUTRON CAPTURE, FISSION, AND DYNAMICS OF THE LIQUID DROP   Meanwhile, Niels Bohr (14) had suggested an interpretation of the newly discovered resonance capture of slow neutrons in nuclei in terms of nuclear oscillations very similar to the oscillations of a liquid drop that had been described in detail by Rayleigh. The same model turned out to be a fruitful starting point for the description of fission (15). Frenkel (16), on the basis of similar considerations, arrived at a precise criterion for the occurrence of spontaneous fission:

$$E_c/E_s > 2,$$ 1.3.

where E_c and E_s denote the fourth and second terms in Equation 1.1. Frenkel also surmised that under certain conditions the most stable shape of a nucleus may be deformed. Jensen (17), in his Nobel Prize lecture in 1963, related that to Niels Bohr, the "hot" liquid drop model, which was so successful in the interpretation of compound nucleus formation, seemed incompatible with shell structure. Because of this, and the intervening war, shell model investigations that had been vigorously pursued in the early 1930s were more or less shelved.

1.5 THE INDIVIDUAL PARTICLE SHELL-MODEL   Finally, after the war, the independent particle shell-model could not be refuted any longer, once the difficulties concerning the interpretation of the magic numbers for the somewhat heavier nuclei were resolved by the hypothesis of strong spin-orbit coupling (18). Briefly, this model, for an odd-A nucleus, consists of a spherically symmetric, central potential in whose field the single odd nucleon moves. Both neutrons and protons independently are predicted to fill shells with the magic numbers 2, 8, 20, 28, 50, 82, and 126. From this model followed not only the systematics of binding energies, relative and absolute abundances and ground-state spins (19), and magnetic moments, but also the sequence of spins and magnetic moments of excited states in odd-A nuclei.

1.6 RELATION BETWEEN ENERGY AND TRANSITION RATE FOR SINGLE PARTICLE TRANSITION   Weisskopf computed the rate T for a single particle transition from an initial state J' to a final state J. For an electric quadrupole transition, the γ transition rate for a single proton is

$$T_\gamma[(\text{sec})^{-1}] = 7.23 \times 10^7 A^{4/3} E_\gamma^5(\text{MeV}) = 2.45 \times 10^{60} B(E2)_{\text{sp}}(\text{cm}^4) E_\gamma^5(\text{MeV}),$$

1.4.

where $T_\gamma = [\tau(\text{sec})]^{-1}[1+\alpha]^{-1}$, τ denotes the mean lifetime, E is the transition energy (in MeV), α the total internal conversion coefficient, and $B(E2)_{\text{sp}}$ (in cm^{-4}) denotes the reduced single-particle transition rate (a "Weisskopf unit").

1.7 ODD-A QUADRUPOLE MOMENTS, ENHANCED $E2$ TRANSITIONS IN EVEN-EVEN NUCLEI, AND ROTATIONAL BANDS. THE BOHR-MOTTELSON COLLECTIVE MODEL AND THE EXPANSION OF ROTATIONAL ENERGIES IN TERMS OF $J(J+1)$. The shell model stimulated new interest in the systematic study of electric quadrupole moments of ground states of odd-A nuclei. An earlier result had given an interesting clue: in 1940 Schmidt (20) reported a systematic deviation, as a function of Z, of nuclear quadrupole moments from values expected for a sphere, and pointed out that minima of this function occurred near the "magic numbers" $Z = 50$ and 82. Several authors (21) presented compilations of Q vs the number of the odd nucleons. The results showed a strong dependence of the magnitude and sign of the quadrupole moment on shell structure. The quadrupole moment was found to become larger away from closed shells and to attain negative values just after a shell was filled. This clearly implied a deformation of the nucleus between closed shells. Townes, Foley & Low (22), from a more detailed analysis of the data, then concluded that protons and neutrons tend to be "oriented or polarized to allow maximum overlap between proton and neutron distributions." This result was explained by Rainwater (23), who showed that once a slight deformation exists due to the presence of several nucleons beyond a closed shell, the addition of more extra-core nucleons will lead to increasing spheroidal distortion, assuming the volume remains constant. He thus put Frenkel's hypothesis (16) on a more solid basis. A. Bohr went on to develop the theory of the quantization of angular momentum in the field of a nucleus with spheroidal symmetry, independent of the origin of deformation (24). He pointed out that since in a spheroidal potential the degeneracy in the magnetic quantum number will be removed, the energies of the m substates will vary as functions of the axially symmetric deformation. Later, Nilsson (25) calculated these functions $E(K\pi Nn_z\Lambda)$ using a harmonic oscillator potential, both for positive (prolate) and negative (oblate) deformations. Carefully revised on the basis of empirical data over the years, such "Nilsson diagrams" have become an essential tool in nuclear physics.

Bohr explained the absence of low-lying rotational levels arrived at by Teller & Wheeler (13) by assuming that the nucleus does not rotate like a rigid structure: "It seems quite possible that a rotation of the nucleus involves the motion of only a fraction of the nuclear particles, due to an incomplete rigidity of the nuclear structure, or due to exchange effects of the type considered by Teller & Wheeler."

The successes of the shell model directed the interest of nuclear physicists toward the study of excited states of odd-A nuclei, which in a very short period gave abundant proof of the simplicity and predictive power of the shell model. The classification of isomeric transitions by Goldhaber & Sunyar (26) provided valuable tests for the new individual particle theory of electromagnetic transition rates. In most cases the observed rates were lower. There were, however, some surprising exceptions. While even-even nuclei were generally thought to be rather featureless structures that simply provided a spherical potential for the odd nucleon, there were among the isomeric transitions studied a small number of low-energy ($\sim 80 \rightarrow$ 100 keV) electric quadrupole ($E2$) transitions in even-even rare-earth nuclei that had transition rates two orders of magnitude higher than expected. The authors attributed this "enhancement" to the cooperative nature of these transitions. They pointed out that they occurred exactly in a region where the largest quadrupole moments in odd-A nuclei had been observed, indicating that both phenomena must have the same origin. Further, a survey of the spin-parity of first excited states in even-even nuclei showed that they are predominantly 2+. The implications of this empirical rule were considered in detail from a theoretical point of view by A. Bohr (27).

The low-lying 2+ states in rare earth even-even nuclei depopulated by enhanced $E2$ ($2+ \rightarrow 0+$) transitions, were recognized by Bohr & Mottelson (28) as the missing lowest levels of rotational bands expected from Equation 1.2. They were soon able to carry their analysis further (29). In a review of isomeric transitions (30), they found that the four γ rays following the 5.5-hr transition in ^{180}Hf reported by Burson et al (31) obeyed almost precisely the $J(J+1)$ rule for a cascade $0+ \leftarrow 2+ \leftarrow 4+ \leftarrow 6+ \leftarrow 8+$. A coincidence measurement (32) proved that indeed all four γ's are in cascade, and an angular correlation measurement (33) showed that they are all $E2$ transitions. Further supporting the evidence for rotational bands (29) were 3 triple cascades, in ^{176}Hf, and in two actinide nuclei, ^{226}Ra and ^{228}Th. Although in some of these nuclei the $J(J+1)$ rule was more or less violated, the authors believed that it should suffice to add a correction term

$$\Delta E_J = -BJ^2(J+1)^2, \qquad\qquad 1.5.$$

of the type known from the theory of molecular spectra, attributed to rotation-vibration interaction. It therefore became customary to list $\mathscr{I}$ and B as the parameters characterizing a rotational band.

The properties of rotational bands strengthened Bohr & Mottelson's conviction that a deformed nucleus has cylindrical symmetry, so that the component K of the total angular momentum along the axis of symmetry is a good quantum number. The deformation parameter β provides a measure of the departure from spherical symmetry. The deformed nucleus was assumed to be an essentially incompressible spheroid of constant density, in which neutrons and protons are uniformly distributed (34). In addition to rotation, two types of quadrupole vibrations of considerably higher frequency than that of the rotations ($\hbar\omega \approx 1$ MeV) were thought to be most prominent: beta vibrations around the equilibrium deformation, for which $K = 0$, and for which the spin sequence of the superimposed rotational band is 0, 2, 4, . . . , even parity, and the asymmetry producing gamma vibrations, with $K = 2$ and spin sequence 2, 3, 4, 5, . . . , even parity. (The asymmetry parameter

γ is an angular coordinate. For $\gamma = 0$ or π the deformation is symmetric about the axis of symmetry, and of prolate and oblate shape, respectively. For $\gamma = 30°$, the asymmetry is at a maximum.)

1.8   RELATION BETWEEN MOMENT OF INERTIA AND TRANSITION QUADRUPOLE MOMENT: THE HYDRODYNAMICAL MODEL LEADS TO DISCREPANCY   The hydrodynamical expression proposed for the nuclear moment of inertia defined by Equation 1.2 was the irrotational one (34):

$$\mathscr{I} = \frac{9}{8\pi} AMR_0^2\beta^2,$$ 1.6.

where β is the deformation parameter, M the nucleon mass, and R_0 the nuclear radius. The intrinsic quadrupole moment Q_0, which is related to the reduced transition probability $B(E2)$ by

$$Q_0^2 = \frac{16\pi}{5} B(E2)(0 \to 2),$$ 1.7.

is given by

$$Q_0 = 3/(5\pi)^{1/2}ZR_0^2\beta.$$ 1.8.

By eliminating β from Equations 1.6 and 1.8, one obtains

$$\mathscr{I} = \frac{5M}{8} \frac{A}{Z^2R_0^2}Q_0^2 = \frac{5M}{8R_0^2} \frac{A^{1/3}}{Z^2}Q_0^2.$$ 1.9.

However, this equation led to a discrepancy.

A detailed analysis by Sunyar (35) of the measured energies and lifetimes of $2+$ states in even-even nuclei ranging from ^{152}Sm to ^{198}Hg resulted in the conclusion that the moments of inertia deduced from E_2 values are 4 to 5 times larger than the irrotational values. This is 2 to 3 times smaller than $\mathscr{I}_{\text{rigid}}$. The problem of the origin of the nuclear moment of inertia, to which we soon return, is the main concern of this review.

1.9   FIRST $2+$ STATE ENERGIES IN EVEN-EVEN NUCLEI: ENERGY GAP AND PAIRING FORCE   Soon after the discovery of enhanced $E2$ transitions of $80 \to 100$ keV to the ground states of even-even nuclei in the rare-earth region, there started a largely model-independent phenomenological study of level schemes of all known even-even nuclei, both stable and radioactive. The motivation was the expectation that this procedure might yield some insight into the structure of the even-even core of an odd-A nucleus, which had so far been considered only as the source of the potential in which the odd nucleon moves. This study was based partly on literature reviews and partly on experiments performed with this objective in mind. It went through several stages, which are described below.

The questions asked were: what are the properties of excited states of even-even nuclei, their spins and parities, energies, lifetimes, and electromagnetic moments? What symmetry properties do they reveal? How do they vary with

neutron and proton number? Can analytic expressions be found that describe the
level energies as functions of other level properties such as the angular momentum?

The first stage consisted of a review (36) of the spins, parities, and energies of the
three lowest states in even-even nuclei. In contrast to the variety of the spin-

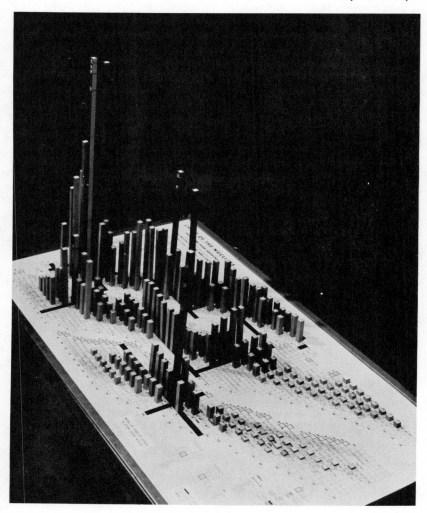

Figure 1   Three-dimensional model of the energies of the first 2+ states of even-even
nuclei vs proton and neutron numbers. Magic numbers are indicated by heavy bars in the
$Z - N$ plane. The model shows that the excitation energies are very large at magic numbers,
but fall off steeply and attain the smallest values midway between magic numbers. The
lowest energies are attained in the larger shells (rare earths and actinides).

parities of levels in odd-A nuclei, it was found that even-even nuclei show astonishingly regular features: without exception the ground states were found to be 0+, the first excited states, with very few exceptions,[4] were 2+, and the overwhelming majority of second excited states were either 4+, 2+, or 0+. These findings were in agreement both with the collective model (27) and the extended shell model (37), which had evolved simultaneously.

The excitation energy of the first excited (2+) state reflects the nuclear shell structure in great detail (Figure 1). From the highest values at doubly magic numbers [$\sim$20 MeV at ^{4}He (not shown), $\sim$6 MeV at ^{16}O] it decreases first rapidly, then more gradually to reach the lowest value approximately midway between the closed neutron and proton shells. The lowest values, $\sim$40 keV, are attained for the heaviest nuclei known (e.g. ^{250}Cf).[5] An unsuccessful attempt was made to represent the energy of the first 2+ state as a function that is separable into a neutron and proton part. This failure implied that neutron-proton interactions contribute to the 2+ energy.

The fact that the spacing between the ground-state (g.s.) and the first excited state near magic numbers is approximately an order of magnitude larger than in odd-A nuclei, instead of being equal or even smaller, presented a puzzle. Also, no "intrinsic" states, i.e. states due to the promotion of one or two nucleons to a higher orbit at energies $\lesssim 1$ MeV are populated in even-even nuclei: this "energy gap" was attributed (39) to the effect of "residual" forces acting between nucleons, i.e. forces not included in the average single-particle potential. It was concluded that in addition to the force that tends to deform the nucleus, there must be a force between like particles that tends to preserve the spherical shape. Also, such a force must be responsible for the reduction of the moment of inertia compared with the rigid-body value, since the Inglis cranking model (40), assuming *noninteracting* particles contained in a rotating external potential, yields $\mathscr{I} = 0$ for closed-shell nuclei, and $\mathscr{I} = \mathscr{I}_{\text{rigid}}$ for nuclei with any number of nucleons beyond

---

[4] The exceptions to the 2+ rule for the first excited state now known are the following:

Nucleus	$I\pi$	E
^{4}He	0+	20.2
^{14}O	(1−)	5.17
^{14}C	1−	6.0932
^{16}O	0+	6.052
^{40}Ca	0+	3.348
^{72}Ge	0+	0.690
^{90}Zr	0+	1.75
^{98}Mo	0+	0.7349
^{208}Pb	3−	2.6145

[5] The shell dependence of the quadrupole excitations is in sharp contrast with the behavior of the giant electric dipole vibrations, which are ascribed to vibrations of protons against neutrons and whose energy, in first approximation, depends only on the mass of the nucleus (38).

closed shells.[6] A force fulfilling these requirements is the pairing force. First, Bohr & Mottelson (42) suggested "a greatly simplified model, in which the whole effect of nucleons outside of closed shells is represented by two interacting nucleons in p-states." As interaction (pairing-force) parameter between the nucleons they chose $v = U/h\omega$, where U is the energy difference between the $J = 0$ and $J = 2$ states of the two nucleons and $h\omega$ the configuration spacing. They were able to fit approximately the empirical $\mathscr{I}$ values [taken to be $(h^2/2E_2)$] as a function of β [as deduced from $B(E2)$] by setting $v = 1/3$. The nuclei included in this survey ranged from $^{150}\mathrm{Nd}_{90}$ to $^{188}\mathrm{Os}_{112}$, with β ranging from 0.2 to 0.45.

1.10 ANALOGY WITH A SUPERCONDUCTOR: PAIRING-PLUS-QUADRUPOLE MODEL AND MOMENT OF INERTIA   An important step forward was made in 1958. Bardeen, Cooper & Schrieffer, and, independently, Bogoliubov had succeeded in explaining superconductivity on the assumption that a weak attractive force exists between two electrons. Following this, Bohr, Mottelson & Pines (43) pointed out that the mathematical approach developed for the understanding of superconductivity might be applied to the "pairing force" in nuclei, while nuclear deformation might be ascribed to a quadrupole-quadrupole interaction between two nucleons. Following their suggestion, Belyaev developed a theory based on the "pairing-plus-quadrupole (PPQ) force." Griffin & Rich (44) and Nilsson & Prior (45) applied the same theory to the problem of the moment of inertia and arrived at fairly satisfactory values for a few nuclei in the rare earth and actinide regions.

Mottelson & Valatin (46) carried the analogy with a superconductor one step further. They argued that just as superconductivity vanishes when an external magnetic field has reached a critical magnitude, so the energy gap vanishes at a critical angular velocity ω_c due to the action of the Coriolis forces.[7] At that point the nuclear moment of inertia will abruptly approach the rigid-body value $\mathscr{I}_{\mathrm{rigid}}$. They stated that the critical angular momentum J_c is given by the equation $J_c = \omega_c \mathscr{I}_{\mathrm{rigid}}$ and estimated that $J_c \simeq 12$ in the rare-earth region and $J_c \simeq 18$ in the actinide region. A search for this effect was thought to be valuable to test the description of the pairing correlations in nuclei. As we show in Section 4 and later sections, an effect of this kind was indeed later discovered, and the estimates for J_c were in remarkably good agreement with observation. However, the phenomena observed above J_c are more complex than Mottelson & Valatin had foreseen.

In Section 2 a phenomenological approach to the problem of the moment of inertia, applicable not only to strongly deformed but to all nonmagic nuclei is discussed.

1.11 FIRST 2+ STATE ENERGY TREATED AS A SCALE FACTOR: THE VIBRATIONAL MODEL   The second stage in the phenomenological study occurred long before

---

[6] A lucid presentation of the cranking model approach is included in Sorensen's review article entitled "Nuclear Moment of Inertia at High Spin" (41).

[7] The role of the Coriolis force was first discussed by Bohr (27).

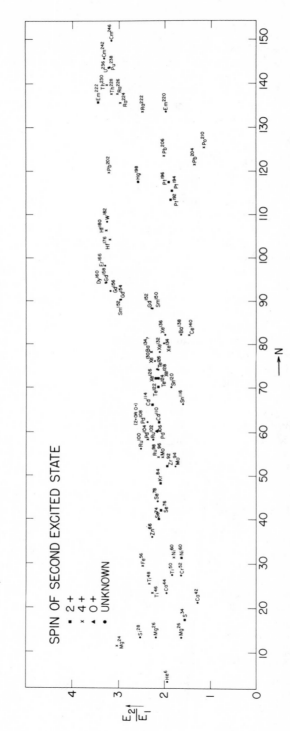

Figure 2 Ratios of E_2/E_1 as function of neutron number N. The character of the first excited state is $2+$ in all cases. The character of the second excited state is denoted by triangles for $0+$, squares for $2+$, crosses for $4+$, and dots for states of unknown spin (unless given in parentheses). An abrupt increase occurs between 88 and 90 neutrons. Low values occur at magic numbers 50 and 82. The figure appeared in (47). The character of the second excited state in all nuclei included with $N \geqq 90$ was later found to be $4+$.

the pairing-plus-quadrupole model was conceived. It was concerned with the variation of the energy (E_2) of the second excited state (47), established to be in general
either $2+$, $0+$, or $4+$. The excitation energy E_1 of the first $2+$ state was taken
as a scale factor. As we have seen, E_1 depends not only on the nuclear volume,
but reflects details of the shell structure, including neutron-proton interactions.
Figures 2 and 3, reproduced from the original paper, demonstrate that the ratios,
E_2/E_1, both as functions of N and of the energy E_1, form definite patterns: they
can be divided into a rotational group for which $E_2/E_1 \simeq 3.33$ and into a second
group, for which $E_2/E_1 \simeq 2.2$, referred to as vibrational or near-harmonic. This
fact, together with previously established rules for the transition rates between
levels of the vibrational group, suggested as a model a spherical nucleus whose
surface is capable of undergoing one, two, or more phonon excitations of the
quadrupole type; hence the first excited state is $2+$, the second a $0+$, $2+$, $4+$
triplet, etc. For the "ideal" vibrator, $E_2/E_1 = 2$, and $[B(E2)(2 \to 1)]/[B(E2)(1 \to 0)] =$
2. By a shell-model calculation involving four nucleons above a closed shell core,
it could be shown that the degeneracy of the triplet state is lifted and the ratio
is increased from 2 to ~ 2.2. An abrupt transition from the vibrational to the
rotational pattern is observed between 88 and 90 neutrons, both for Sm ($Z = 62$)
and Gd ($Z = 64$). This appears to be due to the fact that the $h_{11/2}$ shell "breaks
up" as a function of deformation (48). A similar abrupt change takes place in
the actinide region, in this case between 86 and 88 protons (49). The vibrational
model is still widely used for the analysis of level schemes, although we now know

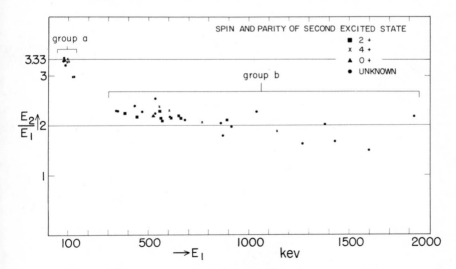

Figure 3   Ratios of E_2/E_1 as function of E_1 for $36 < N < 108$. The same symbols are used
as in Figure 2. A gap in the energy E_1 is seen between the high (i.e. rotational) values
and the low (vibrational) values. At the highest energies the ratio decreases below the
vibrational value. Figure taken from (47).

that it is oversimplified, since already the 2+ state deviates from the spherical shape (see Section 1.13). On the strength of the same empirical evidence, an alternative model was proposed by Wilets & Jean (50), namely that of a "γ-unstable" nucleus. The empirical evidence on level energies then presented was recently confirmed (51), on the basis of the much more numerous data accumulated meanwhile.

1.12   "TRANSITIONAL" LEVEL SCHEMES IN EVEN-EVEN Os NUCLEI AND MALLMANN CURVES   The abruptness in the transitions between the vibrational and rotational regions suggested the possibility that there might be no intermediate solution, i.e. that either the forces that bring about the spherical shape take over, or, if several nucleon pairs of both kinds are present, the aligning forces will give the nucleus an appreciable deformation. An "intermediate coupling" calculation to answer this problem seemed out of the question. However, a survey of the energies of 2+ states as a function of N and Z suggested (Figure 4) that on the high-A side of the rare-earth region a much more gradual transition occurs than between 88 and 90 neutrons. In particular, the Os $(Z = 76)$ nuclei seemed suitable for systematic study via the positron and K-capture decay of the odd-odd Ir $(Z = 77)$ nuclei. A series of experiments was undertaken at BNL, which led to a surprising result (52): a gradual transition from the rotational to the vibrational level pattern occurred with increasing neutron number. Indeed, a cascade 0+, 2+, 4+, 6+, 8+ was found in a nucleus that had an almost equidistant level pattern, namely ^{190}Os. Such a cascade is now referred to as a ground-state band. During the succeeding years a number

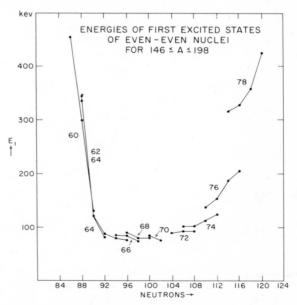

Figure 4   Excitation energies of the first 2+ state for the rare-earths region. The energies in Os $(Z = 76)$ nuclei partly bridge the gap occurring between 88 and 90 neutrons.

of experimental studies of the decay schemes of odd-odd Ir nuclei were carried out (53–55). The results confirmed and greatly extended the early conclusions. Results from several other laboratories, notably at Berkeley (56), further expanded the knowledge of the striking regularities in the excitation patterns of "transitional nuclei." Figure 5 presents a survey of Os level schemes compiled several years ago.

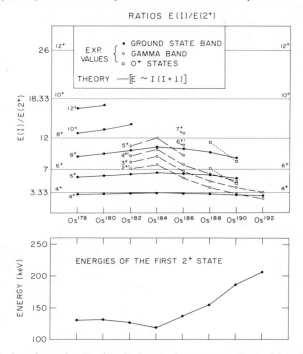

Figure 5   Ratios of energies E_J of excited states in even-even Os nuclei to E_2 (energy of first 2+ state). In the lower part of the figure the energies of 2+ states in Os nuclei with $102 \leq N \leq 116$ are shown. The upper part presents the ratios E_J/E_2 for ground-state bands, denoted by solid circles. These may be compared with the horizontal lines indicating the "perfect rotor" values $J(J+1)$. From the peak values at ^{184}Os, the most neutron-deficient stable isotope of Os (with only 0.018% relative abundance) for which E_2 is at a minimum, the ratios fall off slowly both toward the neutron-deficient side and toward the heavier isotopes. The open circles denote ratios of states in the gamma band. In ^{184}Os the $2'+$ ground state of the ($K = 2$) gamma band (2+, 3+, 4+, 5+) lies above the 6+ state of the ground-state band. An analogous band was populated in ^{182}Os, which exhibits only a slight drop in the ratios. On the neutron-rich side, however, the ratios for the ($K = 2$) bands, while preserving the same spin sequence and general appearance, fall rapidly to lower values so that in ^{192}Os the $2'+$ state lies just below the $4+$ ($K = 0$) state. The ratios R_4 and $R_{2'}$ strongly resemble the corresponding ratios known from two-phonon states of "vibrators." (Two sets of O+ states are known in ^{188}Os and ^{190}Os. By extrapolating the ratios for the lower set to ^{192}Os, the approximate value expected for the O+ member of the triplet results. However, no such O+ state is known in ^{192}Os.) Thus the Os level schemes represent an almost complete transition from rotor (^{184}Os) to vibrator (^{192}Os).

The lower part of the figure presents the energies of the first 2+ states in even-even Os nuclei vs N, and the upper part the ratios E_J/E_2, $K = 0$ (ground-state bands) and E'_J/E_2, $K = 2$ (now referred to as "quasi-gamma bands") vs N, and also a few ratios for excited 0+ states. It is seen that the ground-state band ratios decrease slowly from almost rotational values (indicated by horizontal lines) at ^{184}Os toward the vibrational ones at ^{192}Os. They also decrease toward the neutron-deficient side, where the isotopes become more and more beta-unstable. The ratios for the "quasi-gamma bands" decrease precipitously as the neutron number increases. One interesting feature, not apparent in Figure 5, is the odd-even staggering, particularly well displayed by the quasi-gamma band in ^{186}Os (55).

TRANSITION REGION

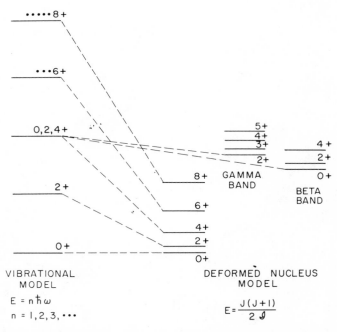

Figure 6   Schematic level schemes of spherical and deformed nuclei. At the right, the ground-state, gamma- and beta-bands of a "good rotor" are indicated. Their components along the symmetry axis are $K = 0, 2$, and 0, respectively. At the left the first three levels predicted for vibrational nuclei are shown. The 0, 2, 4 triplet, here shown as degenerate, is thought to be split by coupling one or two nucleon pairs to the spherical core. The dashed lines indicate the gradual transitions observed in the level schemes of Os nuclei from the states of the two-phonon triplet to the 4+ state of the ground-state band and the 2+ and 0+ ground states of the gamma- and beta-bands. The Mallmann curves (Figure 7) imply the scaling property of the lines connecting the one, two, ... phonon levels with highest spin with the levels of the ground-state band of the good rotor.

The conclusions from this empirical study are summarized in schematic form in Figure 6. The states of the rotational band go over into the highest spin states of the vibrational multiplets, whereas the 2+ ground state of the γ band goes over into the 2+ state of the vibrational two-phonon triplet, and the 0+ ground state of the β band into the 0+ triplet state. In the intervening years, in addition to ground-state bands, "quasi-gamma" and "quasi-beta" bands have been populated in many even-even nuclei. A valuable tabulation of these bands has been provided by Sakai (57).

About two years after the first report on the Os level schemes, Mallmann (58) made an important contribution. Impressed by the similarity of the curves E_J/E_2 vs N for $J = 4, 6, 8$ in the Os nuclei, he plotted the ratios E_6/E_2 and E_8/E_2 vs E_4/E_2 for all known rotational bands, bands for transitional nuclei (Os), and a few other ground-state bands. The latter pertained to nuclei with lower E_4/E_2 values, some of them being singly magic (Figure 7). It turned out that these ratios formed two "universal" curves, i.e. the energies in ground-state bands seemed to "scale!" No existing nuclear theory explained this puzzling fact. Mallmann showed further that the expansion in $J(J+1)$ (Equation 1.5) diverges quite badly already very close to the rotational region (*dashed curves*).

Meanwhile, the detailed analysis of the Os level schemes (55) showed that the ground-state bands, but not the quasi-gamma bands, were in rather good agreement with an axially asymmetric rotor model suggested by Davydov & Filipov (59). The axial asymmetry, according to this interpretation, increased from 16° to 25° between ^{186}Os and ^{192}Os. The fit was not much improved by the inclusion of the interaction of β-vibrational motion with the rotational motion according to Davydov & Chaban (60).

Baranger, who had become intrigued by the interesting regularities and symmetry properties displayed by the even-even Os nuclei, devised a microscopic Hartree-Fock-Bogoliubov theory to explain this behavior (61) (see Section 8). In collaboration with Kumar (62), he analyzed the degree of validity of this method, and was able to give a fairly detailed account of energies and electromagnetic moments of excited states and the transition probabilities between these states. However, the "scaling" behavior did not follow from this theory.

1.13   ELECTRIC QUADRUPOLE MOMENTS OF 2+ STATES IN SPHERICAL NUCLEI AND ENERGIES OF EXTENDED GROUND STATE BANDS IMPLY INCREASE OF MOMENT OF INERTIA WITH ANGULAR MOMENTUM   Following a suggestion of Breit, electric quadrupole moments of 2+ states of several nuclei whose level schemes correspond to the vibrational pattern (Section 1) were determined by means of the "reorientation effect" (63) and found to be rather large. Although the accuracy of determination of static quadrupole moments is not as good (~ 30–100%) as that for transition quadrupole moments (~ 2–10%), the main result was beyond doubt: the vibrational model had to be modified.

A more recent compilation by Christy & Häusser (64) is based on several different methods of determining electric quadrupole moments of 2+ states. The comparison with the transition quadrupole moments shows close correlation in

256

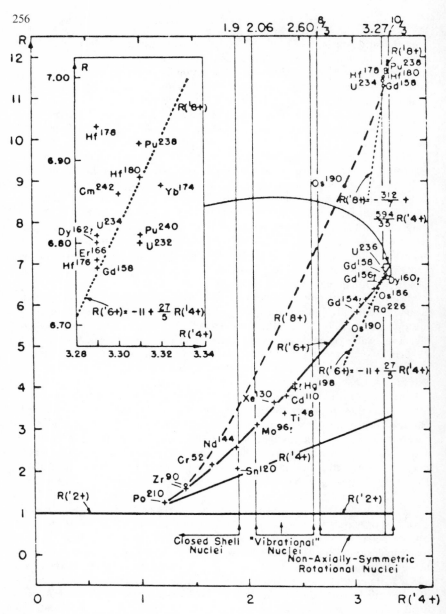

Figure 7 Empirical ratios $E_6/E_2[R('6+)]$ and $E_8/E_2[R('8+)]$ as functions of E_4/E_2. The points lie on two "universal curves." The two-term expansion in $J(J+1)$ (*dotted lines*) is shown to diverge from these curves already close to the rotational limit. Later results showed that the curves do not continue to $E_4/E_2 = 1$, but stop at ~1.8 (see Figure 16). The figure is reproduced from (58).

general. The general behavior of $Q(0+ \rightarrow 2+)$ values is discussed in detail in Sections 3.1 and 3.2.

A new chapter for the exploration of nuclear structure opened when Morinaga & Gugelot (65) devised a method for populating ground-state bands in even-even nuclei by means of $(\alpha, 2n)$ and $(\alpha, 4n)$ reactions. This approach permitted the study of these bands to higher J values than the observation of the decay of natural or artificial radioactive isotopes. The latter is limited to $J \lesssim J'+1$, where J' denotes the spin of the parent nucleus. The new method made use of the high energy and angular momentum imparted to the nucleus by the incoming particle. Soon afterwards it was extended to heavier ions, which permitted the study of even higher angular momenta in the product nucleus, and increased the range of these product nuclei to more unstable species at the neutron-deficient side of the valley of stability (66). Moreover, the study of nuclei undergoing spontaneous fission, in particular of ^{252}Cf (67), and of spallation or fission products obtained by bombardment with very high energy particles, in particular protons, extended the range to the neutron-rich side of the valley of stability. In this way g.s. bands up to $J = 22$ were populated. More recently, Coulomb excitation experiments succeeded in populating very high spin states, e.g. by bombarding ^{238}U with Kr and Xe ions, a 3067.2-keV $(22+)$ state was populated in ^{238}U (68). In all these cases the energy spacings at higher J are smaller than required by the $J(J+1)$ rule. The bands could not generally be fitted by adding a term to the energy quadratic or cubic in $J(J+1)$. Morinaga (69) was the first to suggest that the decrease in spacing was due to an increase in the moment of inertia $\mathscr{I}$. He proposed the term *softness* for the percentage increase of $\mathscr{I}$ per unit change of J, $\Delta\mathscr{I}/\mathscr{I}\Delta J$ and discussed the form of the dependence of this quantity on J as a function of N and Z.

1.14 BETA-STRETCHING MODEL AND SELF-CONSISTENT FOURTH-ORDER CRANKING MODEL FIT BANDS IN STRONGLY DEFORMED NUCLEI   An attempt to give a physical explanation for the increase in $\mathscr{I}$ was made by Diamond, Stephens & Swiatecki (70). They suggested a semiclassical model based on "the idea of a spinning nucleus being stretched out under the influence of the centrifugal force," the so-called beta-stretching model. Here the energy is given by

$$E_J(\beta) = \tfrac{1}{2}C(\beta_J - \beta_0)^2 + \frac{J(J+1)}{2\mathscr{I}(\beta)},$$ 1.10.

where $\mathscr{I}$ is the moment of inertia in terms of $\hbar^2$. The first term represents a kind of rotational Hooke's Law. The equilibrium condition

$$\partial E_J/\partial \beta = 0$$ 1.11.

fixes the value of β_J, if $\mathscr{I}(\beta)$ is a known function. β_0 denotes the g.s. deformation parameter, and C the stiffness parameter. The authors assumed that the relation $\mathscr{I} \propto \beta^2$ holds. Their comparison of Equation 1.10 with g.s. bands in strongly deformed, neutron-deficient nuclei showed good agreement in general.

An important advance was made by Harris (71), who first proposed a consistent expansion of the energy and moment of inertia in powers of ω^2, as an alternative to the $J(J+1)$ expansion of Bohr & Mottelson. He found excellent agreement with the energies in g.s. bands of deformed nuclei. His formalism is a generalization of the Inglis cranking model. As a special case, the Harris model with two parameters may be written

$$[J(J+1)]^{1/2} = \omega(\mathscr{I}_0 + \omega^2/2C), \tag{1.12.}$$

$$E = \frac{1}{2}\left(\mathscr{I}_0 + \frac{3\omega^2}{4C}\right)\omega^2. \tag{1.13.}$$

Elimination of ω leads to an expression for the level energy E in terms of the two parameters $\mathscr{I}_0$ and C.

2   The VMI Model

2.1   AN ANALYTICAL EXPRESSION FOR THE MALLMANN CURVES IN TERMS OF AN EFFECTIVE VARIABLE MOMENT OF INERTIA   Although the energies $E_J(N, Z)$ of states in the g.s. bands of even-even nuclei could not be given in analytical form (Section 1), the time seemed ripe to attempt a derivation of Mallmann's "universal curves" for E_J/E_2 in closed form. As we have seen in Section 1, in such an expression E_2 would have the role of a scale factor. The plan was to include, in addition to strongly deformed nuclei, the transitional, vibrational, and perhaps even magic nuclei (see Figure 7). In searching for such an expression the following considerations were taken into account:

1. The expansion of g.s. band energies in terms of $J(J+1)$ diverges (58) from the observed behavior already close to the "adiabatic limit" (34) ($R_4 = 10/3$).
2. The "moment of inertia" $\mathscr{I}$ deduced from energy spacings in ground-state bands increases steadily with increasing J (69).
3. The static quadrupole moments found in "spherical" nuclei for 2+ states show that these nuclei must also have nonvanishing moments of inertia in the 2+ state.
4. The energies of g.s. bands in even-even Os nuclei, which vary so regularly with neutron number (Figure 5), are best described by assuming that these nuclei have appreciable axial asymmetry. Nevertheless, they are in agreement with the Mallmann scaling law. Thus the analytical expression should not necessarily require axial symmetry.
5. Since the scaling law depends on only two quantities, E_2 and E_4, it should be possible to find a two-parameter expression for the curves E_J/E_2 as a function of E_4/E_2.[8]

The β-stretching formalism (70) appeared to be a promising starting point. However, it presupposes axial symmetry and requires knowledge of the dependence of the moment of inertia on the deformation parameter β. In order to meet these objections, β was replaced (72) by a general variable t, which might include the effects not only of deformation, but also of the effective pairing forces (see also

[8] This approach then seemed to violate two "basic" assumptions of the collective model (34): weak rotation-vibration mixing, and axial symmetry. However, in retrospect these assumptions had no general basis in nuclear dynamics.

Section 5). Least-squares fits to the data were attempted using the relation $\mathscr{I} =$ const t^n with $n = 1, 2, 3$. For g.s. bands ranging from $3.33 \geq E_4/E_2 \geq 2.34$, the best fits were obtained for $n = 1$. Hence, t may be replaced by the effective moment of inertia (73). The level energy is thus given by

$$E_J(\mathscr{I}) = \tfrac{1}{2}C(\mathscr{I} - \mathscr{I}_0)^2 + \tfrac{1}{2}[J(J+1)]/\mathscr{I}. \tag{2.1.}$$

The equilibrium condition

$$\partial E(\mathscr{I})/\partial \mathscr{I} = 0 \tag{2.2.}$$

determines the moment of inertia $\mathscr{I}_J$ (given in units of $\hbar^2$). Equations 2.1 and 2.2 are referred to as the variable moment of inertia (VMI) model.

The parameters C and $\mathscr{I}_0$ are the stiffness parameter and g.s. moment of inertia (for $\mathscr{I}_0 > 0$), respectively. From Equations 2.1 and 2.2 one obtains

$$\mathscr{I}_J^3 - \mathscr{I}_0\mathscr{I}_J^2 = [J(J+1)]/2C. \tag{2.3.}$$

This equation has one real root for any finite positive value of $\mathscr{I}_0$ and C. A least-squares fitting procedure was applied to all measured values E_J in a given band. However, confidence in the VMI model was considerably strengthened when it was found that fitting the two lowest-level energies (E_2, E_4) alone also permits a good fit to higher levels.

2.2   RANGE OF VALIDITY OF THE MODEL   In (73) a softness parameter σ was introduced, which measures the relative initial variation of $\mathscr{I}$ with respect to J. This quantity, which is somewhat analogous to the "initial susceptibility" in ferromagnetism, is obtained from Equation 2.3 as

$$\sigma \equiv [\mathscr{I}^{-1}\,d\mathscr{I}/dJ]_{J=0} = 1/2C\mathscr{I}_0^3. \tag{2.4.}$$

If one defines $r_J = \mathscr{I}_J/\mathscr{I}_0$ and divides Equation 2.3 by $\mathscr{I}_0^3$, one obtains

$$r_J^3 - r_J^2 = \sigma J(J+1). \tag{2.5.}$$

In the "adiabatic limit," i.e. for a perfect rotor, $\sigma = 0$, and hence $r_J = 1$. $R_J = E_J/E_2$ then becomes

$$R_J(\sigma = 0) = \tfrac{1}{6}J(J+1). \tag{2.6.}$$

On the other hand, as $\sigma \to \infty$

$$R_J(\sigma \to \infty) = [\tfrac{1}{6}J(J+1)]^{2/3}. \tag{2.7.}$$

In (73) this was taken as the lower limit of validity of the VMI model. From Equation 2.7 one deduces for this limit $R_4 = 2.23$. Figure 8 presents σ as a function of R_4 for the interval $2.23 \leq R_4 \leq \tfrac{10}{3}$. $\mathscr{I}_0$ vanishes at $R_4 = 2.23$.

2.3   PHYSICAL MEANING OF PARAMETERS   Table 1 of (73) presents experimental values for $E_J(J = 0, 2, 4, \ldots)$ for 88 bands ranging from $A = 108$ to 248, which are compared with the energy values deduced from the VMI law by least-squares fitting. These bands span the range of R_4 from 2.3 to 3.33. For each nucleus the quantities

C, $\mathscr{I}_0$, σ, and $\mathscr{I}_J$ are listed. Figure 9 presents a projective view of $\mathscr{I}_0$ vs N and Z. It is seen that the nuclei nearest closed neutron shells have the smallest $\mathscr{I}_0$ values. As more neutrons and protons are added to closed shells, $\mathscr{I}_0$ rises first steeply and rather smoothly, then less steeply and reaches a peak value approximately at the center of the rectangle defined by $82 \leqq N \leqq 126$ and $50 \ll Z \leqq 82$.

There is, however, no complete particle-hole symmetry in $\mathscr{I}_0$. $\mathscr{I}_0$ does not fall off as steeply on approaching the closed shell at $N = 126$ as it rises above $N = 82$. A behavior of $\mathscr{I}_0$ similar to that in Pt is observed in the Hg ($Z = 80$) isotopes, whose g.s. bands have been studied only recently (74).

The rise in $\mathscr{I}_0$ with increasing number of extra-shell nucleons is further observed for the region $50 \leqq N \leqq 82$, $50 \ll Z \ll 82$, and for $N > 126$, $Z > 82$. It is clear from Figure 9 that the magnitude of $\mathscr{I}_0$ not only reflects the number of extra-shell nucleons, but also the polarization these bring about in the core. The maximum values of $\mathscr{I}_0$, for nuclei ranging from $^{12}\mathrm{C}$ to the actinides, are roughly proportional to $A^{5/3}$.

In Figure 10, the stiffness parameter C is plotted vs A. This graph contains more up-to-date values and includes bands of spherical nuclei as defined in Section 1.

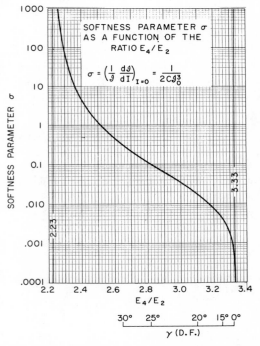

Figure 8   Softness parameter σ plotted vs R_4. (The angular momentum is denoted here by I.) Also shown are the corresponding values of the parameter γ according to the asymmetric model of Davydov & Filippov (59). [This figure was first published in (73).]

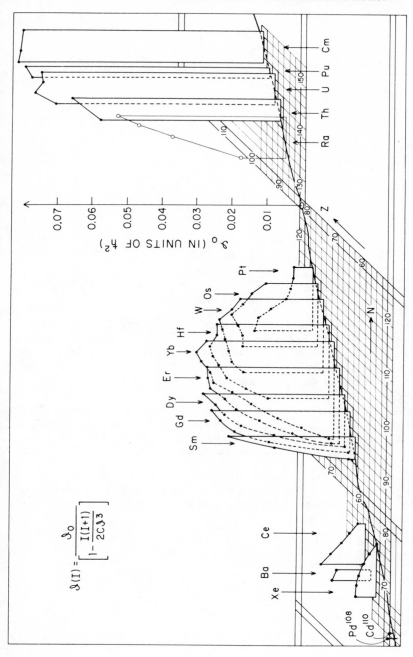

$$\vartheta(I) = \left[\frac{\vartheta_0}{1 - \frac{I(I+1)}{2C\vartheta^3}} \right]$$

Figure 9   Three-dimensional diagram of $\mathscr{I}_0$ (computed from equations 2.1 and 2.3) vs N and Z. [This figure was first published in (73).]

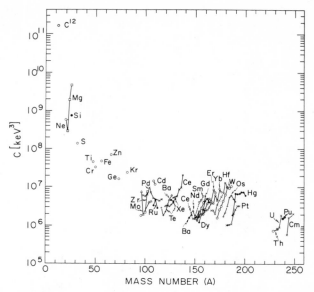

Figure 10 Stiffness parameter C as a function of A. Points for isotopes of the same element are connected by solid lines, which peak at the most stable nuclei.

The stiffness parameter decreases by approximately five orders of magnitude between the nucleus ^{12}C and the heaviest actinides, which, interestingly enough, decay spontaneously. The points for isotopes of the same element are connected by solid lines. For each element one observes a rise of C as the most stable isotope is approached, beyond which point a decrease occurs. It is of interest to note that for the region $82 \ll N \leq 126$, $50 \leq Z \ll 82$, the highest C value occurs for ^{180}Hf. As was mentioned in Section 1.7, this is the nucleus in which the first rotational band was discovered, and which showed very little deviation from $J(J+1)$. In the actinide region, one observes a correlation (75) between the maximum value of C for a given element and the longest spontaneous fission half-life.

2.4 EQUIVALENCE OF VMI MODEL WITH HARRIS'S CRANKING MODEL In (73) the equivalence of the VMI model with Harris's fourth-order cranking model is proven. This complete equivalence is best understood if we write the two sets of equations in the form given below.

$$\text{I.} \begin{cases} E = \mathbf{J}^2/2\mathscr{I} + C(\mathscr{I} - \mathscr{I}_0)^2/2, \\ \left(\dfrac{\partial E}{\partial \mathscr{I}}\right)_J = 0, \end{cases}$$

2.8.

$$\text{II.} \begin{cases} \dfrac{dE}{d\mathbf{J}} = \omega, \\ \text{with} \quad \mathbf{J}^2 = (\mathscr{I}\omega)^2 = J(J+1), \\ \text{and} \quad \mathscr{I} = \mathscr{I}_0 + \omega^2/2C. \end{cases}$$

2.9.

(I) is here analogous to the Newtonian version, and (II), to the Hamilton's equation expression of the same physical model.

2.5   EXTENSION OF VMI MODEL TO NEGATIVE VALUES OF $\mathscr{I}_0$ An extension of the VMI model toward magic nuclei, in the spirit of the "Mallmann curves" (Figure 7), was arrived at (76) on the following grounds: inspection of Figure 9 reveals that $\mathscr{I}_0$ decreases steeply as the magic numbers are approached (e.g. in the rare-earth region as the neutron number decreases toward lower values), and tends to become negative before the magic number (e.g. $N = 82$) is reached. Also, R_J ratios for a number of ground-state bands with $R_4 < 2.23$ had just become available and could be compared with computed values. In general the agreement was found to be good (although not as precise as for higher R_4 values), and the parameters were shown to be physically meaningful. A definite contrast between the computed curves and the original Mallmann curves was found. While the latter (based on very scanty data) continued smoothly toward $R_4 = 1$, the "extension" stops at $R_4 = 1.82$. The solution of the problem is indicated in Figure 11, which presents the left side of Equation 2.5 as a function of $r = \mathscr{I}/\mathscr{I}_0$. The branch on the right corresponds to deformed nuclei for which $\mathscr{I}_0$, and therefore r, is positive; the left, negative, branch corresponds to nuclei for which $\mathscr{I}_0$ is negative. The right side of the equation corresponds to a set of horizontal lines for every value of σ. For the

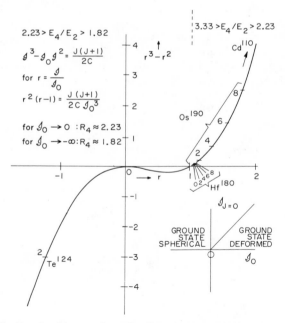

Figure 11   Left side of cubic equation $(r^3 - r^2)$ as function of $r(= \mathscr{I}/\mathscr{I}_0)$. The negative branch pertains to the extension of the VMI model. The inset at the lower right portrays the relation between ground state moment of inertia $\mathscr{I}_{J=0}$ and $\mathscr{I}_0$.

deformed branch, for $J = 0$, one obtains $r = 1$, i.e. $\mathscr{I} = \mathscr{I}_0$. For the spherical branch, however, $r_{J=0} = 0$ and hence $\mathscr{I}_{J=0} = 0$. The relation between the ground-state moment of inertia $\mathscr{I}_{J=0}$ and $\mathscr{I}_0$ is shown in the inset in the right lower part of the figure. The solution for the $2+$ state of a nucleus (^{124}Te) with a spherical ground state is indicated on the left branch. On the right branch, r_J values for a good rotor (^{180}Hf) and a soft rotor (^{190}Os) are shown. No solution exists for $0 < r < 1$. This part of the cubic curve resembles the "Maxwell instability" in the van der Waals equation. Mariscotti (77) attempted to fit bands of closed shell nuclei, using two different roots of the cubic equation for the description of a single band with the transition from one root to the other between the ground state and the first excited state. In this treatment the angular momentum and the angular velocity assigned to the $2+$ state have opposite sign, i.e. $\mathscr{I}(2)$ is negative.

We can now compare the increase of the moment of inertia for a "spherical" nucleus with increasing J, with that of deformed nuclei (Figure 12). One finds that for well-deformed, stable nuclei, e.g. for ^{180}Hf, $\mathscr{I}$ is almost constant as J increases. For nuclei with the *most* deformed ground states, namely those occurring approximately halfway between proton and neutron shells, one finds $\mathscr{I}_0 \propto A^{5/3}$. This relation does not hold, however, for the transition nuclei, as is immediately evident from the coincidence of the curves for ^{120}Xe ($Z = 54$, i.e. four protons beyond the closed shell) and ^{194}Pt ($Z = 78$, four proton holes). For these nuclei the moments of inertia increase steeply from their value at $J = 0$. The most dramatic relative increase of $\mathscr{I}$ with J, however, is found in nuclei with very small positive or with negative values of $\mathscr{I}_0$, e.g. ^{110}Cd and ^{120}Te ($R_4 = 1.99$).[9] The realization that nuclei with $R_4 \sim 2.23$ are extremely "soft" (73) removes the difficulties encountered by the simplest form of vibrational model (47) (which assumes a spherical equilibrium shape for all J), such as the appreciable static quadrupole moments of 2^+ states, (Section 1.13) and the smooth variation of Q_{02} (see Section 3) and of ratios of $B(E2)$'s in the transition from spherical to rotational. As was pointed out above, the parameter C, which determines the steepness of the "potential" $\frac{1}{2}C(\mathscr{I} - \mathscr{I}_0)^2$, is largest, in a given element, for the most stable nuclei. Even for stable isotopes, C takes on smaller values for Xe and Pt, which have relatively low-lying intrinsic odd-parity states, and also for the heavy elements, which have large symmetry energies.

2.6   LOWER LIMIT OF VALIDITY OF EXTENDED VMI MODEL   This limit was deduced as follows: Equation 2.5 may be written in the form:

$$r^2(r-1) = -X, \tag{2.10}$$

[9] For nuclei with $R_4 < 2.23$, the definition (2.4) of the softness parameter is not valid.

$\longrightarrow$

Figure 12   Using Equations 2.1 and 2.3, moments of inertia $\mathscr{I}(J)$ have been computed by least-squares fitting of the energies of ground-state-band levels. Some representative examples of rotational bands (^{180}Hf, ^{170}Hf, ^{238}U, ^{242}Cm, ^{248}Cm); bands in transition nuclei (^{120}Xe, ^{194}Pt); and "vibrational" bands (^{110}Cd, ^{120}Te) are shown. Values for the stiffness parameter C in units of 10^6 keV^3 are given in parentheses. This figure was first published in (76).

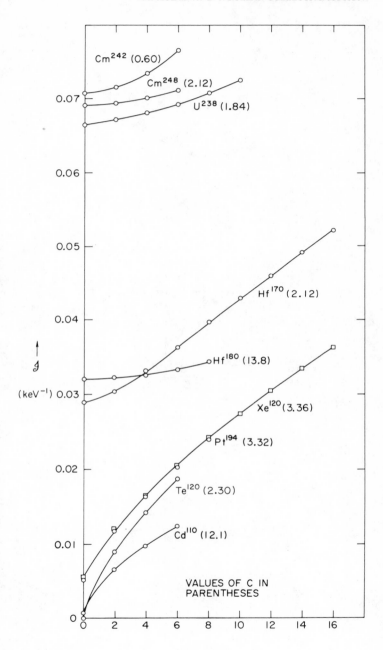

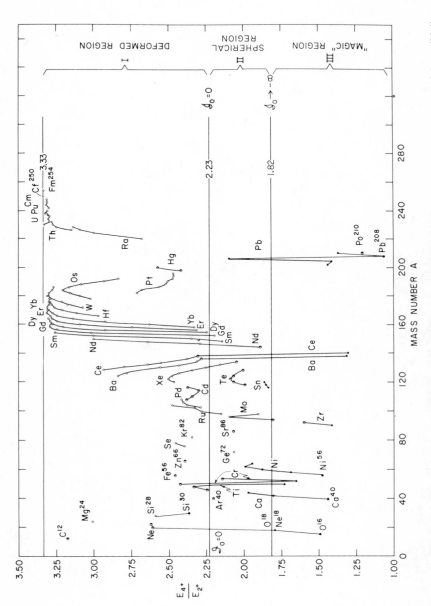

Figure 13 Empirical values of E_4/E_2 as a function of A. Points for isotopes of the same element are connected by solid lines. The horizontal line at 3.33 indicates the perfect rotor limit. The line at 2.23 divides the deformed from the spherical region. At this value the parameter $\mathscr{I}_0$ vanishes. The points lying below this line and above 1.82 (where $\mathscr{I}_0$ reaches $-\infty$), refer to nuclei whose ground state is

with

$$r = \mathscr{I}/\mathscr{I}_0; \quad X = |J(J+1)/2C\mathscr{I}_0^3|.$$

For $r \neq 0$ one can write $r = 1 - X/r^2$ and find graphic solutions for r by plotting each side of the equation separately. For all $X > 0$ there is one real negative root that goes smoothly from $r \approx -(X)^{1/3}$ to $r \approx -(X)^{1/2}$ as X goes from large to small values. For this region one finds $\mathscr{I}_{J=0} = 0$ and therefore, according to Equation 2.1, $E_0 = \frac{1}{2}C\mathscr{I}_0^2.$[10] The limiting value for R_4 is again 2.23 as $\mathscr{I}_0 \to 0$. For large negative $\mathscr{I}_0$ one obtains $R_4 \to (20/6)^{1/2} = 1.82$.

Figure 13 presents the ratio R_4 as a function of A. The interval $2.23 < R_4 < 3.33$ is denoted as the deformed region, the interval $1.82 < R_4 < 2.23$ as the spherical region, and the interval $1 < R_4 < 1.82$, which contains only singly and doubly magic nuclei, as the magic region. In the deformed region the ground-state moment of inertia is $\mathscr{I}_{J=0} = \mathscr{I}_0$, while in the spherical region, as mentioned above, $\mathscr{I}_{J=0}$ vanishes. This means that as $\mathscr{I}_0$ changes sign, $\partial\mathscr{I}_{J=0}/\partial\mathscr{I}_0$ has a discontinuity, which is reminiscent of a second-order phase transition. Negative values for $\mathscr{I}_0$ require the introduction of a new physical concept into the model. This is the notion of "increased resistance to departure from spherical symmetry:" The larger the negative value of $\mathscr{I}_0$, the more firmly the nucleus resists cranking.

2.7   COMPARISON OF EXTENSION OF VMI MODEL WITH EMPIRICAL VALUES   In Figure 14 the established experimental values for R_6 and R_8 are plotted against R_4. Included are all cases with $R_4 < 2.23$ in which γ-ray cascades had been observed, including ^{50}Ti, ^{52}Cr, and ^{90}Zr, whose levels are known from studies of radioactive decay. The solid curves extend the predictions of the VMI model to the spherical region. The experimental points lie along two branches: The first or VMI branch extends leftward from the deformed region. *The points on this branch are well fitted by the extended VMI model and terminate precisely at $R_4 = 1.8$, the natural limit of the model.* The nuclei on this branch in the spherical region are never more than four nucleons from singly magic. The second or magic branch extends to the right from $R_4 = 1$ and consists entirely of doubly and singly magic nuclei for $R_4 < 1.82$.

2.8   THE VMI PHASE   Figure 15 presents R_J ratios vs R_4 for the whole region of validity of the VMI model, and for magic nuclei. The solid lines are computed from Equations 2.1 and 2.3. Ratios for light nuclei are indicated by bold symbols. Although they deviate more from the solid lines than the ratios for heavier ($A \gtrsim 90$) nuclei, they exhibit the same general behavior. This figure was prepared in 1972. In the meantime, ground-state bands have been populated to $J = 22$, and perhaps 24 (in ^{238}U). At high-J states deviations from the VMI ratios have been observed, usually in the direction of smaller energy spacings, i.e. larger moments of inertia. While in some of the ground-state bands the deviations set in gradually, in others they occur abruptly. In the latter case they recalled the "phase transitions" predicted by Mottelson & Valatin (46) at J_c, the "critical angular momentum," at which the

[10] Thieberger (78) proposed an ingenious hydrodynamic analog of the extended VMI model, which explains the meaning of this threshold energy in classical terms.

pairing energy for neutrons or protons suddenly decreases to a low value. Although strictly speaking the concept of phase transition is only applicable to a many-body system, one may define the region in the R_J–R_4 plane below J_c and between $R_4 = 1.82$ and 3.33 as the "VMI phase." J_c is highest at the highest R_4 values and decreases fairly gradually toward $R_4 = 1.82$.

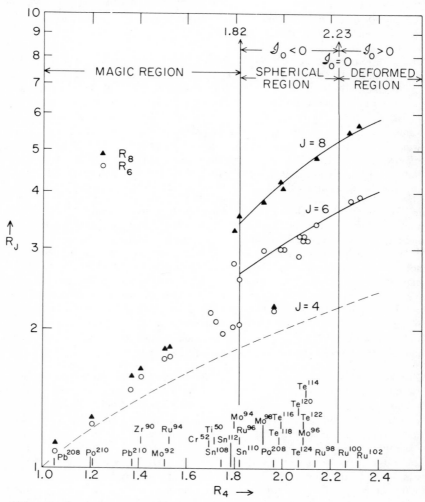

Figure 14   Ratios E_6/E_2 and E_8/E_2 (logarithmic scale) as functions of E_4/E_2 for the magic and spherical regions. The solid curves are computed from Equations 2.1 and 2.3. The dashed line indicates R_4. Only the beginning of the deformed region is shown. In the magic zone the R_4, R_6, and R_8 values are almost degenerate. In ^{208}Po ($R_4 = 1.96$), backbending occurs already above the 4 + state (see Section 4). This figure was reproduced from (76).

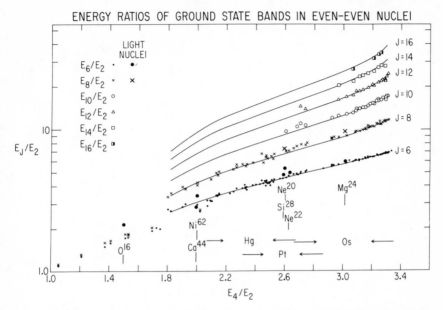

Figure 15 Energy ratios E_J/E_2 plotted vs E_4/E_2 for the whole range of validity of the VMI model: $1.82 \leqq E_4/E_2 \leqq 3.33$. Near the rotational limit, good agreement with the model predictions (*solid curves*) is found in many cases up to $J = 14$ or 16. For lower R_4 values deviations occur at lower J values. The range of isotopes of the transition elements Os and Pt are indicated by arrows. The Hg isotopes range from slightly deformed to spherical. Light isotopes are indicated by bold symbols. At the left ($E_4/E_2 < 1.82$), the almost degenerate $6+$ and $8+$ states of a number of singly or doubly magic nuclei are shown.

2.9   PROTON PARTICLE-HOLE ASYMMETRY IN R_4 AND DEVIATIONS FROM THE VMI MODEL
A striking asymmetry with respect to the $Z = 50$ shell has long been noticed (73), as seen in Figure 13. The R_4 values for the Cd ($Z = 48$) isotopes are above the line separating the deformed from the spherical region, while the values for the Te ($Z = 52$) isotopes lie below. Figure 16 presents more recent data of E_2 and R_4 for the isotopes of the elements $Z = 46$ (Pd), 48 (Cd), 50 (Sn), 52 (Te), and 54 (Xe), as functions of N. The E_2 and R_4 values for Pd and Xe are very similar, but the asymmetry between the two-proton (Te) and two-proton hole nuclei (Cd) appear here to be even more dramatic (79). The non-magic nucleus ^{132}Te ($N = 80$) even violates the rule that only magic nuclei have $R_4 < 1.82$. In Figure 17, the R_6 and R_8 values for Cd and Te are compared with the VMI predictions. A deviation increasing with N is seen for the neutron-rich Te nuclei. This indicates that the phase transition occurs in these nuclei already between $J = 4$ and $J = 6$. The asymmetry has been tentatively attributed to a good overlap between the extra-shell proton pair and an $h_{11/2}$ neutron pair missing from the same major shell ($50 \rightarrow 82$). The deviation of the R_6 and R_8 values in ^{208}Po (Figure 14) appears to be caused by a similar effect.

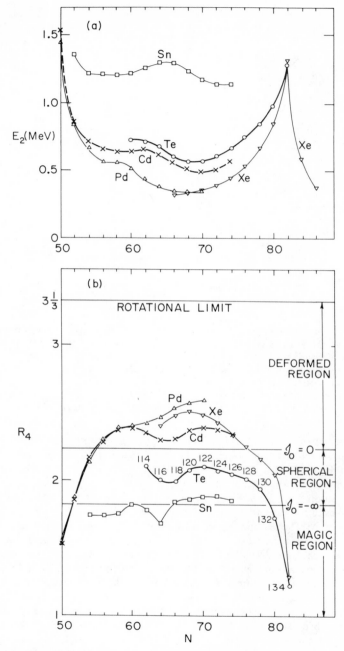

Figure 16 E_2 and R_4 values for $Z = 46$ (Pd), 48 (Cd), 50 (Sn), 52 (Te), and 54 (Xe) are plotted vs N in the top and bottom part of the figure, respectively (see text).

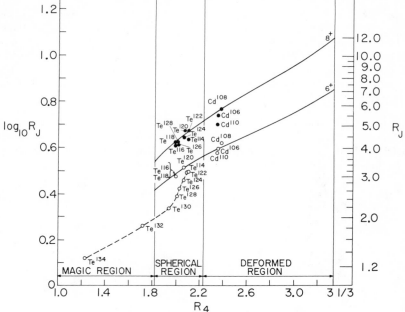

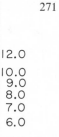

Figure 17 Log plot of R_J vs R_4. The R_J values for the neutron-rich Te nuclei deviate increasingly from the model predictions as $N = 82$ is approached i.e. backbending occurs already above the 4+ state.

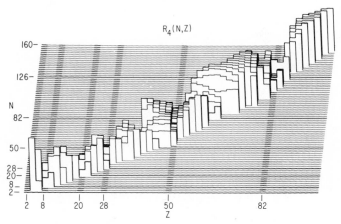

Figure 18 Three-dimensional presentation of R_4 vs N and Z. Magic numbers are indicated by heavier lines. Here R_4 ratios take on minimum values. The abrupt transition from spherical to rotational between 88 and 90 neutrons for $Z = 60$, 62, and 64 is clearly seen. For lower and higher Z values the transition is more gradual. (G.S.G. is indebted to J. K. Kopp for his assistance in the preparation of this figure.)

An interaction of neutrons and protons appears to be also responsible for the change from vibrational to rotational spectra in the rare-earth region. The three-dimensional presentation (Figure 18) of R_4 versus the number of neutrons and protons shows that the previously noted abruptness of the transition in equilibrium shape between 88 and 90 neutrons depends sensitively on the proton number. The transition becomes more gradual on the proton-deficient and proton-rich side of the valley of stability.

2.10   CORRESPONDENCE BETWEEN MODEL PARAMETERS C, $\mathscr{I}_0$, AND h, AND THE ENERGIES E_2 AND E_4 Figure 19 presents the relationship of the VMI model parameters and the energies E_2 and E_4. For a unified treatment of the VMI phase, it is advantageous to replace the softness parameter σ, which has a pole at $R_4 = 2.23$ (see Figure 8), by the "hardness parameter"

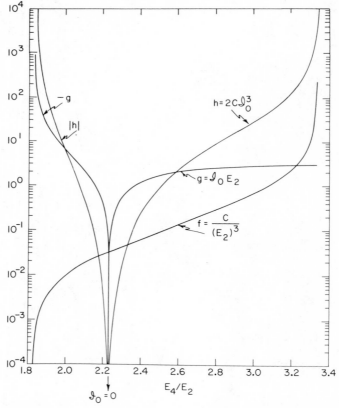

Figure 19   The functions $f = C/(E_2)^3$, $g = \mathscr{I}_0 E_2$, and $h = |2C\mathscr{I}_0^3|$ are plotted vs E_4/E_2. A plot of this type allows the determination of VMI energy values for $J > 4$ using Eq. 2.1. $\mathscr{I}_J$ values can be determined using Figure 11 and the value for h from this figure.

$$h = 1/\sigma = |2C\mathcal{I}_0^3|. \hspace{6cm} 2.11.$$

For $\mathcal{I}_0 < 0$, the original definition given in Equation 2.4 does not hold; nevertheless the parameter provides a useful measure of the increasing resistance to departure from spherical symmetry as $\mathcal{I}_0$ approaches $-\infty$.

We note that starting from the rotational limit the function $g = \mathcal{I}_0 E_2$ falls off quite slowly until it approaches the spherical limit at 2.23. This behavior is responsible for the stability of $\mathcal{I}_0$ discussed in Section 7. The function $f = C/(E_2)^3$, on the other hand, varies by many orders of magnitude in the same region. The effect on the stiffness parameter C of this variation is partly compensated by the variation of $(E_2)^3$ in the opposite direction. Near the rotational limit, the variation of C is relatively unimportant, since there the contribution of the term $C/2(\mathcal{I} - \mathcal{I}_0)^2$ to the energy is very small. However, at lower R_4 values, a relatively small lack in precision of the measured energies can alter the value of C appreciably. Approximate graphical solutions of the VMI parameters for a given ground-state band can be obtained from Figure 19. With the value of $\sigma = 1/h$ so determined, $\mathcal{I}_J$ values may be deduced from Figure 11.

2.11   VMI FITS OF $K = 2$ BANDS IN EVEN-EVEN NUCLEI AND OF GROUND-STATE BANDS IN ODD-ODD NUCLEI   Table 1 lists VMI parameters for $K = 2$ and ground-state bands in a deformed nucleus (^{166}Er) and a transitional nucleus (^{186}Os). For ^{166}Er the two sets of parameters are very similar; the $K = 2$ band is slightly more deformed and less soft. For this nucleus the excitation energy of the ground state of the $K = 2$ band $[E(2'+)]$ is 785.9 keV. The band in ^{186}Os, with $E(2'+) = 767.5$ keV, could not be fitted well because of the odd-even staggering mentioned in Section 1.12. It is, however, less deformed and softer than the ground-state band.

Further compared in Table 1 are ground-state bands in two odd-odd nuclei, one well-deformed (^{164}Ho), and one transitional (^{194}Ir), with the ground-state bands

Table 1   Parameters of $K = 2$ bands of even-even nuclei and ground-state (g.s.) bands of odd-odd nuclei compared with those of ground-state bands of appropriate even-even nuclei. (This table is a reproduction of Table III in reference 73.)

Nucleus		$\mathcal{I}_0$	σ
$_{68}$Er^{166}	g.s.	0.0369	0.0024
	$K = 2$	0.0402	0.0021
$_{76}$Os^{186}	g.s.	0.0215	0.0082
	$K = 2$	0.0176	0.037
$_{67}$Ho^{164}		0.0539	0.000001
$_{66}$Dy^{162}		0.0369	0.0019
$_{77}$Ir^{194}		0.0200	0.071
$_{76}$Os^{192}		0.0124	0.091

of their even-even cores. It had previously been shown (80) that the moment of inertia of an odd-odd nucleus, which always appreciably exceeds that of the even-even core, is in agreement with the assumption that the experimentally determined contributions of the odd proton and that of the odd neutron may simply be added to the moment of inertia of the even-even core to obtain the moment of inertia of the odd-odd nucleus. It is seen from Table 1 that the deformed odd-odd nucleus ^{164}Ho is an almost perfect rotor, and that also the transitional nucleus ^{194}Ir is more deformed and less soft than ^{192}Os.

2.12 MODIFICATIONS OF THE VMI FORMALISM, HIGHER-ORDER CORRECTIONS, AND THEORETICAL JUSTIFICATIONS   Since the VMI model had been shown to fit ground-state bands of a wide range of even-even nuclei, numerous attempts have been made to improve the fits even further, either by using different expressions or higher terms in the angular velocity expansion. Among the latter the lucid and thorough study by Saethre et al (81) stands out. It is dealt with in detail in Sections 5 and 7. Also, both macroscopic and microscopic models have been

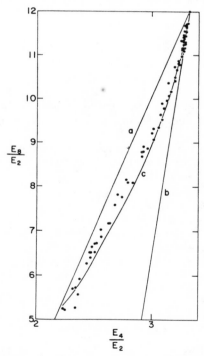

Figure 20 E_8/E_2 is plotted vs E_4/E_2. The solid curves correspond to predictions from (a) the Ejiri formula $E_J = \frac{1}{8}J(J-2)E_4/E_2 - \frac{1}{4}J(J-4)$, which is identical with the prediction from the anharmonic vibrational model; (b) the Bohr-Mottelson two-term expansion in $J(J+1)$; and (c) the VMI model. The figure is reproduced from (83).

proposed in order to explain the observed regularities (Sections 6–8). We wish to discuss here only two energy expressions frequently used by experimentalists for the analysis of their data. The first one is the expansion of E in terms of $J(J+1)$. The second one is a phenomenological expression suggested by Ejiri et al (82), which has the form

$$E_J = aJ + k[J(J+1)]. \qquad 2.12.$$

From Equation 2.12 one obtains the ratios R_J in terms of R_4:

$$R_J = \tfrac{1}{8}J(J-2)R_4 - \tfrac{1}{4}J(J-4). \qquad 2.13.$$

Das, Dreizler & Klein (83), who attempted to fit bands in the "spherical" region by means of an anharmonic vibrational model, arrived at an expression similar to Equation 2.12, except for an additional cubic term in J. These authors then compared three model predictions for R_6 and R_8 as functions of R_4 with the experimental values. Figure 20 presents these comparisons for R_8. Curve a corresponds to Equation 2.13, curve b to the two-term expansion in $J(J+1)$, and curve c to the VMI model. It is evident that b diverges close to the rotational region, as Mallmann had already pointed out, and that c gives a better overall fit for the whole region than curve a, except for $R_4 \approx 2$, where curves a and c give equally good fits.

The Ejiri formalism continues smoothly toward $R_4 = 1$. In the interval between $R_4 = 2$ and $R_4 = 1.5$ the higher states ($J \geq 6$) go below the $4+$ state; at $R_4 = 1.33$ they become degenerate with the $2+$ state; finally, as $R_4 \to 1$, they approach $E = 0$. This inverted order of states has not been observed.

2.13   RELATIONS BETWEEN $E_{2'}/E_2$, $B(E2)(2' \to 2)/B(E2)(2 \to 0)$, AND R_4 For the ideal vibrational nucleus, the enhancement of the transition from the two-phonon state to the one-phonon state is predicted to be twice that from the one-phonon state to the ground state. However, most measured ratios were found to be somewhat smaller than two. The VMI model makes it possible to study the ratio of reduced transition probabilities as a function of the softness parameter σ. First studied (73) was the ratio $[B(E2)(2' \to 2)]/[B(E2)(2 \to 0)]$, which provides a measure of the coupling between the quasi-γ band and the ground-state band. Since σ is a monotonic function of R_4 (Figure 8), the ratio of reduced transition probabilities has been correlated with R_4, (Figure 11, bottom). The top part of Figure 21 presents, on a logarithmic scale, the ratio between the ground-state energy E_2' of the quasi-γ band and the energy E_2. This ratio, of the order of 10 at the rotational limit, rapidly decreases and soon approaches the value ~ 2, which corresponds to the ideal vibrator. In the bottom part, the open circles at $R_4 = 10/3$ and 2 indicate the theoretical predictions ($\ll 1$ and 2) for the rotational and vibrational limits. Starting from the rotational limit, the experimental points lie on a straight line corresponding to the equation:

$$\frac{B(E2)(2' \to 2)}{B(E2)(2 \to 0)} = 5 - \frac{3}{2}R_4. \qquad 2.14.$$

Near the vibrational limit, the statistical errors of the experimental points are

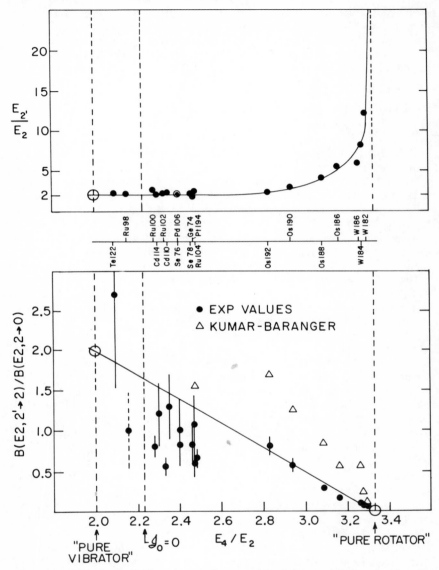

Figure 21 This figure displays the coupling strength of the quasi-gamma band to the ground-state band vs E_4/E_2. The top part shows a logarithmic plot of the ratio of the energy $E_{2'}$ (ground state of the quasi-gamma band) to E_2. In the bottom part the ratio of the $B(E2)$ values for the transitions $2' \rightarrow 2$ to $2 \rightarrow 0$ are presented. The straight line connects the limiting values 2 for $E_4/E_2 = 2$ and $\ll 1$ for $E_4/E_2 = 3.33$. The triangles give Kumar & Baranger's theoretical predictions. This figure is reproduced from (73).

considerably larger. However, with one exception (^{122}Te), there appears to be a tendency for the ratios to fall below the straight line. The regularity exhibited by the experimental points contrasts with the values predicted by the pairing-plus-quadrupole model indicated by triangles. Further experimental and theoretical investigation of the relation of the enhancement ratios, including also the 0+ and 4+ states of the two-phonon triplet, would be of great value.

3   Relation Between Moment of Inertia and Transition Quadrupole Moment

3.1   DEPENDENCE OF RELATION BETWEEN $\mathscr{I}$ AND Q ON A AND Z In Section 1.8 we have seen that the hydrodynamical model predicted a relation between these two quantities of the form

$$\mathscr{I} \propto Q_0^2 A^{1/3}/Z^2. \qquad 3.1.$$

Here $\mathscr{I}$, in units of $\hbar^2$, was taken to be $\mathscr{I} = 3/E_2$. The mass range covered by Sunyar (35) was not extensive enough to allow any definite conclusion on the validity of the A dependence. Grodzins (84) later compared the A and Z dependence of $E_2 \times B(E2)$ for a much larger range of even-even nuclei, namely $12 \leq A \leq 240$, with empirical data and found a deviation of the A dependence from the hydro-dynamical prediction. He proposed for all but a few exceptions adjacent to doubly magic nuclei, the formula

$$1/E_2 \propto B(E2)A/Z^2, \qquad 3.2.$$

whereas, as he pointed out, the A and Z dependence predicted by the hydro-dynamical model (Equation 3.1) as well as by the asymmetric nucleus model (59) and also by the vibrational model (47), is $A^{1/3}/Z^2$. However, for the latter the constant of proportionality is almost three times larger than for a deformed nucleus. Thus an abrupt change would have been expected to occur between 88 and 90 neutrons, but none was found.

3.2   DEPENDENCE OF $\mathscr{I}$ ON Q IN THE VMI MODEL   The VMI model attributes a moment of inertia $\mathscr{I}(J)$ to each state of the ground-state band. We define an "average" moment of inertia $\mathscr{I}_{02} = \frac{1}{2}[\mathscr{I}(0) + \mathscr{I}(2)]$, which may be related to the transition quadrupole moment $Q_{02} = [(16\pi/5)B(E2)(0 \rightarrow 2)]^{1/2}$ for all non-magic even-even nuclei. [It should be kept in mind that in contrast to the previous approaches where $\mathscr{I} \propto 1/E_2$, both $\mathscr{I}(0)$ and $\mathscr{I}(2)$ are deduced from the spectrum of the ground-state band, i.e. from E_2 and E_4.]

A study (73) covering the range $120 < A < 186$ resulted in the relation

$$\mathscr{I}_{02} = Q_{02}^2/k^2, \qquad 3.3.$$

where $k = (39.4 \pm 2.6) \times 10^{-24}$ cm^2 keV$^{1/2}$ is *independent* of A.

After the extension of the VMI model to negative values of $\mathscr{I}_0$, it was possible to correlate data for nuclei with spherical ground states as well as for deformed nuclei. Moreover, light nuclei were included in the survey, which covered a range $12 \leq A \leq 252$.

Figure 22 shows a log-log plot of $\mathscr{I}_{02}$ vs Q_{02} (85, 86), based on recent literature values. It is at once apparent that a strong correlation between the two quantities exists over the whole range of transition quadrupole moments. Closer analysis reveals a strong dependence on shell structure, as is shown presently. (Figure 23 is identical to Figure 22, with the exception of the nuclear symbols.) The plot resembles a sea horse, consisting of three distinct parts:

1. In the "tail," where the $(Q, \mathscr{I})$ values are very small and the points refer to nuclei that possess only one or two neutron or proton pairs (particles or holes) beyond a closed shell. These are either light deformed nuclei or heavier "spherical" nuclei. In this region $\mathscr{I}$ is approximately proportional to Q.
2. The "main body" corresponds to larger $(Q, \mathscr{I})$ values, pertaining to nuclei with several neutron and proton pairs beyond closed shells, i.e. nuclei in the transition and strongly deformed regions. Here $\mathscr{I}$ is proportional to Q^2, as was stated above.

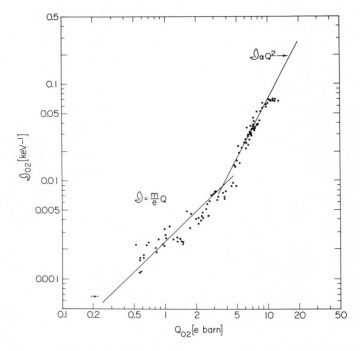

Figure 22 Log-log plot of the average moment of inertia $\mathscr{I}_{02} = [\mathscr{I}(0) + \mathscr{I}(2)]/2$ vs the transition quadrupole moment Q_{02}. A linear part and a quadratic part can be clearly distinguished. The horizontal part for the highest $\mathscr{I} - Q$ values refer to spontaneously fissioning actinides. The linear part is interpreted by the alpha-particle dumbbell model, the quadratic part by a macroscopic two-fluid model.

3. The "head" consists of the very largest $(Q, \mathscr{I})$ values found for the heaviest actinides $[U(A \geqq 234), \mathrm{Pu}, \mathrm{Cm}, \mathrm{Cf}]$ (87). Here $\mathscr{I}$ is roughly constant, while Q continues to increase. All nuclei in this region fission spontaneously.

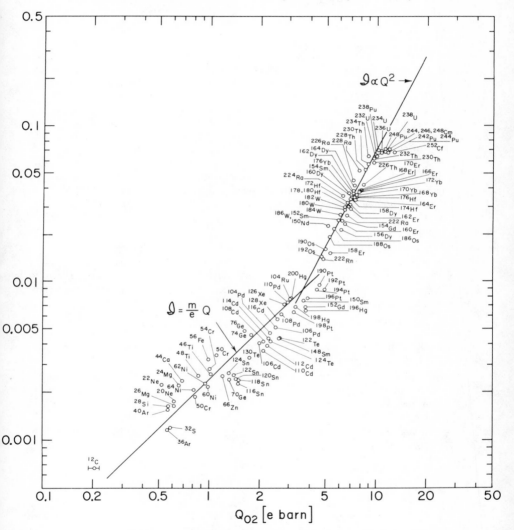

Figure 23 Same as Figure 22, with nuclear symbols added. The linear part consists of nuclei with one or two extra-shell nucleon pairs of one kind. It includes both light deformed nuclei and heavier spherical nuclei. The quadratic part corresponds to the deformed nuclei in the rare-earth and actinide regions.

3.3   THE ALPHA-PARTICLE DUMBBELL MODEL   The straight line traced through the *tail* (Figure 23) is given by a dumbbell model.

$$\mathscr{I} = 2M_\alpha R^2 \approx 2.4 \times 10^{-4} A^{2/3} \text{ (keV)}^{-1},$$
$$Q = 4 e_\alpha R^2 \approx 9.6 \times 10^{-2} A^{2/3} (e \, 10^{-24} \text{ cm}^2),$$
$$R = 1.1 A^{1/3} \text{ fm},$$
$$Q/\mathscr{I} = 415 (e \, 10^{-24} \text{ cm}^2 \text{ keV}).$$

3.4.

In this expression R is the half-density radius of the nucleus. The naive picture is one of two alpha particles connected by a massless rod, whose length is equal to the nuclear half-density diameter.

3.4   MACROSCOPIC TWO-FLUID MODEL FOR DEFORMED NUCLEI   The $\mathscr{I} \propto Q^2$ rule, which covers a wide range of deformations, has so far not found a microscopic explanation. However, one possible derivation of the rule follows from a macroscopic two-fluid model, based on the assumption that a nucleus governed by this rule is deformed into a spheroid characterized by a single significant deformation parameter ε. We assume that the spheroidal nuclear potential acts as a "bucket;" rotating the bucket causes some of the material to be dragged along, while part of the material is in a superfluid state (due to the pairing force), and therefore is not dragged along.

We assume that at every point in the nucleus the local mass density ρ is the sum of the superfluid mass density, which does not participate in the rotation, and an inertial density ρ_{in}, which does participate as in normal rigid rotation. The ratio $f = \rho_{\text{in}}/\rho$ is assumed to be a constant, independent of position in the nucleus. This constant is determined according to the relation

$$f = \xi \int d^3 r [\rho(\mathbf{r}) - \bar{\rho}(\mathbf{r})]^2 / \rho(0);$$

3.5.

where ξ is a dimensionless constant determined by fitting with experiment, $\bar{\rho}(r) = (1/2\pi) \int d\varphi \rho(r)$, and the azimuthal integration over φ is about the axis of rotation of the nuclear spheroid perpendicular to the axis of symmetry. It is at once evident that $\mathscr{I} = f \mathscr{I}_{\text{rigid}}$ and that the fraction f vanishes if the deformation vanishes. Since we have $f \propto \varepsilon^2$ to first order, we get $\mathscr{I} \propto Q^2$ if we assume that $Q \propto \varepsilon$.

It was assumed that the neutrons and protons are both distributed according to the same deformed Woods-Saxon function

$$\rho(\mathbf{r}) = \rho_0 / \{1 + \exp[r - R(1 + \varepsilon P_2(\cos \Theta)]/a\},$$

3.6.

with $\int d^3 r \rho(r) = A$, and $a = 0.5$ fm.

In Figure 24 the experimental results for $\mathscr{I}$ vs Q^2 (*open circles*) are compared with the model predictions (*solid circles*). Excellent agreement is obtained with $\xi = \frac{1}{20}$, *independent of A*, up to and including ^{232}U.

3.5   DEVIATION FROM THE TWO-FLUID MODEL PREDICTION FOR SPONTANEOUSLY FISSIONING ACTINIDES   The sudden deviation for the heaviest actinides is not explained by the model. One would have expected the $\mathscr{I}$ vs Q^2 curve to fall below the model predictions as $\mathscr{I}$ approached the rigid-body value, which is a natural limit, but the actual departure is much more striking than such considerations would suggest. Several possible causes for the departure come to mind. The large hexadecapole moments occurring in this region (87), the near degeneracy of prolate and oblate valence-shell orbits (which might be responsible for the constancy of the 2+ state and 4+ state energies of the heaviest actinide nuclei) (88, 89), or the possibility that due to the Coulomb repulsion the protons might be slightly more concentrated near the poles of these most elongated nuclei than the neutrons.

3.6   THE $\mathscr{I}$ VS Q RELATION FOR HIGHER TRANSITIONS   A study of the $\mathscr{I}$ vs Q relation for higher transitions in the band would be desirable. Here $Q_{J \to J+2}$ is related to $B(E2)(J \to J+2)$ by

$$Q_{J \to J+2}^2 = \frac{32\pi(2J+3)(2J+5)}{15(J+1)(J+2)(2J+1)} B(E2)(J \to J+2). \qquad 3.7.$$

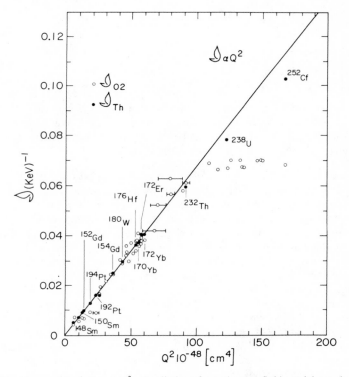

Figure 24   Linear plot of $\mathscr{I}_{02}$ vs Q_{02}^2. Predictions from the two-fluid model are shown by solid circles, empirical values by open circles. The straight line corresponds to $\mathscr{I}_{02} = (1/k^2)Q_{02}^2$. The Q_{02}^2 values for the heaviest actinide nuclei increase by almost 50%, while the $\mathscr{I}_{02}$ values remain approximately constant.

For strongly deformed nuclei a preliminary survey shows that the $\mathscr{I} \propto Q^2$ relation still holds for $2+ \rightarrow 4+$ transitions. For the $4+ \rightarrow 6+$ transitions one finds that the empirical $\mathscr{I}$ values fall above the predicted ones by $\sim 10\%$, for $6+ \rightarrow 8+$ transitions by $\sim 20\%$. For transitional nuclei this lag appears to become more pronounced. Very few $B(E2)$ values are so far known for $(2+ \rightarrow 4+)$ transitions in the vibrational region.

4   Transitions from the VMI Phase

4.1   THE $\mathscr{I}$ VS ω^2 PLOT   According to Equation 2.9,

$$\mathscr{I} = \mathscr{I}_0 + \omega^2/2C. \qquad\qquad 4.1.$$

Measured transition energies within a ground-state band presented in this form yield a very sensitive criterion for deviations from the Harris-VMI model predictions. (For convenience Equation 4.1 is multiplied by 2.) For a transition $J \rightarrow J-2$ the ordinate is given by

$$2\mathscr{I} = \frac{\Delta(J(J+1))_{J \rightarrow J-2}}{\Delta E_{J \rightarrow J-2}} = \frac{4J-2}{\Delta E_{J \rightarrow J-2}} \qquad\qquad 4.2.$$

and the abscissa by

$$\omega^2 = \left[\frac{\Delta E_{J \rightarrow J-2}}{[J(J+1)]^{1/2} - [(J-2)(J-1)]^{1/2}} \right]^2. \qquad\qquad 4.3.$$

According to Equations 4.2 and 4.3, the initial part of the curve consists of an almost straight line, which usually curves slightly upward. Were the quantities plotted differentials instead of energy differences, the line would be perfectly straight. The intercept at $\omega^2 = 0$ is $2\mathscr{I}_0$ and the initial slope is $1/C$.

4.2   PHASE TRANSITIONS IN DEFORMED AND SPHERICAL NUCLEI   By the use of heavier ions as bombarding particles, ground-state bands have been populated during recent years to very high J states. The results have been compiled in various articles and reviews (41, 66, 81). The most extensive compilation (66) ranges from ^{72}Se to ^{242}Pu. It includes near-magic, vibrational, and strongly deformed nuclei. As the magic-number limit is approached, the slope becomes increasingly steeper. The shapes of the $\mathscr{I}$ vs ω^2 curves after deviation from VMI are varied: besides "backbending," i.e. a triple-valued curve in ω^2, a more gradual increase, with ultimate flattening or even downbending has been observed. In Figure 25 we compare an $\mathscr{I}$ vs ω^2 plot for the deformed nucleus ^{158}Dy determined by Thieberger et al (90) up to $J = 22$, with an $\mathscr{I}$ vs ω^2 plot for the "vibrational" nucleus ^{104}Pd, whose ground-state band was populated by Cochavi et al (91) up to $J = 14$. For ^{158}Dy, $J_c = 14$, while for ^{104}Pd $J_c = 6$ or 8. However, the ω^2 value at which back-bending takes place is higher for ^{104}Pd. For this nucleus, double backbending is observed. A systematic study of neutron-deficient even-even Pd nuclei (^{104}Pd$\rightarrow^{98}$Pd) revealed a bewildering number of cascades besides the ground-state band (92), especially in ^{102}Pd and ^{100}Pd. These cascades are frequently interconnected and they all terminate in states $J \geq 2$ of the ground-state band. Among these cascades

quasi-rotational bands, of both odd and even parity, have been established. Iachello, partly in collaboration with Arima (93), has developed an interacting boson model, whose predictions were compared with results for these and other vibrational nuclei. Quadrupole and octopole bosons were considered.

While a number of features in the observed level schemes are in agreement with the model predictions, e.g. the general features of the odd-parity bands, it is too early to judge whether coupling of quasi-particles to quadrupole bosons do not yield equally good or better agreement. As is shown in Sections 6 and 7, the interactions between bands strongly affect the backbending behavior of the ground-state band.

4.3   HEURISTIC VALUE OF THE VMI MODEL FOR THE SPHERICAL REGION   In Section 2.9 we discussed a deviation from VMI predictions occurring in the neutron-rich Te nuclei, which implies $J_c = 4$. This phenomenon was attributed to strong overlap of a proton pair with a neutron pair, i.e. α clustering.

In Figure 14, which presents R_J vs R_4 values for $1.82 < R_4 \leq 2.23$, more scatter of the empirical values is observed than for the deformed region (Figure 15). In

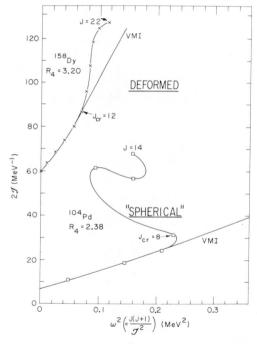

Figure 25   Comparison between an $\mathscr{I}$ vs ω^2 plot for a neutron-deficient rotor (^{158}Dy); $[E_2 = 98.94 \text{ keV}; C = 2.64 \, (10^6 \text{ keV}^3)]$ with that of a "vibrator" (^{104}Pd): $[E_2 = 555.8 \text{ keV}; C = 10.23 \, (10^6 \text{ keV}^3)]$. The VMI predictions are shown by solid lines. For the rotor $J_c = 12$, while for the vibrator $J_c = 6$ or 8.

the case of the Te nuclei (not included in Figure 14), instead of the scatter, one finds with increasing N an increasing but regular deviation from the R_6 curve (Figure 17). It appears that as more nuclei in this region are studied, regularities in the deviations from the VMI model are observed. This kind of "fine structure" may also be connected with α clustering. This, in turn, appears to throw an interesting light on the relation between $\mathscr{I}_{02}$ and Q_{02} found for the spherical region (Section 3.3), which agrees so well with a naive alpha dumbbell model plus a superfluid core.

4.4 VMI BEHAVIOR OF QUASI-ROTATIONAL BANDS IN ODD-A NUCLEI   In this review we have limited our attention to the structure of even-even nuclei. However, it is worth mentioning that excellent fits were obtained for quasi-rotational bands in odd-A nuclei, based on the assumption that the even-even core obeys the VMI law. Such calculations were carried out both for spheroidal (94) and for triaxial (95) nuclei.

5   Phenomenological Generalizations of the VMI Model for the Low-Spin Region

In the previous sections, the phenomenological basis for the VMI model has been reviewed. The success of the model led to a number of attempts to improve upon it phenomenologically or to provide it with a microscopic derivation. In this section, we discuss the VMI model in terms of a more general treatment of the low-spin region. We reserve for Section 6 a treatment of models for both the low-spin and backbending regions and defer the discussion of microscopic models to Sections 7 and 8.

As remarked by Harris and others (71, 96, 97), the energy E must satisfy the classical self-consistency condition

$$\frac{\partial E}{\partial \omega} = \omega \frac{\partial \langle J_x \rangle}{\partial \omega},$$ 5.1.

where x is the axis of rotation perpendicular to the symmetry axis. In analogy to the variational solution of the "cranking" problem, we now assume that E is a function of $\langle J_x \rangle$ and a generalized set of "stretching variables" x_i ($i = 1$ to N). The quantities x_i can be taken to represent quadrupole deformations of the nucleus, neutron and proton pairing parameters, etc. We now require that E be a minimum with respect to variations of x_i at fixed $\langle J_x \rangle$, i.e. $(\partial E/\partial x_i)_{\langle J_x \rangle} = 0$. If we now define the moment of inertia $\mathscr{I}(x_i)$ as $\langle J_x \rangle/\omega$, we can use Equation 5.1 and the above minimum condition to obtain (97)

$$\left(\frac{\partial E}{\partial \langle J_x \rangle}\right)_{x_i} = \langle J_x \rangle/\mathscr{I}(x_i).$$ 5.2.

Integrating Equation 5.2 and taking $\langle J_x \rangle = [J(J+1)]^{1/2}$, we find that E must always have the general form

$$E = \frac{J(J+1)}{2\mathscr{I}(x_i)} + f(x_i),$$ 5.3.

where $f(x_i)$ is an arbitrary "potential function" (98). Further dynamical input is now necessary to specify the dependence of f on the variables x_i.

In microscopic theories (see Sections 1 and 8), deviations in energy from the rigid rotor formula arise mainly from Coriolis anti-pairing of neutrons and protons, centrifugal stretching (near transition nuclei), and higher-order corrections to the cranking model (99, 100). To the extent that these different effects can be de-coupled and treated in the harmonic approximation, the problem reduces to a standard normal-mode problem with N modes, for which

$$f(x_i) = \frac{1}{2} \sum_{i=1}^{N} C_i(x_i - x_{i0})^2, \qquad\qquad 5.4.$$

where x_{i0} is the equilibrium value of the variable x_i. The calculation of the coefficients C_i in a microscopic theory is the subject of Section 7. In this section, we are content to use the *form* (Equation 5.4) suggested by microscopic theories as a vehicle for developing a set of phenomenological models, the VMI model being among them.

It was shown by Ma & Rasmussen (99) that Equation 5.4 together with the conditions $\partial E/\partial x_i = 0$ lead to an energy $E(J)$ expressed in terms of a "generalized stretching variable" t:

$$E(J) = \frac{J(J+1)}{2\mathscr{I}(t)} + \tfrac{1}{2}C't^2, \qquad\qquad 5.5.$$

subject to the minimization condition

$$\partial E(J)/\partial t = 0. \qquad\qquad 5.6.$$

Equations 5.5 and 5.6 are the starting point for a unified treatment of most existing phenomenological models due to Mantri & Sood (101). These authors write $\mathscr{I}(t)$ in the form

$$\mathscr{I}(t) = \mathscr{I}_0 f(t), \qquad\qquad 5.7.$$

where $\mathscr{I}_0$ is the ground-state moment of inertia. Physically acceptable choices for $f(t)$ exhibit the properties: (a) $f(0) = 1$ and (b) $f(t)$ is an increasing function of t. Various models that have been advanced for the description of ground-state bands can now be obtained by particular choices of $f(t)$. Suppose we start from a Taylor series

$$f(t) = (1 + c_1 t + c_2 t^2 + c_3 t^3 + \cdots). \qquad\qquad 5.8.$$

If we make the simplest assumption beyond having a constant moment of inertia, and keep only the term linear in t, we have $f(t) = 1 + t$ and hence

$$t = (\mathscr{I} - \mathscr{I}_0)/\mathscr{I}_0. \qquad\qquad 5.9.$$

Substituting Equation 5.9 in Equation 5.5, we obtain the VMI or Harris model. Note that we can take $c_1 = 1$ in Equation 5.8 in full generality, since Equations 5.5 and 5.6 are invariant in form under a scale transformation $c_1 t \to t$. The VMI model is a special case ($n = 1$) of a more general class of two-parameter models (101), where

$$f(t) = (1 + t)^n.$$

5.10.

The choice $n = 2$ yields the centrifugal stretching model (70, 102, 103). The choice $n \to \infty$ leads to the Draper model (104) if we scale t by $1/n$, i.e. the choice (104)

$$f(t) = \exp(t).$$

5.11.

All of these formulas are equivalent in the region of low t, i.e. when $\mathscr{I}$ is permitted only small excursions from $\mathscr{I}_0$. However, they can differ appreciably in their predictions for high-spin states. In the case where t refers to the "pairing-stretch" degree of freedom, Ma & Rasmussen (105) have provided a microscopic justification of the ansatz (Equation 5.11); see Section 7.

A detailed comparison of formulas like (5.10) for various n can be found in (104, 106, 107). One way to exhibit the differences between these models is to look at the $\mathscr{I}$ vs ω^2 plot. Defining $r = \mathscr{I}/\mathscr{I}_0$, we may write

$$
\begin{aligned}
r &= 1 + \omega^2/\omega_0^2 && (n = 1), \\
r &= (1 - \omega^2/2\omega_0^2)^{-2} && (n = 2), \\
\ln r/r &= \omega^2/\omega_0^2 && (n = \infty),
\end{aligned}
$$

5.12.

where ω_0 is a constant. For $\omega^2/\omega_0^2 \ll 1$, the three formulas in Equation 5.12 are equivalent. For larger values of ω^2/ω_0^2, the VMI curve ($n = 1$) remains linear in ω^2, while the cases $n = 2$ (70, 103, 106) and $n = \infty$ (104) correspond to curves r that are concave upward for increasing ω^2.

The comparison of models like Equation 5.10 can also proceed via a study of the Mallmann curves (58) for R_J vs R_4, where $R_J = E(J)/E(2)$. Such a study has been carried out by Wood & Fink (107). For $2.4 < R_4 < 3.0$, i.e. the "quasi-rotational" region, they find that $n = \frac{1}{2}$ gives a better fit to R_6/R_4 than $n = 1, 2$, or ∞. For $R_4 > 3.25$, i.e. the rotational region, the best model value for n is about 1 for low J, and increases to ∞ for J approaching 16 or so. Their general conclusion (107), also supported by Stockmann & Zelevinsky (106), is that none of the models (5.10) with different n is clearly superior to the others.

A disadvantage of all the two-parameter models discussed above (including the VMI model), is that the energy ratios R_J depend on only one parameter. The other parameter of the theory merely fixes the energy scale and drops out of R_J. Hence the plots of R_J vs R_4 are smooth curves, i.e. one cannot account for nuclei that have essentially the same R_4 values but different R_J values. As seen in Section 2, there is some scatter in the Mallmann plots around the smooth curve predicted by the VMI model. To account for this, one would have to introduce at least a third parameter into the theory. This could be done by including more terms in the series (Equation 5.8), where c_2, c_3, etc are taken to be independent parameters. One could also add an anharmonic term proportional to t^3 to the potential term in Equation 5.5.

Most other phenomenological models for the description of ground-state rotational bands can also be derived (101) within the unified framework of Equations 5.5–5.7. These include the Bohr-Mottelson formula discussed in Section 2, as well as the work of Das, Dreizler & Klein (83), Ejiri (82), Gupta (108), Holmberg & Lipas (109), Sood (110), Bose & Varshni (111), and Warke & Khadkikar (112). The Ejiri formula (81), which in addition to a rotational term includes a contribution proportional to J, also emerges from the exactly soluble model of Arima & Iachello (113).

As shown earlier (see Section 2), the VMI model is equivalent to the angular velocity expansion of Harris (71)

$$E(J) = \alpha\omega^2 + \beta\omega^4 + \gamma\omega^6 + \delta\omega^8 + \cdots,$$
$$\mathscr{I} = 2\alpha + \tfrac{4}{3}\beta\omega^2 + \tfrac{6}{5}\gamma\omega^4 + \tfrac{8}{7}\delta\omega^6 + \cdots,$$

5.13.

if we truncate the expansion at terms of order ω^4 in $E(J)$ or ω^2 in $\mathscr{I}$. An extensive phenomenological analysis for deformed nuclei based on Equation 5.13 has been conducted by Saethre et al (81), keeping terms to order ω^8 in $E(J)$. We compare their results with the VMI model in Section 7.

The angular velocity expansion (Equation 5.13) gives very good fits with four parameters to the energies of deformed nuclei all the way up to the start of the backbending region. Of course, no expansion such as Equation 5.13, even if carried to arbitrarily high order in ω^2, can account for backbending. However, for the low- and moderate-spin region the ω^2 expansion is clearly preferable (81) to an expansion in powers of $J(J+1)$.

6 Phenomenological Band Mixing Models for the VMI and Backbending Regions

In Section 5, we examined a variety of phenomenological approaches that focus attention primarily on the low-spin region. We now consider a class of models capable of dealing simultaneously with the low-spin (VMI) and backbending regimes. In these models the backbending phenomenon results from some sort of band mixing or hybridization. This class of models is still primarily phenomenological in nature, but at least tempered with a certain amount of theoretical motivation.

Before discussing any of the more quantitative calculations that have been performed, it is pedagogically useful to consider the simplest prototype of a band mixing calculation, in which two rotational bands are allowed to interact via a spin-independent interaction. This skeletal model is characterized by two constant moments of inertia $\mathscr{I}_1$ and $\mathscr{I}_2$ for the unperturbed bands, the energy separation E_0 of the bands at zero spin, and the strength V of the band coupling. At least four parameters are needed to describe the low-spin and backbending regions simultaneously. Two parameters are needed for the VMI region and two additional parameters are required to describe the two critical values of $\mathscr{I}$ for which $d\omega^2/d\mathscr{I} = 0$. Additional parameters are in general required to describe the shape of the $\mathscr{I}$ vs ω^2 curve near backbending and the behavior of $\mathscr{I}$ above the backbending region, i.e. the possibility of downbending. The present model corresponds to the minimal complexity necessary to deal with even the qualitative features of the low- and high-spin regions simultaneously.

We start with two unperturbed rotational bands with energies $E_{1,2}(J)$ given by

$$E_1(J) = J(J+1)/2\mathscr{I}_1,$$
$$E_2(J) = E_0 + J(J+1)/2\mathscr{I}_2.$$

6.1.

The energy $E(J)$ of the lowest-lying mixed band is given by

$$E(J) = \tfrac{1}{2}\{E_0 + \alpha J(J+1) - [(E_0 - \beta J(J+1))^2 + 4V^2]^{1/2}\},$$
$$\alpha = \tfrac{1}{2}(1/\mathscr{I}_1 + 1/\mathscr{I}_2),$$
$$\beta = \tfrac{1}{2}(1/\mathscr{I}_1 - 1/\mathscr{I}_2).$$

6.2.

Note that $E(J)$ contains terms to *all* orders when expanded in terms of $J(J+1)$.

The moment of inertia $\mathscr{I}$ of the lowest coupled band can be obtained using Equation 5.2 as

$$\mathscr{I}^{-1} = \alpha + \beta[E_2(J) - E_1(J)]/\xi(J),$$
$$\xi(J) = \{[E_2(J) - E_1(J)]^2 + 4V^2\}^{1/2}.$$

6.3

Equation 6.3 can now be used to write Equation 6.2 in the form

$$E(J) = \frac{J(J+1)}{2\mathscr{I}} + f(\mathscr{I}),$$

6.4.

where $f(\mathscr{I})$ is the "potential" term given by

$$f(\mathscr{I}) = E_0(1-\gamma)/2 - V(1-\gamma^2)^{1/2},$$
$$\gamma = (\mathscr{I}^{-1} - \alpha)/\beta.$$

6.5.

The form of Equation 6.4 may now be used to extract the VMI parameters $\mathscr{I}_0$ and C:

$$\mathscr{I}_0^{-1} = \alpha + \beta\gamma_{\mathrm{m}},$$
$$C^{-1} = \beta^2\mathscr{I}_0^4(1-\gamma_{\mathrm{m}}^2)^{3/2}/V,$$
$$\gamma_{\mathrm{m}} = E_0/(E_0^2 + 4V^2)^{1/2}.$$

6.6.

The parameters of the two-band model can always be chosen so as to match any empirical VMI description of the low-spin states, as per Equation 6.6. We now investigate under what conditions the model also yields backbending. To do this, we first express $\mathscr{I}$ in terms of ω. We obtain

$$\omega^2 = \frac{2V}{\beta\mathscr{I}^2}[g(\mathscr{I}_0) - g(\mathscr{I})],$$
$$g(\mathscr{I}) = \left[\frac{(\beta\mathscr{I})^2}{(1-\alpha\mathscr{I})^2} - 1\right]^{-1/2}.$$

6.7.

Note that this model gives ω^2 as a single-valued function of $\mathscr{I}$. This is in contrast to perturbation theory, i.e. the higher-order cranking formula (71), which would give $\mathscr{I}$ as a Taylor series in ω^2, and hence never yield backbending. If we expand the right-hand side of Equation 6.7 in a Taylor series in $(\mathscr{I} - \mathscr{I}_0)$ of the form

$$\omega^2 = \sum_{n=1}^{\infty} C_n (\mathscr{I} - \mathscr{I}_0)^n \qquad\qquad 6.8.$$

and keep only terms up to $n = 3$, we recover the model of Das & Banerjee (114). It is then clear that we must have $C_1 > 0$, $C_2 < 0$, and $C_3 > 0$ to achieve backbending. In principle, Equation 6.7 can be inverted to yield $\mathscr{I}$ as a function of ω^2, but this will yield a multi-valued function in general. The condition that Equation 6.7 correspond to a backbending situation is that the first derivative of $[g(\mathscr{I}_0) - g(\mathscr{I})]/\mathscr{I}^2$ with respect to $\mathscr{I}$ have two zeroes in the physical region of $\mathscr{I}$. A simple analysis shows that for $\mathscr{I}_2 < \mathscr{I}_1$, i.e. no crossing of the unperturbed bands, the backbending phenomenon cannot occur. For $\mathscr{I}_2 > \mathscr{I}_1$, the unperturbed bands cross, and backbending will occur only if the condition

$$V/E_0 < \frac{\mathscr{I}_2 - \mathscr{I}_1}{\mathscr{I}_2 + \mathscr{I}_1} \qquad\qquad 6.9.$$

is satisfied; thus backbending arises if V is sufficiently *weak*.

We have developed this simple model in some detail, since it displays in an analytic way some of the essential features of more detailed studies. However, a quantitative comparison with data requires a less schematic approach. We now review some of the more sophisticated band mixing calculations that have been performed.

The recent work of Goodman & Goswami (115) is based on a two-band model as outlined above, except that the "unperturbed" bands are not pure rotational bands. Instead of Equation 6.1, we have

$$E_1(J) = E_{\text{VMI}}(J),$$
$$E_2(J) = E_{\text{VMI}}(J) + E_{\text{exc}}(J), \qquad\qquad 6.10.$$

where $E_{\text{VMI}}(J)$ is the VMI energy and $E_{\text{exc}}(J)$ is a four-quasiparticle intrinsic excitation energy parameterized by

$$E_{\text{exc}}(J) = E_0 \left[1 + \exp\left(\frac{x - x_0}{\delta}\right) \right]^{-1},$$
$$x = [J(J+1)]^{1/2}. \qquad\qquad 6.11.$$

In general, this model has six parameters, the VMI parameters $\mathscr{I}_0$ and C; the parameters E_0, x_0, and δ describing the excited band; and the mixing strength V. The procedure used (115) was to fix $\mathscr{I}_0$ and C to the low-spin data alone, choose $E_0 = 4$ MeV and $V = 0.9$ MeV as average values over a range of nuclei, and determine x_0 and δ by a final least-squares fit to all the data, including the backbending region. Very nice fits to both the low- and high-spin data can be obtained in this manner. The model is also capable of producing backbending followed by "downbending," i.e. cases where, for large ω^2, $\mathscr{I}$ decreases after the backbending. In this case, ω^2 is not a single-valued function of $\mathscr{I}$, as per Equation 6.7, so the simplest two-band model of Equation 6.1 cannot explain this effect. In this sense, the Goodman-Goswami model (115) is equivalent to a three (or more) interacting band model.

Some two- and three-band models like Equation 6.1 were first investigated in detail by Molinari & Regge (116). They first used a two-band model, but with a spin-dependent coupling term V. A fair fit to the spectrum of ^{162}Er was obtained. A three-band model (116) with interacting $K = 0$, 2, and 3 bands improved the situation somewhat for ^{162}Er, but suggested that a still more complicated dynamical situation prevails.

Multiband calculations were subsequently carried out by Broglia, Molinari, Pollarolo & Regge (117, 118) for the cases of ^{154}Gd (117) and ^{156}Dy (118). These authors diagonalize a model Hamiltonian $\mathscr{H}$ of the form

$$\mathscr{H} = \mathscr{H}_0 + \mathscr{H}_c \qquad\qquad 6.12.$$

in an n-dimensional vector space (n bands with various K quantum numbers). They take as unperturbed Hamiltonian

$$\mathscr{H}_0 = A(K)\vec{J}^2 + \mathscr{H}_{int}(K) \qquad\qquad 6.13.$$

with eigenvalues $E = A(K)J(J+1) + B(K)$, where $B(K)$ is the eigenvalue of the intrinsic Hamilton $\mathscr{H}_{int}(K)$. The coupling term $\mathscr{H}_c$ is chosen to be

$$\mathscr{H}_c = h_0(J_1^2 + J_2^2) + h_1(J_+ + J_-) + h_2(J_+^2 + J_-^2). \qquad\qquad 6.14.$$

The three terms in Equation 6.14 correspond to the centrifugal, Coriolis, and asymmetry interactions, respectively. The procedure is now to treat the intrinsic matrix elements of $\mathscr{H}_c$ as well as $A(K)$ and $B(K)$ as free parameters to be determined by a least-squares search on all ground-state and excited-band data.

The application of this approach to ^{154}Gd (117) and ^{156}Dy (118) leads to some interesting results. For ^{154}Gd (117), at least three bands were required to give a good fit to the existing data. The two so far observed $K = 0$ bands (ground-state and β-vibrational bands) display considerably different rates of change of $\mathscr{I}$ in the backbending region of the $\mathscr{I}$ vs ω^2 plot. These rates of change are controlled by the positions of the poles of the energy $E(J)$ in the complex J plane (117, 118). For the two-band model of Equations 6.1–6.9, the backbending behavior of both bands is governed by a single pole. It is clear that at least one additional band (pole) is necessary in order to reproduce two distinct types of backbending curve, as well as the phenomenon of downbending.

The experimental situation for ^{156}Dy is even more complete. Here, the ground-state ($K = 0$), beta ($K = 0$), gamma ($K = 2$), and a $K = 1$ band are known experimentally (119). Using the framework of Equations 6.12–6.14, a multiband calculation is reported in (117). Five bands are needed to obtain a good fit to the high-spin region, while a four-band model is sufficient to fit the low-spin region alone. For ^{156}Dy, even the low-spin VMI region cannot be fitted without the mixing of $K \neq 0$ bands; in particular, the mixing of the γ band ($K = 2$) is important for low spins.

The fact that no acceptable fit can be found for either ^{154}Gd or ^{156}Dy using only $K = 0$ bands suggests that a Mottelson-Valatin (46) type of phase transition (uniform breakdown of pairing) is unlikely for cases where backbending is observed (117, 118). Even if such a fit could be obtained, this type of phase transition would not account for the downbending phenomenon. On the other hand, the high-lying $K = 1$ (or $K = 3$) bands which appear in these calculations (117, 118) have rather large moments of inertia. They thus have a natural interpretation in terms of pairs of high-spin particles decoupling from the rotating core, i.e. as *two-quasiparticle* bands. Nilsson orbitals with large spin are indeed available in this mass region, and can give the large moment of inertia required. The Stephens-Simon model (120) and subsequent similar treatments (121, 122) of the backbending phenomenon are a particular realization of the two-quasiparticle band hypothesis, in which a pair of $i_{13/2}$ neutrons are involved. The evidence of this model is treated in detail in (120–122), so we will not repeat it here. The main point here is that the band mixing calculations (117, 118) are consistent with the idea of decoupling pairs of particles, and hence provide support for the general ideas of Stephens & Simon (120). A study of the competition between the Coriolis antipairing (46) and pair decoupling (120) mechanisms as a function of mass number has been given by Sheline (123). As we see in Section 8, the decoupled-pair idea also emerges from self-consistent HFB calculations in cases where backbending is predicted. Thus there is a certain consistency in the physical content of these various approaches, even though they may appear to be quite different.

The band mixing model also enables one to predict the $B(E2)$ transition probabilities for interband and intraband transitions (117). Rather little experimental information is available. However, the $B(E2)$ values would provide an even more stringent test of the model than the energy levels themselves.

In summary, the band mixing approach is largely phenomenological, but the form of Equations 6.12–6.14 of the Hamiltonian has some physical basis. The number of data that are fitted (4 bands for ^{156}Dy) considerably exceeds the number of adjustable parameters. Also, the model makes a number of additional predictions regarding transition rates and the energy levels of higher bands that in principle can be tested experimentally. The disadvantage of such models is that there is as yet no firm connection between the parameters of the phenomenological Hamiltonian and the underlying microscopic theory.

A number of other phenomenological approaches, closely related to the band mixing idea, have been advanced to unify the description of the high-spin and VMI regions. Several authors (124–128) have argued that symmetric rotating nuclei must be unstable with respect to asymmetric deformations for high spins, i.e. the γ degree of freedom becomes "unfrozen." Smith & Volkov (128) have shown from a rather general variational approach how to generalize the VMI model to include the backbending region. They arrive at an energy formula

$$E(J) = \frac{R(\gamma)}{2\mathscr{I}} + \tfrac{1}{2}C_0(\mathscr{I} - \mathscr{I}_0)^2 + C_1\gamma(\mathscr{I} - \mathscr{I}_0) + \tfrac{1}{2}C_2\gamma^2 \qquad\qquad 6.15.$$

subject to the variational conditions $\partial E/\partial\mathscr{I} = 0$ and $\partial E/\partial\gamma = 0$. Here $R(\gamma)$ is obtained by diagonalizing the operator

$$R = \frac{3}{4}\sum_{v=1}^{3}\frac{J_v^2}{\sin^2(\gamma - 2\pi v/3)}. \qquad\qquad 6.16.$$

For many cases, the condition $\partial E/\partial \gamma = 0$ affords two solutions for each spin, one corresponding to $\gamma \approx 0$ (VMI band, no backbending) and another to nonzero values of γ. Above some critical value $J = J_c$, the latter solution lies lower in energy and becomes part of the ground-state band, which then exhibits backbending in this region of J. A very similar phenomenological model was proposed somewhat earlier by Das, Dreizler & Klein (127) for the simultaneous analysis of ground-state and β or γ bands. A formula identical in form to Equation 6.15 was obtained, except that $R(\gamma)$ is replaced by

$$R(\gamma) \approx J(J+1) + n\gamma. \qquad 6.17.$$

where $n = 0$ or 1.

Another method for dealing with the backbending region is that of Wahlborn & Gupta (129) who express $\mathscr{I}$ as a ratio of polynomials in $J(J+1)$. This enables one to reproduce the VMI behavior at low spins and also ensure that $\mathscr{I}$ approaches the proper limit for large J.

Various three-parameter models (125, 126) have been proposed for fitting the low-spin properties of the ground-state and γ bands simultaneously. These models replace the $J(J+1)$ term in the VMI model of Equation 2.11 by various forms of the energies of an asymmetric rotor. No backbending behavior is predicted, however. Note that the usual VMI model of Equation 2.11 was also applied to γ bands in (73).

7 Microscopic Calculations of the VMI Parameters $\mathscr{I}_0$ and C

In this section we review several attempts to calculate the VMI parameters $\mathscr{I}_0$ and C, which characterize the low-spin properties of the ground-state band. We first outline the theoretical basis for these calculations. We then examine the question of whether the values of $\mathscr{I}_0$ and C determined empirically by a VMI-model fit can be sensibly compared to the theoretical values, or whether they represent only phenomenological *effective* parameters. In cases where the comparison of theoretical and empirical VMI parameters is appropriate, we evaluate the extent to which the microscopic theory provides a quantitative account of the data.

A considerable amount of theoretical effort has gone into trying to understand the deviations of rotational energy spectra from the rigid-rotor formula. One general type of approach has been based on the cranking model of Inglis (40). Some early calculations (130–134) take into account only the centrifugal stretching and Coriolis antipairing effect (46). More complete calculations, which include the important fourth-order corrections to the cranking model, have been carried out by Marshalek (135), Ma & Rasmussen (99), and Ma & Tsang (100). We focus our attention on these latter calculations here.

We first outline the essential features of the microscopic cranking formalism, following Ma & Rasmussen (99). We consider a Hamiltonian $\mathscr{H}(\lambda_i^{(0)})$ describing the nucleons moving in some average nonrotating deformed potential [a Nilsson potential (25), for example] and interacting through some specified residual interaction. The quantities $\lambda_i^{(0)}$ ($i = 1, N$) label collective degrees of freedom (the shape of the deformed potential) as well as characteristics of the residual interaction (pairing strengths, etc). The Schrödinger equation describing this nonrotating system is

$$\mathscr{H}(\lambda_i^{(0)})\Psi^{(0)}(\lambda_i^{(0)}) = E^{(0)}(\lambda_i^{(0)})\Psi^{(0)}(\lambda_i^{(0)}). \qquad 7.1.$$

Now consider the situation where the deformed potential is made to rotate around the x axis with frequency ω, through the action of an external force. The Schrödinger equation for this rotating system in the intrinsic frame is

$$[\mathscr{H}(\lambda_i^{(\omega)}) - \omega J_x]\Psi^{(\omega)}(\lambda_i^{(\omega)}) = E^{(\omega)}(\lambda_i^{(\omega)})\Psi^{(\omega)}(\lambda_i^{(\omega)}), \qquad 7.2.$$

where the quantities $\lambda_i^{(\omega)}$ assume different values from the $\lambda_i^{(0)}$ of Equation 7.1. Note that the potential is time-dependent when expressed in laboratory coordinates. In transforming the Schrödinger equation to the intrinsic frame (body-fixed axes) to obtain a time-independent potential, we must consider the Coriolis and centrifugal inertial forces that exist in a rotating reference frame. These forces give rise to the term $-\omega J_x$ in Equation 7.2. Using the semi-classical relationship

$$\langle \Psi^{(\omega)}(\lambda_i^{(\omega)}) | J_x | \Psi^{(\omega)}(\lambda_i^{(\omega)}) \rangle = [J(J+1)]^{1/2}, \qquad 7.3.$$

We can evaluate the energy difference $E(J)$ between the rotational state of spin J and the ground state as

$$E(J) = \langle \Psi^{(\omega)}(\lambda_i^{(\omega)}) | \mathscr{H}(\lambda_i^{(\omega)}) | \Psi^{(\omega)}(\lambda_i^{(\omega)}) \rangle - \langle \Psi^{(0)}(\lambda_i^{(0)}) | \mathscr{H}(\lambda_i^{(0)}) | \Psi^{(0)}(\lambda_i^{(0)}) \rangle. \qquad 7.4.$$

We can make further progress by decomposing $E(J)$ as follows:

$$E(J) = E_1(J) + E_2(J),$$
$$E_1(J) = \langle \Psi^{(\omega)}(\lambda_i^{(\omega)}) | \mathscr{H}(\lambda_i^{(\omega)}) | \Psi^{(\omega)}(\lambda_i^{(\omega)}) \rangle - \langle \Psi^{(0)}(\lambda_i^{(\omega)}) | \mathscr{H}(\lambda_i^{(\omega)}) | \Psi^{(0)}(\lambda_i^{(\omega)}) \rangle, \qquad 7.5.$$
$$E_2(J) = \langle \Psi^{(0)}(\lambda_i^{(\omega)}) | \mathscr{H}(\lambda_i^{(\omega)}) | \Psi^{(0)}(\lambda_i^{(\omega)}) \rangle - \langle \Psi^{(0)}(\lambda_i^{(0)}) | \mathscr{H}(\lambda_i^{(0)}) | \Psi^{(0)}(\lambda_i^{(0)}) \rangle,$$

where the wave functions $\Psi^{(0)}(\lambda_i^{(\omega)})$ are defined by

$$\mathscr{H}(\lambda_i^{(\omega)})\Psi^{(0)}(\lambda_i^{(\omega)}) = E^{(0)}(\lambda_i^{(\omega)})\Psi^{(0)}(\lambda_i^{(\omega)}). \qquad 7.6.$$

In Equation 7.5, $E_1(J)$ is the energy difference between the rotating and static systems when the parameters $\lambda_i^{(\omega)}$ are held fixed at the values appropriate to frequency ω. $E_2(J)$ is the contribution arising solely from changes in the parameters λ_i, without the external cranking term.

We now proceed to evaluate $E_1(J)$ and $E_2(J)$. Treating ωJ_x as a perturbation, one can evaluate $E_1(J)$ by a method similar to that of Harris (71). Keeping only second- and fourth-order terms in perturbation theory, we find (99)

$$E_1(J) = J(J+1)/2\mathscr{I}(x_i) + \tfrac{1}{2}C_\eta \eta^2, \qquad 7.7.$$

where $\eta \equiv \omega^2$ and $x_i \equiv \{\eta, \lambda_i^{(\omega)}\}$. The moment of inertia $\mathscr{I}(x_i)$ is given by

$$\mathscr{I}(x_i) = \mathscr{I}_0 + 2C_\eta \eta. \qquad 7.8.$$

In Equation 7.8, $\mathscr{I}_0$ is given by the second-order cranking formula of Inglis (40) and Belyaev (136)

$$\mathscr{I}_0 = 2 \sum_{m \neq 0} \frac{|\langle 0 | J_x | m \rangle|^2}{E_m - E_0}, \qquad 7.9.$$

where $|m\rangle = |\Psi_m^{(0)}(\lambda_i^{(\omega)})\rangle$ is a two-quasiparticle state and $E_m = E_m^{(0)}(\lambda_i^{(\omega)})$ is the corresponding energy. The constant C_η is given in fourth-order perturbation theory by (99, 100)

$$C_\eta = 2 \sum_{i,j,k \neq 0} \frac{\langle 0|J_x|i\rangle \langle i|J_x|k\rangle \langle k|J_x|j\rangle \langle j|J_x|0\rangle}{(E_i - E_0)(E_j - E_0)(E_k - E_0)}$$
$$-2 \sum_{i,j \neq 0} \frac{|\langle i|J_x|0\rangle|^2 |\langle j|J_x|0\rangle|^2}{(E_i - E_0)^2 (E_j - E_0)}, \qquad \qquad 7.10.$$

where $|0\rangle$ is the quasiparticle vacuum, $|i\rangle$ and $|j\rangle$ are two-quasiparticle states, and $|k\rangle$ is a two- or four-quasiparticle state; E_0, E_i, E_j, and E_k are the corresponding energies.

We now turn to the evaluation of $E_2(J)$ of Equation 7.5. Physically, $E_2(J)$ registers changes in energy due to changes in the variables λ_i. The λ_i represent collective degrees of freedom, such as quadrupole (β_2), hexadecapole (β_4), and triaxial (γ) deformations, the number-displacement degree of freedom (137) (changes of moment of inertia induced by shifting nucleons between orbits of opposite parity); and parameters v_p and v_n characterizing the proton and neutron pairing strengths, for instance. The simplest approximation is to assume that these various modes are uncoupled, and that the $\lambda_i^{(\omega)}$ do not deviate strongly from their equilibrium values $\lambda_i^{(0)}$. In this harmonic approximation we would write

$$E_2(J) = \tfrac{1}{2} \sum_i C_i (\lambda_i^{(\omega)} - \lambda_i^{(0)})^2, \qquad \qquad 7.11.$$

where the spring constants C_i correspond to the stiffness of the system with respect to the various kinds of deformation.

Combining Equations 7.7 and 7.11, we find the total energy

$$E(J) = \sum_i \tfrac{1}{2} C_i (x_i - x_{i0})^2 + \frac{J(J+1)}{2\mathscr{I}(x_i)} \qquad \qquad 7.12.$$

subject to the equilibrium conditions

$$[\partial E(J)/\partial x_i]_J = 0. \qquad \qquad 7.13.$$

The application of Equation 7.13 to Equation 7.12 yields

$$x_i - x_{i0} = \frac{\omega^2}{2C_i} \left(\frac{\partial \mathscr{I}}{\partial x_i}\right)_{x_{i0}}. \qquad \qquad 7.14.$$

We can now identify the VMI constant C of Equation 2.21 by expanding $\mathscr{I}$ to first order in $x_i - x_{i0}$ and using Equation 7.14. We find

$$C^{-1} = \sum_i (C_i)^{-1} (\partial \mathscr{I}/\partial x_i)_{x_{i0}}^2. \qquad \qquad 7.15.$$

Equations 7.9 and 7.15 form the basis for a comparison of the microscopic theory with the VMI model.

It should be noted that the small amplitude approximation in Equation 7.11 is not expected to be generally valid. In fact, for well-deformed nuclei far from closed shells, Ma & Rasmussen (105) have shown that the dependence of $\mathscr{I}$ on the pairing correlation parameter v is nearly exponential, i.e. $\mathscr{I} = \mathscr{I}_0 \exp(-\gamma v)$ for $0.3\Delta \leq v \leq \Delta$, where Δ is the pairing-gap parameter. This provides a partial microscopic justification for the Draper model of Equation 5.11.

Before discussing the detailed comparison of microscopic and empirical values of $\mathscr{I}_0$ and C^{-1}, we must first establish the conditions under which this comparison is meaningful. An important step in this direction is to assess the stability of the VMI parameters $\mathscr{I}_0$ and C to the addition of further terms in the expansion of the energy powers of ω. This has been done by Saethre et al (81), who start from the generalized Harris expansion (71)

$$E(J)/E_{\text{rot}}(J) = 1 + \gamma_J + \varepsilon_1 \gamma_J^2 + \varepsilon_2 \gamma_J^3, \qquad\qquad 7.16.$$

where $E_{\text{rot}}(J) = J(J+1)/2\mathscr{I}_0$ and $\gamma_J = 3\omega^2/4\mathscr{I}_0 C$. Saethre et al (81) have determined the parameters $\mathscr{I}_0$, C, ε_1, and ε_2 by a least-squares fit to the experimental E2 transition energies for ground state-bands in the mass region from Ce to Pt. For most cases, the lowest five transition energies are included in the four-parameter fits; thus the emphasis is on the low- to moderate-spin region ($J \leq 10$) below the onset of backbending.

The critical question in appraising the significance of the parameters $\mathscr{I}_0$ and C is whether or not the higher-order corrections ($\varepsilon_1, \varepsilon_2$) in Equation 7.16 are small. The size of the corrections is dependent on the spin J, and increases monotonically with increasing J. From the work of Saethre et al (81), we are able to calculate these corrections, which are displayed in Table 2 for $J = 4$. If these corrections were small for $J = 4$ then the parameters $\mathscr{I}_0$ and C^{-1} obtained by fitting only $E(2)$ and $E(4)$ could be more meaningfully compared with the microscopic expressions 7.9 and 7.15 than the values of $\mathscr{I}_0$ and C^{-1} resulting from fitting all data up to $J = 10$, say. However, from Table 2, we see that the higher-order corrections $\varepsilon_1 \gamma_J$ and $\varepsilon_2 \gamma_J^2$ relative to the second term are significant in almost all cases, even for $J = 4$. Only for very few nuclei (^{162}Dy, ^{164}Er, ^{168}Er, ^{166}Yb, ^{174}W, ^{178}Os) are the corrections from *both* third- and fourth-order terms less than 20% in magnitude. Since the third- and fourth-order terms are almost always opposite in sign, they tend to cancel. However, their combined contribution for $J = 4$ is still greater than 20% of the second-order term for half of the nuclei listed in Table 2.

In Equation 7.16 the value of γ_J registers the deviation of the energy spectrum from the rigid-rotor limit $E_{\text{rot}}(J)$. In general, the third- and fourth-order corrections are larger for "soft" nuclei for which γ_J is large. Thus a VMI description, which sets $\varepsilon_1 = \varepsilon_2 = 0$ in Equation 7.16, is most appropriate for very good rotors (small γ_J). For a typical nucleus, the value of ω^2 for $J = 10$ is about 2–3 times that for $J = 4$. Hence, for many nuclei in Table 2, the expansion 7.16 is already slowly convergent, if not divergent, for $J = 10$. If we thus fit the transition energies with a *truncated* Taylor series in ω^2, as per Equation 7.16, the resulting parameters will in general only have significance as *effective* parameters.

The extent to which the above considerations represent a real difficulty can be assessed by determining the VMI parameters $\mathscr{I}_0$ and C^{-1} by several different methods, and observing whether or not the same values emerge. We consider three methods here:

1. The two-parameter VMI fit of Mariscotti, Scharff-Goldhaber & Buck (73), giving $\mathscr{I}_0^{\text{VMI}}$ and C_{VMI}^{-1}.
2. The four-parameter fit of Saethre et al (81), giving $\mathscr{I}_0^s$, C_s^{-1}.
3. A two-parameter VMI fit to $E(2)$ and R_4 only, giving $\mathscr{I}_0(2,4)$ and $C^{-1}(2,4)$.

In method 3, we used the experimental $E(2)$ and $E(4)$ values of Saethre et al (81). For $\mathscr{I}_0$, the agreement between the methods 1–3 is excellent, indicating that the value of $\mathscr{I}_0$ is stable to variations in the method used to extract it from the data. For good rotors, the values of $\mathscr{I}_0$ vary by 1% or less. Even for soft nuclei such as the Ce isotopes the variation in $\mathscr{I}_0$ is less than 20%.

The situation is somewhat different for C^{-1}. In Table 3, we show the values of C^{-1} determined by methods 1–3. For about one half of the nuclei in Table 3, significant differences ($>20\%$) exist between the three determinations of C^{-1}, often a discrepancy of a factor of two or three. These correspond to the relatively "soft"

Table 2 Higher-order terms in the angular velocity expansion Equation 7.16 for $J = 4$ (81)

Nucleus	γ_4	$\varepsilon_1\gamma_4$	$\varepsilon_2\gamma_4^2$	Nucleus	γ_4	$\varepsilon_1\gamma_4$	$\varepsilon_2\gamma_4^2$
^{128}Ce	1.24	−0.59	0.14	^{176}Yb	0.04	−0.21	1.45
^{130}Ce	2.14	−0.64	0.09	^{166}Hf	0.96	−0.53	0.18
^{132}Ce	3.90	−0.66	0.05	^{168}Hf	0.41	−0.29	0.19
^{134}Ce	5.38	−0.59	0.03	^{170}Hf	0.19	−0.06	0.22
^{152}Sm	0.70	−0.34	0.09	^{172}Hf	0.12	−0.09	0.23
^{154}Gd	0.71	−0.38	0.11	^{174}Hf	0.09	−0.07	0.35
^{156}Gd	0.14	−0.15	0.23	^{176}Hf	0.07	−0.05	0.24
^{158}Gd	0.08	−0.38	1.94	^{178}Hf	0.06	−0.06	0.50
^{156}Dy	0.99	−0.40	0.09	^{180}Hf	0.04	−0.09	0.37
^{158}Dy	0.20	−0.20	0.24	^{172}W	0.45	−0.21	0.24
^{160}Dy	0.09	−0.09	0.28	^{174}W	0.23	−0.08	0.17
^{162}Dy	0.05	−0.03	0.13	^{176}W	0.20	−0.23	0.38
^{158}Er	2.43	−0.58	0.07	^{178}W	0.15	−0.15	0.32
^{160}Er	0.46	−0.34	0.19	^{180}W	0.11	−0.19	0.65
^{162}Er	0.15	−0.15	0.25	^{182}W	0.06	−0.12	0.28
^{164}Er	0.08	−0.03	0.12	^{178}Os	0.52	−0.13	0.11
^{166}Er	0.06	+0.02	0.33	^{180}Os	0.53	−0.55	0.42
^{168}Er	0.03	−0.02	0.10	^{182}Os	0.45	−0.67	0.52
^{160}Yb	4.30	−0.65	0.05	^{184}Os	0.24	−0.36	0.31
^{162}Yb	1.33	−0.65	0.16	^{186}Os	0.35	−0.44	0.29
^{164}Yb	0.34	−0.26	0.23	^{188}Os	0.56	−0.47	0.20
^{166}Yb	0.15	−0.12	0.19	^{182}Pt	3.34	−0.67	0.05
^{168}Yb	0.09	−0.06	0.37	^{184}Pt	3.53	−0.63	0.05
^{170}Yb	0.05	−0.04	0.20	^{186}Pt	6.00	−0.60	0.02
^{172}Yb	0.04	−0.08	0.60	^{188}Pt	6.49	−0.56	0.02
^{174}Yb	0.03	+0.11	−1.87				

Table 3 Empirical values of C^{-1} determined by different methods (in units of 10^2 MeV^{-3})

Nucleus	C_{VMI}^{-1}	C_s^{-1}	$C_{(2,4)}^{-1}$	Nucleus	C_{VMI}^{-1}	C_s^{-1}	$C_{(2,4)}^{-1}$
^{128}Ce	1.84	3.61	2.89	^{176}Yb	1.28	1.20	1.10
^{130}Ce	1.72	3.59	2.08	^{166}Hf	3.97	6.24	4.15
^{132}Ce	1.64	3.15	1.86	^{168}Hf	3.85	4.94	3.95
^{134}Ce	1.07	2.06	1.23	^{170}Hf	4.72	3.91	3.77
^{152}Sm	5.95	9.84	7.45	^{172}Hf	2.84	2.78	3.00
^{154}Gd	5.44	9.52	7.03	^{174}Hf	2.16	2.30	2.23
^{156}Gd	3.38	4.05	3.64	^{176}Hf	1.70	1.85	1.80
^{158}Gd	2.45	2.85	2.24	^{178}Hf	1.35	1.36	1.31
^{156}Dy	6.25	10.04	7.36	^{180}Hf	0.73	0.85	0.79
^{158}Dy	3.79	4.34	3.78	^{172}W	6.10	5.99	5.35
^{160}Dy	2.19	2.65	2.50	^{174}W	3.79	3.64	3.52
^{162}Dy	1.95	1.90	1.87	^{176}W	3.09	3.31	2.86
^{158}Er	4.90	9.66	5.94	^{178}W	2.23	2.49	2.29
^{160}Er	3.68	5.37	4.11	^{180}W	1.88	1.96	1.74
^{162}Er	2.55	2.95	2.72	^{182}W	0.98	1.16	1.06
^{164}Er	1.97	1.91	1.89	^{178}Os	5.68	5.86	5.54
^{166}Er	2.40	2.09	2.13	^{180}Os	4.39	5.43	3.76
^{168}Er	1.10	1.16	1.14	^{182}Os	2.94	4.75	2.74
^{160}Yb	4.03	8.14	4.62	^{184}Os	1.80	2.92	2.19
^{162}Yb	3.85	7.50	4.37	^{186}Os	1.62	2.90	2.05
^{164}Yb	3.33	4.19	3.64	^{188}Os	1.95	3.53	2.43
^{166}Yb	2.55	2.90	2.67	^{182}Pt	9.62	24.66	13.40
^{168}Yb	2.58	2.66	2.58	^{184}Pt	9.26	22.54	12.59
^{170}Yb	1.60	1.66	1.73	^{186}Pt	8.93	21.40	12.12
^{172}Yb	2.14	1.47	1.39	^{188}Pt	4.72	8.57	5.18
^{174}Yb	1.08	1.21	1.27				

nuclei for which $\sigma_s > 0.01$. On the other hand, the nuclei for which the three C^{-1} values are the same to within 20% correspond in most cases to relatively "hard" nuclei (good rotors) for which $\sigma_s \leqq 0.01$. For these nuclei, the VMI correction to the rigid-rotor energy formula is also small at low spins, i.e. $\gamma_4 \leqq 0.2$ in Table 2. The four nuclei ^{164}Yb, ^{172}W, ^{174}W, and ^{178}Os are borderline cases; they have C^{-1} differences of only 5–20%, but fairly large σ_s values in the range 0.01–0.03.

A rather general conclusion is now clear: nuclei that have small VMI corrections $\gamma_4 \leqq 0.2$ to the rigid-rotor energy formula also tend to have small higher-order corrections *relative* to γ_4, and hence C^{-1} values that are relatively stable (to 20%) to the method used to extract them from the data. For these nuclei, one can make a sensible comparison of the microscopic expression 7.15 with the empirical value of C^{-1}. For soft nuclei, however, such a comparison does not seem meaningful; this includes most nuclei for which $\sigma_s > 0.01$. For these nuclei, the VMI model remains a valid phenomenological procedure for fitting the data, as shown in Sections 2–4. However, for these nuclei, the parameters $\mathscr{I}_0$ and C^{-1} are to be interpreted as *effective* parameters that cannot be identified with $\mathscr{I}_0$ and C^{-1} as given by the microscopic cranking formalism outlined here.

We now proceed with a comparison of the VMI parameters with the microscopic theory for those nuclei where this comparison is reasonable. We restrict our attention to the results of Marshalek (135) and Ma & Tsang (100), which are the most complete.

In (100) the following procedure was used to calculate $\mathscr{I}_0$ and C^{-1}. One first generates a deformed single-particle basis using the deformed oscillator potential of Nilsson et al (138). The residual interaction is taken to be a pairing force of strength G given by

$$G = \left[g_0 - \tau_z g_1 \left(\frac{N-Z}{A} \right) \right] \bigg/ A, \qquad\qquad 7.17.$$

where $\tau_z = +1$ for neutron and -1 for proton pairing, $g_0 = 19.2$ MeV and $g_1 = 7.4$ MeV. Using Equation 7.17, one now solves the BCS equations numerically (100) to obtain the usual occupation probabilities U_k and V_k and the quasiparticle energies E_k. The pairing correlation parameter v is now defined by

$$\left. \begin{array}{c} U_k^2 \\ V_k^2 \end{array} \right\} = \frac{1}{2} \left\{ 1 \pm \frac{(\varepsilon_k - \lambda)}{[(\varepsilon_k - \lambda)^2 + v^2]^{1/2}} \right\}, \qquad\qquad 7.18.$$

where ε_k is the Nilsson single-particle energy and λ is the chemical potential.

In evaluating C^{-1} from Equation 7.15, we need to know both the spring constants C_i and the derivatives $\partial \mathscr{I} / \partial x_i$. In (100) the parameters x_i are taken to be the neutron and proton pairing parameters v_n and v_p, the quadrupole and hexadecapole deformation parameters ε_2 and ε_4, and the cranking parameter η. Triaxial deformation and number displacement degrees of freedom (137) among other possibilities, are not considered in (100). Note that the inclusion of such additional degrees of freedom can only *increase* the theoretical value of C^{-1} as per Equation 7.15. For the case at hand, Equation 7.15 can be written more explicitly as

$$C^{-1} = 4C_\eta + \frac{1}{C_{v_n}} \left(\frac{\partial \mathscr{I}_0}{\partial v_n} \right)^2_{v_n^{(0)}} + \frac{1}{C_{v_p}} \left(\frac{\partial \mathscr{I}_0}{\partial v_p} \right)^2_{v_p^{(0)}} + \frac{1}{C_{22}} \left(\frac{\partial \mathscr{I}_0}{\partial \varepsilon_2} \right)^2_{\varepsilon_2^{(0)}} + \frac{1}{C_{44}} \left(\frac{\partial \mathscr{I}_0}{\partial \varepsilon_4} \right)^2_{\varepsilon_4^{(0)}} \quad 7.19.$$

The derivatives of $\mathscr{I}_0$ with respect to v_n and v_p are evaluated for fixed G and particle number $n = \sum_{k>0} 2V_k^2$. The spring constants $C_i = \{ C_{v_n}, C_{v_p}, C_{22}, C_{44} \}$ are obtained from the ground-state energy ε_0 via

$$C_i = (\partial^2 \varepsilon_0 / \partial x_i^2)_{n, G}. \qquad\qquad 7.20.$$

The calculations of Marshalek (135) are very similar in spirit to the method of Tsang & Ma (100) outlined above. Outside of the details of the choice of deformed-potential and residual-interaction parameters, the main difference is that Marshalek (135) neglects some corrections due to particle-number conservation and considers the triaxial degree of freedom γ, and not ε_4. The γ degree of freedom was found to be of minor importance in most cases (135).

The results of the theoretical calculations of $\mathscr{I}_0$ and C^{-1} are shown in Tables 4 and 5. The empirical values $\mathscr{I}_0^s$ of Table 4 and C_s^{-1} of Table 3 are from the analysis

Table 4 Comparison of empirical and theoretical values of $\mathscr{I}_0$ (in units of MeV^{-1})

Nucleus	$\mathscr{I}_0^s$	$\mathscr{I}_0^{TM}$ (ref. 100)	$\mathscr{I}_0^M$ (ref. 135)
^{152}Sm	22.6	20.2	18.2
^{154}Gd	22.3	19.1	17.3
^{156}Gd	33.2	25.4	25.9
^{158}Gd	37.5	26.6	30.7
^{160}Dy	34.2	24.8	26.3
^{162}Dy	37.0	26.7	29.9
^{162}Er	28.9	22.6	—
^{164}Er	32.6	25.1	28.8
^{166}Er	37.0	27.0	31.9
^{168}Er	37.5	28.3	32.8
^{160}Yb	7.3	29.6	—
^{166}Yb	28.9	23.0	—
^{168}Yb	33.9	25.1	—
^{170}Yb	35.4	26.8	32.1
^{172}Yb	38.0	28.7	34.8
^{174}Yb	39.1	29.7	35.9
^{176}Yb	36.4	29.1	34.7
^{174}Hf	32.7	25.9	—
^{176}Hf	33.7	28.1	34.3
^{178}Hf	32.0	27.4	31.5
^{180}Hf	32.0	29.0	28.0
^{180}W	28.6	25.0	—
^{182}W	29.8	26.6	28.6
^{184}Os	24.4	24.3	—
^{186}Os	21.1	21.6	—
^{188}Os	18.2	18.8	—

of Saethre et al (81), based on Equation 7.16. In Table 5, we have included only those nuclei for which the empirical value of C^{-1} is empirically determined to 20% or better (see earlier discussion). If one decomposes the theoretical value of C^{-1} into contributions from the various mechanisms, one finds:

1. The fourth-order cranking contribution C_η and the neutron Coriolis anti-pairing contribution are comparable in size and generally represent the largest contributions to C^{-1}.
2. The proton antipairing term usually contributes only 10–20% of the total C^{-1}.
3. The contributions of quadrupole and hexadecapole stretching to C^{-1} are comparable but usually small, except for nuclei with $N = 90$ or 92 (among those calculated).

From Table 4, we see that the trends of $\mathscr{I}_0$ with N and Z are well reproduced, but the calculated magnitudes of $\mathscr{I}_0$ are too small (except for ^{160}Yb), generally by 10–40%. This level of disagreement is not very serious relative to the situation for C^{-1} values. We note that $C_M^{-1} > C_{TM}^{-1}$ except for ^{156}Gd and ^{160}Dy. This is mostly due (99) to the sizable reduction in the Coriolis antipairing effect $(C_{v_{n,p}}^{-1})$ due to the inclusion of number conservation corrections to the BCS energy, which were included by Ma & Tsang (100) and omitted by Marshalek (135). We also note that C_{TM}^{-1} is neither systematically larger nor smaller than the empirical values C_s^{-1} (80). For only a few nuclei (^{160}Dy, ^{164}Er, ^{168}Yb, ^{170}Yb, ^{178}Hf) is the agreement between C_{TM}^{-1} and C_s^{-1} better than 20%; in several cases the discrepancy is of the order of a factor of two.

It is clear that the calculations done thus far with the microscopic cranking perturbation formalism do not seem to provide a very quantitative description of the VMI parameter C^{-1}. However, it should be noted that C^{-1} is much more sensitive than $\mathscr{I}_0$ to the details of the single-particle levels. Preliminary studies (100) indicate that the C^{-1} values can be considerably improved by an adjustment of the neutron single-particle Nilsson levels around $N = 104$ and 108. Of course, these adjustments must be done in such a way as not to disturb the description of other nuclear properties. It is still not clear whether the discrepancies in Table 5 reflect the inadequacies of the cranking formalism or an inappropriate choice of a single-particle basis and residual interaction. These questions deserve further study.

Table 5   Theoretical values of C^{-1} (in units of 10^2 MeV^{-3})

Nucleus	C_{TM}^{-1} (ref. 100)	C_M^{-1} (ref. 135)
^{156}Gd	2.88	2.31
^{158}Gd	2.18	2.64
^{160}Dy	2.49	2.21
^{162}Dy	2.35	2.86
^{162}Er	1.68	—
^{164}Er	1.94	2.82
^{166}Er	2.54	3.27
^{168}Er	1.74	2.89
^{166}Yb	1.79	—
^{168}Yb	2.36	—
^{170}Yb	1.79	3.18
^{172}Yb	2.28	3.94
^{174}Yb	2.63	4.26
^{176}Yb	1.63	3.46
^{174}Hf	1.78	—
^{176}Hf	2.77	4.59
^{178}Hf	1.59	3.55
^{180}Hf	1.97	2.27
^{180}W	1.49	—
^{182}W	1.80	2.82

8   The Hartree-Fock-Bogoliubov Approach to the VMI Region and Backbending

8.1   INTRODUCTION   In previous sections, we have considered phenomenological models for the low-spin region and also approaches that can account for the backbending behavior observed at high spins. The microscopic cranking model developed in Section 7 provides a method of calculating the VMI parameters $\mathscr{I}_0$ and C^{-1}, which characterize the low-spin behavior of the nucleus. However, neither the cranking model in perturbation theory nor the VMI model can account for the properties of all nuclei at high spins. We thus need a microscopic model capable of dealing simultaneously with the VMI and backbending regions. The Hartree-Fock-Bogoliubov (HFB) method is one such microscopic model. We devote the present section to an evaluation of its successes and limitations.

We first derive the HFB equations in a compact fashion. A more extensive discussion is given in (139–141). We note that the HFB method includes all of the physical effects considered in the cranking-model formalisms of Ma & Rasmussen (99) and Marshalek (135), treated in Section 7. However, the HFB method is not restricted to small oscillations of each collective variable around its equilibrium value. Since the HFB equations represent a nonlinear system that is solved self-consistently, they are capable of describing large oscillations, coupling between normal modes of the system, and phase transitions. We show that the HFB method includes the important physical mechanisms of (a) Coriolis antipairing of neutrons and protons, (b) axially symmetric as well as nonaxial deformations, (c) pair re-alignment effect, and (d) possibility of shape transitions. The HFB equations are sufficiently flexible to allow the nuclear dynamics to choose which of these mechanisms will dominate in a given situation.

8.2   DERIVATION OF HFB EQUATIONS   We start from the many-body Hamiltonian

$$H_\omega = H - \omega J_x,$$
$$H = \sum_{ij} T_{ij} C_i^\dagger C_j + \frac{1}{4} \sum_{ijkl} v_{ijkl} C_i^\dagger C_j^\dagger C_l C_k, \qquad \qquad 8.1.$$

where ωJ_x is the external cranking potential (see Equation 7.2), $\langle J_x \rangle$ is given by Equation 7.3, $C_i^\dagger$ and C_i are particle creation and destruction operators, T is the kinetic energy, and v is an effective nucleon-nucleon interaction. If an inert core is assumed, then T includes the spherical single-particle potential which describes the effect of the core on a valence particle.

It is impractical to diagonalize H_ω when many active nucleons are present. Consequently we seek a canonical transformation such that the transformed quasi-particles are approximately noninteracting. When the quasiparticle interactions are neglected, the resulting Hamiltonian may not conserve particle number. Consequently H_ω must be replaced by

$$H' = H_\omega - \lambda N, \qquad \qquad 8.2.$$

where $N = \sum_i C_i^\dagger C_i$ is the number operator and the Fermi energy λ is chosen so that N has the correct expectation value.

The goal is to express H' as

$$H' = H'_0 + \sum_i E_i a_i^\dagger a_i + H_{\text{int}},\qquad\qquad 8.3.$$

where $H'_0 + \lambda \langle N \rangle + \omega \langle J_x \rangle$ is the energy of the quasi-particle vacuum $|\Phi_0\rangle$, the E_i are the quasi-particle energies, and H_{int} is the neglected quasi-particle interaction. H' is a function of ω, so the quasi-particle states, energies, and vacuum are ω-dependent.

The HFB quasi-particle creation operators $a_i^\dagger$ are obtained from the unitary transformation

$$a_i^\dagger = \sum_j (U_{ij} C_j^\dagger + V_{ij} C_j),\qquad\qquad 8.4.$$

subject to the constraints

$$UU^\dagger + VV^\dagger = I, \quad U\tilde{V} + V\tilde{U} = 0.\qquad\qquad 8.5.$$

Equation 8.5 also follows from the requirement that the quasi-particle operators obey Fermion commutation rules. The quasi-particle vacuum $|\Phi_0\rangle$ is defined by $a_i|\Phi_0\rangle = 0$. Since the quasi-particles are fermions, a solution is $|\Phi_0\rangle = \Pi_i a_i |0\rangle$.

The HFB equations may be derived by choosing the quasi-particle transformations (Equation 8.4) so that Equation 8.2 is cast into the form of Equation 8.3. The derivation employs Wick's theorem: any product of operators can be expressed as a sum of normal products$^{\cdot\cdot}$ $:\mathcal{O}:$ containing all possible contractions $\langle\mathcal{O}\rangle$. For example,

$$C_i^\dagger C_j = \langle C_i^\dagger C_j \rangle + :C_i^\dagger C_j:$$
$$C_i^\dagger C_j^\dagger C_l C_k = \langle C_i^\dagger C_k \rangle \langle C_j^\dagger C_l \rangle - \langle C_i^\dagger C_l \rangle \langle C_j^\dagger C_k \rangle + \langle C_i^\dagger C_j^\dagger \rangle \langle C_l C_k \rangle + \langle C_i^\dagger C_k \rangle :C_j^\dagger C_l:$$
$$+ \langle C_j^\dagger C_l \rangle :C_i^\dagger C_k: - \langle C_i^\dagger C_l \rangle :C_j^\dagger C_k: - \langle C_j^\dagger C_k \rangle :C_i^\dagger C_l:$$
$$+ \langle C_l C_k \rangle :C_i^\dagger C_j^\dagger: + \langle C_i^\dagger C_j^\dagger \rangle :C_l C_k: + :C_i^\dagger C_j^\dagger C_l C_k:.$$

$$8.6.$$

The expectation values $\langle\mathcal{O}\rangle$ are with respect to the quasi-particle vacuum. The density matrix ρ and the pairing tensor t are defined by

$$\rho_{ij} = \rho_{ji}^* = \langle \Phi_0 | C_j^\dagger C_i | \Phi_0 \rangle = (V^\dagger V)_{ij},\qquad\qquad 8.7.$$

$$t_{ij} = -t_{ji} = \langle \Phi_0 | C_j C_i | \Phi_0 \rangle = (V^\dagger U)_{ij},\qquad\qquad 8.8.$$

where the expectation values are evaluated by inverting Equation 8.4 and applying $a_i|\Phi_0\rangle = 0$ and the fermion commutation relations. The Hamiltonian H' may be re-expressed by substituting Equation 8.6 into 8.2, so that

$$H' = H'_0 + H'_2 + H'_4,\qquad\qquad 8.9.$$

where the subscripts denote the number of uncontracted operators, and

$$H'_0 = \sum_{ij} (T - \lambda - \omega J_x + \tfrac{1}{2}\Gamma)_{ij} \rho_{ji} + \tfrac{1}{2}\sum_{ij} \Delta_{ij} t_{ji}^\dagger,\qquad\qquad 8.10.$$

$$H'_2 = \sum_{ij} (H - \lambda - \omega J_x)_{ij} : C_i^\dagger C_j : + \tfrac{1}{2} \sum_{ij} \Delta_{ij} : C_i^\dagger C_j^\dagger : + \tfrac{1}{2} \sum_{ij} \Delta_{ij}^\dagger : C_i C_j : , \qquad 8.11.$$

$$H'_4 = \tfrac{1}{4} \sum_{ijkl} v_{ijkl} : C_i^\dagger C_j^\dagger C_l C_k : . \qquad 8.12.$$

The HF Hamiltonian H, the HF potential Γ, and the pair potential Δ are defined by

$$H = H^\dagger = T + \Gamma, \quad \Gamma_{ij} = \Gamma_{ij}^\dagger = \sum_{kl} v_{ikjl} \rho_{lk}, \quad \Delta_{ij} = -\Delta_{ji} = \tfrac{1}{2} \sum_{kl} v_{ijkl} t_{kl}. \qquad 8.13.$$

Expectation values of Wick normal products $: \mathcal{O} :$ with respect to $|\Phi_0\rangle$ vanish by construction. Consequently the vacuum energy $\langle \Phi_0 | H' | \Phi_0 \rangle$ is H'_0. The first sum in Equation 8.10 is the HF energy, whereas the second sum is the pairing energy. The neglected quasi-particle interactions are described by H'_4.

The Hamiltonian (Equation 8.9) assumes the desired form (Equation 8.3) if H'_2 is an independent quasi-particle Hamiltonian

$$H'_2 = \sum_i E_i a_i^\dagger a_i. \qquad 8.14.$$

The commutator $[H'_2, a_i^\dagger]$ may be evaluated with Equation 8.14:

$$[H'_2, a_i^\dagger] = E_i a_i^\dagger = E_i \sum_j (U_{ij} C_j^\dagger + V_{ij} C_j). \qquad 8.15.$$

or with Equation 8.11:

$$[H'_2, a_i^\dagger] = \sum_{jk} [H - \lambda - \omega J_x)_{jk} U_{ik} + \Delta_{jk} V_{ik}] C_j^\dagger$$
$$+ \sum_{jk} [-\Delta_{jk}^* U_{ik} - (H - \lambda - \omega J_x)_{jk}^* V_{ik}] C_j. \qquad 8.16.$$

Equating the coefficients of $C_j^\dagger$ and C_j in Equations 8.15 and 8.16, we obtain the HFB eigenvalue equations

$$\begin{pmatrix} (H - \lambda - \omega J_x) & \Delta \\ -\Delta^* & -(H - \lambda - \omega J_x)^* \end{pmatrix} \begin{pmatrix} U_i \\ V_i \end{pmatrix} = E_i \begin{pmatrix} U_i \\ V_i \end{pmatrix}. \qquad 8.17.$$

This energy matrix (κ) is hermitian, so the quasi-particle energies E_i are real.

The HFB equations are nonlinear, and solutions are obtained by an iterative procedure. A guess is made for the quasi-particle transformations (Equation 8.4), subject to the constraints of Equation 8.5. The wave function (ρ, t) is calculated from Equations 8.7 and 8.8. Then the potentials (Γ, Δ) are constructed from Equation 8.13. The HFB energy matrix (Equation 8.17) is diagonalized to obtain a new set of quasi-particles. This procedure is repeated until the same wave function is obtained on successive iterations. At each iteration λ and ω are adjusted to satisfy the number constraint and the cranking constraint (Equation 7.3).

8.3 PHYSICAL EFFECTS INCLUDED IN HFB-CRANKING THEORY In order to explain both the low- and high-spin behavior of the moment of inertia, a microscopic theory must be general enough to encompass a great variety of physical effects, of which the HFB method includes the following:

8.3.1 *Deformations* The nuclear shape may be characterized by the multipole moments (Q_{LM}) of the quasi-particle vacuum. If j_{max} is the largest single-particle angular momentum in the model space, then Q_{LM} can be nonzero for $L = 0, 2, \ldots, 2j_{max}$ and $M = 0, 2, \ldots, L$. (For the realistic calculations described below, $j_{max} = \frac{13}{2}$.) The quadrupole (β_2), the hexadecapole (β_4), and the nonaxial (γ) deformation parameters are only a few of the degrees of freedom that the nuclear shapes are permitted to have. All deformations result from the nucleon-nucleon effective interaction.

8.3.2 *Pairing field* In HFB theory Δ is a matrix constructed from the same two-body interaction as is the HF field. (The BCS approximation retains only the diagonal elements of Δ and neglects self-consistent rearrangements in the HF single-particle basis due to the presence of pairing correlations.) The pair field is level-dependent, since each pair of orbits $|ab\rangle$ has a potential Δ_{ab}, in contrast to simpler models where a single order parameter Δ is assumed to represent all pairs of levels.

8.3.3 *Inertial forces* The Coriolis and centrifugal forces are described by $-\omega J_x$ and are treated exactly in HFB, as opposed to perturbation methods, which treat them only in lowest orders. At high spins these inertial forces are strong, and perturbation techniques are not appropriate.

8.3.4 *Self-consistency* The HF field, the pair field, and the inertial forces are all treated simultaneously, on an equal footing, and self-consistently. That is, any change in one of these three potentials results in rearrangements in the other two. Consequently, all deformation parameters and pairing gaps are J-dependent. Self-consistency of these various effects is ensured since the potentials (Γ, Δ) are determined by the wave function (ρ, t) and vice versa.

8.3.5 *VMI law* At low spins $\mathscr{I}_{exp}$ is approximately linear in ω^2, and depends upon both deformation and pairing properties. Since HFB treats the response of these properties to the inertial forces, it is of considerable interest to see whether this linearity can be reproduced and explained.

8.3.6 *Coriolis antipairing effect* In 1960 Mottelson & Valatin (46) observed that the Coriolis force acts in opposite directions on each member of a time-reversed pair and therefore tends to decouple the correlated pair. They predicted that the pairing gap Δ would decrease with increasing angular velocity ω. At a critical value of angular momentum J_c, the nucleus would suddenly become normal with $\Delta = 0$ for all pairs, and $\mathscr{I}$ would rise to $\mathscr{I}_{rigid}$. The HFB theory is suitable for investigating this effect.

8.3.7 *Realignment effect* Stephens & Simon (120) have proposed that the effect of the Coriolis force may be to decouple a single neutron or proton pair from the pairing field with subsequent alignment of this pair along the rotation axis. Consequently the nucleus would not be axially symmetric at high spins. The re-aligned pair should be in a high-j orbital, where the Coriolis force is the strongest.

Since HFB theory permits a state-dependent pair gap as well as nonaxial deformations, it includes the possibility of such a realignment effect.

8.3.8   *Shape transitions*   Two rotational bands with different moments of inertia and deformations (such as spherical and deformed, prolate and oblate, or axial and nonaxial) could cross at a high spin (116, 118, 142, 143). At the J corresponding to the band intersection, $\mathscr{I}$ would suddenly rise, producing a backbending curve. Similarly, an axially symmetric rotating nucleus may become unstable with respect to asymmetric deformations at some critical angular momentum (128, 144). HFB predicts not only absolute minima (ground-state intrinsic bands) but also relative minima (excited intrinsic bands). At each J several shapes may be produced. Therefore a band crossing of different shapes or a sudden transition from axial to nonaxial shape could result from an HFB calculation.

8.3.9   *Gapless superconductivity*   The order parameter (Δ) and the energy gap of the excitations (quasi-particle energy E_{qp}) are distinct. Goswami, Lin & Struble (145) showed that the Coriolis force could cause the smallest E_{qp} to vanish, even though Δ is nonzero. Such gapless excitations may be connected with backbending (146, 147) and they are permitted by HFB.

8.3.10   *Negative quasi-particle energies*   The eigenvalues of the HFB equations (E_{qp}) occur in pairs ($+E_i, -E_i$). In conventional applications of HFB theory the positive eigenvalues ($+E_i > 0$) are selected as quasi-particle energies. As was noted above in Section 8.3.9, at a critical angular velocity ω_a the lowest quasi-particle energy E_1 can go to zero. Banerjee, Ring & Mang (148) have proven that if the positive branch is chosen for $\omega > \omega_a$, then the quasi-particle vacuum changes from even to odd number parity. A wave function has even (odd) number parity if all components have an even (odd) number of particles. To preserve the number parity for $\omega > \omega_a$, the negative branch of E_1 must be followed. If $E_1 + E_2 = 0$ at $\omega_b > \omega_a$, then the signs of both quasi-particles can be changed at ω_b, and number parity will be conserved. This prescription guarantees that a two-quasi-particle excitation (which is the lowest noncollective excited state of an even nucleus) will never have lower energy than the quasi-particle vacuum.

8.3.11   *Symmetries in a rotating frame*   The symmetries of the Hamiltonian H are extremely useful in simplifying ground-state ($\omega = 0$) HFB calculations. Unfortunately the inertial forces, $-\omega J_x$, violate most of these symmetries, such as time-reversal, rotations about the y and z axes, and reflections through the xy and xz planes. Consequently, cranked wave functions contain neither time-reversal, nor axial or triaxial symmetry, and they are therefore difficult to calculate.

Goodman (149) has used the one remaining reflection symmetry σ_x (reflection through the yz plane) to reduce the dimension of the cranked HFB equations, and hence make a realistic calculation possible. The σ_x basis is shown (149) to block diagonalize J_x, $\mathscr{H}$, Δ, and consequently κ. Also, the HFB quasi-particle vacuum $|\Phi_0\rangle$ is reduced to canonical form. For all ω, we have

$$|\Phi_0\rangle = \prod_\alpha C_\alpha^\dagger \prod_{\beta \neq \alpha} (u_\beta + v_\beta C_\beta^\dagger C_{\hat\beta}^\dagger)|0\rangle, \qquad\qquad 8.18.$$

where $|\alpha\rangle$, $|\beta\rangle$ and $|\hat\beta\rangle$ are eigenvectors of σ_x with eigenvalues, respectively, of $\pm i$, $-i$ and $+i$. $|\hat\beta\rangle = T|\beta\rangle$ only when $\omega = 0$. The orbits $|\beta\hat\beta\rangle$ are pair-correlated, whereas the orbits $|\alpha\rangle$ are decoupled from the pair field.

8.4 HFB-CRANKING CALCULATIONS

8.4.1 *Exact tests of HFB in soluble models* How well does the HFB method approximate the exact eigenstates of the cranking Hamiltonian H_ω of Equation 8.1? Chu et al (150) studied this question by applying HFB to the exactly soluble two-level $R(8)$ model of Krumlinde & Szymanski (151). This model consists of 2 Ω identical fermions interacting through a pairing force, and populating two single-particle levels ($m = \frac{1}{2}, \frac{3}{2}$), which are 2 Ω-fold degenerate, separated in energy, and coupled to an external rotor with fixed $\mathscr{I}$. The Hamiltonian can be exactly diagonalized by group theoretical methods.

Self-consistent cranking (HFB) is extremely accurate below and above the transition region (which encompasses only two or three yrast states). HFB is more accurate when sharp backbending occurs than when there is no backbending. The relative lack of accuracy of HFB in the transition region is largely caused by the vanishing of Δ when only one pair of particles has their spins aligned. This behavior is intrinsic to the two-level model and would not occur in a realistic calculation. It is concluded that the many-body techniques (HFB) will be even more accurate for studies with a larger number of particles in a more realistic model space.

Although backbending is often accompanied by gapless superconductivity and one negative quasi-particle energy, the correspondence is not one to one. The existence of a negative E_{qp} is due entirely to the matrix element $\langle m = \frac{1}{2}|J_x|m = -\frac{1}{2}\rangle$. In the two-dimensional $R(5)$ model (152), this matrix element is omitted, so that negative quasi-particle energies cannot occur. Nevertheless, sharp backbending is produced in the $R(5)$ model. Finally, it is argued that backbending does not necessarily imply instability in the HFB equations.

8.4.2 *Reid interaction* Goodman & Vary (153, 154) have attempted a realistic description of the spectra of rare-earth nuclei. The Brueckner **G** matrix (ladder-series summation) was calculated (155) using the Reid soft-core nucleon-nucleon potential (156). The valence space contains essentially one major shell of each parity for both neutrons and protons and can accommodate 52 protons and 66 neutrons. The spherical single-particle energies (T_i) of the inert core (40 protons and 70 neutrons) are fixed by equating average spherical HF single-particle energies with spherical Nilsson energies. The HF potential and the pair potential are both determined by the **G** matrix, and include matrix elements of **G** between two particle states $|j_1 j_2 JT\rangle$ of all permissible couplings.

The experimental and HFB spectra of [168,170,172] Yb and [174]Hf are presented in Figures 26a and 26b. The essential characteristics of the experimental spectra are (a) slow variations in $\mathscr{I}$ at low spins, (b) rapid increase in $\mathscr{I}$ at high spins, (c) $\mathscr{I}(\omega^2)$ remains single-valued for [168]Yb and [174]Hf, and (d) $\mathscr{I}(\omega^2)$ is triple-valued (backbends) for ^{170}Yb. All of these features are reproduced by the HFB moment of inertia.

The low-spin behavior of $\mathscr{I}$ is governed by the VMI rule Equation 2.9. The

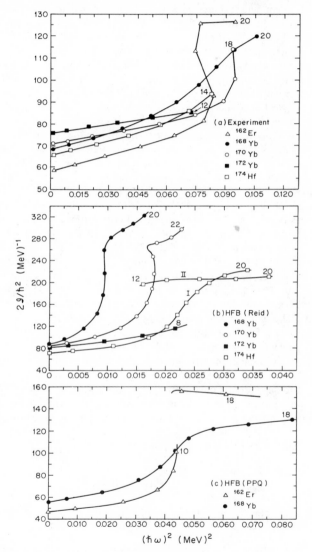

Figure 26 $\mathscr{I}$ vs ω^2 curves for (*a*) experiment (66), (*b*) HFB with **G** matrix derived from Reid potential (153, 154), and (*c*) HFB with pairing-plus-quadrupole interaction (157). Note that the scales differ.

HFB values $\mathscr{I}_0$ of $\mathscr{I}$ at $\omega \to 0$ (Table 6) agree with the VMI values of $\mathscr{I}_0$ (Table 4) to about 20%. However the slopes $1/2C$ of the HFB $\mathscr{I}$ vs ω^2 curves at $\omega \to 0$ (Table 6) are much larger than the VMI slopes of Table 3. Furthermore, the HFB curves deviate from linearity at too small a value of J. These discrepancies are correlated with the magnitudes of the HFB $J = 0$ prolate-oblate energy differences E_{po} (0.49, 1.45, 2.46, and 2.20 MeV), which are too small by several MeV, thereby making these nuclei too soft with regard to nonaxial deformations and increases in $\mathscr{I}$. As E_{po} increases towards more reasonable values (4–5 MeV), the HFB slope decreases, and the extent of the linear region in $\mathscr{I}(\omega^2)$ increases. At high spins $\mathscr{I}_{HFB}$ rises more rapidly than $\mathscr{I}_{exp}$. As E_{po} increases towards more physical magnitudes, $\mathscr{I}_{HFB}$ $(J = 20)$ decreases rapidly towards $\mathscr{I}_{exp}$ $(J = 20)$.

The triple-valued portion of the ^{170}Yb curve cannot be obtained by performing HFB for fixed ω, since all physical quantities are multivalued in ω, thereby creating apparent instabilities. However, all observables are single-valued in J. Consequently, when J is held fixed (and ω varied) at each iteration, these artifical "instabilities" disappear, and convergence in $\langle H' \rangle$ and ω is rapid.

To decide which of the proposed mechanisms for the rapid rise of $\mathscr{I}$ is operative in each nucleus, a brief analysis of the wave functions is presented in Table 7. The properties of four neutron pairs near the Fermi surface are given for various $\hbar\omega$. These neutrons are primarily in the $i_{13/2}$ shell. In the ground state ($\hbar\omega = 0$) $\alpha = 1, 2, 3, 4$ corresponds to $m_z = 7/2, 5/2, 3/2, 1/2$.

Consider ^{168}Yb. At $\hbar\omega = 0$ the state-dependent pair gap $\Delta_{\alpha\hat{\alpha}}$ is nearly constant. The members of each pair are related by time-reversal, so that each pair produces a net $\langle J_x \rangle = 0$. The rapid rise in $\mathscr{I}$ begins at $\hbar\omega = 0.095$ MeV and ends at 0.100 MeV. It is apparent that $\Delta_{\alpha\hat{\alpha}}$ decreases to zero over this interval very uniformly for all orbits. The angular momentum is spread over several pairs, and no single pair predominates. The maximum for any one pair is 5.3 units, whereas a fully aligned $i_{13/2}$ pair would have $\langle J_x \rangle = 13/2 + 11/2 = 12$. The nucleus ^{168}Yb therefore presents a clear example of the Mottelson-Valatin Coriolis antipairing effect (46).

Consider ^{170}Yb. At $\hbar\omega = 0.130$ MeV, $\mathscr{I}(\omega^2)$ is triple-valued. For $\langle J_x \rangle = 11$, $\Delta_{\alpha\hat{\alpha}}$ is not very state-dependent. One pair has $\langle J_x \rangle = 6.2$, which is only half the aligned value. The solution at $\langle J_x \rangle = 17.332$ is drastically altered. While the entire nucleus gained 6.3 units of angular momentum, the fourth pair has acquired 9.9 units.

Table 6 HFB values of $\mathscr{I}_0$ and C^{-1}

Nucleus	$\mathscr{I}_0$ (MeV^{-1})	C^{-1} (10^2 MeV^{-3})	Ref.
^{162}Er	24	3.4	157
^{168}Yb	28	3.7	157
	44	34.1	153, 154
^{170}Yb	42	15.8	153, 154
^{172}Yb	41	10.0	153
^{174}Hf	36	9.2	153

Consequently, all of the other nucleons actually "slow down." Furthermore, this pair has decoupled from the pairing field. The average of the state-dependent pairing gap for the first three pairs is 45% of the $\hbar\omega = 0$ strength, while $\Delta_{\alpha\hat{\alpha}}$ is rapidly vanishing for the last pair.

What is the nature of this decoupled pair $|\alpha\hat{\alpha}\rangle$ in ^{170}Yb? Employment of the σ_x basis makes possible a precise identification. The HFB orbital $|\alpha\rangle$ and the J_x eigenstate $|i_{13/2}, m_x = 13/2\rangle$ have a 98% overlap at $\hbar\omega = 0.14$ MeV. The overlap of $|\hat{\alpha}\rangle$ and $|i_{13/2}, m_x = 11/2\rangle$ is 94%. The decoupled pair is therefore aligned along the x axis of rotation. Whereas m_z is a good quantum number in the ground state, for high spins m_x is essentially a good quantum number for this pair. The nucleus

Table 7   HFB with **G** matrix derived from Reid potential (153, 154)[a]

$\hbar\omega$ (MeV)	$\langle J_x\rangle$	α	v_α^2	$\Delta_{\alpha\hat{\alpha}}$ (MeV)	J_α
^{168}Yb					
0	0	1	0.206	0.520	0
		2	0.731	0.522	0
		3	0.907	0.498	0
		4	0.942	0.483	0
0.095	7.961	1	0.078	0.297	−0.231
		2	0.851	0.301	+1.538
		3	0.964	0.305	+1.771
		4	0.987	0.265	+1.156
0.100	13.844	1	0	0.007	0
		2	1	0.008	−1.216
		3	1	0.009	+4.001
		4	1	0.008	+5.291
^{170}Yb					
0	0	1	0.438	0.517	0
		2	0.883	0.483	0
		3	0.955	0.450	0
		4	0.969	0.437	0
0.130	11.000	1	0.245	0.249	−1.076
		2	0.981	0.277	+0.046
		3	0.987	0.212	+6.171
		4	0.993	0.252	+1.232
0.130	17.332	1	0.055	0.225	+0.045
		2	0.984	0.205	−0.622
		3	0.996	0.205	+0.536
		4	1.000	0.045	+11.106

[a] The occupation probabilities v_α^2, the pairing gap $\Delta_{\alpha\hat{\alpha}}$, and the contribution to the angular momentum, $J_\alpha = [(J_x)_{\alpha\alpha} + (J_x)_{\hat{\alpha}\hat{\alpha}}]v_\alpha^2$, are given for four neutron pairs near the Fermi surface for various angular velocities ω. The single-particle basis α diagonalizes the density matrix ρ. At each ω the states α are ordered with increasing v_α^2. The total angular momentum of the nucleus is $\langle J_x\rangle$.

^{170}Yb is therefore a clear illustration of the realignment effect. In the backbending region a single quasi-particle energy goes to zero and becomes negative, evoking the picture of gapless superconductivity.

The richness of HFB is manifested in ^{174}Hf. There are two single-valued bands, one $(0 \leq J \leq 20)$ resembling the experimental band, and a second $(12 \leq J \leq 20)$ with a nearly constant $\mathscr{I}$. The first band has Δ_n slowly vanishing, an $i_{13/2}$ neutron pair aligning along the x axis, and one negative quasi-particle energy. The second band has a nearly constant Δ_n, a realigned pair, and one gapless excitation. This clearly demonstrates that there is not a one-to-one correspondence between any particular physical mechanism and the existence (or nonexistence) of backbending.

Finally, we consider the spin dependence of the shapes. For $J = 0$ these nuclei are prolate and axially symmetric $(\gamma = 0°)$. At high spins β_2 fluctuates by only 4%. The nonaxial deformation parameter γ rises rapidly in the transition region to about 15° for ^{168,170}Yb and 10° for ^{174}Hf.

8.4.3   *Pairing-plus-quadrupole interaction*   Banerjee, Mang & Ring (157) have solved the HFB cranking equations with the phenomenological PPQ force. The model space is identical to that described above for the Reid interaction. The parameters of the PPQ interaction are adjusted to produce reasonable ground-state properties.

The experimental and HFB spectra of ^{162}Er and ^{168}Yb are given in Figure 26a and 26c. The experimental $\mathscr{I}(\omega^2)$ curves vary slowly at low spins, and rise rapidly at high spins. The ^{162}Er curve is multivalued, whereas the ^{168}Yb curve is single-valued. These essential characteristics are reproduced by HFB. However the backbending section of ^{162}Er was not obtained.

The HFB values of $\mathscr{I}_0$ (Table 6) compare favorably with the VMI values of Table 4. The slopes of the HFB curves at $\omega \to 0$ (Table 6) differ from the VMI slopes $(2C)^{-1}$ (Table 3) by about 40%. However, significant deviations from linearity begin at $J = 4$ in the HFB curves. This may be caused by the absence of higher multipoles of the interaction in the construction of Δ.

The primary mechanism for the rapid rise of $\mathscr{I}$ in ^{162}Er is the decoupling of an $i_{13/2}$ neutron pair from the pairing field concurrent with the alignment of this pair along the x rotation axis. This realignment effect is accompanied by one negative quasi-particle energy. The increase in $\mathscr{I}$ in ^{168}Yb is essentially caused by the sudden reduction of the neutron pairing gap. The Reid and the PPQ interactions give very similar explanations for the ^{168}Yb curve.

The HFB-cranking equations have also been solved for the odd-mass nuclei ^{155,159}Dy (158, 159). The agreement with experiment for both negative and positive parity bands is excellent. In particular, the inversion of J levels at high spins in the positive parity bands is reproduced exactly. This level inversion is known to result from the Coriolis force. However, in other theories the Coriolis force must be attenuated by an arbitrary parameter R $(0.4 < R < 0.9)$ in order to fit such spectra.

In the HFB-cranking theory no attenuation factor was needed. The reason is as follows: the Coriolis force is included in $-\omega J_x$, where $\omega \sim 1/\mathscr{I}$. In HFB $\mathscr{I}$ is calculated self-consistently and is strongly dependent on J. In particular $\mathscr{I}_{\text{HFB}}$

includes the effect of decoupling the odd particle. For low J this particle is coupled to the core, and it is easy to gain angular momentum in the x direction by decoupling the particle. Consequently $\mathscr{I}_{\text{HFB}}$ is large, and the Coriolis force is strongly "attenuated." At higher spins where this particle is becoming x-aligned, its contribution to $\mathscr{I}_{\text{HFB}}$ is greatly reduced, so that $\mathscr{I}_{\text{HFB}}$ decreases as J increases. Correspondingly, there is less "attenuation" of the Coriolis force.

This self-consistency effect may also be explained by noting that the quasi-particle vacuum $|\Phi_0(J)\rangle$ is a function of angular momentum. Consequently, the one quasi-particle state $a^\dagger_{\alpha(J)}|\Phi_0(J)\rangle$ that describes odd nuclei contains three, five, ... $J = 0$ quasi-particle $(a^\dagger_{\alpha(0)})$ admixtures into $|\Phi_0(0)\rangle$. That the HFB-cranking theory explains the "attenuation" of the Coriolis force and the level inversions in odd nuclei without the conventional adjustable parameter is a remarkable achievement.

8.4.4 *Effect of $J \neq 0$ multipoles of the interaction on the pairing field* The PPQ interaction contains only a $J = 0$ pairing force. Krumlinde & Szymanski (160) included a quadrupole pairing interaction in their two-dimensional two-level model. The qualitative features of the backbending curves are not altered by this inclusion. However the critical angular momentum of the transition is slightly decreased.

Bhargava & Thouless (161, 162) have used the Sussex interaction in a model calculation containing only the $1h_{11/2}$ shell. The effect of including the $J = 2$ and $J = 4$ components of the particle-particle (pairing) channel is studied. They find that whereas the effective strength of the $J = 0$ pairing force rapidly diminishes at high spins, the higher multipole components retain some strength and considerably alter the shapes of the $\mathscr{I}(\omega^2)$ curves. It should be noted that in the work of Goodman & Vary (154) all components ($0 \leq J \leq 12$) of the **G** matrix are included.

8.4.5 *Assessment of HFB-cranking* The HFB-cranking theory (a) compares very well with exact solutions in soluble models, (b) reproduces the essential qualitative features of $\mathscr{I}_{\text{exp}}(\omega^2)$, (c) exhibits the ability to include and clearly differentiate the various mechanisms leading to high-spin anomalies, and (d) explains the attenuation of the Coriolis force in odd nuclei.

Quantitative defects of the predicted spectra include deviations from the VMI model at low spins. $\mathscr{I}_0^{\text{HFB}}$ and $\mathscr{I}_0^{\text{VMI}}$ differ by 20%. C_{HFB}^{-1} and C_{VMI}^{-1} differ by 40% for the PPQ interaction and by an order of magnitude more for the Reid **G** matrix. Improvements in the effective interaction may lead to a more quantitative agreement with experiment.

8.5 PROJECTION OF ANGULAR MOMENTUM AND PARTICLE NUMBER   Faessler et al (163–165) have minimized the pairing-plus-quadrupole Hamiltonian with respect to the set of projected trial wave functions $|\Delta_p\Delta_n\beta JM\rangle$, which are constructed as follows: given β, axially symmetric Nilsson single-particle states are calculated. Given Δ_p and Δ_n, a BCS state is formed with the Nilsson levels. Angular momentum and partial number projection on these BCS states provide the trial wave functions. There are three variational parameters Δ_p, Δ_n, and β. Kumar (166) has employed the same degrees of freedom, but he used the cranking constraint without any projection.

Since the realignment effect requires the possibility of one pair of particles de-coupling from the pair field, Δ must be state-dependent (as in HFB) for realignment to occur. However in this model Δ_n is the same for all neutron Nilsson levels, and similarly for protons. Furthermore, realignment of particles along the rotation axis requires that the nuclear shape be nonaxial. Triaxial degrees of freedom are expected to be significant at high spins (128, 144, 167–170). In this model the nuclear shape is restricted to be axial at all J. Consequently the realignment effect and transitions to triaxial shapes are excluded from the model space. The moment of inertia of the core is a free parameter.

The nuclei investigated include the backbending ^{162}Er, ^{166,170}Yb, ^{168}Hf, and ^{182}Os and the non-backbending ^{158}Dy and ^{168}Yb. The theory predicts backbending only for ^{158}Dy and ^{170}Yb. The cause in ^{158}Dy is a sudden jump in deformation due to level crossings at the Fermi surface. For ^{170}Yb the cause is Coriolis anti-pairing of the neutrons. In general the high-spin anomalies cannot be reproduced. Partial-number projection considerably alters the unprojected spectra. This is also suggested by the similar model calculations of Dalafi et al (171, 172).

9   Conclusions

The VMI-Harris model has replaced the appealing simplicity of the early collective model by an equally succinct and more general description. The collective model distinguished between a "strong coupling region" halfway between the larger closed shells, in which each nucleus was characterized by an intrinsic, deformation-dependent quadrupole moment and moment of inertia, derived from the transition probability and energy of the lowest $2+$ state, and a weak coupling region in which nuclei were thought to be spherical and undergoing only vibrational motions related to the values of $B(E2)$ and $E2$. The discovery of the energy gap, as well as the rigid moment of inertia result for the independent particle cranking model, later led to the introduction of the pairing-plus-quadrupole model. While this model is more realistic, some of the intuitive appeal of the early collective model was lost.

In the VMI model, each non-magic nucleus is characterized by two parameters, which can be deduced from the spectrum (E_2, E_4) only: a ground-state moment of inertia $\mathscr{I}_0 = \mathscr{I}_{J=0}$ and a stiffness parameter C. In the case of $\mathscr{I}_0 < 0$, $\mathscr{I}_{J=0}$ vanishes and $(C/2)\mathscr{I}_0^2$ measures the resistance to deformation. In both cases, for $J > 0$ the moment of inertia $\mathscr{I}(J)$ is positive and increases monotonically. Stable, strongly deformed nuclei are nearly perfect rotors—the spectra deviate only slightly from $J(J+1)$. With increasing β-ray instability and with decreasing deformation the softness of the spectra becomes more marked, i.e. the moment of inertia increases more rapidly with angular momentum, producing a quasi-vibrational spectrum.

The VMI model fits ground-state bands up to a limiting value $J = J_c$ for the whole range of non-magic nuclei $(1.82 \leq R_4 \leq 3.33)$. R_J ratios for ground-state bands in magic nuclei, for which $R_4 \leq 1.82$, abruptly become almost degenerate.

The empirical relation of $\mathscr{I}$ to Q is essentially independent of A and Z, but can be divided into two clearly distinguishable domains, which roughly coincide with the previous weak-coupling and strong-coupling regions. For nuclei with one or two extra-shell nucleon pairs of one kind, a linear relationship exists between $\mathscr{I}$ and Q.

described by a dumbbell model. For nuclei with several extra-shell nucleon pairs a quadratic $\mathscr{I}$ vs Q relationship exists, described by a macroscopic two-fluid model. For the heaviest, spontaneously fissioning nuclei $\mathscr{I}$ remains essentially constant while Q increases.

In the region in which the quadratic relationship holds for the $0 \to 2$ transition, $\mathscr{I}$ increases somewhat faster than Q^2, as J increases. The relationship of $\mathscr{I}$ and Q at higher J for all three groups requires further analysis.

The knowledge of the mode of coupling of other bands to the ground-state band is still at a primitive stage; e.g. an empirical relationship not yet well understood is the coupling of the γ band, with a strength decreasing linearly with increasing R_4, while the ratio of the vibrational to rotational frequency first remains almost constant at ~ 2, and increases steeply by approximately an order of magnitude as the rotational limit is approached.

The onset of deviations from the VMI predictions at the critical angular momentum J_c can be studied best by means of the $\mathscr{I}$ vs ω^2 representation (where ω denotes the angular velocity). In many cases the deviations set in suddenly at J_c, and in a number of cases "backbending" occurs, i.e. the curve becomes triple-valued in ω^2. The process taking place in the nucleus near J_c is frequently referred to as a phase transition. It now appears that several mechanisms, either singly or combined, are involved in this transition, such as decoupling of a nucleon pair with high j, as well as a sudden decrease in pairing.

Numerous authors have attempted to find improvements on the VMI approach, either by somewhat different two-parameter expressions, or by adding higher-order terms in the angular velocity expansion. Attempts have further been made to develop more detailed nuclear theories in order to fit the R_J vs R_4 curves.

In cases where the angular velocity expansion converges, we obtain an identification of the phenomenological VMI parameters $\mathscr{I}_0$ and $(2C)^{-1}$ with corresponding expressions in the microscopic cranking perturbation formalism. This provides a test of the adequacy of our theories of nuclear structure, in particular the correctness of our choice of a deformed single-particle basis and residual interaction. The close connection of the VMI parameterization and the microscopic cranking formalism enables us to predict which physical effects (Coriolis antipairing, fourth-order cranking, centrifugal stretching, etc) contribute to deviations from a constant moment of inertia for a given nucleus.

Even for very good rotors ($\sigma \lesssim 0.01$), for which the ω^2 expansion converges, the microscopic perturbative cranking formalism does not give a quantitative account of either $\mathscr{I}_0$ or C^{-1}. The values of $\mathscr{I}_0$ are systematically too low by 10–40%, while the C^{-1} values often suffer a discrepancy of a factor of two. The existing cranking perturbation calculations do not use realistic nucleon-nucleon interactions, are not fully self-consistent, and are of course not capable of dealing with the backbending phenomenon. A promising approach seems to be provided by the HFB-cranking method with realistic residual forces; this method is nonperturbative and self-consistent. The HFB method relates the rotational spectrum directly to the effective nucleon-nucleon interaction (**G** matrix). It is sufficiently flexible to accommodate exactly the physical mechanisms whose effect can only be estimated perturbatively in the usual cranking model. The HFB method is thus capable of achieving simultaneously a description of both the low-spin VMI region and the high-spin backbending regions in nuclei. Only relatively few HFB-cranking calculations with either realistic or phenomenological forces have been performed, but the preliminary results are very encouraging. For the few cases studied, the sudden rise in $\mathscr{I}$ in backbending nuclei is caused by the realignment of a single neutron pair along the rotation axis, whereas the rapid increase in $\mathscr{I}$ in non-backbending nuclei results from the sudden disappearance of the neutron pairing gap. Further work employing this method may ultimately provide a quantitative understanding of rotational spectra in nuclei.

ACKNOWLEDGMENTS

G. Scharff-Goldhaber would like to thank many of her colleagues for helpful discussions, and A. S. Goldhaber for his constant interest and understanding support. She further owes particular gratitude to M. McKeown for his resourceful and enthusiastic collaboration during many years of research, and to W. F. Piel, Jr. for his participation in the more recent phases of the work described here. Two of us (G.S.G. and C.B.D.) should like to acknowledge the generous advice and incisive criticism of Joseph Weneser. A. L. Goodman is indebted to R. Sorensen for a critical reading of Section 8.

Literature Cited

1. Pauli, W. 1924. *Naturwissenschaften* 37:741–43
2. Casimir, H. B. G. 1963. On the interaction between atomic nuclei and electrons. San Francisco & London: Freeman. 2nd ed. *Teyler's Tweede Genootschap, Haarlem.* Originally published in 1936.
3. Schüler, H., Schmidt, T. 1935. *Z. Phys.* 94:452; 95:265
4. Heitler, W., London, F. 1927. *Z. Phys.* 44:455–72
5. Gamow, G. 1930. *R. Soc. London Proc. A* 126:632–44
6. Rutherford, E., Chadwick, J., Ellis, C. D. 1930. *Radiations from Radioactive Substances,* p. 530. Cambridge, Engl: Univ. Press
7. von Weizsäcker, C. F. 1935. *Z. Phys.* 96:431–58
8. Bohr, A., Mottelson, B. R. 1969. *Nuclear Structure,* 1:141, 142. New York & Amsterdam: Benjamin
9. Bethe, H. A. 1971. *Ann. Rev. Nucl. Sci.* 21:93–244
10. Clark, J. W., Lam, P. M., ter Louw, W. J. 1975. *Nucl. Phys. A* 255:1–12 and references therein
11. Flowers, B. 1952. *Progress in Nuclear Physics,* 2:235–70. London: Pergamon

12. Thibaud, J. 1930. *CR Acad. Sci. Paris* 191:656–58
13. Teller, E., Wheeler, J. 1938. *Phys. Rev.* 53:778–89
14. Bohr, N. 1936. *Nature* 137:344–48
15. Bohr, N., Wheeler, J. A. 1939. *Phys. Rev.* 56:426–50
16. Frenkel, J. 1939. *Phys. Rev.* 55:987
17. Jensen, J. H. D. 1963. Zur Geschichte der Theorie des Atomkerns, pp. 153–64. *Les Prix Nobel en 1963.* Stockholm: Norstedt
18. Mayer, M. G. 1948. *Phys. Rev.* 74:235–39
19. Feenberg, E. 1949. *Phys. Rev.* 75:320–22; Nordheim, L. W. 1949. *Phys. Rev.* 75:1894–1901
20. Schmidt, T. 1940. *Naturwissenschaften* 28:565
21. Feenberg, E., Hammack, K. 1949. *Phys. Rev.* 75:1877–93; Gordy, W. 1949. *Phys. Rev.* 76:139–40; Hill, R. D. 1949. *Phys. Rev.* 76:998
22. Townes, C. H., Foley, H. W., Low, W. 1949. *Phys. Rev.* 76:1415–16
23. Rainwater, J. 1950. *Phys. Rev.* 79:432–34
24. Bohr, A. 1951. *Phys. Rev.* 81:134–38
25. Nilsson, S. G. 1955. *Kgl. Dan. Vidensk. Selsk. Mat. Fys. Medd.* 29(16):1–68

26. Goldhaber, M., Sunyar, A. W. 1951. *Phys. Rev.* 83:906–18
27. Bohr, A. 1952. *Kgl. Dan. Videsnk. Selsk. Mat. Fys. Medd.* 26:1–40
28. Bohr, A., Mottelson, B. R. 1953. *Phys. Rev.* 89:316–17
29. Bohr, A., Mottelson, B. R. 1953. *Phys. Rev.* 90:717–19
30. Goldhaber, M., Hill, R. D. 1952. *Rev. Mod. Phys.* 24:179
31. Burson, S. B., Blair, K. W., Keller, H. B., Wexler, S. 1951. *Phys. Rev.* 83:62–68
32. Mihelich, J. W., Scharff-Goldhaber, G., McKeown, M. 1953. *Phys. Rev. A* 94:794
33. Mihelich, J. W., Scharff-Goldhaber, G., McKeown, M. Unpublished
34. Bohr, A. Mottelson, B. R. 1953. *Kgl. Dan. Vidensk. Selsk. Mat. Fys. Medd.* 27(16):1–174
35. Sunyar, A. W. 1955. *Phys. Rev.* 98:653–57
36. Scharff-Goldhaber, G. 1953. *Phys. Rev.* 90:587–602. Energies of the lowest 2+ states were independently compiled by Preiswerk, P., Stahelin, P. 1952. *Helv. Phys. Acta* 24:623
37. Kurath, D. 1952. *Phys. Rev. A* 87:218; Flowers, B. H. 1952. *Phys. Rev.* 86:254–55
38. Goldhaber, M., Teller, E. 1948. *Phys. Rev.* 74:1046–49
39. Preface to 2nd ed. Ref. 34. 1957, pp. 3–6
40. Inglis, D. R. 1956. *Phys. Rev.* 103:1786–95
41. Sorensen, R. A. 1973. *Rev. Mod. Phys.* 45:353–77
42. Bohr, A., Mottelson, B. R. 1955. *Kgl. Dan. Vidensk. Selsk. Mat. Fys. Medd.* 30(1):1–24
43. Bohr, A., Mottelson, B. R., Pines, D. 1958. *Phys. Rev.* 110:936–38
44. Griffin, J. J., Rich, M. 1960. *Phys. Rev.* 118:850–54
45. Nilsson, S. G., Prior, D. 1961. *Kgl. Dan. Vidensk. Selsk. Mat. Fys. Medd.* 32(16):1–61
46. Mottelson, B. R., Valatin, J. G. 1960. *Phys. Rev. Lett.* 5:511–12
47. Scharff-Goldhaber, G., Weneser, J. 1955. *Phys. Rev.* 98:212–14
48. Mottelson, B. R., Nilsson, S. G. 1955. *Phys. Rev.* 99:1615–17
49. Scharff-Goldhaber, G. 1956. *Phys. Rev.* 103:837–38
50. Wilets, L., Jean, M. 1956. *Phys. Rev.* 102:788–96
51. Hadermann, J., Rester, A. C. 1974. *Nucl. Phys. A* 231:120
52. Scharff-Goldhaber, G. 1957. *Proc. Pitts-burgh Conf.*, ed. S. Meshkov, pp. 447–79
53. Scharff-Goldhaber, G., Alburger, D. E., Harbottle, G., McKeown, M. 1958. *Phys. Rev.* 111:913–19
54. Kane, W. R., Emery, G. T., Scharff-Goldhaber, G., McKeown, M. 1960. *Phys. Rev.* 119:1953–69
55. Emery, G. T., Kane, W. R., McKeown, M., Perlman, M. L., Scharff-Goldhaber, G. 1963. *Phys. Rev.* 129:2597–2621
56. Diamond, R. M., Hollander, J. M., Horen, D. J., Naumann, R. A. 1961. *Nucl. Phys.* 25:248–58
57. Sakai, M. 1975. *At. Data Nucl. Data Tables* 15:513–42
58. Mallmann C. 1959. *Phys. Rev. Lett.* 2:507–9
59. Davydov, A. S., Filippov, G. F. 1958. *Nucl. Phys.* 8:237–49
60. Davydov, A. S., Chaban, A. A. 1960. *Nucl. Phys.* 20:499–508
61. Baranger, M. 1961. *Phys. Rev.* 122:992–96
62. Kumar, K., Baranger, M. 1968. *Nucl. Phys. A* 110:490–528, 529–54
63. de Boer, J., Eichler, J. 1968. *Adv. Nucl. Phys.* 1:1–65 and references therein. See also Kleinfeld, A. M., Mäggi, G., Werdecker, D. 1975. *Nucl Phys. A* 248:342–55
64. Christy, A., Häusser, O. 1972. *At. Nucl. Data Tables* 11:281–98
65. Morinaga, H., Gugelot, P. C. 1963. *Nucl. Phys.* 46:210–24
66. Sayer, R. O., Smith, J. S. III, Milner, W. T. 1975. *At. Nucl. Data Tables* 15:85–110
67. Cheifetz, E., Jared, J. C., Thompson, S. G., Wilhelmy, J. B. 1970. *Phys. Rev. Lett.* 25:38–43; Wilhelmy, J. B., Thompson, S. G., Jared, J. C., Cheifetz, E. 1970. *Phys. Rev. Lett.* 25:1122–25
68. Grosse, E., de Boer, J., Diamond, R. M., Stephens, F. S., Tjøm, P. 1975. *Phys. Rev. Lett.* 35:562
69. Morinaga, H. 1966. *Nucl. Phys.* 75:385–95
70. Diamond, R. M., Stephens, F. S., Swiatecki, W. J. 1964. *Phys. Rev. Lett.* 11:315–18
71. Harris, S. H. 1965. *Phys. Rev. B* 138:509–13
72. Scharff-Goldhaber, G. 1967. *Proc. Int. Conf. Nucl. Struct., Tokyo, 1967*, pp. 150–59
73. Mariscotti, M. A. J., Scharff-Goldhaber, G., Buck, B. 1969. *Phys. Rev.* 178:1864–87
74. Proetel, D., Diamond, R. M., Stephens, F. S. 1974. *Phys. Lett. B* 48:102–4; Lieder, R. M., Beuscher, H., Davidson,

W. F., Neskakis, A., Mayer-Böricke, C. 1975. *Nucl. Phys. A* 248:317–41
75. Garrett, J. D., Scharff-Goldhaber, G., Vary, J. P. 1974. *Bull. Am. Phys. Soc.* 19:59
76. Scharff-Goldhaber, G., Goldhaber, A. S. 1970. *Phys. Rev. Lett.* 24:1349–53
77. Mariscotti, M. A. J. 1970. *Phys. Rev. Lett.* 24:1242–45; Gorfinkiel, J. I., Mariscotti, M. A. J., Pomar, C. 1974. *Phys. Rev. C* 9:1243–52
78. Thieberger, P. 1970. *Phys. Rev. Lett.* 25:1664–66
79. Scharff-Goldhaber, G. 1974. *J. Phys. A: Math. Nucl. Gen.* 7:L212–14
80. Scharff-Goldhaber, G., Takahashi, K. 1967. *Bull. Acad. Sci. USSR Phys. Ser.* 31:42
81. Saethre, Ø., Hjorth, S.-A., Johnson, A., Jagare, S., Ryde, H., Szymanski, Z. 1973. *Nucl. Phys. A* 207:486–512
82. Ejiri, H., Ishihara, M., Sakai, M., Katori, K., Inamura, T. 1968. *J. Phys. Soc. Jpn.* 24:1189
83. Das, T. K., Dreizler, R. M., Klein, A. 1970. *Phys. Rev. C* 2:632–38
84. Grodzins, L. 1962. *Phys. Lett.* 2:88–91
85. Scharff-Goldhaber, G., Goldhaber, A. S. 1974. *Proc. Int. Conf. Nucl. Struct. Spectrosc., Amsterdam, The Netherlands,* 2:182–85
86. Goldhaber, A. S., Scharff-Goldhaber, G. To be published
87. Ford, J. L. C. Jr., Stelson, P. H., Bemis, C. E. Jr., McGowan, F. K., Robinson, R. L., Milner, W. T. 1971. *Phys. Rev. Lett.* 27:1232–34
88. Griffin, R. E., Jackson, A. D., Volkov, A. B. 1971. *Phys. Lett. B* 36:281–86
89. Van Rij, W. I., Kahana, S. H. 1972. *Phys. Rev. Lett.* 28:50–54
90. Thieberger, P., Sunyar, A. W., Rogers, P. C., Lark, N., Kistner, O. C., der Mateosian, E., Cochavi, S., Auerbach, E. H. 1972. *Phys. Rev. Lett.* 28:972–74
91. Cochavi, S., Kistner, O., McKeown, M., Scharff-Goldhaber, G. 1972. *J. Phys. France* 33:102
92. Scharff-Goldhaber, G., McKeown, M., Lumpkin, A. H., Piel, W. F. Jr., 1973. *Phys. Lett. B* 44:416–20; Piel, W. F. Jr., Scharff-Goldhaber, G., Lumpkin, A. To be published
93. Arima, A., Iachello, F. 1975. *Phys. Lett. B* 57:34–43
Iachello, F. 1974. *Proc. Int. Conf. Nucl. Struct. Spectrosc., Amsterdam, Sept. 1974,* 2:163–81
94. Krien, K., Naumann, R. A., Rasmussen, J. O., Rezanka, I. 1973. *Nucl. Phys. A* 209:572–88

Wisshak, K., Klewe-Nebenius, H., Habs, D., Faust, H., Nowicki, G., Rebel, H. 1974. *Nucl. Phys. A* 247:59–73
95. Toki, H., Faessler, A. 1975. *Nucl. Phys. A* 253:231–52
96. Klein, A., Dreizler, R. M., Das, T. K. 1970. *Phys. Lett. B* 31:333–35
97. Draper, J. E. 1970. *Phys. Lett. B* 32:581–82
98. Thieberger, P. 1970. *Phys. Lett. B* 45:417–20
99. Ma, C. W., Rasmussen, J. 1970. *Phys. Rev. C* 2:798–819
100. Ma, C., Tsang, C. F. 1975. *Phys. Rev. C* 11:213–27
101. Mantri, A. N., Sood, P. C. 1973. *Phys. Rev. C* 7:1294–1305
102. Moszkowski, S. A. 1966. *Proc. Conf. Nucl. Spin Parity Assignments, Gatlinburg, Tenn., 1965,* ed. H. B. Gove, R. L. Robinson. New York: Academic
103. Sood, P. C. 1968. *Can. J. Phys.* 46:1419–23; 1969. *J. Phys. Soc. Jpn.* 26:1059
104. Draper, J. E. 1972. *Phys. Lett. B* 41:105–9
105. Ma, C. W., Rasmussen, J. O. 1974. *Phys. Rev. C* 9:1083–90
106. Stockmann, M. I., Zelevinsky, V. G. 1972. *Phys. Lett. B* 41:19–22
107. Wood, J. L., Fink, R. W. 1974. *Nucl. Phys. A* 224:589–95
108. Gupta, R. K. 1971. *Phys. Lett. B* 36:173–78
109. Holmberg, P., Lipas, P. O. 1968. *Nucl. Phys. A* 117:552–60
110. Sood, P. C. 1967. *Phys. Rev.* 161:1063–70
111. Varshni, Y. P., Bose, S. 1972. *Phys. Rev. C* 6:1770–80; Varshni, Y. P. 1974. *J. Phys. Soc. Jpn.* 36:317–25
112. Warke, C. S., Khadkikar, S. B. 1968. *Phys. Rev.* 170:1041–46
113. Iachello, F., Arima, A. 1974. *Phys. Lett. B* 53:309–12
114. Das, T. K., Banerjee, B. 1973. *Phys. Rev. C* 7:2590–92
115. Goodman, A., Goswami, A. 1974. *Phys. Rev. C* 9:1948–53
116. Molinari, A., Regge, T. 1972. *Phys. Lett. B* 41:93–96
117. Broglia, R. A., Molinari, A., Pollarolo, G., Regge, T. 1974. *Phys. Lett. B* 50:295–98
118. Broglia, R. A., Molinari, A., Pollarolo, G., Regge, T. 1975. *Phys. Lett. B* 57:113–16
119. Lieder, R. M. et al 1974. *Phys. Lett. B* 49:161–64
120. Stephens, F. S., Simon, R. S. 1972.

VMI MODEL AND NUCLEAR COLLECTIVE MOTION   317

Nucl. Phys. A 183:257–84
121. Stephens, F. S., Kleinheinz, P., Sheline, R. K., Simon, R. S. 1974. *Nucl. Phys. A* 222:235–51; Grosse, E., Stephens, F. S., Diamond, R. M. 1973. *Phys. Rev. Lett.* 31:840–43; 1974. 32:74–77
122. Stephens, F. S. 1975. *Rev. Mod. Phys.* 47:43–65
123. Sheline, R. K. 1972. *Nucl. Phys. A* 195:321–32; 1973. *Phys. Lett. B* 45:459–62
124. Mottelson, B. R. *Proc. Nucl. Struct. Symp. Thousand Lakes, Joutsa, 1970,* Pt. II, pp. 148–65 (Univ. Jyväskylä Res. Rep. 4/1971)
125. Abecasis, S. M., Hernandez, E. S. 1972. *Nucl. Phys. A* 180:485–96; Abecasis, S. M. 1973. *Nucl. Phys. A* 205:475–80
126. Gupta, R. 1973. *Phys. Rev. C* 7:2476–83; 1972. 6:26–35 Trainor, L. E. H., Gupta, R. 1971. *Can. J. Phys.* 49:133
127. Das, T. K., Dreizler, R. M., Klein, A. 1970. *Phys. Rev. Lett.* 25:1626–28
128. Smith, B. C., Volkov, A. B. 1973. *Phys. Lett. B* 47:193–96
129. Wahlborn, S., Gupta, R. K. 1972. *Phys. Lett. B* 40:27–31
130. Udagawa, T., Sheline, R. K. 1966. *Phys. Rev.* 147:671–84
131. Chan, K. Y., Valatin, J. G. 1966. *Nucl. Phys.* 82:222–40
132. Chan, K. Y. 1966. *Nucl. Phys.* 85:261–72
133. Bes, D. R., Landowne, S., Mariscotti, M. A. J. 1968. *Phys. Rev.* 166:1045–51
134. Krumlinde, J. 1968. *Nucl. Phys. A* 121:306–28
135. Marshalek, E. R. 1965. *Phys. Rev. B* 139:770–89; 1967. *Phys. Rev.* 158:993–1000
136. Belyaev, S. T. 1959. *Kgl. Danske Videnskab Selskab, Mat. Fys. Medd.* 31(11):1–54
137. Rasmussen, J. O., Ma, C. W. 1973. *Phys. Rev. Lett.* 31:317–19
138. Nilsson, S. G., Tsang, C. F., Sobiczewski, A., Szymanski, Z., Wycech, S., Gustafson, C., Lamm, I. L., Moller, P., Nilsson, B. 1969. *Nucl. Phys. A* 131:1–66
139. Mang, H. J. 1975. *Phys. Rep.* 18(6):325–68
140. Ring, P., Beck, R., Mang, H. J. 1970. *Z. Phys.* 231:10–25
141. Beck, R., Mang, H. J., Ring, P. 1970. *Z. Phys.* 231:26–47
142. Pavlichenkov, I. M. 1974. *Phys. Lett. B* 53:35–38
143. Perazzo, R. P. J., Reich, S. L. 1975. *Phys. Lett. B* 55:354–56
144. Turner, R. J., Kishimoto, T. 1973. *Nucl. Phys. A* 217:317–41
145. Goswami, A., Lin, L., Struble, G. L. 1967. *Phys. Lett. B* 25:451–54
146. Grin, Y. T. 1974. *Phys. Lett. B* 52:135–37
147. Goswami, A. To be published
148. Banerjee, B., Ring, P., Mang, H. J. 1974. *Nucl. Phys. A* 221:564–72
149. Goodman, A. L. 1974. *Nucl. Phys. A* 230:466–76
150. Chu, S. Y., Marshalek, E. R., Ring, P., Krumlinde, J., Rasmussen, J. O. 1975. *Phys. Rev. C* 12:1017–35
151. Krumlinde, J., Szymanski, Z. 1973. *Ann. Phys. NY* 79:201–49
152. Bose, S., Krumlinde, J., Marshalek, E. R. 1974. *Phys. Lett. B* 53:136–40
153. Goodman, A. L. 1976. *Nucl. Phys.* 113–41
154. Goodman, A. L., Vary, J. P. 1975. *Phys. Rev. Lett.* 35:504–7
155. Goodman, A. L., Vary, J. P., Sorensen, R. A. 1976. *Phys. Rev. C* 13:1674
156. Reid, R. V. 1968. *Ann. Phys.* 50:411
157. Banerjee, B., Mang, H. J., Ring, P. 1973. *Nucl. Phys. A* 215:366–82
158. Ring, P., Mang, H. J., Banerjee, B. 1974. *Nucl. Phys. A* 225:141–56
159. Ring, P., Mang, H. J. 1974. *Phys. Rev. Lett.* 33:1174–77
160. Krumlinde, J., Szymanski, Z. 1974. *Phys. Lett. B* 53:322–24
161. Bhargava, P. C. 1973. *Nucl. Phys. A* 207:258–72
162. Bhargava, P. C., Thouless, D. J. 1973. *Nucl. Phys. A* 215:515–24
163. Faessler, A., Lin, L., Wittmann, F. 1973. *Phys. Lett. B* 44:127–30
164. Faessler, A., Grümmer, F., Lin, L., Urbano, J. 1974. *Phys. Lett. B* 48:87–90
165. Gruemmer, F., Schmid, K. W., Faessler, A. 1975. *Nucl. Phys. A* 239:289–300
166. Kumar, K. 1973. *Phys. Rev. Lett.* 30:1227–30
167. Sugawara-Tanabe, K., Tanabe, K. 1973. *Nucl. Phys. A* 208:317–32
168. Meyer ter Vehn, J. 1975. *Phys. Lett. B* 55:273–76
169. Meyer ter Vehn, J. 1975. *Nucl. Phys. A* 249:111–40
170. Meyer ter Vehn, J. 1975. *Nucl. Phys. A* 249:141–65
171. Dalafi, H. R., Banerjee, B., Mang, H. J., Ring, P. 1973. *Phys. Lett. B* 44:327–29
172. Dalafi, H. R., Mang, H. J., Ring, P. 1975. *Z. Phys. A* 273:47–55

Ann. Rev. Nucl. Sci. 1976. 26:319–50
Copyright © 1976 by Annual Reviews Inc. All rights reserved.

FOSSIL NUCLEAR REACTORS   ×5576˙

Michel Maurette

Laboratoire René BERNAS du Centre de Spectrométrie Nucléaire
et de Spectrométrie de Masse, Orsay, France

CONTENTS

1   INTRODUCTION

The first man-made nuclear chain reactor occurred in a closely guarded laboratory of the University of Chicago on December 2, 1942. This reactor was a mass, the "pile," containing uranium in some suitable arrangement throughout a block of graphite, which acted as moderator for slowing down fast fission neutrons. According to the nuclear-reactor theory (1, 2) the condition for maintaining a chain reaction is that on the average at least one neutron created in a fission process causes another fission. In a more quantitative way this condition of "criticality" can be stated by expressing that the neutron-multiplication factor of the reactor, $k = k_\infty \cdot P$, is equal to 1. In this expression k_∞ is the "infinite" multiplication factor of the reactor computed for an infinite mass, and P is a geometrical factor that expresses that $k < k_\infty$ because of the loss of neutrons by leakage through the finite surface of the reactor.

Quantitative studies of "fossil" fission products trapped in uranium minerals were performed from about 1953 to 1963 by the groups of Thodes, Wheterill, and Kuroda.

319

The main results of these investigations have already been summarized elsewhere (3, 4). They mainly show that uranium minerals, when extracted from old and well-preserved uranium ore deposits, contain fission products due both to the spontaneous fission of ^{238}U ($^{238}U_{sf}$) and to a lesser extent to the slow neutron fission of ^{235}U ($^{235}U_{nf}$). However, the small proportion ($\sim 10\%$) of fission events attributed to $^{235}U_{nf}$ was compatible with very small neutron fluences ($\lesssim 10^{15}$ n cm^{-2}), mostly due to $^{238}U_{sf}$, and showed no evidence for a marked multiplication of the neutrons.

During the course of these studies, Wheterill (5) and then Kuroda (6) suggested the possible occurrence of natural chain reactions in old ($\gtrsim 2$ billion years) uranium-rich ore deposits. In particular, Kuroda was the first to evaluate k_∞ for uranium mines considered as assemblages of fuel (^{235}U), moderator (water), and neutron-absorbing "poisonous" impurities. The striking conclusion of this evaluation was that criticality ($k_\infty = 1$) could have been reached in old uranium ores both having an age, $T \gtrsim 2$ billion years, and a water-to-uranium mass ratio $\gtrsim 0.5$. Kuroda even suggested that the absence of uranium ores older than 2 billion years could well result from their destructive disintegration during the reactor phase.

The first evidence of a fossil nuclear reactor was accidentally discovered in 1972 following an extraordinary combination of chance, extreme technical care, and scientific ingenuity by a team of the French Atomic Energy Commission (CEA). Bouzigue et al (7) were controlling on a routine basis the input of uranium at the French uranium gas-diffusion plant at Pierrelate. Suddenly, in June 1972, secondary natural uranium standards were found to be 0.4% lower in the ratio $^{235}U/^{238}U$ than primary standards, when this ratio was considered constant in nature to within 0.1%. This discrepancy, although very small, was taken seriously and was further investigated. The resulting "detective" investigation was then run backward in time, through the complex chain of processes involved in the preparation of uranium in several centers of CEA, and up to the original source of uranium used to prepare the suspect standards, some ores from the Oklo deposit in Gabon. To their amazement, the CEA team found that samples extracted from a uranium-rich zone within the ore bodies were as much as 50% depleted in ^{235}U, and various hypotheses were then advanced to account for this marked depletion. Subsequently, a search for rare-earth fission products (Nd) showed that they were abundant in the Oklo ores. Therefore in September 1972, Neuilly et al (8) felt sufficiently confident to announce that the Oklo deposit had been the site of a natural chain reaction in the distant past. Since then, R. Naudet has been leading the so-called Franceville project established by CEA to investigate the Oklo phenomenon. This project has initiated many studies, partially reviewed by Naudet (9, 10), and recently reported at the first international meeting on the Oklo phenomenon held in Libreville in June 1975.

So far the Oklo phenomenon is quite unique. In fact, the six distinct reaction zones found in the Oklo mine are the only fossil reactors discovered in nature and only two of these zones have been carefully investigated to date. Therefore this review cannot be a critical comparison of fossil nuclear reactors, intended to draw general conclusions about their characteristics (types, frequency of occurrence, association with peculiar geological features, etc). Instead, in reviewing the multidisciplinary work performed to date (January 1976), we try to illustrate how the Oklo phenomenon challenges our understanding of a wide variety of complex problems (aging of geological and geochemical features over a long time scale, stability of radiation-damage features in insulators, radiation-induced corrosion, disposal of man-made radioactive wastes, composition and circulation of deep ground waters, etc), as well as the most refined analytical methods recently used in the fields of earth sciences, irradiation effects in solids, radiochemistry, and so forth.

In Section 2 we briefly present a plausible scenario for the birth of the Oklo deposit, in which we point out a few of the formidable problems underlying the formation of uranium-ore deposits. This background in earth sciences is indeed necessary to question whether or not the Oklo deposit is already unique with respect to its geological and geochemical features, as well as to identify the "initial" configuration of the ore bodies at the time of the chain reactions.

In Section 3 we report on recent studies concerning both uranium- and neutron-induced isotopes, which have confirmed the occurrence of fossil reactors in the Oklo deposit. The major difficulties attached to these studies are first discussed, then attempts to infer important reactor-operating conditions (neutron fluences, duration of the chain reactions, etc) from the isotope measurements are reviewed.

In Section 4 we describe still-unpublished searches for a variety of radiation-damage effects expected in the reaction zones and their borders, and we briefly analyze their relevance pertinent to the understanding of the Oklo phenomenon and to the long-term behavior of material exposed to high neutron fluences.

In Section 5 the basic assumptions underlying the simple static reactor models, mostly used to date to infer the reactor-operating conditions from isotope measurements, are discussed. The first serious attempt to improve these pioneering models is presented, in which the initial conditions prevailing at the time of the chain reactions (uranium distribution, "poison" concentrations, etc) can be varied as a function of time during the functioning of the reactors (dynamic model).

In the last section a summary of the studies performed to date as well as a tentative list of future extensions is presented.

2 THE EARTH SCIENCES BACKGROUND[1]

2.1 *Formation of the Oklo Uranium-Ore Deposit*

The uranium geochemical cycle, as first proposed by Bigotte (11) and then extended by Barnes & Ruzika (12), governs the repeated transport and deposition of uranium from one geological setting to the other, and is responsible for a variety of models concerning the formation and the subsequent "aging" of uranium ores. One of the dominating scenarios,[2] proposed to account for the formation of uranium-ore deposits in sedimentary rocks (13) includes 3 basic steps that concern the formation

---

[1] Throughout this section we extensively use matter extracted from the Proceedings of the 1st Symposium on the Formation of Uranium Ore Deposits, held in Vienna in 1974, but we only refer to a few of the most important papers herein.

[2] Mechanical transport of uranium in placer-type deposit (Blindriver, Witwatersrand) corresponds to a different type of important scenario.

of the ore-forming fluid,[3] the precipitation of uranium[4] in the sediments leading to the "primary" uranium deposit, and the subsequent redistribution of uranium[5] that possibly triggers its overconcentration in local traps within the primary deposit. The pioneering work of Weber (14), performed before the discovery of the Oklo phenomenon, suggests the following comments when applying Figure 1 to the Oklo deposit:

1. Uranium was dissolved as a result of surface leaching, which acted either on volcanic ashes from the Francevillien series or on older igneous rocks from the basement.

2. The ore-forming fluid infiltrated layers of sediments already covered by the sea, and organic matter then played a fundamental role in reducing and precipitating uranium as pitchblende[6] in favorable geological traps.

3. The sediments were probably buried at shallow depths (perhaps < 1000 m) at the time of the primary uranium mineralization, but as the relative level of the sea increased, the ore bodies were gradually buried at great depths ($\gtrsim 5000$ m) by further sedimentation.

---

[3] Uranium present as U^{4+} in "fertile" parent rocks is incorporated in solutions that have a correct pH and redox potential to oxidize U^{4+} to the soluble U^{6+} state (bacterial life can also contribute to the solubility of U). This "leaching" process could be caused by extensive surface weathering due to torrential rains (supergene scenario). The ore-forming fluid could also originate from U-rich waters expelled during the crystallization of a magmatic chamber (hypogene scenario). Whatever the detailed process may be, uranium is transported as uranyl ions (UO_2^{2+}) with metals that have a similar geochemical behavior (V, Mo, etc) and then interacts with the geological environment.

[4] The deposition of ores, which is prepared by the formation of a great variety of geological traps, can include the following scheme: the deformations of the earth crust produce porosity and permeability in the host sedimentary rocks via the formation of brecciation and fractures. When uraniferous solutions entering such tectonic traps suffer a change in their pH and redox conditions, uranium can be reduced to U^{4+} and precipitates as a fine-grained mineral coating breccia fragments and filling up or lining the walls of the pore space. The nature of the minerals can depend on a variety of other factors such as the presence of other metals and heat. This precipitation of uranium can take place either simultaneously or subsequently to the formation of the sediments (syngenetic versus epigenetic scenario).

[5] Uranium readily fluctuates between its oxidized water-soluble state and its reduced insoluble state as a consequence of changes in the redox conditions of the sedimentary environment. Each one of these successive cycles acts as a distinct chemical-fractionation step that can function both to decrease the concentration of other elements (V, Mo, etc), generally associated with uranium in the ore-forming fluid, and to gradually enrich uranium in the most favorable traps of the geological environment. These processes, which tend to increase the uranium concentration found in the fertile rock by a factor $\sim 10^4$, are necessarily long and repetitive.

[6] Uranium has been identified in some 115 different species among which 8 form most of the ores throughout the world. Pitchblende is a Th-poor variety of uraninite (UO_2) that crystallizes as a colloidal solution. With coffinite [$U(SiO_4)_{1-x}(OH)_{4x}$], uranium vanadates (carnotite, etc) and uraninite, pitchblende constitutes the major uranium mineral.

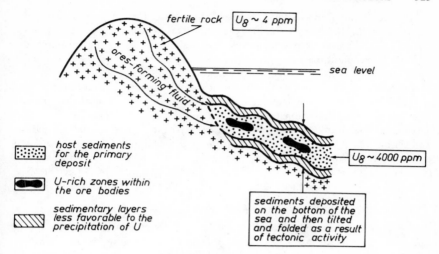

Figure 1 Schematic representation of a plausible scenario for the formation of the Oklo uranium deposit.

4. During the subsequent geological evolution (basin fill-up, tectonic activity induced by deformation of the earth's crust) the relative level of the sea decreased and the U-rich sediments became exposed to erosional processes acting at an estimated rate of $\sim 10 \, m/10^6$ years, which brought the Oklo ores into a near-surface position quite recently.

In this earlier scenario it was frequently stressed that the Oklo site, in contrast to the other Francevillien uranium deposits, was a remarkably quiet place with respect to tectonic activity and for this reason the U mineralization was not attributed to traps (brecciation, fractures) induced by tectonic activity. After the discovery of the Oklo phenomenon this hypothesis was further scrutinized by the group of Weber at Strasbourg, in order to explain the additional observation of the 6 uranium-rich lenses dispersed within the "normal" ore bodies that have been functioning as natural reactors. Gauthier-Lafaye et al (15) thus conducted a careful field analysis from which they concluded that contrary to previous thoughts, tectonic traps have played an important role in the formation of such lenses. In particular, the observation that the sediments were generally not deformed beyond their limit of plasticity during the production of such traps would imply that they were deeply buried ($\gtrsim 5000$ m) during the formation of the U-rich lenses. In this "deep-burial" model ground water was probably overcritical[7] during the chain reactions; this conclusion has profound implications concerning the functioning of the reactors (cf Section 5).

[7] For pressures, $p \gtrsim p_c = 225$ bars, observed at depths $\gtrsim 1000$ m, water is above its critical point and cannot experience any phase transformation.

However, none of the scenarios proposed to account for the formation and the subsequent aging of specific uranium ore deposits is far from controversy. Indeed, they are connected to formidable problems in the earth sciences. For example, the pH and redox conditions that govern the uranium geochemical cycle are modulated by a variety of parameters including climatic conditions, biological activity and worldwide continental-drift motion (inducing tectonic movements that generate pore space in the ore bodies), which are mostly unknown for precambrian environments (age $\gtrsim 0.6$ aeons). In addition, experts do not necessarily agree on the interpretation of geological features. For example, Samana (16) recently suggested that the indices of tectonic activity reported by the Strasbourg group imply a very shallow depth of burial of the sediments ($\lesssim 500$ m) during the formation of the U-rich lenses. We note here that this last model would fit better with the expectations of geochemists who infer that the age of the reaction is similar to that of the sediments (17, 18).

2.2   Specific Features of the Oklo Deposit

Nothing looks particularly unique in the formation and subsequent evolution of the Oklo deposit except the relatively good shielding of the ore bodies from any severe redistribution of uranium. Indeed, the Oklo mine was buried most of the time under a thick layer of sediments. In addition, it was quite distant (~ 150 km) from the nearest magmatic chamber so far identified that could both trigger the metamorphism of the sediments and inject into them hydrothermal solutions. Finally, the ores were only brought into a near-surface position in recent times, where they were further shielded from extensive surface weathering by top layers of impermeable sediments (pellites).

To complement this brief search for peculiar features of the Oklo deposits we report in Table 1 various characteristics of other uranium ore deposits that are prerequisites to the finding of well-preserved fossil reactors. To identify favorable conditions in the size, age, and uranium concentration of the deposits we have retained a rule of thumb roughly similar to that proposed by Cowan (19), where the deposits are thicker than ~ 50 cm (this minimum value corresponds to the size of a "minipile" found in the Oklo ores), more than 1 aeon ($= 10^9$ years) old,[8] and have at least approximately 20% of their weight in uranium. The deposits listed in Table 1 include typical members of the major families of sedimentary-type deposits (for example the Nabarlek mine has over 100 analogs in Australia), as well as 6 other mines from the Francevillien series that are in the near vicinity of Oklo. One of the major difficulties in establishing this table is that data concerning very rich uranium deposits are considered confidential by private companies. In spite of this limitation, the data allowed the following tentative conclusions:

[8] As explained in footnote 5, the high uranium overconcentrations found at Oklo are the end products of a long chain of repetitive fractionation processes, where oxygen played a dominating role as an oxidizing agent. It is generally assumed that this element was injected only 2 aeons ago in the earth's atmosphere, by a new generation of living organisms. Consequently, the high uranium overconcentration ($\gtrsim 20\%$) required for triggering nuclear chain reaction were probably never achieved in uranium-ore deposits older than 2 aeons, and the occurrence of fossil nuclear reactors was probably limited to a relatively narrow period of time ranging from about 1 to 2 aeons.

1. Within the Oklo ore bodies the uranium-rich lenses appear as very marked anomalies (see Section 3.2). In particular, they show large thicknesses (~ 1 m) and average uranium concentrations that are $\gtrsim 50$-fold those measured in the normal ores ($\sim 0.5\%$). However, such uranium-rich zones (column 5) are also found in other uranium deposits (Cluff, Nabarlek, etc).

2. Uraninite appears as a remarkably stable mineral (cf Sections 3 and 4) that contains most of the rare-earth elements and behaves as a closed chemical system with respect to the long-term redistribution of most neutron-induced isotopic anomalies. This mineral, which also occurs in other uranium deposits (column 8), can thus keep in its "memory" the signature of a chain reaction over a time scale of ~ 2 aeons.

3. Among the Francevillien deposits the Oklo ores look particularly free of minerals containing high concentrations of poisonous elements, which have high capture–cross sections for neutrons. Therefore it could be argued that a geological "filter" was active during the formation of the Oklo ore bodies to eliminate highly poisonous elements, but this relative "cleanliness" is not necessarily unusual (column 6). In particular, the notion of a geological filter was apparently introduced for the first time to account for the cleanliness of the Rabbit Lake uranium ores (20).

4. The advanced age of the Oklo ores is not unique and there is a clustering of the ages of old uranium mines at about 2 aeons.

5. The Oklo ores, as well as other uranium deposits, were quite well shielded against a severe redistribution of uranium (column 7), and this point is even further strengthened by the remarkable stability of uraninite.

Therefore, none of the characteristics of the Oklo deposit listed in Table 1 appear particularly unique when examined individually. However, they all combine favorably to increase the probability of finding well-preserved fossil nuclear reactors within these old ore bodies. Since such a favorable trend seems to be delineated for other uranium deposits, we believe that the Oklo phenomenon was probably not unique in the distant past. It could be that the only real unique feature of the Oklo deposit was the very careful routine analysis of uranium performed by Bouzigue et al (7). Indeed, the relative mass of uranium cycled through the chain reactions at Oklo has been evaluated at about 10^{-3}. Therefore the detection of fossil reactors in other uranium ores probably requires the frequent control of the isotopic composition of uranium with an accuracy better than $\sim 0.1\%$. Unfortunately it is not certain that such good routine analyses have been and/or will be conducted in other uranium deposits. Indeed, they are costly and the finding of "anomalous" abundance, if advertised, could perturb the economical exploitation of the ore deposits.

3   NEUTRON-INDUCED ISOTOPES AND REACTOR OPERATING CONDITIONS

3.1   Selection of "Useful" Isotopes

In a neutron reactor there is a mixture of thermal ($E_0 \cong kT$, where k is the Boltzmann constant, and T the temperature in °K), epithermal ($E \gtrsim 10E_0$), and fast

Table 1 Characteristics of various types of uranium-ore deposit

Ore deposits	Host rocks[a]	Age (10^6 yr)	U_8^b (%)	U-rich lens	Poisons	Shielding	Uraninite
Oklo	sandstones	~1.800 ■	0.5	50 ■	■	■	■
Mounana	sandstones	~1.800 ■	0.4				
Okelobondo	sandstones	~1.800 ■					
Boyindzi	sandstones	~1.800 ■				?	
Mikeloungou	sandstones	~1.800 ■					
Kaya-Kaya	sandstones	~1.800 ■					
Powder river basin	sandstones	~65	0.2–1				
Ambrosia Lake	sandstones	~140	0.15				
Nabarlek	sandstones	~1.800 ■	? ■	>10 ■			
Jabiluka	sandstones	~1.800 ■	? ■	?			
Witwatersrand	conglomerates	≥2.000 ■	0.04			■	■
Blind river	conglomerates	~1.700 ■	0.1			■	■
Vinaninkarena	impregnations of lacustrine sediments	≤1.8	0.01	0.2	?		
St. Pierre du Cantal	sediments	~20	0.2	10			

Deposit	Type of mineralization	3	4	5	6	7	8
Schaentzel		~300	0.2				
Chattanooga	impregnations of shales	~350	<0.01				
Billingen		~550	0.03				
South Dakota	impregnations of lignite	~60	0.33				
Phosphoria Formation		~250	0.01 → 0.065				
Bone valley	marine phosphates	~5	≲0.015				
Khouribga		~65	0.015				
Shinkolobwe	dolostones	~630	■	■	■	■	■
Sud Madagascar	placers	~0	?	?			
Port Radium		■ ~1.400	1	50	■	■	
Rabbit lake	other types of host rocks	■ ~1.100	■		■	■	■
Cluff		■ ~1.200	■	■	■	■	■

[a] The first 2 columns identify the deposits without referring to the type of the uranium mineralization.

[b] In columns 3 to 8, favorable characteristics for the finding of fossil nuclear reactors have been identified with a dark-lined frame containing a black square. In column 3, which gives either the age of the sediments or that of the uranium mineralization, favorable ages are greater than 1 aeon ($=10^9$ years). When possible, the average uranium concentration has been reported as weight percent in columns 4 and 5 for the normal-ore bodies and the U-rich zones, respectively. In column 5 the favorable conditions refer to the existence of U-rich lenses with sizes and uranium concentrations in excess of ~50 cm and ~20‰, respectively. The favorable "poison" conditions indicated in column 6 correspond to the lack of major uranium minerals such as monazite and carnotite, which contain highly poisonous elements such as V, Th, etc. The shielding is quoted in column 7 evaluates the protection of the ore bodies against weathering (a good degree of shielding is attributed to ore bodies that show a small amount of secondary uranium minerals resulting from the alteration of the primary ones). In column 8 deposits in which uraninite (cf footnote 6) is a major mineral have been listed as favorable.

($E \gtrsim 1$ MeV) neutrons. Neutrons of all energies can induce the fission of ^{235}U as well as that of various nuclei (^{233}U, ^{239}Pu, ^{241}Pu, etc) resulting from neutron capture on ^{238}U, ^{235}U, and ^{232}Th, but only fast neutrons can induce fission in ^{238}U and ^{232}Th. The resulting fission fragments are distributed on two-bumped mass yield curves. Their usefulness in the understanding of the Oklo phenomenon results first from the variations of their fission yields with respect to the mass of the fissioning nuclei (^{235}U, ^{238}U, ^{239}Pu) and the neutron energy, and/or from their high neutron capture cross sections, which contribute further to modify their abundance during the chain reactions. Indeed, in the Oklo reaction zones the isotopic composition of elements in the mass range corresponding to the maxima of the mass yield curves (~ 100 to 150) can be interpreted as a mixture of natural elements and fission fragments both perturbed by neutron capture.

After an appropriate selection discussed hereafter, the neutron-induced isotopes observed in the Oklo samples can be correlated to both the reactor operating conditions defined in Table 2 and the concentration of their parent nuclei by using sets of Bateman equations (21). As these equations have been described elsewhere (22, 23), we just remind here that the concentration of a stable fission fragment induced during the thermal fission of ^{235}U, which is further destroyed by neutron capture, can be approximated by the following expression:

$$N = \frac{N_{235}\sigma_F Y}{\sigma_c^*} \left| 1 - \exp\left(-\sigma_c^* F\right) \right|,$$

where N and N_{235} are the number of atoms of the fission isotope and fissile nuclei in a given volume, respectively; σ_F is the fission cross section of ^{235}U; σ_c^* and Y are the effective capture cross section[9] and the fission mass yield of the fission isotope, respectively, and F is the neutron fluence (cf Table 2).

The major difficulties attached to the determination of the reactor operating conditions at Oklo are briefly summarized below. They mainly concern the choice of useful isotopes, the definition of a meaningful technique for extracting samples from the reaction zones, the uncertainties attached to neutron cross sections and fission mass yields, and assumptions about the mode of operation of the Oklo reactors. Although these last two points are more appropriately discussed in Section 5, we already note here that most of the reactor operating conditions listed in Table 2 have been deduced from "static" models of the reactor where the variation of important characteristics of the ore bodies (U concentration, organic matter-to-uranium ratio, "poison" concentration, etc) are only attributed to the neutron fluences and natural radioactivity.

---

[9] The concept of an effective cross section, $\sigma^* = \sigma_0(g + r \cdot s)$ has been proposed by Wescott (24). In this expression σ_0 is the cross section corresponding to neutrons with a speed of 2200 m sec^{-1}; g is a temperature factor that must be used if the cross section is not exactly inversely proportional to the neutron speed s, and can be identified under specific circumstances with the resonance integral, $I = \int \sigma(E) \, dE/E$, taken between a low limit just above thermal energy and an upper limit equal to the average energy of the fission neutron; and r is the neutron epithermal index.

Table 2 Reactor operating conditions

Symbol	Definition	Range of values[a]
α and β	Proportion of fission events induced during the fast neutron fission of ^{238}U (α) and the slow neutron fission of ^{239}Pu (β)	0–4% 4–10%
δ	Duration of the chain reaction	50,000 up to 1,500,000 years
$F = nv_0\delta$	Thermal neutron fluence (n is the neutron density and v_0 the speed of neutron in thermal equilibrium at temperature T_0)	$\sim 10^{21}$ n cm^{-2}
C	Conversion factor giving the proportion of ^{235}U atoms resulting from the α decay of ^{239}Pu	0.2–0.6
r	Epithermal index of the neutron energy spectrum (cf footnote 9)	0.10–0.20
N_{fs}/N_8	Total number of ^{235}U fission per residual uranium atom	2–3%

[a] These values are estimated for U-rich zones where the present-day uranium concentration is $\gtrsim 10\%$.

The choice of neutron-induced isotopes is restricted by the two following conditions:

1. These isotopes have to be easily corrected for the abundance of natural elements (background correction). Therefore, they should be either much more abundant than the natural elements, or the corresponding element should have at least one isotope shielded from fission (Figure 2). In the specific case of fission products this condition already limits their selection to the maximum of the mass-yield distribution.

2. The daughter and parent nuclei, especially when they have different atomic numbers, should not have suffered any marked redistribution either during or after the occurrence of the chain reactions.

The sampling of uranium ores at Oklo was complicated by their heterogeneity down to a microscale ($\lesssim 0.1$ cm). To avoid this difficulty, Naudet et al (26) decided to "homogenize" samples with a minimum size of ~ 3 cm by crushing. In this manner the precious information stored in the Oklo ores as a marked heterogeneity in their properties is lost, but on the other hand the intercomparison of results obtained in various laboratories is now possible. In addition, the random sampling technique used in the earlier studies was only of a marginal interest (cf Section 3.2). Spectacular progress in the understanding of the Oklo reactors was only achieved

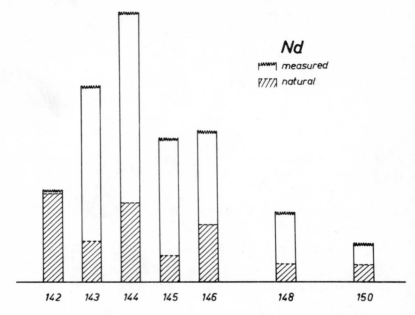

Figure 2 Isotopic composition of neodynium in a sample (Oklo-321) extracted from one of the reaction zones. As several isotopes of neodynium (143–146) are abundant fission products, this diagram clearly reflects the occurrence of a chain reaction. In addition, ^{142}Nd is an isotope shielded from fission that can be used as the natural reference to easily perform the background correction (Courtesy C. Allegre).

recently by extracting samples along traverse profiles throughout the reaction zones (Figure 3). In particular, with this technique it was possible to study the borders of the reactors where the isotope profiles are expected to be modulated by a variety of factors including steep gradients in the neutron fluence, the injection of depleted uranium from the reactor cores, etc. In addition, a good concordance between the traverse profiles of the parent and daughter nuclei contributes to single out useful elements, namely those that were not redistributed on a macroscopic scale of a few centimeters.

Essential progress in selecting couples of non-redistributed isotopes was achieved by using the ion microprobe[10] to map on a microscale of a few microns the distribution of uranium isotopes, and to look for the isotopic composition of both major and trace elements generally associated with uranium. The major results of this investigation (31) were to demonstrate the unique characteristics of uraninite for the retention of fission and (n, γ) reaction products and to reveal a drastic

[10] In this powerful instrument a primary beam of low-energy ions ($E \lesssim 0.1$ keV/amu) with a diameter of a few microns sputters away the surface of the sample, and the ionized species thus released are analyzed with a mass spectrometer.

decrease in the concentration of carbon compounds in the reaction zones. In addition, the ion microprobe runs showed that a small number of other isotopes have been partially lost from the uraninite grains. In particular, radiogenic lead mostly derived from the α decay of uranium is found in the clay matrix; this explains the difficulties met in applying the uranium-lead method to determine the age of the reactions (17, 18).

3.2   Distribution of Uranium Isotopes

As shown by Geoffroy (27), uranium is distributed in two very different ways in the Oklo ore bodies: First, outside the reaction zones the uranium mineralization occurs in both a conglomerate constituted of quartz grains embedded in organic matter and in clay mineral. In these two types of sediments, uranium appears mostly as microveinlets of pitchblende and the corresponding average uranium weight concentration is $U_8 \sim 0.5\%$. Second, the reaction zones represent a drastic discontinuity in this normal type of mineralization as the conglomerate has now been completely replaced by clay minerals. In addition, uranium appears as clusters of tiny and well-crystallized uraninite grains, and U_8 shows an approximately 50-fold increase over the normal value of 0.5%. In addition, "hot" spots with local U concentrations of up to 60% are frequently observed, and the concentration of organic matter is much smaller than in the normal zones. So far the six uranium-rich zones found at Oklo have all been functioning as neutron reactors. They are lens-shaped along the general direction of the sedimentation and their thicknesses and surfaces are about 1 m and 100–200 m^2, respectively (Figure 4). Recently a "minipile" with a smaller thickness of $\sim$ 40 cm (28) has even been found at the end of one of the reaction zones.

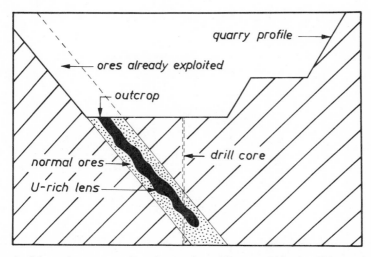

Figure 3   Schematic representation of a uranium-rich zone within the Oklo ore bodies. Drill-core tubes provide samples for traverse profile analysis (Courtesy R. Naudet).

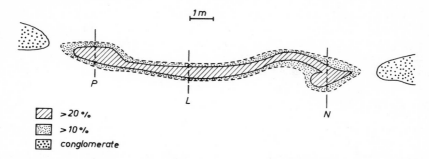

Figure 4 Outcrop of one of the uranium-rich lenses. The reactor thickness is about 1 m and the continuity of the conglomerate has been strikingly interrupted in this lens (Courtesy R. Naudet).

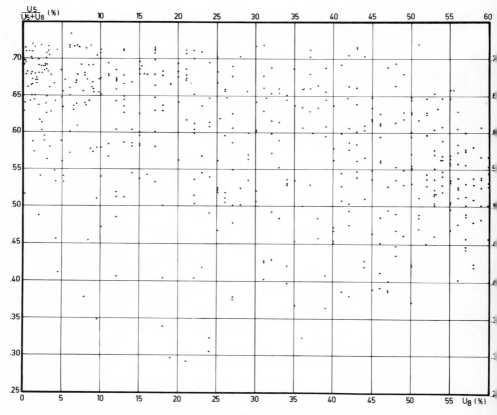

Figure 5 Correlation plots between the ^{235}U isotopic abundance and the uranium concentration for $\gtrsim 500$ different samples.

In Figure 5 we summarize the uranium isotopic analysis reported for about 500 samples extracted from 3 of the reaction zones (29). This correlation plot between the ^{235}U and ^{238}U abundances, which corresponds somewhat to a random selection of samples, clearly shows the predicted trend of finding a marked depletion of ^{235}U in the uranium-rich samples. However, there is a wide scatter of the experimental points that cannot be easily interpreted on this type of diagram. This disturbing feature was partially resolved by Naudet et al (26), who plotted their results along traverse profiles throughout the reaction zones. A very clear-cut correlation now appears on the plots (Figure 6). The depletion of ^{235}U generally increases with the U concentration, but departures from this correlation are generally observed near the center and the edges of the reaction zones. When analyzed in terms of the neutron theory (30), these departures do in fact give further support for the occurrence of chain reactions. For example, the negative correlation between the ^{235}U and ^{238}U abundances observed near the reactor center simply reflect the high value of the conversion factor at this location. In addition, the overshoot in the depletion of ^{235}U at the reactor boundaries would represent the effects of the natural reflectors active at the edges of the reactor to reinject neutrons into the reaction zones. One of the important results of these analyses is the extreme variety of the profiles so far obtained, which suggests a great variety of local modes of reactor functioning within each one of the reaction zones. This in turn complicates the modeling of the Oklo reactors, but also explains most of the scatter observed in Figure 5.

3.3   Fission Fragments and (n, γ) Products

Although isotope measurements have been reported by several groups (32), we mainly discuss hereafter the work of Hagemann et al (33), which gives a good illustration of the potential of isotope studies for determining the Oklo reactor operating conditions in the framework of a static model of the reactor.

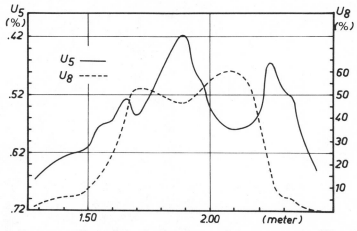

Figure 6 Traverse profile analysis, showing the correlation between the ^{235}U isotopic abundance (*solid line*) and the uranium concentration (*dotted line*) (Courtesy R. Naudet).

Hagemann et al simultaneously used 19 isotopes corresponding to 3 rare-earth elements (Nd, Gd, Sm), as well as 3 isotopes of ruthenium, 3 isotopes of palladium, and the 2 uranium isotopes, to obtain sets of consistent values for the reactor

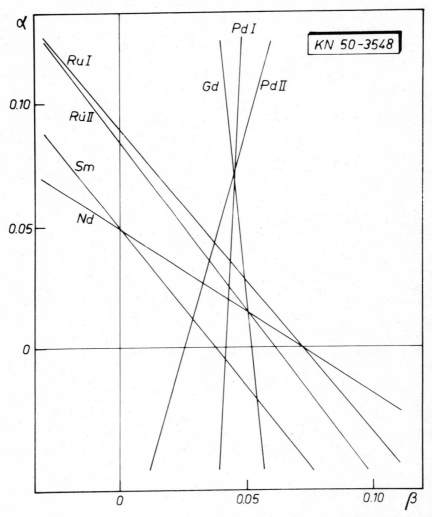

Figure 7 Graphical determination of the proportion of fission events due to the fast and thermal neutron fission of ^{238}U and ^{239}Pu, respectively. For this purpose Hagemann et al (21) measured 7 distinct isotope ratios identified as Nd = ^{150}Nd/(^{143}Nd + ^{144}Nd + ^{145}Nd + ^{146}Nd); Sm = ^{154}Sm/(^{147}Sm + ^{148}Sm); Gd = (^{157}Gd + ^{158}Gd)/(^{155}Gd + ^{156}Gd); RuI = ^{104}Ru/^{101}Ru; RuII = ^{104}Ru/^{102}Ru; PdI = ^{110}Pd/^{105}Pd; and PdII = ^{110}Pd/^{106}Pd (Courtesy R. Hagemann).

operating conditions defined in Table 2. The results of this investigation concern six main areas, described below.

1. The proportion of fission events, α and β, due to $^{238}U_{nf}$ and $^{239}Pu_{nf}$ respectively. The separation of ^{239}Pu and ^{238}U fissions can be best performed by using isotopes near both the high- and low-mass wings of the distributions; this is a difficult task in view of the abundances of natural elements (cf Section 3.1). For this purpose Hagemann et al measured 7 different isotope ratios, identified in Figure 7, which can be simply expressed as a function of α, β, and the corresponding fission-mass yields. Finally, the technique of concordant diagrams illustrated in Figure 7 is used to solve the $\alpha = f(\beta)$ equations, and this provides a set of consistent values for α ($\sim 3\%$) and β ($\sim 4\%$).

2. The thermal neutron fluence F and the epithermal index r. These parameters are simultaneously derived by similar graphical methods (Figure 8) from equations relating special pairs of parent-daughter isotopes where the parent nucleus is a fission product from which the daughter nucleus is derived by neutron capture. The daughter nucleus is subsequently destroyed by neutron capture, but σ_c (daughter) $\ll$ σ_c (parent). Since the sensitivity to the epithermal index increases from ^{143}Nd to ^{145}Nd and ^{147}Sm, the (^{143}Nd, ^{144}Nd) pair gives F values ($\sim 10^{21}$ n cm^{-2}) that are roughly independent of r, and the (^{147}Sm, ^{148}Sm) pair provides the best evaluation of r (~ 0.15).

3. The conversion factor C. The values of this parameter ($0.4 < C < 0.6$) can be deduced from an equation relating F, r, and Δr to the uranium depletion factor, $U_5/(U_5 + U_8)$, where U_5 and U_8 are the weight concentrations of ^{235}U and ^{238}U, respectively.

4. The total number of ^{235}U fission, N_{f5}. For samples extracted from the reaction zones, $N_{f5}/N_8 \sim 2\%$, where N_8 is the residual number of uranium atoms. This value is derived from 5 groups of rare-earth fission products that depend only slightly on α and β, and which have well-known fission mass yields—($^{143}Nd + ^{144}Nd$), ($^{145}Nd + ^{146}Nd$), (^{150}Nd), ($^{147}Sm + ^{148}Sm$), and (^{154}Sm).

5. The duration of the chain reaction, δ. In contrast to the other determinations, the evaluation of δ requires the use of radioactive isotopes that decay with a half-life comparable to δ. Hagemann et al tentatively used ^{239}Pu (half-life $\approx$ 24,000 years) and ^{99}Tc, which both decay to ^{99}Ru with a half-life $\approx$ 213,000 years, and lead to the formation of ^{100}Ru by neutron capture acting either directly on ^{99}Tc or on ^{99}Ru. In the ^{239}Pu method, which gives the most reliable results, $\delta \approx$ 600,000 years is inferred from the β parameter, as determined by the graphical method reported in Figure 7. On the other hand, the ^{99}Tc method, which is based on the measurement of the $^{100}Ru/^{99}Ru$ ratio, yields higher values ($\sim$ 1,400,000 years) that are less reliable as a result of both uncertainties attached to the σ_c values of Ru and the possible redistribution of this element.

6. The age of the reaction, Δ_r. Δ_r figures explicitly in the expression of N_{f5}/N_8 and N_{232}/N_8, where N_{232} is the total number of ^{232}Th atoms in a given volume resulting from the α decay of ^{236}U formed by neutron capture on ^{235}U. N_{f5} is inferred as indicated above. By applying both methods, the Δ_r values are found to be roughly similar to the age of the uranium mineralization ($\Delta_r \sim \Delta_U \sim 1.8$ aeons).

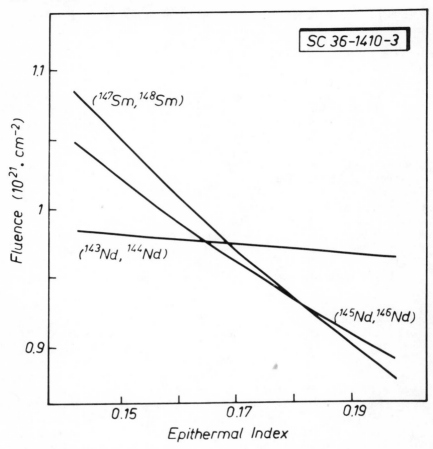

Figure 8 Graphical determination of the thermal neutron fluence and the neutron epithermal index (Courtesy R. Hagemann).

The results of these isotope studies look very coherent, but this feeling is indeed challenged by various observations, which follow below.

1. Isotope methods give incoherent results for samples extracted from the borders of the reactors. In particular, by taking the most reasonable assumption, $\Delta_r \approx \Delta_s \approx \Delta_U \approx 1.8$ aeons, the N_{f5} values inferred from rare-earth fission products can be either much higher ($\times 10$) or even slightly lower than the value directly obtained from the neutron fluence F, by applying the following relationship: $|(N_5)_0/(N_5+N_8)_{now}| \times \sigma_{f5} \times F$, where $(N_5)_0$ and $(N_5+N_8)_{now}$ are the number of ^{235}U atoms at the time of the chain reaction and the total number of residual uranium atoms, respectively (34). These fission anomalies, which are apparently not observed in the U-rich

zones ($U_8 \gtrsim 20\%$) of the reactor cores, have still to be quantitatively explained. It is likely (35) that they reflect the injection of a small amount of fission products and/or depleted uranium from "hot" spots ($U_8 \gtrsim 20\%$) to "cold" spots ($U_8 \lesssim 5\%$) either within the reaction zones or on their borders. If so, it would be difficult to infer any meaningful conclusions about the functioning of the Oklo reactors from isotopic measurements conducted on samples extracted from the reactor borders.

2. The concordant diagrams used by Ruffenach et al (36) to infer the values of α and β give consistent values for only one sample (KN-50-3548), but not for the 3 other samples also investigated in this work. Furthermore, the β value reported by the CEA team has been criticized by Drodzt (37), who still claims that there is no evidence for ^{238}U fission in the fission spectra of xenon measured for core samples (38). Consequently, Drodzt evaluated β values much higher ($\sim 10\%$) than those quoted by Ruffenach et al, which in turn implies much smaller values for δ ($\sim 50,000$ years). These rare-gas studies are based on very powerful methods that have been already highly successful in predicting the isotopic composition of xenon resulting from the spontaneous fission of ^{244}Pu in a complex mixture of meteoric xenon (39). In addition, Drodzt believes that the severe loss of xenon suffered by the samples cannot affect his high β value. Therefore we do think that there is a discrepancy between the rare-gas and rare-earth elements (REE) data that is still not explained.

3. All the reactor operating conditions reported in Table 2 have been deduced in the framework of a simple approximation for the Oklo reactors, where the possible redistribution of uranium and fission products is not taken into consideration. Although this model was well-adapted to the preliminary exploration of the Oklo phenomenon, its validity appears now to be questioned by a variety of observations (cf Section 5.1).

4   RADIATION-DAMAGE STUDIES

4.1   *Introduction*

We felt that the discussion of radiation-damage features stored in the Oklo minerals should be discussed in detail. Indeed, in striking contrast to expectations based on the high neutron fluences quoted in the previous section, no high concentration of radiation-damage defects has so far been observed in insulator grains extracted from the reactor cores. Although this result severely contradicts the very marked isotopic anomalies observed in the same zones, no specific discussion of these conflicting observations has been reported in previous review papers (9, 10). Finally, it is interesting to question whether or not these studies can help in improving our understanding of the long-term behavior of neutron-irradiated materials.[11]

Radiation-damage features in the Oklo minerals have been mostly analyzed to

---

[11] A working group has recently (40) emphasized how little is known about this problem, which contributes to limit the development of the new generations of fusion and fission reactors.

date by J. C. Petit in a still-unpublished work (41). The observations were made on both insulator grains and uraninite, which has semiconductor properties. The insulator grains included quartz, a variety of mica still poorly defined, which behaves

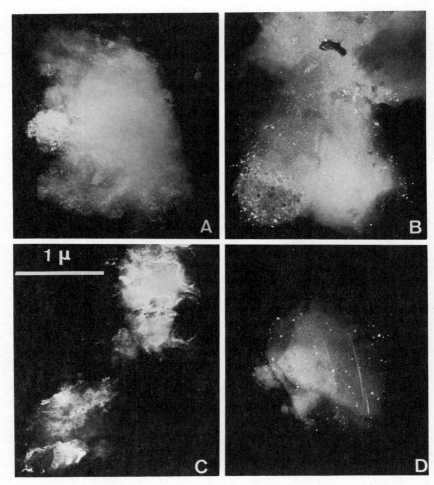

Figure 9 1-MeV dark field micrographs of clay minerals from the Oklo ores. When directly observed with the HVEM the grains extracted from a core sample (Oklo-321) (*A*) show no evidence for a high degree of lattice disordering or track crystallites, and they are very similar to those (*C*) extracted from a normal sample (Oklo-330). In addition, after a thermal annealing of 2 hr at 500°C, these grains become loaded with tiny crystallites (*B*) that look similar to those induced by krypton tracks artificially registered in grains from Oklo-330, and similarly annealed (*D*). By contrast, only very few crystallites are observed after heating in nonirradiated grains from Oklo-330, which contains an average uranium concentration (0.1%) at least 200 times lower than that estimated for Oklo-321.

like biotite upon chemical etching, and two minerals (illite and chlorite) that are the major constituents of the clay matrix of the Oklo ores.

Most of the radiation-damage features expected in insulator grains from the Oklo ores are nuclear-particle tracks, which are produced by a variety of heavy ions, either emitted during the radioactive decays of heavy elements (spontaneous fission of ^{238}U; α decay of ^{235}U, ^{232}Th, etc), or induced by the neutron fluences in the reaction zones and on their borders. Therefore, these ions will include fission fragments with a distribution of masses and energies (~ 100 MeV), α recoils with masses $\gtrsim 200$ amu and energies of about 100 keV, and lighter recoils mostly induced during the elastic scattering of fast neutrons with the constituent major elements of the grains. For more details concerning either the registration of nuclear-particle tracks in solids or the formula applied to compute their densities (in nb cm^{-2}) as a function of the track origins, we refer the reader to the work of Fleischer, Price & Walker (42, 43). We recall below a few features of latent and etched tracks necessary to the understanding of this section.

Latent tracks are the continuous trails of radiation-damaged material with a diameter of a few tens of angstroms, which are believed to be only formed along the path of heavy ions in insulators. Upon heating at a temperature generally much lower than the fusion temperature (thermal annealing), the tracks act as nucleation sites and are transformed into tiny "track crystallites" (TCs) that can only be conveniently observed with a high-voltage electron microscope (HVEM) (Figure 9D). The TCs are very stable in the electron beam of the HVEM and their sizes can be limited to those of single dots. They are thus particularly useful as indicators of an irradiation history when the latent tracks cannot be observed either because their densities (in nb cm^{-2}) are too high ($> 10^{11}$ cm^{-2}) to be directly resolved or when their contrast has already been altered as a result of partial annealing in nature. In addition, as the sizes of the TCs markedly increase with the annealing temperature, the observation of "fossil" TCs in natural minerals gives clues concerning their thermal history in nature (44). Upon appropriate chemical etching, non-annealed latent tracks can be etched out at a much faster rate than that of the host grain material. They are thus revealed as etched tracks (Figure 10). Above a critical temperature ξ_c,[12] which depends very much on the nature of the mineral as well as on the type of ions, the length and density of the etched tracks markedly decrease with the annealing temperature.

4.2   Expected Effects

The constituent minerals of the Oklo reaction zones have been subjected to fluences of thermal and fast neutrons of about 10^{21} cm^{-2} and $\gtrsim 10^{20}$ cm^{-2}, respectively. During fast-neutron irradiations, the lattice structures of most insulators are more readily damaged than those of semiconductors and metals. In particular, quartz appears to be one of the most sensitive minerals, showing an amorphous structure when exposed to fast-neutron fluences of about 10^{20} n cm^{-2} (45). In addition, if

---

[12] ξ_c is currently defined as the temperature at which the track density has dropped to about 50% of its initial value, after a one-hour annealing. ξ_c is generally 100°C to 200°C lower than the temperature required to form observable TCs with sizes $\gtrsim 100$ Å.

the radiation-damage stability scale of minerals inferred from heavy-ion irradiations (46) can be extrapolated to fast neutrons, then mica and the clay minerals are expected to be even more sensitive than quartz to a fast-neutron irradiation. Therefore, on the basis of their exposure to fast neutrons, most insulator grains extracted from the reaction zones should show a highly disordered crystalline lattice when examined by X-ray and electron-diffraction techniques ("metamict" state).

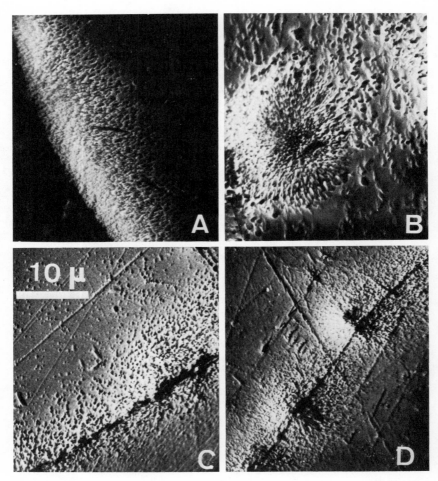

Figure 10 Etched fission-track micromapping observed with a scanning electron microscope (SEM) in a quartz grain extracted from a normal sample (Oklo-329). This grain was first subjected to a high fluence of thermal neutrons (10^{19} n cm^2), which induced $^{235}U_{nf}$ tracks, then etched in sodium hydroxide. The tracks are very clearly clustered in edge zonings (*A*), fission stars (*B*), and "microcrack" zonings (*C* and *D*).

Fast neutrons also contribute to the density of fission-fragment tracks in the same minerals, by inducing the fission of ^{235}U, ^{238}U, and ^{232}Th atoms. But the work described in Section 3 clearly indicates that their contribution to the number of fissions ($\lesssim 10\%$) is much smaller than that of thermal neutrons, which will be mostly responsible for a fission track "excess," $\Gamma = \rho_{nf}/\rho_{sf}$, where ρ_{nf} and ρ_{sf} are the fission-track densities expected from the induced fission of ^{235}U and the spontaneous fission of ^{238}U, respectively. The value of Γ linearly increases with the thermal neutron fluence F, and ranges from about 1 to 10^6 when F increases from $\sim 10^{16}$ up to 10^{21} cm^{-2}. As a fission-track excess of $\sim 50\%$ can be easily measured (cf footnote 14), Γ should be a very useful parameter to determine the relatively weak thermal neutron fluences active in the borders[13] of the reaction zones.

In the normal zones where $U_8 \approx 0.5\%$, the thermal and fast-neutron fluences have probably been smaller than 10^{15} n cm^{-2} (47). Consequently, $\Gamma_{normal} \lesssim 0.1$, and no lattice disorder should be observed in minerals extracted from these zones. In addition, the fission-track densities will be mostly due to the spontaneous fission of ^{238}U and the so-called fission-track method is potentially useful to obtain a variety of ages. In this method the ρ_{sf} value and the uranium concentration, U_8 (i), are related to a "model" age, Δ_i, depending on the type of parent uranium distribution from which the tracks originate. More specifically, if uranium is trapped within a tiny inclusion formed at the time of crystallization of the mineral, then the tracks will form a typical fission "star" (Figure 10B) and $\Delta_i = \Delta_{incl}$ will give the formation age of the mineral. On the other hand, the fission tracks can originate from the uranium-rich matrix in which the grains become embedded at the time of the uranium mineralization, and they will be registered in the external layers of the grain, thus forming a very clear superficial "zoning" of the track density (Figure 10A) if the uranium concentration in the matrix is much higher than in the grain. In this case $\Delta_i = \Delta_{zoning}$ will give the apparent age of the uranium mineralization, which can be significantly different than Δ_{incl} if uranium becomes redistributed in the clay matrix.

4.3   Observations

HVEM observations (48, 49) very clearly indicate that clay minerals from the reaction zones (core clays) look very similar to those from normal zones (normal clays) (Figures 9A, C). In particular, they are not metamict and show tiny crystallites only after an artificial thermal annealing (Figure 9B) at a low temperature ($\sim 500°C$). These crystallites look then like artificial TCs (Figure 9D) and their densities, which are correlated to the average uranium concentration, are much smaller than the high values ($\gtrsim 10^{11}$ cm^{-2}) expected from the thermally induced recrystallization of a metamict lattice. These HVEM observations were independently confirmed by X-ray diffraction studies performed by Weber et al (50), which were used to evaluate an index of crystallinity that was then applied to scale the

---

[13] However, it is not possible to measure the high Γ values expected in the reactor cores with this technique because the tracks are then registered in minerals already heavily damaged by fast neutrons.

clays from 1 (well-ordered lattice) to 8 (highly disordered lattice). These X-ray studies firstly showed a clustering of the indices around a value of about 6, which is independent of the clay origin, and secondly revealed that core clays were slightly different than normal clays in showing a preferential orientation along the main direction of sedimentation.

Quartz grains have been found to date only in the normal zones and in the reactor borders, and both their crushing strength and size markedly decrease in the near vicinity of the reactor edges (51). So far they have been the most useful track detector discovered in the Oklo ores, and this is mostly due to their exceptional track-retentivity against thermal annealing ($\xi_c \approx 1300°C$). Indeed, laboratory experiments of a type described elsewhere (43) show that fission tracks registered in the Oklo quartz should be stable at 300°C, over a time scale of 10 aeons, and this is indeed a world record for solid-state track detectors.

A detailed fission-track micromapping was performed in this mineral (Figure 10), indicating three distinct types of uranium distributions from which the tracks originate, which include tiny inclusions generating fission stars (Figure 10B), the U-rich clay matrix in which the grains become embedded (Figure 10A), and a variety of microcracks loaded with uranium (Figures 10C, D). In normal quartz, fission-track ages for the inclusions[14] were found to agree with the Rb-Sr ages of igneous rocks from the basement (~ 2.6 aeons), which demonstrates that the Oklo quartz originates from such rocks. On the other hand, quartz grains, extracted at a distance of about 30–50 cm from the edge of one of the reactors, clearly showed a fission-track excess of ~ 10 compatible with a residual neutron fluence of about 10^{17} n cm^{-2} at this location. In all quartz grains so far studied, the fossil tracks were markedly shortened, with the average track lengths being about 3 times shorter than the expected value of $\sim 10 \mu m$.

Other mineral species were also investigated with the HVEM for samples extracted from the reactor cores. The tiny ($\sim 50 \mu m$) flakes of mica showed no evidence for a severe lattice disordering or for high TC densities. On the other hand, uraninite grains were loaded with very high densities ($\gtrsim 10^{11}$ cm^{-2}) of tiny crystallites that look like TCs.

4.4 Implications

There are two tentative explanations for the lack of a severe radiation-damage disordering in clay minerals and mica flakes extracted from the reaction zones.

1. If the Oklo ores were indeed buried most of the time at great depths ($\gtrsim 5000$ m), then in the water-rich, high-temperature ($\gtrsim 150°C$) environment of the

[14] In the technique recently developed by Dran et al (52) the grains are first heated to shorten the fossil track, if necessary. Then they are exposed to a known fluence of thermal neutrons that induce long $^{235}U_{nf}$ fission tracks and subsequently etched on internal surfaces obtained by polishing, fracture, etc. The track densities are finally measured with electron-microscope replicas (Figure 11) for both the "short" tracks, ρ (short), which are mostly fossil fission tracks, and the "long" tracks, ρ (long), which weight the uranium concentration from which ρ (short) originates. After various corrections, Δ_{inc} is directly deduced from the ρ (short)/ρ (long) ratio by using the appropriate formula.

ore bodies, an unusual long-term annealing process has suppressed the tracks without leaving their characteristic TCs remnants, which usually indicate a period of marked heating in the grain history.

2. The clay and mica grains found to date in the reaction zones have not been exposed to high neutron fluences. This conclusion either implies that the uraninite grains, where most of the isotopic anomalies are trapped, were processed by high neutron fluences elsewhere, prior to their deposition in the Oklo sediments or that the clay minerals initially present in the reaction zones get drastically redistributed during or after the cessation of the chain reaction while staying confined within these zones.

This last hypothesis appears to be the most satisfactory because the quartz grains in the borders of the reactors have been clearly exposed to a gradient of neutron fluence that is quite compatible with a thermal neutron fluence of about 10^{21} cm^{-2} active at the center of the reactors.

Although much more work in the field of simulation experiments is required before definitively excluding the occurrence of a peculiar track-annealing process at Oklo, the redistribution hypothesis is progressively gaining a wider acceptance (50) because it is compatible with additional observations including the preferential orientation of the core clays as detected by X-ray diffraction techniques; the marked degradation of the quartz grains in the near vicinity of the reactor edges; and the lack of isotopic anomalies for the constituent "poisonous" elements of the clays (Li, B), which suggests that the expected anomalies have simply been washed out

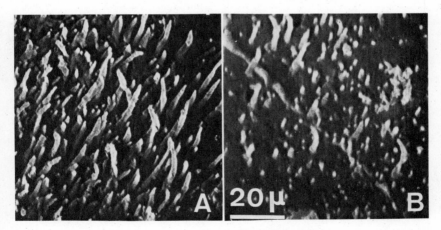

Figure 11 SEM micrographs of electron-microscope replicas of a normal quartz grain (Oklo-329) that was artificially irradiated in a fluence of thermal neutrons of about 10^{17} n cm^2. Micrograph *A* represents an edge zoning corresponding to Figure 10*A*, where most of the tracks are long and result from $^{235}U_{nf}$. In contrast, a high proportion of the tracks in the "microcrack" zoning reported in micrograph *B* corresponds to short fossil tracks due to $^{238}U_{sf}$. This marked excess of short tracks most likely reflects the fact that uranium has been washed out of the microcrack in recent times.

and diluted with normal elements. If so, this redistribution was probably triggered by the preferential etching of tracks in deep ground waters, which was further activated by the heat evolved during the chain reactions.[15] Then a new generation of minerals, similar to the previous ones, precipitated from the resulting radiation-damaged "mud," which was remarkably confined within the reaction zones (49).

Besides suggesting a remobilization scenario for the minerals of the reaction zones that was quite unexpected 3 years ago, the radiation-damage studies summarized in this section appear to be useful both in improving our understanding of the Oklo phenomenon and in challenging our knowledge of radiation-damage effects in insulators.

The following conclusions are particularly noteworthy:

1. The marked shortening of fossil tracks in quartz, when coupled with the exceptional track retentivity of this mineral, clearly shows that the Oklo ores were buried most of the time at great depths ($\gtrsim 5000$ m) in order to achieve the relatively high temperatures ($> 300°C$) required for track annealing. Therefore, this observation probably rules out any formation model assuming that the Oklo ores have resided a very long time in a near-surface position.

2. The tiny uranium-rich inclusions in quartz grains extracted from the borders of the reactor have generated fission-track excesses that are modulated by a steep gradient in the thermal neutron fluences. This gradient, which cannot be obtained from isotope measurements, should be helpful in improving the modeling of the Oklo reactors.

3. If the tiny crystallites present in uraninite grains are indeed TCs produced with a small efficiency of registration, then the detailed study of their growth kinetics upon heating could contribute in fixing interesting limits for the temperature attained at the reactor centers.

4. About both irradiation effects and the long-term behavior of neutron-irradiated material the following comments can be made: (a) the quartz variety found in the Oklo ores has the highest track retentivity ever reported for an insulator. Indeed, freshly induced as well as fossil tracks are still stable at a temperature, $\xi_c \approx 1300°C$, which is only slightly lower than the fusion temperature ($\sim 1500°C$) of quartz. This astonishing stability constitutes an interesting solid-state problem that challenges our understanding of the annealing of radiation-damage effects in insulators. Indeed, in very pure varieties of synthetic quartz grown by the fusion-zone technique, $\xi_c \approx 400°C$. Therefore it can be postulated that the much higher ξ_c value of the Oklo quartz is related to peculiar impurities that act as stabilizing agents, which will be very interesting to identify in the general context of learning how to stabilize defects in solids. (b) When examined with the HVEM the Oklo uraninite shows only tiny crystallites firmly embedded in the grains, and no evidence for microcracks, naturally etched features, voids, etc is found. This observation agrees

---

[15] It is difficult to estimate the composition of ground waters at great depths, but the pH of water issuing from deep drillings usually ranges from about 6 to 8.5. If this radiation-induced corrosion was indeed active at Oklo, then naturally etched tracks should soon be observed in grains extracted from the borders of the reactors.

well with the remarkable stability of this mineral with respect to the redistribution of neutron-induced isotopes. (*c*) In contrast to uraninite, all insulator grains that were irradiated in the Oklo reactors had a poor long-term stability with respect to the corrosion due to ground water.

5 REACTOR MODELING

5.1 *Static Modeling*

Naudet et al (53) have developed the concept of a static model for the Oklo reactors. In this type of model, with the exception of water, which is considered an adjustable parameter, the present-day characteristics of the Oklo ore bodies are used to evaluate the values of the neutron multiplication factor k expected in the U-rich zones at the time of the chain reactions ($\Delta r \approx 1.8$ aeons). The reactor parameters considered in these computations are the isotopic abundance, $U_5/(U_5+U_8)$ and the uranium concentration U_8 at the time of the chain reactions; the macroscopic absorption cross section Π (in $mm^2\ g^{-1}$) of the clay matrix evaluated for the major constituent elements now present in the clays (Fe, Si, Al, Mg, etc); the thickness ϕ of the U-rich lenses; and the water-to-uranium mass ratio, H_2O/U. The computations show that criticality was indeed easily achieved at Oklo, as $k > 1$ with the following set of average parameters estimated from a compilation of experimental data: $U_5/(U_5+U_8) \approx 3\%$, $U_8 \approx 30\%$, $\Pi \approx 1\ mm^2\ g^{-1}$, $\phi \approx 1$ m, and $H_2O/U \approx 0.15$. Surprisingly enough, this set of natural values is very similar to that observed for man-made reactors using natural water as a moderator. From these computations Naudet has proposed a spectacular model for the functioning of the Oklo reactors, in which the chain reactions occurred first as a series of local "fires" confined to the richest uranium "hot" spots within the reaction zones. Then, as poisonous elements such as boron get progressively burned, this fire could spread to nearby areas containing smaller uranium concentrations. In this scenario the very efficient control of the chain reactions is attributed to the removal of water during boiling, which causes a sudden drop in the value of k. Then, as the shut-down reactors cool down, liquid water is reinjected into their cores and the chain reactions can start again.

In the static models uranium is neither lost nor injected into the uranium-rich zones, the porosities of the ores keep constant values, the distribution of the uraninite lumps within the moderator is fixed for once and for all, etc. Although these approximations have indeed been useful, it is interesting to question their validity by evaluating the extent to which the following present-day characteristics of the Oklo reactors could have been perturbed during both their functioning and subsequent aging. (We again emphasize that the understanding of the aging of Precambrian sediments is a formidable problem in the earth sciences, and consequently this part of the discussion is quite speculative.)

1. MODERATOR The moderator properties of the ores result from their content of water and organic matter. In previous sections evidence for a marked perturbation of the reactor configuration during the chain reaction was presented. In

particular, neither high concentrations of radiation damage nor isotopic anomalies for poisonous constituent elements (Li, B) has so far been found in the clay minerals from the reaction zones. Therefore the clay matrix that certainly played a dominating role in the control of the chain reactions was probably drastically redistributed during the chain reactions, while still being confined within the actual reactor edges. Such a perturbation was perhaps further enhanced if the U-rich sediments contained organic matter subsequently driven away during the reactions, as a result of processes (pyrolysis and/or radiolysis) that formed volatile species (CO, CO_2, N_2, H_2O, etc).

2.   DEPTH OF BURIAL OF THE REACTORS D_r DURING THE CHAIN REACTIONS   The indices of tectonic activity observed in the Oklo ore bodies (cf Section 2.2), as well as mineralogical studies (50), strongly suggest that $D_r \gtrsim 5000$ m. However, this value is still uncertain, which is very unfortunate. Indeed, D_r determines the initial temperature and pressure of the ore bodies during the functioning of the reactors. These parameters in turn strongly affect the physical and thermodynamic properties of water (54), which played a dominating role as a moderator and coolant, as well as the temperature factor g of effective cross sections (cf footnote 7).

3.   POISON CONCENTRATIONS   The macroscopic absorption cross section of the ores is dominated by high-σ_c elements present in the clay matrix and in the uraninite grains. However the poisons now trapped in the clay minerals (Li, B) do not show anomalous isotopic concentrations, which is not astonishing in view of the postulated redistribution of the clay minerals within the reaction zones. Therefore the poison distribution active at the time of the reactions is still not defined.

4.   REFLECTOR   The reflector properties of the reactor borders, which depend on both their uranium and carbon contents, also appear to be poorly defined. Indeed, these quantities have probably been perturbed by the injection of uranium from the reactor cores, as well as by the possible redistribution of organic matter exposed to the temperature gradient of the reactors.

5.   URANIUM DISTRIBUTION   The clear-cut traverse profiles are generally considered as indicative that the geometrical configuration of the reaction zones have been well preserved. However, there was certainly a slight injection of uranium from "hot" to "cold" spots, either in the reaction zones or in their borders (cf Section 3.3). In addition, even by supposing that uraninite was formed before the reactions, the detailed configuration of the uraninite lumps in the moderator is correlated with the ill-defined remobilization history of the clay minerals and organic compounds in the same zones.

6.   COOLANT   Water was probably the active coolant. However, a detailed heat budget cannot be evaluated at the present time as a result of uncertainties attached to the determination of both the duration of the chain reactions and the depth of burial of the reactors during criticality.

Static models have indeed contributed to the understanding of the Oklo phenomenon. In particular, one of their most spectacular achievements was to qualitatively explain the complex correlations between the ^{235}U depletion factor and the uranium concentration along traverse profiles (26). However, they now have to be improved by using more elaborate "dynamic" approximations, intended to account for possible variations in the characteristics of the Oklo ore bodies during or after the cessation of the chain reactions.

5.2 Dynamic Modeling

Cowan et al (55) reported an interesting attempt to develop a dynamic model for the Oklo reactors. Their idea was to compute the variations of n isotope ratios, R_i, by using as variables the neutron fluences (thermal, epithermal, fast), the uranium distribution prevailing during the chain reactions, and the duration of the reactions. The computations are performed in many neighboring regions on a closely spaced grid. The deviations from unity of the calculated-to-experimental ratios at each point in the grid, $\varepsilon_i = [R_i(\text{calc})/R_i(\text{exp})] - 1$, are then used to compute the quantity, $(1/n) \sum_{i=0}^{n} \varepsilon_i^2$, which is finally minimized with respect to the variables.

So far the only "dynamic" injected in this model was to consider a simmering water reactor, where a water reflux action would reconcentrate the uranium from the borders to the center of the reactor. This model, which is still in its infancy, opens the way to a better understanding of the Oklo phenomenon. In particular, the computations already indicate that the $\Sigma \varepsilon_i^2$, used as criteria for quality of fit, rapidly increase for $\Delta r \gtrsim 2.1$ aeons and $\delta \lesssim 200,000$ years, and also when the redistribution of uranium involves more than $\sim 10\%$ of the total amount of uranium.

One of the interesting features of fossil reactors that emerged from the work of Cowan (19) is that as the age of the initial deposits decreases, the ratios of ^{235}U to ^{238}U drop to the point where the critical configuration will breed rather than deplete its fissile material. The crossover point between a burner and a breeder should be observed at an age smaller than 1 aeon, and in this case the net result might have been to enrich ^{235}U, rather than to deplete it.

At the present time, the results of Cowan et al (55) are only slightly better than those previously reported by Naudet et al (53). Indeed, dynamic models should also be seriously improved by taking into consideration other variables such as the poison distribution active during the chain reaction, as well as a variety of corrections including a careful investigation of all parameters involved in the evaluation of the resonance integrals (cf footnote 9), the consideration that overcritical water was probably the active coolant, etc.

6 SUMMARY AND FUTURE PROSPECTS

With the possible exception of a good shielding against the remobilization of isotopic anomalies and of a relatively good "cleanliness," the geological and geochemical features of the Oklo uranium ore deposits do not look exceptional. However, they all combine favorably to increase the probability of finding well-preserved fossil reactors within the deposit. The discovery of fossil reactors in other old ore bodies is dependent on the willingness of private companies to communicate the confidential characteristics of uranium deposits, as well as on both the frequent analysis of the uranium isotopic abundance and the identification of local "hot" spots of sufficient size within these deposits. In this context the use of methods developed by Gingrich et al (56) to detect α-recoil tracks formed during the α decay of radon emanating from such "hot" spots could greatly help in their detection.

One of the most formidable challenges in the earth sciences is to extrapolate in a meaningful way the present-day configuration of the Oklo ore bodies back to the time of the chain reactions, and in particular to evaluate the initial depth of burial of the reactors. In this context the study of the reactor "ashes" (neutron-induced isotopes and radiation-damage features) injected into the Oklo sediments ~ 2 aeons ago will give the first serious opportunity to solve this very difficult problem.

Despite the presence of a complicated mixture of isotopes, geochemical methods have been successfully applied to the most abundant fission fragments produced during the chain reactions. Some of these fission-product isotopes were shown to be only slightly redistributed during the geological aging of the reactors and are therefore very useful probes. However, fission fragments near the valley of the mass distribution still cannot be used, and the interpretation of the isotope data in terms of reactor parameters has mainly been conducted to date by using a static configuration for the ore bodies. This approximation must be improved, although the extent of remobilization of most neutron-induced isotopes is astonishingly limited, thanks to the remarkable stability of uraninite. In addition, there are still conflicting reports about the proportion of fission due to ^{239}Pu, and consequently about the duration of the chain reaction. This prevents any meaningful evaluation of the temperature attained during the chain reaction. Therefore it is very difficult to pinpoint the mode of operation of the reactor (pulsed versus continous mode) and to accurately correct the effective cross sections that are used to infer the reactor parameters. Any marked progress in the understanding of the Oklo phenomenon will now come by using dynamic models of the reactor based on a realistic treatment of water above the critical point, on good evaluations of neutron cross sections (57), and on other such improvements.

The homogenization of samples by crushing, although allowing the intercomparison of data between different laboratories, has certainly introduced artifacts that are poorly understood at the present time. Indeed, one of the major characteristics of the Oklo ore samples is their extreme heterogeneity down to a microscale, which should be further investigated by using a variety of microprobe techniques. One of the most challenging problems in this type of analysis is to understand whether or not uraninite grains were formed before the chain reactions.

The most powerful methods used in carbon geochemistry should be applied as soon as possible to the Oklo ores in order to understand the marked depletion of carbon compounds in the reaction zones. If these studies show that large amounts of carbon compounds have indeed been lost from the reactor cores by pyrolysis and/or radiolysis, then this important finding would invalidate any static model for the Oklo reactors.

Radiation-damage studies have been useful in suggesting for the first time a drastic redistribution of the clay matrix during the chain reaction, and in requiring a long residence time of the ore bodies at great depths ($\gtrsim 5000$ m). In addition, they already have the potential of measuring several important reactor-operating conditions in the borders of the reactors (gradients in the fluence of thermal and fast neutrons, temperature gradients, etc), where the injection of depleted uranium from the reactor cores severely perturbs isotope measurements.

Finally the following by-products of the Oklo studies are worth mentioning:

1. These studies challenge in many ways analytical methods used in a variety of fields. In particular, only the most refined methods used in geochronology, rare-gas determinations and ion-microprobe analysis can be tentatively exploited in the "non-closed" geochemical system of the Oklo reactors, where in addition to other processes, neutron fluences have been active to further modify isotopic abundances.

2. They contributed to important problems dealing with the redistribution of elements in nature over a very long time scale. This in turn has a direct bearing on the long-term disposal of man-made radioactive wastes (57, 58, 59), in identifying those fission products that have been stable over a time scale of about 2 aeons. In particular, the Oklo studies already suggest that with the exception of fissiogenic rare gases, the safe storage of radioactive waste in natural geological settings is feasible.

3. The formation of an international consortium of scientists for the study of the Oklo samples has been encouraged by the CEA. This consortium approach, in catalyzing the partnership between the earth sciences and the whole of chemistry and physics, has already been very fruitful in the study of lunar samples, deep-sea sediments, and Antarctic ices. As pointed out by Cowan (19), a similar type of effort for the study of the Oklo phenomenon will certainly be useful in giving new clues concerning the formation of uranium ore deposits in the sedimentary rocks (wherein are contained about 75% of the uranium reserves of the world), as well as their rates of survival or destruction, thus contributing substantively to the solution of the energy problem.

ACKNOWLEDGMENTS

Dr. Massignon initiated the author's interest in the study of the Oklo phenomenon and made valuable comments about this manuscript. The constant input and interest of J. C. Petit was decisive in the preparation of this manuscript. The author is also very grateful to the following colleagues: J. Geoffroy and F. Weber for their enthusiastic expertise in the earth science aspect of the Oklo phenomenon; R. Hagemann and E. Roth for comments about isotope studies; R. Naudet for illuminating discussions, and J. Audouze, C. Cowan, R. Klapisch, and G. Raisbeck, who constructively reviewed the present manuscript.

Literature Cited

1. Fermi, E. 1947. *Science* 105:27–31
2. Glasstone, S., Sensonske, A. 1963. *Nuclear Reactor Engineering.* Princeton: Van Norstrand. 830 pp.
3. Rankama, K. 1963. *Progress in Isotope Geology,* pp. 560–61. New York: Interscience. 679 pp.
4. Kuroda, P. K. 1975. *The Oklo phenomenon,* pp. 479–87. Vienna: IAEA. 646 pp.

5. Wheterill, G. W., Inghram, M. G. 1953. *Proc. Conf. Nucl. Processes Geol. Settings*, pp. 30–32. Washington DC: Natl. Res. Counc.
6. Kuroda, P. K. 1956. *J. Chem. Phys.* 25:781–82
7. Bouzigues, H., Boyer, R. J. M., Seyve, C., Teulieres, P. See Ref. 4, pp. 237–43
8. Neuilly, M. et al 1972. *CR Acad. Sci. Paris* 275:1847
9. Naudet, R. 1974. *Bull. Inf. Sci. Tech. Paris* 183:7–46
10. Naudet, R. 1975. *Recherche* 6:508–18
11. Bigotte, G. 1964. *Int. Geol. Congr. 22nd, New Delhi*, Sect. 5, pp. 351–62
12. Barnes, F. Q., Ruzicka, V. 1972. *Int. Geol. Congr., 24th, Montreal*, Sect. 4, p. 159
13. Adler, H. H. 1974. *Formation of Uranium Ore deposits*, pp. 141–67. Vienna: IAEA. 728 pp.
14. Weber, F. 1971. *Rapp. CEA-R-4054*. Paris: Commis. Energ. At. 328 pp.
15. Gauthier-Lafaye, F., Ruhland, M., Weber, F. 1975. See Ref. 4, pp. 103–18
16. Naudet, R. Personal communication
17. Devillers, C. et al. See Ref. 4, pp. 293–99
18. Lancelot, J. R., Vitrac, A., Allegre, J. C. 1975. *Earth Planet. Sci. Lett.* 25:189–96
19. Cowan, G. A. 1975. *Los Alamos Rep.* LA-UR-75-2040
20. Knipping, H. D. See Ref. 13, pp. 531–48
21. Ruffenach, J. C. et al. See Ref. 4, p. 378; Hagemann, R. et al 1975. See Ref. 4, p. 416; Neuilly, M., Naudet, R. 1975. See Ref. 4, p. 542
22. Loubet, M., Allegre, J. 1976. *Geochim. Cosmochim. Acta*. Submitted for publication
23. Evans, R. D. 1955. *The Atomic Nucleus*. New York: McGraw-Hill. 560 pp.
24. Wescott, C. H. 1958. *AECL n° 670* (CRRP-787)
25. Naudet, R. 1973. *DPRMA n°8*. Gif-sur-Yvette: Commiss. Energ. At.
26. Naudet, R., Filip, A., Renson, R. 1975. See Ref. 4, pp. 83–101
27. Geoffroy, J. See Ref. 4, pp. 133–49
28. Naudet, R. Personal communication
29. Naudet, R., Renson, C. See Ref. 4, pp. 265–89
30. Naudet, R., Filip, A. See Ref. 4, pp. 557–64
31. Havette, A., Naudet, R., Slodzian, G. See Ref. 4, pp. 463–78
32. 1975. *The Oklo Phenomenon*, Sects. 3–5. Vienna: IAEA. 679 pp.
33. Hagemann, R. et al. See Ref. 4, pp. 293–99, 371–84, 415–23
34. Bassière, H., Cesario, J., Poupard, D., Naudet, R. See Ref. 4, pp. 385–99
35. Hagemann, R. Personal communication
36. Ruffenach, J. C. et al. See Ref. 4, pp. 371–84
37. Drozd, R. J. Personal communication
38. Drozd, R. J., Hohenberg, C. M., Morgan, C. J. 1974. *Earth Planet. Sci. Lett.* 23:28–33
39. Alexander, E. C., Lewis, R. S., Reynold, J. H., Michel, M. C. 1971. *Science* 172:837–40
40. Vook, F. L. et al 1975. *Rev. Mod. Phys* 47:S1–S14 (suppl. 3)
41. Petit, J. C. 1976. *Thèse Doct. 3è Cycle*. Orsay: Faculté Sci. In press
42. Fleischer, R. L., Price, P. B., Walker, R. M. 1965. *Ann. Rev. Nucl. Sci.* 15:1–28
43. Fleischer, R. L., Price, P. B., Walker, R. M. 1975. *Nuclear Tracks in Solids*. Berkeley: Univ. Calif. Press. 605 pp.
44. Durrieu, L., Jouret, C., Le Roulley, J. C., Maurette, M. 1971. *Jernkontorets Ann.* 155:535–40
45. Comes, R., Lambert, M., Guinier, A. 1967. *Interaction of Radiation with Solids*, pp. 319–37. London: Plenum
46. Seitz, M. et al 1970. *Radiat. Effects* 5:143–50
47. Drozd, R. J. 1974. PhD thesis. Washington Univ., St. Louis, Mo.
48. Dran, J. C., Duraud, J. P., Maurette, M. 1974. *Bull. Inf. Sci. Tech. Paris* 183:59–63
49. Dran, J. C. et al. See Ref. 4, pp. 223–33
50. Weber, F., Geoffroy, J., Le Mercier, M. See Ref. 4, pp. 173–92
51. Geoffroy, J. Personal communication
52. Dran, J. C., Langevin, Y., Maurette, M., Petit, J. C. 1976. *Earth Planet. Sci. Lett.* In press
53. Naudet, R. et al. See Ref. 4, pp. 527–40, 557–64, 565–71, 573–85
54. Langevin, Y. Personal communication
55. Cowan, G. A., Bryant, E. A., Daniels, W. R., Maeck, W. J. See Ref. 4, pp. 341–55
56. Gingrich, J. E., Lowett, D. B. 1972. *Trans. Am. Nucl. Soc.* 15:118
57. Maeck, W. J., Spraktes, F. W., Tromp, R. L., Keller, J. H. See Ref. 4, pp. 319–39
58. Walton, R. D., Cowan, G. A. See Ref. 4, pp. 499–507
59. Frejacques, C. et al. See Ref. 4, pp. 509–24

Ann. Rev. Nucl. Sci. 1976. 26: 351–83
Copyright © 1976 by Annual Reviews Inc. All rights reserved.

POSITRON CREATION IN SUPERHEAVY QUASI-MOLECULES

×5577

Berndt Müller

Department of Physics, University of Washington,
Seattle, Washington 98195[1]

CONTENTS

1 INTRODUCTION

During the last century atomic physics has always been a primary source of fundamental physical insight. Two of the most successful theories, quantum mechanics and quantum electrodynamics, have drawn their bases from novel atomic physics experiments. The common feature of all atoms encountered in nature (including also mesonic atoms) is that the binding energy per particle is much less than the total rest energy of this particle. This would be drastically changed if there were higher islands of stability for nuclei, perhaps around $Z = 114$ or 164. These superheavy

[1] Present address: Institut für Theoretische Physik, Robert-Mayer-Strasse 8–10, Frankfurt am Main, West Germany.

nuclei were first speculated upon by Werner & Wheeler (1), who also calculated the electronic states up to $Z = 164$. The stability of superheavy nuclei was investigated in the context of deformed-shell models by Mosel & Greiner and by Nilsson, Thompson & Tsang (1). In addition, the present status of heavy-ion accelerators has made it possible to perform collision experiments with a projectile energy of more than 5 MeV/nucleon. This allows one to bring all existing nuclei so close to each other that they will simulate for a very short moment a superheavy nucleus, at least from the atomic physicist's point of view (22, 24). Thus, experiments with atomic systems up to $Z = 184$ ($= 92 \times 2$) or even 190 or 196 seem to be feasible in the near future.

This prospect has led to an unprecedented interest in questions of strong binding in the years since 1968. Most of this work has been done in Frankfurt (4, 11–13, 15, 18) and by the Russian group (9, 14, 23, 41, 42). It turns out that superheavy atomic systems open up a fundamentally new process: at $Z = 172$ a K-shell electron is bound by twice its rest mass and beyond this point spontaneous, energyless pair production is possible. We shall see that a K vacancy in such a system (which we call supercritical) will be emitted as a positron and the K shell will be filled. This is more than just another possible experiment. When one really tries to describe the supercritical atom, one is forced to give up the old idea of the vacuum as being neutral and without particles. Instead, a new, charged vacuum has to be introduced. This constitutes a kind of dynamical symmetry breaking, but in this case of a fermion field.

This review bears witness to the two aspects of the physical problem: Sections 2 and 5 are devoted to a discussion of the new concepts that have been introduced in recent years, and Sections 3 and 4 contain a review of the phenomenological and experimental aspects. During the entire paper we use natural units ($\hbar = c = 1$) and measure energies in units of the electron mass $m_e = 511.0041$ keV.

2 SPONTANEOUS POSITRON EMISSION FROM SUPERHEAVY ATOMS

2.1 The Electronic Spectrum of Superheavy Atoms— the Critical Charge Z_{cr}

Let us start with an examination of the electronic binding energies of the presently known elements ($Z \leq 106$), as shown in Figure 1. Compared with the 1.022-MeV gap between the positive and negative energy solutions of the Dirac equation (electron and positron scattering states), all bound states are seen to lie in a small region close to the top of the gap. A simple extrapolation of the nonrelativistic Z^2 scaling of the K-shell binding energies from hydrogen would predict the $1s$ state would traverse the gap and reach the bottom at $Z \approx 275$, which is out of reach for any presently conceivable experiment.

However, it is obvious that for binding energies E_B comparable to the electron rest energy m_e the relativistic Dirac equation must be solved,

$$\mathscr{H}_D \psi(\mathbf{r}) \equiv [\boldsymbol{\alpha} \cdot \mathbf{p} + \beta m_e + V(\mathbf{r})] \psi(\mathbf{r}) = E\psi(\mathbf{r}), \qquad 2.1.$$

where

$$\alpha = \begin{pmatrix} 0 & \sigma \\ \sigma & 0 \end{pmatrix} \quad \text{and} \quad \beta = \begin{pmatrix} I & 0 \\ 0 & -I \end{pmatrix} \qquad 2.2.$$

are the four anticommuting Dirac matrices and E is the total electron energy, related to the binding energy by $E = E_B + m_e$. $V(\mathbf{r})$ is the nuclear Coulomb potential and may also include electron-electron interaction terms. The general spectrum of Equation 2.1 is determined from the following consideration: for any asymptotically vanishing potential V, the form of the wave function at large r is $\psi \approx \exp[i(E^2 - m_e^2)^{1/2}r]$. This is oscillating for $E > m_e$ for electrons and for $E < -m_e$ for positrons, and there is a region $m_e > E > -m_e$ where all eigenstates are bound states.

It is a general property of relativistic equations of motion, classically as well as quantum mechanically, that the singular $-(Z\alpha/r)$ potential can support a stationary motion only for not-too-large coupling constants $(Z\alpha)$. This can easily be seen as follows: the relativistic energy-momentum relation

$$p^2 + m^2 - (E - V)^2 = 0 \qquad 2.3.$$

gives rise to an effective potential

$$V_{\text{eff}} \approx (J + \tfrac{1}{2})^2/r^2 + 2EV - V^2, \qquad 2.4.$$

which depends on the angular momentum J. V_{eff} is repulsive for small distances r as long as $(Z\alpha)^2 \lesssim (J + \tfrac{1}{2})^2$. For potentials stronger than this value, the wave function (or orbit, classically) will collapse to the center of the potential, unless some deviation occurs from the $1/r$ behavior at very small distances.

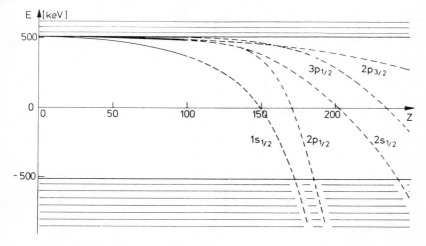

Figure 1   The energies of the lowest atomic states versus the nuclear charge Z. Solid lines give the experimentally known region, dashed lines represent Hartree-Fock-Slater calculations. Observe the critical point for the $1s$, $2p_{1/2}$, and $2s$ states.

Mathematically, this means that the relativistic Hamiltonian $\mathscr{H}_D$ loses its hermiticity when $(Z\alpha) > (J + \frac{1}{2})$. In order to give meaning to the eigenvalue problem, one has to specify an additional parameter that characterizes the solution. This was first done—with little respect to physical situations, however—by Case (2). From physical arguments it is clear that the potential is created by an extended source, so that it is weakened in a region $r \lesssim R$. Then for $Z\alpha > J + \frac{1}{2}$ the eigenstates of Equation 2.1 will critically depend on R:

$$\psi(\mathbf{r}) = \psi(\mathbf{r}, Z\alpha, R).$$

In superheavy atoms, R is the nuclear radius. Of course, the wave function ψ and the energy E depend not only on R but also on the precise form of the nuclear charge distribution. It turns out, however, that any realistic model will do, unless precision experiments can be made. In the calculations reported below, either a homogeneous charge distribution or a Fermi-type model was used.

The energy eigenvalues of the Dirac equation with a realistic potential were first investigated by Pomeranchuk & Smorodinskii (3a); later, the precise values were given by Werner & Wheeler (1) and by Voronkov & Kolesnikov (3b). The binding energies of electrons for all elements up to $Z = 170$ were obtained by Pieper & Greiner (4) and by Rein (5) in the one-electron approximation. Electron screening was first taken into account in the Hartree-Fock calculations of Fricke et al (6). The results due to Soff & Fricke (7) are shown in Figure 1 by dashed lines. The $1s_{1/2}$ state, which is of main interest to us, quickly gains binding energy when the nuclear charge is increased beyond $Z = 100$. The binding energy reaches $2m_e = 1022$ keV at $Z_{cr} \approx 172.5$, which we call the critical nuclear charge Z_{cr} (8, 9). The second noteworthy feature of Figure 1 is the drastic change in the properties of the $2p_{1/2}$ state beyond $Z \approx 140$. At $Z = 170$ its binding energy is more than twice that of the $2s_{1/2}$ state and the radial part of its wave functions resembles more and more that of the $1s_{1/2}$ state. The binding energy of the $2p_{1/2}$ states reaches $2m_e$ at $Z_{cr}(2p_{1/2}) \approx 185$ and for the $2s_{1/2}$ state we find $Z_{cr}(2s_{1/2}) \approx 245$ (11).

We have said before that for E below $-m_e$ no bound-state solution of Equation 2.1 can be constructed. On the other hand, it cannot correspond to the physical situation that the two $1s$ electrons would simply disappear if the nuclear charge is increased beyond Z_{cr}. Pieper & Greiner (4), and later, independently, Rein (5) and Gershtein & Zeldovich (22), suggested that beyond Z_{cr} spontaneous pair creation would become possible if the K shell was vacant. The K hole, so to say, would be emitted as a positron. This idea is elaborated below.

2.2   The Scattering States of Supercritical Atoms

We have seen in the preceding paragraph that there is a critical charge Z_{cr} in the superheavy atom region beyond which the Dirac equation does not support a $1s_{1/2}$ bound state. This happens when the $1s$ bound state becomes energetically degenerate with the positron continuum states of the Dirac equation. It seems natural to investigate the properties of these scattering states for $Z > Z_{cr}$ and to see how they are influenced by the bound state.

We write the Dirac wave function ψ with quantum numbers $\kappa = \pm(j + \tfrac{1}{2})$ and $\mu = j_z$ as follows (10):

$$\psi = \begin{pmatrix} f\chi_\kappa^\mu \\ ig\chi_{-\kappa}^\mu \end{pmatrix} = \frac{1}{r} \begin{bmatrix} (E + m_e)^{1/2} & (\phi + \phi^*) & \chi_\kappa^\mu \\ -(E - m_e)^{1/2} & (\phi - \phi^*) & \chi_{-\kappa}^\mu \end{bmatrix}. \qquad 2.5.$$

For a potential $V = -(Z\alpha/r)$ it can be shown that ϕ satisfies the second-order differential equation,

$$\frac{d^2\phi}{dx^2} + \frac{1}{x}\frac{d\phi}{dx} - \left[\frac{1}{4} + \left(\frac{1}{2} + iy\right)\frac{1}{x} - \frac{\gamma^2}{x^2}\right]\phi = 0, \qquad 2.6.$$

where we have introduced the dimensionless variable $x = 2ipr$, and $p^2 = E^2 - m_e^2$, $y = (Z\alpha E/p)$, $\gamma^2 = (Z\alpha)^2 - \kappa^2$. γ has already been defined in view of the $s_{1/2}$ and $p_{1/2}$ states, for which $\kappa = \pm 1$, and therefore $Z\alpha > |\kappa|$. The general solution of Equation 2.5 is given (11) in terms of the Whittaker function $M_{a,b}(x)$:

$$\phi = Nx^{-1/2}\left[\exp(i\eta)M_{-(iy+1/2),iy}(x) + \exp(-i\eta - \pi y)\frac{i\gamma - iy}{\kappa + iym_e/E}\right.$$

$$\left. \times M_{-(iy+1/2),-i\gamma}(x)\right]. \qquad 2.7.$$

Here N is a normalization factor that is usually chosen such that for large r,

$$\phi \to \frac{1}{2(\pi p)^{1/2}}\exp[i(pr + \Delta)] \qquad 2.8.$$

and η is an arbitrary phase.

For $Z\alpha < |\kappa|$, the situation in all known atoms, η can be determined from the requirement that ϕ be regular at the origin. For $Z\alpha > |\kappa|$, the case of present study, this is not possible. For every choice of η, ϕ has an essential singularity at the origin, but stays bounded. In order to determine η, the potential has to be replaced by a nonsingular one for $r < R$ (inside the nucleus) and the external solution must be matched to the internal solution at $r = R$ (11). Of course, η obtained from this procedure is strongly dependent on the nuclear radius R. The phase shift Δ from Equation 2.8 becomes a function of R in this way, too (12):

$$\Delta(R) = y\ln(2pr) + \arg\left\{\exp[i\eta(R)]\frac{\Gamma(2i\gamma + 1)}{\Gamma(i\gamma + 1 + iy)}\right.$$

$$\left. + i\exp[-i\eta(R) - \pi\gamma]\frac{\gamma - y}{\kappa + iym_e/E}\frac{\Gamma(-2i\gamma + 1)}{\Gamma(-i\gamma + 1 + iy)}\right\}. \qquad 2.9.$$

The radii of physical nuclei are much smaller than the electron Compton wavelength ($R \ll m_e^{-1}$) and in this limit

$$\eta \to -\gamma\ln(2pR) + \eta_0 + \mathcal{O}(R), \qquad 2.10.$$

where η_0 depends slightly on the details of the nuclear charge distribution.

The physical scattering phase shift δ is defined from Δ by requiring $\delta \to 0$ for small momenta p. In (11) it was shown that this is taken care of, if an angular momentum–independent part is subtracted:

$$\delta = \Delta - \frac{\pi}{4} - y[\ln(2pr/|y|) + 1].$$ 2.11.

This phase shift has been evaluated as a function of energy in the negative-energy continuum for various supercritical atoms. Figure 2 shows a typical example for the $s_{1/2}$ phase shift. The phase shift δ varies smoothly from zero at $E = -m_e$ to a finite limit at $E \to -\infty$, except that there is a sharp resonance around a certain energy E_{res}. This resonance can be identified as a bound state imbedded in a continuum of scattering states of the same symmetry.

By numerical computation this resonance energy can be traced as a function of Z. The result is shown in Figure 3 for the $s_{1/2}$ and the $p_{1/2}$ resonance. In both cases the value of Z where the resonance enters the positron continuum is identical with the critical charge Z_{cr} where the equivalent bound state disappears. This shows that the bound-state solution becomes a continuum resonance as soon as its binding energy reaches $2m_e$, a result that was obtained independently by Peitz, Müller, Rafelski & Greiner (13) and by Zeldovich & Popov (14). For Z up to 250 the resonance energy has been parameterized as a function of the supercritical part of the nuclear charge $Z' = Z - Z_{cr}$:

$$E_{res} \approx -m_e - Z'\delta - Z'^2\tau.$$ 2.12.

The values of the parameters δ and τ for the $1s_{1/2}$ and the $2p_{1/2}$ states are listed in Table 1. They can be understood in terms of the simple model of the subsequent paragraph.

If we fit the resonance shape of Figure 2 by a simple Breit-Wigner formula,

$$\tan \delta(E) = \tfrac{1}{2}\Gamma(E - E_{res})^{-1},$$ 2.13.

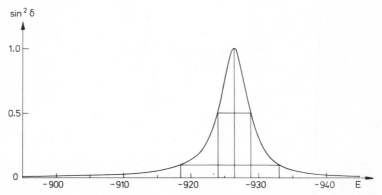

Figure 2 The $s_{1/2}$ phase shift in the negative-energy Dirac continuum for the element $Z = 184$ (U + U). The sharp resonance at $E = -926.4$ keV with a width of 4.8 keV represents the 1s electron state, which is admixed to the positron-scattering states.

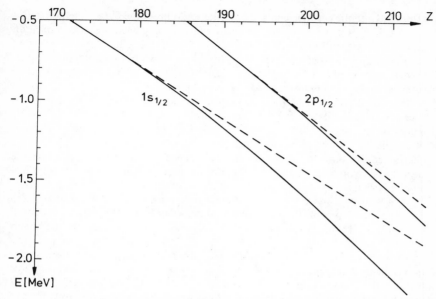

Figure 3   The resonance energies obtained by $s_{1/2}$ and $p_{1/2}$ phase-shift analysis (*full lines*). The dashed lines indicate the linear approximation discussed in Section 2.3.

we can extract a width Γ of the resonance. When Γ is plotted for various $Z > Z_{cr}$ and also different nuclear radii R, it turns out to depend only upon the location of the resonance. For reasons that will become clear in the next paragraph, we have plotted the dimensionless quantity $\gamma = m_e\Gamma/(E_{res}+m_e)^2$ in Figure 4 for the $1s_{1/2}$ resonance as a function of E_{res} (15). In the threshold region $E_{res} \approx -m_e$ it is strongly energy-dependent, whereas it becomes almost constant as the resonance moves deeper into the continuum. If the nuclear radius is kept constant, it turns out that Γ is proportional to Z'^2 in the threshold region.

2.3   The Auto-Ionization Model

The results of the last paragraph lead to the conclusion that by going into the supercritical region $Z > Z_{cr}$, the bound state, e.g. the $1s_{1/2}$ electronic state, becomes

Table 1   The parameters δ and τ for the energy location $E_{res} = -m_e - Z'\delta - Z'^2\tau$ of the supercritical bound-state resonances

State	$1s$	$2p_{1/2}$
Z_{cr}	172	185
δ (keV)	29.0	37.8
τ (keV)	0.33	0.22

embedded into the negative-energy scattering solutions of the Dirac equation and is converted into a sharp resonance. This situation is analogous to the auto-ionization process that is well known in atomic and nuclear physics. In this case, a bound state (usually a two-particle excitation) is degenerate with a continuum (e.g. one-particle scattering states). Due to configuration interaction the bound state is admixed to the continuum and becomes a resonance.

The main difference from the usual auto-ionization process is the fact that in our case the scattering states have to be considered as occupied. Indeed, the Dirac hole model of antiparticles tells us that the whole body of negative-energy electron states should be viewed as occupied and any hole created in this "Dirac sea" interpreted as a positron. This is intimately connected with the quantization of the electron-positron field and the definition of the vacuum state (ground state) of the atomic system.

At present we choose to postpone these questions to the next paragraph and instead proceed to apply the auto-ionization theory of Fano (16) to the supercritical atom (13, 17). In our model we assume that it is sufficient to consider the bound state $|n\rangle$ which becomes supercritical (e.g. the $1s_{1/2}$ state) together with the negative-energy scattering states $|E\rangle$ of the same symmetry (e.g. the $s_{1/2}$ states). This assumption is justified because all other states of same symmetry are separated from these by more than 500 keV, and because in a situation of spherical symmetry, states of different angular momentum should not influence each other (see Figure 1).

We start from the just-critical atom with the bound state $|n_{cr}\rangle$ and scattering states $|E, Z_{cr}\rangle$, such that

$$\mathscr{H}(Z_{cr})|n_{cr}\rangle = -m_e|n_{cr}\rangle, \qquad\qquad 2.14a.$$

$$\mathscr{H}(Z_{cr})|E, Z_{cr}\rangle = E|E, Z_{cr}\rangle, \qquad\qquad 2.14b.$$

$$\langle E', Z_{cr}|E, Z_{cr}\rangle = \delta(E'-E). \qquad\qquad 2.15.$$

The supercritical Hamiltonian can be split into the critical one plus the potential increase due to the additional charge $Z' = Z_{cr} - Z$ inside the nucleus:

$$\mathscr{H}(Z) = \mathscr{H}(Z_{cr}) + V(Z'). \qquad\qquad 2.16.$$

In addition, for small Z' we can assume that the shape of the potential remains unchanged by the additional charge, since the nuclear radius R grows very slowly with Z. Thus,

$$V(Z') \approx Z'U(r). \qquad\qquad 2.17.$$

The matrix elements of the supercritical Hamiltonian (Equation 2.16) in the reduced basis (Equation 2.14) are

$$\langle n_{cr}|\mathscr{H}(Z)|n_{cr}\rangle = -m_e + \Delta E(Z') \approx -m_e - Z'\delta, \qquad\qquad 2.18a.$$

$$\langle E_1 Z_{cr}|\mathscr{H}(Z)|n_{cr}\rangle = V_E(Z') \approx Z'v_E, \qquad\qquad 2.18b.$$

$$\langle E_1' Z_{cr}|\mathscr{H}(Z)|E, Z_{cr}\rangle = E\,\delta(E-E') + U_{E'E}. \qquad\qquad 2.18c.$$

This system can be solved analytically for the eigenstates of $\mathscr{H}(Z)$ when the continuum-continuum interaction $U_{E'E}$ is neglected. We shall do this in order to obtain a transparent model of the supercritical atom. The deviations from the exact phase-shift analysis then can be used to determine the influence of the rearrangement of the continuum states. The eigenstates can be expanded in terms of the basis (Equation 2.14):

$$|E, Z\rangle = a(E)|n_{cr}\rangle + \int_{E' < -m_e} dE'\, b_{E'}(E)|E', Z_{cr}\rangle \qquad 2.19.$$

with the coefficients (11, 18)

$$a(E) = V_E^* [E + m_e - \Delta E - F(E) + i\pi|V_E|^2]^{-1}, \qquad 2.20a.$$

$$b_{E'}(E) = \delta(E - E') + a(E)V_{E'}(E - E' + i\varepsilon)^{-1}. \qquad 2.20b.$$

Here $F(E)$ is the principal value integral

$$F(E) = P\int_{E' < -m_e} dE'\, |V_{E'}|^2/(E - E'). \qquad 2.21.$$

We see that we may interpret

$$E_n(Z) = -m_e + \Delta E(Z') \qquad 2.22.$$

as the location of a resonance and

$$\Gamma_E = 2\pi|V_E|^2 \qquad 2.23.$$

as the (slightly energy-dependent) width of this resonance, which becomes even more obvious if we write:

$$|a(E)|^2 = \frac{(1/2\pi)\Gamma_E}{\{E - [E_n + F(E)]\}^2 + \frac{1}{4}\Gamma_E^2}. \qquad 2.24.$$

Equation 2.24 gives the probability with which the previously bound state $|n_{cr}\rangle$ becomes admixed to the continuum states in the supercritical atom.

In the next paragraph we show that this result, which is analogous to the broadening of an unstable auto-ionizing state, means that a vacant supercritical state can be filled spontaneously via pair production. This provides a simple explanation of the additional shift $F(E)$ of the center of the resonance: it is the level shift because the state is damped by some coupling to a decay channel. By analogy to radiation theory (19), the shift is small.

We now can compare the prediction of our model with the phase-shift calculation of Section 2.2. The proper connection between the two descriptions is provided by Fano's formula (16):

$$|a(E)|^2 = 2\sin^2(\delta_E)/\pi\Gamma_E. \qquad 2.25.$$

One may check easily that Equations 2.25 and 2.13 allow this identification. If we use the scaling laws (Equation 2.18) of our matrix elements with Z', we find that $E_n(Z)$ corresponds to the straight dashed lines in Figure 3. The numerical values of

δ from Equation 2.18a and from the resonance fit (Equation 2.12) coincide. The additional quadratic term $Z'^2\tau$ in Equation 2.12 is partly due to a change in the nuclear radius as Z' becomes large, and partly to the influence of the continuum-continuum interaction neglected in our model. The width of the resonances, Γ_E (2.23), is proportional to Z'^2 as long as the matrix element can be considered independent of the energy E. Close to the threshold $E = -m_e$ this is not the case because the penetration factor of positron-scattering states

$$|\psi_E(r=0)|^2 \approx \exp(\pi Z\alpha E/p) \qquad\qquad 2.26.$$

rapidly goes to zero for $E \to -m_e$. Since the wave function $|n_{cr}\rangle$ is localized close to the nucleus, the resonance width behaves like $\exp(-\pi Z\alpha m_e/p)$ near the threshold. This explains the steep falloff of $\tilde{\gamma}(E)$ at $E \to -m_e$ in Figure 4.

2.4 Quantization of the Supercritical Atom and Spontaneous Positron Emission

So far we have discussed the solutions of the Dirac equation in a supercritical potential for a single electron. In order to discuss the occupation of states, especially what the mixture of electron and positron solutions means, one has to introduce the language of field theory. The usual way to define the operator A for a fermion field ψ, i.e. to antisymmetrize with respect to $\psi : \frac{1}{2}[\bar{\psi}, A\psi]$, is equivalent to the redefinition of negative-energy states as antiparticle states. Thus the "destruction" operator ψ acting on a positive energy state annihilates an electron, but acting on a negative energy state creates a hole, i.e. a positron:

$$\psi(x,t) = \sum_{E_p>0} b_p \varphi_p(x) \exp(-iE_p t) + \sum_{E_p<0} d_p^+ \varphi_p(x) \exp(-iE_p t). \qquad (2.27)$$

This is Dirac's picture of a completely filled "sea" of negative-energy states, the charge of which is renormalized to be zero.

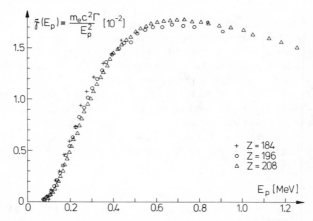

Figure 4   Plot of the invariant $1s$ resonance width (positron emission constant) $\tilde{\gamma}$ versus the positron energy E_p. $\tilde{\gamma}$ is clearly independent of the special supercritical system.

This familiar picture has to be rethought, when one of the unoccupied electron states joins the positron states, as we have seen in the preceding paragraphs. On the one hand it would seem quite natural to add this dissolved bound state to the ensemble of filled states; on the other hand this would either produce a charged vacuum state or it would violate the conservation of charge. First of all, it becomes obvious that the boundary that separates the particle and the antiparticle states cannot be drawn at zero energy. In the language of many-body theory, the boundary in momentum space that separates occupied and vacant states is called the Fermi surface F. We therefore define a generalized Fermi surface F in the same way for the electron field. Equation 2.27 then reads

$$\psi(x,t) = \sum_{p>F} b_p \varphi_p(x) \exp(-iE_p t) + \sum_{p<F} d_p^+ \varphi_p(x) \exp(-iE_p t). \qquad 2.28.$$

For a system where all bound states are filled, we have $E_F = +m_e$, whereas the vacuum state is defined by $E_F = -m_e$. Utilizing a statistical method, this is shown in Section 5.

Next, we have to ask what should be called a vacuum state. Certainly, it has to be the state of lowest energy possible for the system in question (e.g. for the electron field in the presence of a nucleus), but we must also require that this state be stable. We shall now show that the configuration in which the supercritical bound state resonance is vacant is in fact unstable and decays by positron emission (18, 20). For convenience we put the Fermi energy at $E_F = -m_e$ and construct the desired configuration by a vacancy-creation operator. The unitary of the transformation (Equation 2.19) allows us to construct a creation operator for vacancies in the bound state also in a supercritical atom (18):

$$\tilde{d}_n^+ = \int_{E<-m_e} dE'\, a(E')\, d_{E'}^+. \qquad 2.29.$$

The tilde sign on $\tilde{d}_n^+$ is meant to stress that it is not a linearly independent operator in the supercritical situation, i.e. that it really corresponds to the subcritical basis of states. If we prepare a bound-state vacancy in this way, the true state vector of our system can be written:

$$|\Psi(t)\rangle = y(t)\tilde{d}_n^+ |F\rangle + \int_{E<-m_e} dE'\, W_{E'}(t)\, d_{E'}^+ |F\rangle. \qquad 2.30.$$

The state $|F\rangle$ denotes the configuration that is filled according to our prescription of the Fermi energy. The time dependence of $|\Psi(t)\rangle$ is obtained by solving the equation of motion

$$i\frac{\partial}{\partial t}|\Psi(t)\rangle = \mathscr{H}(Z)|\Psi(t)\rangle, \qquad 2.31.$$

where $\mathscr{H}(Z)$ is the (quantized) Hamiltonian from Equation 2.16. Taking matrix elements in the same way as in Section 2.3, one arrives at the differential equation

$$E[y(t)a(E) + W_E(t)] = i\frac{\partial}{\partial t}[y(t)a(E) + W_E(t)]. \qquad 2.32.$$

The solution of this equation corresponding to our initial condition $y(0) = 1$, $W_E(0) = 0$ is (18)

$$y(t) = \exp(-iE_{res}t - \tfrac{1}{2}\Gamma|t|),$$ 2.33a.

$$W_E(t) = a(E)[\exp(-iEt) - \exp(-iE_{res}t - \tfrac{1}{2}\Gamma|t|)].$$ 2.33b.

Here E_{res} has the familiar meaning

$$E_{res} = E_n + F(E_{res})$$ 2.34.

and the decay constant is given by Equation 2.23:

$$\Gamma = 2\pi|V_{E_{res}}|^2.$$ 2.35.

Thus the probability of finding a localized vacancy after time t,

$$|y(t)|^2 = \exp(-\Gamma|t|),$$ 2.36.

shrinks to zero, whereas the energy distribution of positrons approaches

$$|W_E(t \to \infty)|^2 = |a(E)|^2.$$ 2.37.

Now let us consider what we have achieved: the state prepared at $t = 0$ in Equation 2.30 is just the familiar vacuum state, in which all negative-energy scattering states are filled and all localized states are vacant. We have seen that this state is unstable and develops into a configuration with the localized state occupied and a positron distribution given in Equation 2.37. This, of course, is not the lowest possible state. If we remove the positrons we gain the energy

$$\Delta E = \int_{E' < -m_e} dE'(-E' + m_e)|a(E')|^2 = -E_{res} + m_e,$$ 2.38.

thus arriving at a stable state of even lower energy. However, this state is just the one with Fermi energy $E_F = -m_e : |F, E_F = -m_e\rangle$. If we are willing to take our definition of the vacuum state seriously, we have to call this state the new *overcritical vacuum* $|0, Z > Z_{cr}\rangle$.

We want to investigate some of the properties of this new state. As was already mentioned, charge conservation forces us to consider it as charged. Indeed, if the usual vacuum $|0, Z_{cr}\rangle$ is defined as having charge zero, we only have to remark that (because of the two spin projections)

$$|0, Z_{cr}\rangle = \tilde{d}_{n\uparrow}^+ \tilde{d}_{n\downarrow}^+ |0, Z > Z_{cr}\rangle$$ 2.39.

to see that the supercritical vacuum must bear two negative units of charge. We shall therefore call it the *charged vacuum* (13, 14, 18) as distinguished from the familiar vacuum state. Although it is impossible to ascribe the charge distribution to a stationary one-electron state, it is possible to understand it as a collective effect of real vacuum polarization. Using the field theoretic prescription for the charge-density operator

$$\rho(x) = -\frac{e}{2}[\bar{\psi}(x), \gamma_0 \psi(x)],$$ 2.40.

it is possible to calculate the charge density of the charged vacuum state. The usual regularization problems are avoided if we extract the interesting part of $\langle 0|\rho(x)|0\rangle$ by integrating over a small energy interval around the resonance (11). Figure 5 shows the result of such an analysis for various supercritical states in comparison with the charge distribution of the just-critical $1s$ state in $Z = 172$. The shape of the radial distribution of charge is completely identical with that for a bound state out of the point spectrum of the subcritical Dirac Hamiltonian. To illustrate the dramatic change in vacuum polarization between the normal and the charged vacuum, Figure 5 also shows (exaggerated by a factor of 100!) the charge density ($r^2 \Delta\rho_{vac}$) for $Z = 172$ in the same energy interval where the resonance is found for $Z = 184$.

This completes our analysis of a stable supercritical atom, and we can summarize the phenomenology: beyond $Z = 172$ the $1s$ states of the (hypothetical) atoms "dive" into the negative-energy Dirac continuum. In doing so they become sharp resonances that can be detected by their phase shifts. If the $1s$ state was occupied before the "diving" occurred, it could be analyzed by a phase-shift analysis of positron scattering. Or, probably more easily, the photo-absorption of the $1s$ state becomes strongly broadened when it is supercritical. The width is given by Equation 2.35. On the other hand, if the $1s$ state was vacant before "diving" or if it was

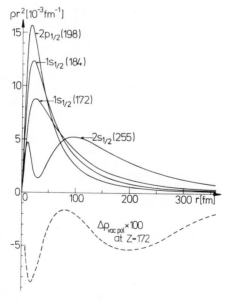

Figure 5 Electron density distributions for various supercritical states are shown in comparison to the subcritical $1s$ state in $Z = 172$. The transition to the supercritical region is seen to be smooth. The virtual vacuum polarization density $\Delta\rho$ (see text) is shown in magnification.

artificially ionized after becoming supercritical, the vacancy decays by emission of a positron. The typical lifetime is 10^{-19} sec, i.e. about a factor 10^2–10^3 shorter than the lifetime against radiative decay (21). The energy distribution of the emitted positrons is given by Equation 2.24.

Although it is impossible to define a bound one-electron $1s$ state in the super-critical atom, there is a charge distribution corresponding to the K shell, which is produced by the vacuum electrons as a collective property. We have called this state the charged vacuum, because it is charged twice (or more when more than one bound state has "dived"). Another difference to the neutral vacuum is that the photon propagator acquires additional poles (besides $k^2 = 0$) at the energies necessary to excite a collective K-shell electron into a higher vacant bound state. These poles have an imaginary part $i\Gamma_E$, i.e. they are off the mass shell and can only be seen as resonances in photo-absorption.

3   SUPERCRITICAL QUASI-MOLECULES

3.1   The Two-Center Dirac Equation and the Critical Distance R_{cr}

In order to test the theoretical results of the preceding chapter we need a super-critical system. The present status of nuclear physics makes it very unlikely that a stable or at least metastable nucleus with charge $Z > 172$ can be produced. On the other hand, the rather short lifetime of a supercritical K-shell vacancy against positron emission ($\tau_{e^+} \approx 10^{-18}$–$10^{-19}$ sec) would allow an experiment even if the supercritical system lives only for a comparable period of time. It has therefore been proposed that the collision of two extremely heavy nuclei, e.g. U and U, could be used to probe the charged vacuum (22–24). An estimate of the order of magnitude shows that this is indeed feasible: the typical collision time of two such nuclei at energies just below the Coulomb barrier is

$$\tau_{coll} \approx \frac{2R_{cr}}{v} \approx 0.25 \times 10^{-20} \text{ sec,} \qquad\qquad 3.1.$$

with $R_{cr} \approx 35$ fm (see below). The emission time for positrons is typically 100 times longer, so that one expects a yield of roughly 1% in this reaction. The theoretical treatment of the process is greatly facilitated by the great mass of the two nuclei: the Sommerfeld parameter $\eta = Z_1 Z_2 \alpha / v > 500$. Hence the classical approximation to the nuclear motion is very good, and only the electrons have to be treated quantum mechanically.

In the scattering system composed of two atoms, the electrons, at least the innermost ones, feel the attraction from both nuclei and form some kind of two-center states. The ratio between electronic and nuclear velocity, to speak in crude classical terms, is on the order of 10:1, which allows for the formation of quasi-stationary electronic states. Because of the similarity to stable or metastable molecules formed by valence-shell electronic binding, these binary systems are called quasi-molecules. The formation and existence of such quasi-molecules have been proven by the observation of X radiation from the transition between molecular states (25–30).

In this paragraph we neglect the dynamics of the collision process and calculate the electronic states as if the scattering system would live infinitely long. Due to the binding energies we expect, the proper wave equation to be solved is the two-center Dirac equation

$$[\mathscr{H}_{\text{TcD}} - E]\psi(\mathbf{x}) = \left[\boldsymbol{\alpha}\cdot\mathbf{p} + \beta m_e + V_1\left(\mathbf{x} - \frac{\mathbf{R}}{2}\right) + V_2\left(\mathbf{x} + \frac{\mathbf{R}}{2}\right) - E\right]\psi(\mathbf{x}) = 0.$$

3.2.

V_1 and V_2 are the potentials exerted by the two nuclei, respectively. To this date no calculation has been made that incorporates electron screening; therefore V_1 and V_2 are taken to be $1/r$ potentials outside the nuclei and, if necessary, quadratically cut off potentials inside the nuclear charge distribution.

Several attempts have been made to solve Equation 3.2 numerically. The variational method of Müller, Rafelski & Greiner (31, 32) and the numerical integration by Rafelski & Müller (33) determine the energy eigenvalues and the wave functions for every distance R, while the variational calculations of Popov et al (34–38) only determine the *critical distance* R_{cr}, at which the binding energy of the lowest molecular state equals $-2m_e$. We shall adopt here the treatment of (31) and (32) and later compare the numerical results with those of the other calculations.

Introducing prolate spheroidal coordinates ξ, η, and φ (39), where R is the distance between the two nuclei, one may separate off the φ coordinate. This is possible because $j_z = l_z + \frac{1}{2}\sigma_z$ commutes with the Hamiltonian (Equation 3.2) and thus represents a constant of motion that will henceforth be called μ. If the wave function is written in the form

$$\psi(\mathbf{x}) = \begin{pmatrix} 1 & 0 \\ 0 & i \end{pmatrix} \exp(i\varphi l_z)\psi'(\xi, \eta),$$

3.3.

a two-dimensional eigenvalue equation for the spinor ψ' is obtained, which may be written as a system of four coupled differential equations. They cannot be further separated nor solved analytically, in contrast to the nonrelativistic two-center equation. The eigenvalues and eigenfunctions where obtained by variation (diagonalization) within the basis

$$\psi_{nls}^{\mu}(\xi_1\eta) = (\xi^2 - 1)^{(\mu + \varepsilon_s)/2} \exp\left(-(\xi - 1)/2a\right) \cdot L_n^{(\mu + \varepsilon_s)}\left(\frac{\xi - 1}{a}\right) P_l^{(\mu + \varepsilon_s)}(\eta)\chi_s$$

3.4.

with $\varepsilon_s = \frac{1}{2}\sigma_z$ and a scaling factor a to approximate the asymptotic behavior $\xi \to \infty$. The L_n^{α} and P_l^{α} are the associated Laguerre and Legendre polynomials, respectively, and χ_s represents the four unit spinors. The basis functions (Equation 3.4) are not orthogonal, but they form a complete set if all values of n and l are allowed. For numerical computation, of course, the basis must be truncated, which was done in general at n, $l = 5$, keeping $5 \times 5 \times 4 = 100$ basis functions.

The eigenstates obtained in this manner are classified according to their quantum number $\mu = \pm\frac{1}{2}$, $\pm\frac{3}{2}$, etc. It can be shown that the states that differ only by the sign of μ are always degenerate (32). For symmetric systems (e.g. U-U) the parity is also a good quantum number. In addition to these exact symmetries the asymptotic quantum numbers n and j^2 are used from the limit $R \to 0$ (the so-called united atom limit). For example, the two lowest-energy molecular states are always the $1s_{1/2,1/2}(1s\sigma)$ and the $2p_{1/2,1/2}(2p_{1/2}\sigma)$ states, respectively. These will be the two states of main interest to us.

Figure 6 shows the complete diagram of the lowest 36 quasi-molecular levels in the symmetric system $_{92}U + _{92}U$, as obtained from the variational calculations of (32). At an internuclear separation R of about 5000 fm, the two $1s$ states of the two U atoms split noticeably, but both gain strongly in binding energy.

The binding energy of the $1s\sigma$ level increases until at $R_{cr} \approx 35$ fm, it equals $-2m_e$ and the quasi-molecule is rendered supercritical, in just the same way as the superheavy atom was at $Z > Z_{cr}$. For a distance $R = 0$ the quasi-molecule should approach the hypothetical atom $Z = 184$, but the nuclear reactions setting in for $R \lesssim 15$ fm do not allow a reliable description of the nuclear charge distribution in this region. The extension of the two nuclei becomes important at this point, where the two-center separation is comparable to the nuclear radius. However, the precise value of the critical radius is affected by less than 1.5 fm.

It is difficult to give an estimate of the precision of the results shown in Figure 6 for the following reason: the Hamiltonian of Equation 3.2 is not bounded below, because it allows for the negative-energy continuum states. Therefore the expectation value $\langle \psi_t | \mathcal{H}_{TcD} | \psi_t \rangle$ with a variational test wave function ψ_t does not provide an upper limit to the exact eigenvalue E. It is therefore hard to tell to what extent negative-energy scattering states are admixed to the best variational wave function.

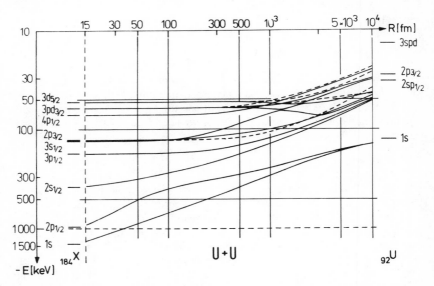

Figure 6 The stationary quasi-molecular states as they develop in a $U + U$ collision. R is the internuclear distance. The lowest state ($1s\sigma$) reaches the negative-energy continuum at $R_{cr} \approx 36$ fm.

The same statement applies to the variational determination of R_{cr} (i.e. the solution of Equation 3.2 for $E = -m_e$) by Popov et al (34–38). The problem, however, is circumvented in the approach of (33), where the eigenstate is obtained by direct numerical integration of the $2 \cdot (2j + 1)$ coupled equations after a multipole expansion up to angular momentum j. By explicitly forcing the wave function into an asymptotic bound-state form, it is orthogonalized against all scattering states. It should be mentioned that the results (especially on R_{cr} in $U + U$) agree very well with those of (31) and (32).

Let us return to the quasi-molecular diagrams of supercritical systems. Figure 7 compares the two lowest states ($1s\sigma$ and $2p_{1/2}\sigma$) for the three systems $_{92}U + _{92}U$, $_{92}U + _{98}Cf$, and $_{98}Cf + _{98}Cf$. Again it is seen that the nuclear extension is of minor importance for the critical radius R_{cr}, but it determines how far the level "dives" into the continuum. Figure 7 also shows that (without any strong compression of nuclear matter, such as shock waves) the $2p_{1/2}\sigma$ level in $U + U$ does not become supercritical, but approaches the threshold $-m_e$ up to only 250 keV. In the two heavier systems there is a critical radius for the $2p_{1/2}\sigma$ state. In Table 2 we have listed the numerical values for R_{cr} (32). Popov et al (34–38) also give values for $R_{cr}(1s\sigma)$, which for the $U + U$ system range from 34 fm up to 51 fm. Thus at present the values of Table 2 are favored, but the remaining uncertainties should be clarified by future work.

The energy function $E_{1s\sigma}(R)$ may be used to obtain cross sections for positron production in the so-called static approximation: the emission process is calculated in the Golden Rule approximation, i.e. one assumes that at every point R the system has an infinitely long time to emit. The time dependence is taken into account

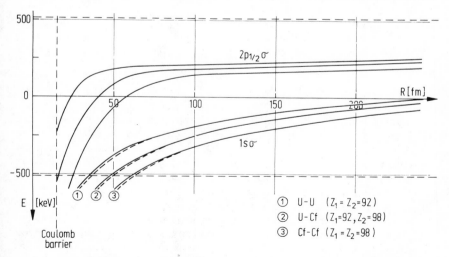

Figure 7   Only the $1s\sigma$ and $2p_{1/2}\sigma$ states are shown for three supercritical collision systems. Dashed lines indicate the effect if point nuclei are used in the calculation. On the heavier systems the $2p_{1/2}\sigma$ states also begin to dive before the Coulomb barrier is reached.

Table 2 The values of the critical distance in three supercritical quasi-molecular systems (in fm). The data of Müller, Rafelski & Greiner (32) are compared with those of Marinov, Popov & Stolin (37)

	$R_{cr}(1s\sigma)^a$	$R_{cr}(2p_{1/2}\sigma)^a$	$R_{cr}(1s\sigma)^b$
U + U	36	—	51
U + Cf	48	16	—
Cf + Cf	61	25	78

[a] From (32).
[b] From (37).

only through the relation $R(t)$. The probability for the escape of a positron with total energy E is (40)

$$p(R, E) = \frac{\Gamma(R)}{h} |a(E, R)|^2 \qquad\qquad 3.5.$$

with Γ and $a(E)$ from Section 2.3. Using $R(t)$ from classical Coulomb trajectories, one can integrate over time and obtain a positron distribution $W(E, \theta)$ depending on the scattering angle θ. The full cross section becomes

$$d\sigma_{pos} = L_0 W(E, \theta) dE \, d\sigma_{Ruth}. \qquad\qquad 3.6.$$

L_0 here denotes the probability of having a vacancy in the $1s\sigma$ level at the point where it becomes supercritical. The value of L_0 is unknown, but it was taken to be 0.01 in the calculations by Peitz et al (40), which are shown graphically in Figure 8. Similar calculations have been performed by Popov (41, 42), giving comparable results.

3.2   Dynamically Induced Positron Emission

The plots shown in Figure 8 exhibit very sharp peaks in the positron distribution. However, the uncertainty relation requires a large energy broadening due to the finite lifetime of the quasi-molecular system. In the preceding section this collision time was estimated to be on the order of 2.5×10^{-21} sec, which corresponds to an energy uncertainty of 400 keV or more. This inevitable broadening can be quantitatively understood as follows: the dynamics of the collision process cause the electric potential exerted by the two nuclei to be time-dependent. The electric field therefore contains rather large Fourier frequencies that can act as (longitudinal) quanta and add to the spontaneous energy of any emitted positron. The general formalism for a dynamical description of quasi-molecular collision processes has been given by Smith, Müller & Greiner (43). We give a brief outline of this theory.

When the nuclear motion is treated classically, the dynamical wave equation for the electrons is

$$\mathcal{H}_{\text{TcD}}[\mathbf{R}(t)]\,\psi(\mathbf{r},t) = i\frac{\partial}{\partial t}\,\psi(\mathbf{r},t). \tag{3.7}$$

The two-center Hamiltonian $\mathcal{H}_{\text{TcD}}$ from Equation 3.2 depends on the vector $\mathbf{R}$ between the two nuclei, but the angular dependence corresponds to a trivial rotation of the coordinate system. It is therefore possible to eliminate the (nuclear) angles by introducing an intrinsic wave function $\psi'(\mathbf{r},t)$:

$$\psi(\mathbf{r},t) = \exp(-i\int dt\,\mathbf{\Omega}\cdot\mathbf{j})\psi'(\mathbf{r},t). \tag{3.8}$$

The Dirac equation (3.7) in the intrinsic co-rotating system becomes

$$\{\mathcal{H}_{\text{TcD}}[\mathbf{R}(t)] - \mathbf{\Omega}\cdot\mathbf{j}\}\,\psi'(\mathbf{r},t) = i\frac{\partial}{\partial t}\,\psi'(\mathbf{r},t), \tag{3.9}$$

where $\mathcal{H}$ and ψ depend only on the scalar R. $\mathbf{\Omega}(t)$ denotes the angular velocity of the nuclear axis and $\mathbf{j}$ the total electronic angular momentum.

The solution of Equation 3.9 is achieved in the stationary two-center Dirac equation (3.2):

$$\psi'(\mathbf{r},t) = \sum_n a_n(t)\,\varphi_n[\mathbf{r},R(t)]\exp(-i\int E_n\,dt). \tag{3.10}$$

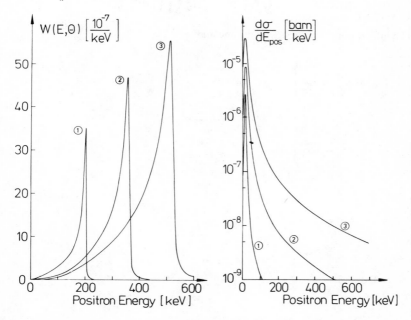

Figure 8 Spontaneous positron emission in 1) U−U, 2) U−Cf, and 3) Cf−Cf collisions. The scattering energy is always at the Coulomb barrier. The Golden Rule approximation was made to describe the transition. (*a*) Emission probability W as a function of positron energy; (*b*) the differential cross section versus positron energy ($L_0 = 0.01$).

Upon substitution of this ansatz one is left with an infinite system of coupled channel equations

$$\dot{a}_n(t) = - \sum_{m \neq n} a_m(t) \exp\left[i \int dt'(E_n - E_m)\right] \cdot \left\langle \psi_n \left| \dot{R} \frac{\partial}{\partial R} + iV(t) \right| \psi_m \right\rangle. \qquad 3.11.$$

This system cannot be solved explicitly in general, but for the calculation of the positron spectrum, first-order perturbation theory is sufficient (remember the estimate that only about 1% of the K vacancies participate in the reaction). Therefore we write (15) $a_m = a_0 \delta_{m1s}$, where $|a_0(t = -\infty)|^2 = L_0$ is the K-vacancy probability before the close encounter between the two atoms. In addition, we assume that $a_0(t)$ is essentially time-independent, and varies only because of positron decay:

$$a_0(t) = \exp\left[-\frac{1}{2} \int_{-\infty}^{t} dt' \Gamma(t') \right]. \qquad 3.12.$$

Then we can solve Equation 3.14 and obtain the amplitudes for emission of a positron with (total) energy E:

$$a_E(t) = -i \int_{-\infty}^{t} dt' \, \tilde{V}_E(t') \exp\left[i \int dt''(E - E_{1s\sigma}) - \tfrac{1}{2} \int dt'' \Gamma \right]. \qquad 3.13.$$

The integration now goes over all time and

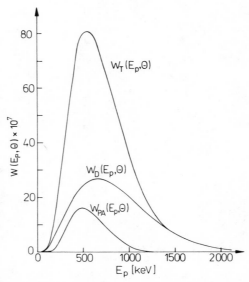

Figure 9 Fully dynamical calculation of the positron emission. The emission probability W in the supercritical region of the trajectory (D), before and after diving (PA), and the total (T). The plot is for $U + U$ at 812-MeV center-of-mass energy.

$$\tilde{V}_E(t) = \langle E, R_{cr} | H(R) - H(R_{cr}) | 1s\sigma, R_{cr} \rangle \qquad (R < R_{cr})$$

$$\tilde{V}_E(t) = -i\dot{R}\langle E, R \left| \frac{\partial}{\partial R} \right| 1s\sigma, R \rangle \qquad (R > R_{cr})$$

3.14.

The probability of producing a positron with energy between E_p and $E_p + dE_p$ is

$$W(E_p, \theta, E_{ion}) dE_p = |a_E(t \to \infty)|^2 dE_p,$$

3.15.

and depends on the scattering energy E_{ion}, which determines the amount of energy broadening. In order to obtain a cross section, again Equation 3.6 must be applied, with the assumption that $L_0 = 0.01$.

Before we discuss the validity of some of our approximations, let us have a glance at the results (15). Figure 9 shows the emission probability as a function of positron energy E_p in the collision U + U. The emission probability W_D during the "diving" (the spontaneous contribution) and the probability W_{PA} before and after the stage of supercriticality (the induced contribution) are listed separately. The two contributions add up coherently and form a much larger W_T. Figure 10 displays

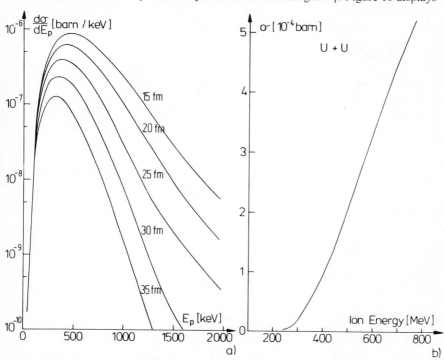

Figure 10 Fully dynamical positron emission cross sections for U + U. (*a*) The differential cross section versus positron energy at five different scattering energies (given by the distance of closest approach). (*b*) The total cross section as a function of center-of-mass bombarding energy ($L_0 = 0.01$).

the corresponding cross sections, $d\sigma/dE_p$ in (a) and σ in (b). It is evident that at high kinetic energies E_p, the cross sections fall off exponentially, which is typical for the high-frequency behavior of the Fourier transform of the potential. The total cross section begins to be measurable at an incident ion energy of $E = 300$ MeV, which corresponds to a distance of closest approach $E_{min} = 40$ fm, i.e. roughly the critical radius. Due to the collision dynamics, a vacancy in the $2p_{1/2}\sigma$ level in U + U (which does not become supercritical) can also be emitted as a positron. It has been calculated, however, that the cross section is smaller by a factor of 20 than the emission from the $1s\sigma$ state (H. Peitz, to be published), so that the contributions from even higher states can be expected to be negligible. With the amount of supercritical charge in the system, $(Z_1 + Z_2 - Z_{cr})$, the cross section increases rapidly. This is seen from Figure 11 in the comparison of the three systems U + U, U + Cf, and Cf + Cf, for which the molecular diagrams were shown in Figure 7.

Figure 12 will become of interest when we discuss background effects in Section 4. The differential positron cross section is plotted as a function of impact parameter b for a fixed positron energy $E_p = 1$ MeV. The maximum yield is encountered at $b = 10$ fm, corresponding to $90°$ collisions.

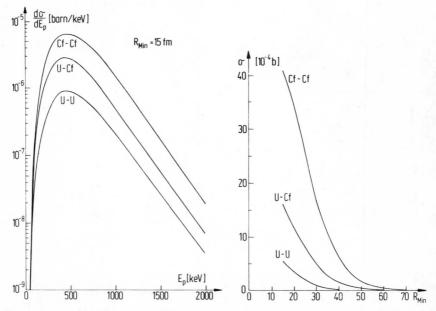

Figure 11   Three supercritical collision systems, U + U, U + Cf, and Cf + Cf, in comparison. The calculations are fully dynamical. (a) Differential cross section versus positron energy E_p for scattering at 15 fm closest approach. (b) Total cross section versus distance of closest approach ($L_0 = 0.01$).

The main assumption that enters into the calculations of Figures 9–12 is the magnitude and form of the $1s\sigma$-vacancy amplitude $a_0(t)$. Its square L_0 was taken to be 0.01. Our use of such an ad hoc assumption is justified only by our complete ignorance of its true value. The standard methods for calculating K-vacancy production cross sections in heavy-ion collisions (44, 45) are inapplicable in the range of Z envisaged here. Estimates for L_0 scatter widely, from the heuristic extrapolations of Meyerhof (46) and Saris et al (47) obtained from relatively light-ion scattering experiments ($L_0 \sim 10^{-5}$) to the predictions of Burch et al (48) in the framework of the binary encounter approximation (BEA) that L_0 could be as high as 0.1.

Very recently Betz, Soff, Müller & Greiner (49) have obtained the first quantitative results from a full calculation of Coulomb ionization in the context of the quasi-molecular model. Excitation of $1s\sigma$ electrons into the continuum (ionization) as well as into high-lying bound states were taken into account. In the positron emission region ($R < 35$ fm), only excitation into the continuum is of importance. At 1600-MeV scattering energy and zero impact parameter, the average ionization rate over this region is 0.08, i.e. larger than assumed in the calculations discussed above. However, this means not only a change in the absolute magnitude of L_0, because the ionization amplitude $a_0(t)$ turns out to be a rapidly varying function of time. This may drastically change the energy spectrum of the positrons. Moreover, the amplitude a_0 depends quite strongly on the impact parameter, with a characteristic falloff at $b = 20$ fm. This again will influence the energy spectrum and the total cross section. The results of the new calculations that are presently being carried out can therefore be expected with interest.

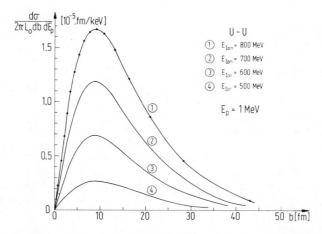

Figure 12 Positron emission cross section versus projectile ion impact parameter for various bombarding energies. The positron window is at $E_p = 1$ MeV. The thick points give every ten degrees of scattering angle, the peak being at 90°.

4   BACKGROUND EFFECTS

In this section we discuss some of the effects that may—or might be thought to— prevent the observation of the spontaneous positron production in a supercritical quasi-molecule. First, we discuss the virtual-vacuum polarization effect on the binding energy of the supercritical bound state. It will be seen that it is only a minor contribution to the total binding energy and does not prevent the state from "diving." Then we turn to the background production of electron-positron pairs from the conversion of excited nuclear states and from direct pair creation out of the time-varying fields.

4.1   Virtual Vacuum Polarization

It is well known that every external charge induces a nonvanishing charge distribution (see Equation 2.40) in the vacuum, which is commonly called vacuum polarization. As long as the system is subcritical ($Z < Z_{cr}$ or $R > R_{cr}$), the total charge contained in this polarization cloud is zero. The dimension of the polarization charge distribution is typically of the order of the electron Compton wavelength $1/m_e \approx 386$ fm. It is obvious that the K electrons, which are localized in this region, must be strongly influenced by the vacuum polarization (VP).

When the VP is calculated in perturbation theory, the dominant term is linear in $(Z\alpha)$. This Uehling potential (50) can be written as a convolution integral:

$$V_{Ueh} = -\frac{2\alpha^2}{3m_e} \int_0^\infty dr' \rho_N(r')[\xi(|r-r'|) - \xi(|r+r'|)], \qquad 4.1.$$

with the structure function

$$\xi(x) = \int_1^\infty \exp(-2m_e xt)\left(1 + \frac{1}{2t^2}\right)(t^2 - 1)^{1/2} \frac{dt}{t^3}. \qquad 4.2.$$

The nuclear charge density $\rho_N(r')$ is normalized to Z. Soff, Müller & Rafelski (8) find that the energy shift of the $1s$ state at $Z = Z_{cr}$ due to the Uehling potential is $\Delta E = -11.834$ keV, i.e. it favors the diving of the $1s$ state.

The higher-order (in $Z\alpha$) VP contributions have been evaluated up to the critical point by Gyulassy (51), and by Rinker & Wilets (52). Both use numerical techniques to evaluate the induced charge density (Equation 2.40). Gyulassy expresses $\rho(x)$ in terms of the exact Green's function in an external Coulomb potential. His result for the additional energy shift due to the higher orders in $(Z\alpha)$ is for $Z = Z_{cr}$: $\Delta E \sim +1.15$ keV. This has the opposite sign of the Uehling correction, but is smaller by an order of magnitude. Rinker & Wilets evaluate Equation 2.40 directly after subtraction of the Uehling term. In the monopole approximation they find that in the U + U system at $R = R_{cr}$ the total VP effect is $\Delta E_{R_{cr}} = -3.98$ keV on the $1s\sigma$ quasi-molecular state.

We conclude that virtual VP does not seriously affect the nature or the precise value of the critical point. The self-energy correction to the bound-state energy has been evaluated by Cheng & Johnson (to be published) with the method of (53). They obtain $\Delta E_{se}(160) = 7.36$ keV, extrapolating to $\Delta E_{se} \approx 12.3$ keV at $Z = 172$.

4.2 Pair Conversion of Nuclear Coulomb Excitations

In order not to encounter various nuclear reactions, the scattering experiment proposed in Section 3 should be carried out at ion energies below the Coulomb barrier. Unfortunately, even at these energies the Coulomb forces between the two nuclei are able to excite collective nuclear (mostly quadrupole) states, a process known as Coulomb excitation (54). In such heavy-collision systems a U + U nuclear state of extremely high energy and angular momentum can be excited sometimes beyond the fission threshold (55). Since many of the states involved are not known from experiment, the process was calculated in the framework of the rotation-vibrational model (56) by Oberacker, Soff & Greiner (57). The calculated excitation of a certain state can be very model-dependent, but one should expect that the overall excitation function is reasonably well represented. The total excitation is on the order of several barns.

Because of the relatively long lifetime of collective nuclear states ($\sim 10^{-12}$–10^{-10} sec), their decay will not take place during the collision. On the other hand, the lifetime is too short to allow experimental separation of their decay products from those produced during the collision. When the excitation energy of a nuclear state is larger than $2m_e = 1022$ keV, it has the possibility to decay via emission of an electron-positron pair. The branching ratio β for electric quadrupole transitions is given in the Born approximation by (58)

$$\frac{d\beta}{dE_p} = \frac{\alpha}{\pi\omega^5}\left\{\omega^2[E_p^2+E_e^2-2m_e^2]\ln\left[\frac{m_e^2+E_pE_e+P_pP_e}{\omega}+\frac{8}{3}P_pP_e(E_eE_p-m_e^2)\right]\right\}.$$

4.3.

Here the indices p and e denote "positron" and "electron," respectively, and ω is the nuclear transition energy. β is typically of the order of 10^{-3}, but is somewhat reduced by Coulomb corrections, especially for small positron momentum P_p.

The total pair-production cross section in a 800-MeV (center-of-mass energy) U + U collision has been calculated (53) to be 1.25×10^{-4} barns and is maximal at backward scattering. If folded with the Rutherford cross section the peak moves to smaller angles, but stays at larger angles than the peak of the positron distribution from the decaying vacuum. The full differential cross section for the two positron-production mechanisms is plotted in Figure 13. At a $30°$ scattering angle, the vacuum positrons are dominant by more than an order of magnitude. Of course, the factor L_0 (again 0.01 was used) causes a large uncertainty for this process. It is safe to say that $L_0 \approx 10^{-3}$ is the critical threshold above which experimental verification of the positron autoionization mechanism seems feasible in a U + U collision.

4.3 Pair Production by Direct Processes

Figure 14 displays the diagrams of four mechanisms to produce positrons in a heavy-ion collision. Part (a) represents the spontaneous decay of a K vacancy as

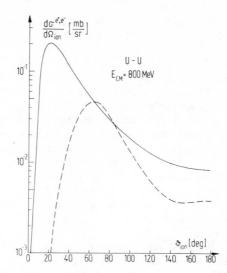

Figure 13 Differential positron-production cross sections versus ion scattering angle for pair conversion of excited nuclear states (*dashed line*) and decay of the vacuum (*solid line*). Again $L_0 = 0.01$ was used. The cross section has not been symmetrized around 90°, as would be necessary for a symmetric collision.

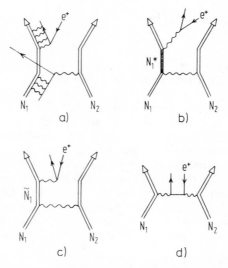

Figure 14 Diagrams of four positron-production mechanisms in a heavy ion collision. (a) Decay of a $1s\sigma$ vacancy; (b) pair conversion of excited nuclear states; (c) pair conversion of bremsstrahlung photons; and (d) direct pair production. Processes a and b are dominant.

discussed in Sections 2 and 3, and (b) shows the pair creation by an excited nucleus N_1^*. The two additional mechanisms (c and d) are called direct because they do not require any internal structure of the scattering atoms and nuclei. Process c, where one of the nuclei ($\tilde{N}_1$) is scattered off the mass shell and returns by pair creation, is equivalent to the pair conversion of a bremsstrahlung photon. Although this diagram is of the same order as diagram (b), its cross section is much smaller: Reinhardt, Soff & Greiner (59) have calculated that its contribution in an 800-MeV U + U collision is approximately 6×10^{-8} barns. It is therefore no major experimental background.

The fourth diagram (d), a type of "trident" graph, is the familiar pair creation in a Coulomb scattering process. It is presently calculated for the first time for non-relativistic scattering energies (J. Reinhardt, G. Soff, to be published).

5   SOME IMPLICATIONS OF THE NEW GROUND STATE

In this final section we return to the discussion of the new charged vacuum state that was introduced as the consequence of the supercritical "diving into the Dirac sea" of a bound state. We have already seen that the vacuum charge is two, later four, six, etc. This successive change is once more illustrated in Figure 15. It becomes obvious that at the various critical points the vacuum state undergoes a

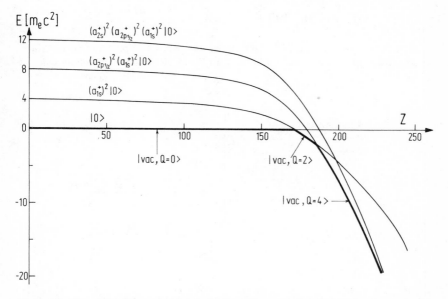

Figure 15   The energy of various atomic configurations as a function of atomic number Z. The vacuum (*solid line*), defined as ground-state configuration, changes abruptly at $Z_{cr} = 172$ and again at $Z_{cr}(2p) = 185$. The change may be viewed as a phase transition of the electron-positron field in which charge symmetry is dynamically broken.

phase transition, whereby each consecutive phase is charged higher. As with charge, the other physical properties of the vacuum state can no longer be renormalized to zero. Take, for example, the energy: because the resonance location changes as a function of Z (or R in the two-center case), one expects the vacuum energy to change in the same way.

5.1   Density of States in the Vacuum

In order to find a unique description to separate the finite contribution from the infinite regularization of vacuum quantities, one has to define the density of scattering states. If zero flux through a (e.g. spherical) boundary at $r = R$ is required, then the artificially discretized continuum states have momentum p with

$$\sin\left[pR + \Delta(p)\right] = 0, \qquad \text{i.e. } pR + \Delta(p) = n\pi. \qquad 5.1.$$

$\Delta(p)$ is the phase shift, which is assumed to be asymptotically independent of R. The density can be defined by

$$\frac{dn}{dp} = \frac{R}{\pi} + \frac{1}{\pi}\frac{d\Delta}{dp}, \qquad \frac{dn}{dE} = \frac{RE}{\pi p} + \frac{1}{\pi}\frac{d\Delta}{dE}. \qquad 5.2.$$

We are now able to define a global quantity A (charge, energy, etc) of the vacuum state by

$$\langle A \rangle_0 = \frac{1}{2\pi}\sum_{\text{spin}}\int_{-\infty}^{-m_e} dE\left[A(E)\frac{d\Delta^\circ(E)}{dE} - A(-E)\frac{d\Delta^\circ(-E)}{dE}\right] - \frac{1}{2}\sum_{\text{b.st.}} A_i. \qquad 5.3.$$

We have taken into account the sign difference between occupied and vacant states, and have already discarded the infinite contribution $RE/\pi p$ to the vacuum expecta-

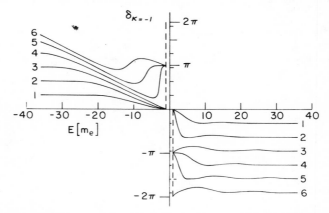

Figure 16   s-phase shifts due to square-well potentials of various strengths. The well depth of the nth line is $n \cdot 5m_e$. The phase jump between lines 2 and 3 at $E = m_e$ signals the appearance of a bound s state; the jump between 3 and 4 at $E = -m_e$ indicates that the bound state has joined the negative-energy continuum. The vacuum of situations 4 through 6 is charged.

tion value which is independent of the potential. The integral in Equation 5.3 is still in general divergent; therefore we have written Δ° instead of Δ. It can be shown that the proper way to regularize $\Delta(E)$ is to subtract successive orders of Born-approximation phase shifts until the integral converges: $\Delta^{\circ} = \Delta - \Sigma_n^N \Delta_n^{Born}$ [Puff (60), B. Müller, to be published]. For the subcritical vacuum all expectation values (Equation 5.3) vanish, but for the supercritical vacuum state they provide a description of its physical properties. As an example we show Figure 16, which shows the relativistic phase of electrons in a square-well potential. When one bound state becomes supercritical (curve 4), a phase jump occurs leading to a nonzero charge expectation value

$$\langle Q \rangle_0 = \frac{1}{2\pi} \sum_{spin} [\pi - (-\pi)] = 2. \qquad 5.4.$$

The first-order Born phase shift must be subtracted to ensure convergence.

5.2   A Statistical Model of the Supercritical State

On theoretical grounds, but also because of speculations about superdense nuclear matter and abnormal states [Bodmer (61), Lee & Wick (62)], one would like to describe the supercritical state for very large nuclear charges Z. This implies also that the ground state is charged many times and screening due to the real vacuum-polarization charge becomes an important effect. Müller & Rafelski (63) have generalized the Thomas-Fermi approximation to include vacuum effects. The charge density is expressed in terms of the generalized Fermi momentum (see Section 2.4) P_F:

$$\rho_{el} = \frac{e}{3\pi^2} P_F^3. \qquad 5.5.$$

With the relativistic energy-momentum relation, P_F is replaced by the Fermi energy E_F, and upon substitution into Coulomb's law one obtains (63):

$$\nabla^2 V(\mathbf{r}) = e\rho(\mathbf{r}) = e\rho_N(\mathbf{r}) - \frac{4\alpha}{3\pi} [(E_F - V)^2 - m_e^2]^{3/2} \theta(E_F - V - m_e). \qquad 5.6.$$

$\rho_N(\mathbf{r})$ is the (external) nuclear charge distribution, and E_F is chosen as $-m_e$ (see below). Equation 5.6 must be solved self-consistently under the boundary condition $\lim_{r \to \infty} [rV(\mathbf{r})] = \gamma$. The eigenvalue γ is interpreted as the asymptotically observable charge of the system: $\gamma = Z - \langle Q \rangle_0$. The variation of γ with Z, as shown in Figure 17, leads to an almost complete saturation of the external charge: asymptotically, $\gamma \approx \ln Z$.

This puts an upper limit to the electrostatic potential and, more importantly (63), limits the Coulomb energy necessary to add another proton to the nucleus to be smaller than

$$\Delta E = 93 \text{ MeV} \cdot \left(\frac{\rho}{\rho_0}\right)^{-1/3} \qquad 5.7.$$

The Coulomb energy therefore would not be able to stop the growth of an abnormal nuclear state. The total energy of the state (including the emitted positrons),

$$\mathscr{H}[E_F] = 2 \int d^3x \int_0^{P_F(x)} \frac{d^3p}{(2\pi)^3} [(p^2 + m_e^2)^{1/2} - E_F$$

$$+ (E_F + m_e)\theta(m_e + E_F)] + \tfrac{1}{2} \int d^3x \, V(x)\rho(x),$$ 5.8.

can be used as a Hamiltonian to deduce Equation 5.6. Upon variation with respect to the Fermi energy E_F, one obtains an energy functional $\mathscr{H}(E_F)$, which is shown in Figure 18. The energy clearly exhibits a minimum at $E_F = -m_e$, which is therefore the stable configuration (J. Rafelski, B. Müller, to be published). This result provides the ultimate justification for the choice $E_F = -m_e$ to define the ground state introduced in Section 2.4.

6   CONCLUSION AND OUTLOOK

When the binding energy of an electron in a stable or quasi-stable atomic system exceeds $2m_e$, a new situation arises. No longer is the vacant bound state a stable configuration, and it decays via emission of a positron into the configuration in which the bound state is occupied. This corresponds to a qualitatively new process: the decay of the old, neutral vacuum into the supercritical charged vacuum. The process becomes accessible to experimental verification, since in a $U - U$ collision the supercritical region is reached at an internuclear separation of 36 fm. The

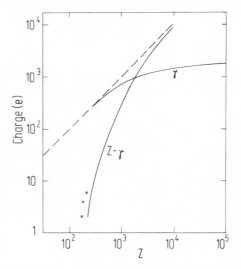

Figure 17   Unscreened charge γ and charge of the vacuum $(Z - \gamma)$ as a function of the bare nuclear charge Z. For very large Z (collapsed nuclei, neutron stars) the screening becomes almost complete, leading to spontaneous overall neutralization.

feasibility of such an experiment depends critically on the K-vacancy production rate in such collisions, which we do not know at all, and it seems to be an important task to investigate this problem theoretically and experimentally. If, after all, the experiment would turn out to be unfeasible in this way, one should think about methods to produce completely stripped, very heavy-ion beams (64).

It may be asked whether similar effects could be seen in exotic atoms. Unfortunately, the radius of a bound state is inversely proportional to the rest mass of the particle, so that even muonic $1s$ states in a superheavy atom would lie entirely within the nucleus. It has been calculated (8) that the critical nuclear charge for muons is approximately $Z_{cr}(\mu^-) = 2200$, out of reach for any laboratory experiment.

However, the supercritical phenomenon is essentially caused by the strong coupling ($Z\alpha > 1$) between the superheavy nucleus (or the two nuclei) and the electron. This case of strong electromagnetic coupling is of particular interest because the interaction is known precisely. We are therefore in a situation to make a reliable theoretical model that may be tested by experiment. However, it seems natural to expect many similar situations in strong-interaction physics where the coupling constant is much larger than one. Due to the formidable problems of field theory with strong coupling, the investigation of supercritical phenomena in hadron physics is just beginning. However, if we can hope at all to solve problems like the confinement of quarks, which are presumably the constituents of hadronic matter, then we probably have to deal with supercritical situations. To this end the ideas put forward in this paper must be generalized to cope with a situation where bosons are also present (65, 66), and to include many-particle phenomena such as bose condensation. Also, if there is no external source that creates a supercritical field but only the interacting particles themselves, it is necessary to include the correlation energy into the model (B. Müller, J. Rafelski, to be published). This will produce a different behavior of the supercritical fermion state which, however, can only be understood on the basis of the theory of supercritical atomic levels.

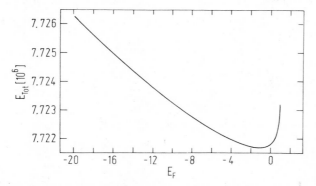

Figure 18 Total energy of a hypothetical nucleus with $Z = 10,000$ versus the generalized Fermi surface E_F. Clearly, the choice $E_F = -m_e$ corresponds to the stable, but charged vacuum.

ACKNOWLEDGMENTS

This work has been financially supported by the US Energy Research and Development Administration, the Bundesministerium für Forschung und Technologie, and the Deutsche Forschungsgemeinschaft. I should like to thank W. Greiner, W. Meyerhof, H. Peitz, J. Rafelski, G. Soff, and L. Wilets for valuable discussions; and Mrs. Knolle and Mrs. Utschig for help in the preparation of the manuscript

Literature Cited

1. Werner, F. G., Wheeler, J. A. 1958. *Phys. Rev.* 109:126–44; Mosel, U., Greiner, W. 1969. *Z. Phys.* 222:261–82; Nilsson, S. G., Thompson, S. G., Tsang, C. F. 1969. *Phys. Lett. B* 28:458–61
2. Case, K. M. 1950. *Phys. Rev.* 80:797–806
3a. Pomeranchuk, I., Smorodinskii, Ya. 1945. *J. Phys. USSR* 9:97
3b. Voronkov, V. V., Kolesnikov, N. N. 1960. *Zh. Eksp. Teor. Fiz.* 39:189–91; 1961. *Sov. Phys. JETP* 12:136–37
4. Pieper, W., Greiner, W. 1969. *Z. Phys.* 218:327–40
5. Rein, D. 1969. *Z. Phys.* 221:423–30
6. Fricke, B., Greiner, W. 1969. *Phys. Lett. B* 30:317–19
7. Fricke, B., Soff, G. 1974. *Dirac-Fock-Slater calculations of the elements fermium (Z = 100) to Z = 173,* Darmstadt: GSI-Bericht T1-74
8. Soff, G., Müller, B., Rafelski, J. 1974. *Z. Naturforsch. Teil A* 29:1267–75
9. Popov, V. S. 1972. *Yad. Fiz.* 15:1069–81; *Sov. J. Nucl. Phys.* 15:595–600
10. Rose, E. M. 1961. *Relativistic Electron Theory,* p. 302. New York: Wiley
11. Müller, B., Rafelski, J., Greiner, W. 1972. *Z. Phys.* 257:183–211
12. Müller, B., Rafelski, J., Greiner, W. 1973. *Nuovo Cimento A* 18:551–73
13. Müller, B., Peitz, H., Rafelski, J., Greiner, W. 1972. *Phys. Rev. Lett.* 28:1235–38
14. Zeldovich, Ya.B., Popov, V. S. 1971. *Usp. Fiz. Nauk.* 105:403–40; 1972. *Sov. Phys. Usp.* 14:673–94
15. Smith, R. K., Peitz, H., Müller, B., Greiner, W. 1974. *Phys. Rev. Lett.* 32:554–56
16. Fano, U. 1961. *Phys. Rev.* 124:1866–78
17. Müller, B., Rafelski, J., Greiner, W. 1972. *Z. Phys.* 257:62–77
18. Rafelski, J., Müller, B., Greiner, W. 1974. *Nucl. Phys. B* 68:585–604
19. Jackson, J. D. 1962. *Classical Electrodynamics,* p. 641. New York: Wiley
20. Fulcher, L. P., Klein, A. 1973. *Phys. Rev. D* 8:2455–57
21. Anholt, R., Rasmussen, J. O. 1974. *Phys.* *Rev. A* 9:585–92
22. Gershtein, S. S., Zeldovich, Ya.B., 1969. *Zh. Eksp. Teor. Fiz.* 57:654–59; 1970. *Sov. Phys. JETP* 30:358–61
23. Popov, V. S. 1971. *Yad. Fiz.* 14:458–68; 1972. *Sov. J. Nucl. Phys.* 14:257–62
24. Rafelski, J., Fulcher, L. P., Greiner, W. 1971. *Phys. Rev. Lett.* 27:958–61
25. Saris, F. W., van der Weg, W. F., Tawara, H., Laubert, R. 1972. *Phys. Rev. Lett.* 28:717–20
26. Mokler, P. H., Stein, H. J., Armbruster, P. 1972. *Phys. Rev. Lett.* 29:827–30
27. Macdonald, J. R., Brown, M. D., Chiao, T. 1973. *Phys. Rev. Lett.* 30:471–74
28. Meyerhof, W. E., Saylor, T. K., Lazarus, S. M., Little, W. A., Triplett, B. B. 1973. *Phys. Rev. Lett.* 30:1279–82
29. Gippner, P., Kaun, K.-H., Stary, F., Schulze, W., Tretyakov, Yu.P. 1974. *Nucl. Phys. A* 230:509–14
30. Davis, C. K., Greenberg, J. S. 1974. *Phys. Rev. Lett.* 32:1215–18
31. Müller, B., Rafelski, J., Greiner, W. 1973. *Phys. Lett. B* 47:5–7
32. Müller, B., Greiner, W. 1976. *Z. Naturforsch. Teil A* 31:1–30
33. Rafelski, J., Müller, B. 1975. *Phys. Rev. Lett.* 36:517–20
34. Popov, V. S., Rozhdestvenskaya, T. I. 1971. *Zh. Eksp. Teor. Fiz. Pisma Red.* 14:267–70
35. Popov, V. S. 1972. *Zh. Eksp. Teor. Fiz. Pisma Red.* 16:355–58; *J. Exp. Theor. Phys. Lett.* 16:251–54
36. Popov, V. S. 1973. *Yad. Fiz.* 17:621–33; *Sov. J. Nucl. Phys.* 17:322–28
37. Marinov, M. S., Popov, V. S., Stolin, V. L. 1973. *Zh. Eksp. Teor. Fiz. Pisma Red* 19:76–80; 1974. *J. Exp. Theor. Phys. Lett.* 19:49–51
38. Marinov, M. S., Popov, V. S. 1975. *Zh. Eksp. Teor. Fiz.* 68:421–31; 1975. *Sov. Phys. JETP* 41:205–9
39. Morse, P. H., Feshbach, H. 1953. *Methods of Theoretical Physics,* Vol. 2. New York: McGraw-Hill
40. Peitz, H., Müller, B., Rafelski, J., Greiner, W. 1973. *Lett. Nuovo Cimento* 8:37–42

41. Popov, V. S. 1973. *Zh. Eksp. Teor. Fiz.* 65:35–53; 1974. *Sov. Phys. JETP* 38:18–26

42. Popov, V. S. 1974. *Yad. Fiz.* 19:155–68; *Sov. J. Nucl. Phys.* 19:81–87; *Yad. Fiz.* 20:1223–28; 1975. *Sov. J. Nucl. Phys.* 20:641–43

43. Smith, R. K., Müller, B., Greiner, W. 1975. *J. Phys. B* 8:75–101

44. Ogurtsov, G. N. 1972. *Rev. Mod. Phys.* 44:1–17

45. Garcia, J. D., Fortner, R. J., Kavanagh, T. M. 1973. *Rev. Mod. Phys.* 45:111–77

46. Meyerhof, W. E. 1974. *Phys. Rev. A* 10:1005–7

47. Foster, C., Hoogkamer, T., Woerlee, P., Saris, F. W. 1975. *Proc. ICPEAC Conf., 9th, Seattle, July 1975.* 1:511–12

48. Burch, D., Ingalls, W. P., Wieman, H., Vandenbosch, R. 1974. *Phys. Rev. A* 10:1245–54

49. Betz, W., Soff, G., Müller, B., Greiner, W. 1976. Direct formation of quasi-molecular $1s\sigma$ vacancies in uranium-uranium collisions. Frankfurt Univ. prepr.

50. Uehling, E. A. 1935. *Phys. Rev.* 48:55–63

51. Gyulassy, M. 1974. *Phys. Rev. Lett.* 33:921–25

52. Rinker, G. A. Jr., Wilets, L. 1975. *Phys. Rev. A* 12:748–62

53. Desiderio, A. M., Johnson, W. R. 1971. *Phys. Rev. A* 3:1267–75

54. Alder, K., Bohr, A., Huus, T., Mottelson, B., Winther, A. 1956. *Rev. Mod. Phys.* 28:432–542

55. Holm, H., Greiner, W. 1972. *Nucl. Phys. A* 195:333–52

56. Faessler, A., Greiner, W., Sheline, R. K. 1965. *Nucl. Phys.* 70:33–88

57. Oberacker, V., Soff, G., Greiner, W. 1976. *Nucl. Phys. A* 259:324–42; *Phys. Rev. Lett.* 36:1024–27

58. Berestetsky, W., Shmushkevich, I. M. 1949. *Zh. Eksp. Teor. Fiz.* 19:591

59. Reinhardt, J., Soff, G., Greiner, W. 1976. *Z. Phys. A* 276:285–93

60. Puff, R. 1975. *Phys. Rev. A* 11:154–59

61. Bodmer, A. 1971. *Phys. Rev. D* 4:1601–6

62. Lee, T. D., Wick, G. C. 1974. *Phys. Rev. D* 9:2291–316

63. Müller, B., Rafelski, J. 1975. *Phys. Rev. Lett.* 34:349–52

64. Betz, W., Heiligenthal, G., Reinhardt, J., Smith, R. K., Greiner, W. 1975. *Proc. ICPEAC Conf., 9th, Seattle, July 1975,* ed. J. Risley, R. Geballe, p. 29. Seattle: Univ. Wash. Press

65. Migdal, A. B. 1972. *Zh. Eksp. Teor. Fiz.* 61:2209–24; *Sov. Phys. JETP* 34:1184–91

66. Klein, A., Rafelski, J. 1975. *Phys. Rev. D* 11:300–11; 12:1194–95

Ann. Rev. Nucl. Sci. 1976. 26 : 385–456

DIFFRACTION OF HADRONIC WAVES

✷5578

U. Amaldi, M. Jacob, and G. Matthiae

European Organization for Nuclear Research (CERN), Geneva, Switzerland

CONTENTS

1 INTRODUCTION

Most of the known fundamental particles are hadrons, particles with strong interactions, and evidence for their composite nature is now overwhelming. Indeed, hadrons are extended in space with typical dimensions of the order of 1 fm. This is the range of strong interactions. In a collision the center-of-mass (c.m.) energy of two hadrons is thus concentrated, for a time of the order of 10^{-23} sec, in a small volume whose transverse dimensions are of the order of 1 fm. Strong interactions are such that in collisions at c.m. energies much larger than the proton rest mass, there are many possible final states that in general contain many newly produced hadrons. Since many inelastic interactions can occur, a non-negligible fraction of the wave function describing the two initial hadrons is absorbed as the collision takes place. The diffraction phenomena discussed in this paper are to a large extent simple consequences of the absorption caused by the open inelastic channels. They are characterized by the dimensions of the region of space in which absorption takes place (1 fm). They are, however, more complicated than simple elastic diffraction because hadrons apparently have internal degrees of freedom that may be excited, a strong evidence for their composite nature.

In reviewing this subject we have in mind the nonspecialist. Thus we build up our arguments starting in Section 2 from elementary considerations of optical diffraction. In Section 3 we consider hadron-nucleus collisions. This we find is the best introduction to hadron-hadron phenomenology, since nuclei have well-defined dimensions. Within the nucleus the incoming hadron wave is absorbed quite independently of the details of the hadron-nucleon dynamics and the observed scattering depends essentially upon the dimensions of the nucleus. Sections 4 and 5 are then devoted to describing elastic and inelastic hadron-hadron diffraction phenomena. A large sample of experimental data is presented and discussed using both the optical-geometrical point of view developed in Sections 3 and 4 (often referred to as the s-channel approach) and a simple form of the phenomenological exchange picture (t-channel approach). The exchange picture of diffraction, with its recent developments and its predictions, is discussed in more detail in the last two sections, where we review some topical questions at ISR and Fermilab energies.

It goes without saying that space does not allow for a complete presentation of this wide subject. We have chosen material that could be fitted into a logical presentation. Complications and further generalization of the optical analogy develop along the way. We hope this will help newcomers to follow and leave them with some self-contained facts, even if they do not read the entire article. The readers who already know the subject will find in the last sections detailed presentations of recent developments in the interpretation of the phenomena and throughout the paper many figures in which recent data are compiled. In order to be systematic with such compilations, we tried whenever possible to stick to a uniform convention in presenting data points in the figures that contain data produced at various high-energy accelerators: ▼ Fermi National Accelerator Laboratory (FNAL), Batavia, Illinois; □ Stanford Linear Accelerator Center

(SLAC), Stanford, California; ∇ Alternating Gradient Synchrotron (AGS), Brookhaven National Laboratory, Upton, New York; ■ Institute for High Energy Physics (IHEP); Serpukhov, USSR; ● Intersecting Storage Ring (ISR), CERN, Geneva; and ○ Proton Synchrotron (PS), CERN, Geneva.

2 DIFFRACTION IN OPTICS AND ITS EXTENSION TO HADRONIC WAVES

2.1 *Fresnel-Kirchhoff Theory*

Historically, the optical analogy has been instrumental in the development of the field reviewed here and many previous presentations have taken it as a starting point (1–4). We follow a similar track using a language suitable to the extension to hadronic physics.

Optics relies on approximations (5). In speaking of diffraction theory one usually refers to the applications of the Huygens-Fresnel-Kirchhoff approximate method. It can be applied to the propagation of a plane wave of light behind an opaque screen with a "hole" in it (Figure 1) if the wavelength λ is much smaller than the dimensions R of the diffracting hole:

$$kR \gg 1 \quad \text{(short wavelength condition),} \qquad\qquad 1.$$

where $k = 2\pi/\lambda$. The propagation of a plane light wave is described by means of a complex scalar function $A(x, y, z)$; the basic formula of the theory connects the value A_0 of this quantity on a plane Σ, to the value of the same quantity at some point P of the detector plane (5):

$$A(x, y, z) = (k/4\pi i) \int_{\Sigma} \mathrm{d}^2 a A_0 S(\mathbf{a})(1 + \cos \theta') \exp{(ikd)}/d. \qquad 2.$$

The quantity $A_0 \cdot S(\mathbf{a})$ may be interpreted as the amplitude of the wave "just behind" the plane Σ. The quantity $S(\mathbf{a})$, which corresponds to the S matrix of scatter-

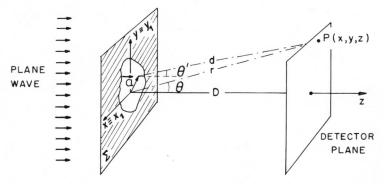

Figure 1 Propagation of a plane wave behind a screen with a hole. Reference systems and definition of the various quantities.

ing theory, describes the variation introduced in the amplitude by the "hole." When the short-wavelength condition is satisfied, the diffracted wave is concentrated at small angles and the "inclination factor" is practically constant: $(1 + \cos \theta') \approx 2$. The main variation of the factor $\exp(ikd)/d$ comes from the exponent whenever the distance D of the detector plane satisfies the inequality

$$R/D \ll 1 \quad \text{(large distance condition)}. \qquad\qquad 3.$$

The exponent kd may then be written as a power series in the coordinates x_1 and y_1 on the plane Σ and, as is well known, two regimes are distinguished according to the relative values of the quantities $kR \gg 1$ and $R/D \ll 1$:

$$kR^2/D \ll 1 \quad \text{(Fraunhofer diffraction)}; \qquad kR^2/D \approx 1 \quad \text{(Fresnel diffraction)} \qquad 4.$$

In the first case the quadratic terms in x_1 and y_1 are negligible and $d \approx r - (xx_1 + yy_1)/r$, while in the second they cannot be neglected.

Since geometrical optics is obtained in the limit $\lambda \to 0$, that is $k \to \infty$, for any choice of R and D (i.e. also when condition 3 is not satisfied), one may say that it corresponds to

$$kR^2/D \gg 1 \quad \text{(geometrical optics)}. \qquad\qquad 5.$$

It is seen that the parameter kR^2/D determines the propagation regime. Since Fraunhofer diffraction is of great importance for our purpose, it is worth rewriting the diffracted amplitude of Equation 2 in this particular case:

$$A(x, y, z) \approx -(ik/2\pi)[A_0 \exp(ikr)/r] \int_\Sigma d^2 a X S(\mathbf{a}) \exp(i\mathbf{q} \cdot \mathbf{a}). \qquad 6.$$

The two-dimensional momentum transfer $\mathbf{q}$ lies in the plane Σ and has by definition components $(kx/r, ky/r)$, so that

$$|\mathbf{q}| = k \sin \theta. \qquad\qquad 7.$$

Let us now consider a diffracting absorbing disc, which is a closer analogy to an absorbing hadron than a screen with a hole. If $S(\mathbf{a})$ is now the S matrix for diffraction by the disc, its "profile function" is defined as

$$\Gamma(\mathbf{a}) = 1 - S(\mathbf{a}) = 1 - \exp[i\Delta(\mathbf{a})]. \qquad\qquad 8.$$

$\Delta(\mathbf{a})$ is the complex phase shift. Im Δ describes the absorption introduced by the disc and Re Δ is the actual phase shift given to the incoming wave. Introducing $S(\mathbf{a}) = 1 - \Gamma(\mathbf{a})$ in Equation 6, one obtains the full amplitude behind the disc. The term proportional to 1 is the unperturbed plane wave, and the one that contains $\Gamma(a)$ describes the diffracted wave.

In Equation 6 the factor $A_0 \exp(ikr/r)$ represents an outgoing spherical wave of amplitude A_0 emanating from the center of the diffracting object to the observation point. The only physically interesting quantity is the factor multiplying this spherical wave, the so-called scattering amplitude $f(\mathbf{q})$. Since $f(\mathbf{q})$ is proportional to the Fourier transform of the profile function, the converse is also true and one can write:

$$f(\mathbf{q}) = (ik/2\pi) \int d^2 a \Gamma(\mathbf{a}) \exp(i\mathbf{q}\cdot\mathbf{a}); \; \Gamma(\mathbf{a}) = (1/2\pi ik) \int d^2 q \, f(\mathbf{q}) \exp(-i\mathbf{q}\cdot\mathbf{a}). \qquad 9.$$

When the profile function is spherically symmetric, Equation 9 simplifies to

$$f(\mathbf{q}) = ik \int_0^\infty da \, a\Gamma(a) J_0(qa). \qquad 10.$$

As many as 140 years ago a similar expression was used by Airy for computing the diffraction of a plane light wave from a circular aperture of radius R. In our notation, he obtained (4)

$$f(\mathbf{q}) = ikR^2 J_1(qR)/(qR). \qquad 11.$$

Equation 10 may be integrated analytically in various cases. Some examples, which will offer a useful guideline later, are presented in Figure 2.

2.2 The Optical Theorem

The optical theorem relates the forward-scattering amplitude to the total cross section σ_t:

$$\sigma_t = 4\pi \, \mathrm{Im}\, f(0)/k. \qquad 12.$$

For future use we sketch here the derivation of the theorem. The amplitude of the incoming wave is modified by the screen with the factor $S(\mathbf{a})$, so that the ratio of the

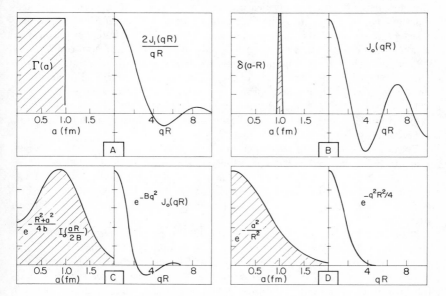

Figure 2 Simple profile functions $\Gamma(a)$ and the corresponding scattering amplitudes computed by means of Equation 10.

energy flux absorbed by the screen to the incident flux is $[1 - |S|^2] = 2\,\mathrm{Re}\,\Gamma - |\Gamma|^2$. In scattering theory its integral is the inelastic cross section:

$$\sigma_{in} = \int d^2a[2\,\mathrm{Re}\,\Gamma(\mathbf{a}) - |\Gamma(\mathbf{a})|^2]. \qquad\qquad 13.$$

Similarly, the differential elastic cross section $d\sigma/d\Omega$ is defined as the outgoing energy density per unit solid angle normalized to the incoming flux, so that $d\sigma/d\Omega = |f(\mathbf{q})|^2$. Its integral over the solid angle (with $d\Omega = d^2q/k^2$) gives the elastic cross section:

$$\sigma_{el} = \int d^2a|\Gamma(\mathbf{a})|^2. \qquad\qquad 14.$$

The sum of these two quantities is the total cross section σ_t. By comparing it with Equation 9a one obtains the optical theorem of Equation 12.

2.3   Application of Optical Concepts to Hadronic Waves

Hadrons propagate in free space according to relativistic wave equations where the wave number k is related to the momentum p, the total energy E, and the mass m by the usual relation $k = p/h = (E^2 - m^2c^4)^{1/2}/hc$. Numerically $k \approx 5.10^{13}\, p(\mathrm{cm}\,\mathrm{GeV}/c)^{-1}$. As pointed out in the introduction, hadrons as well as nuclei are extended objects. In electron-scattering experiments the rms radii of the charge distributions are actually measured (6). The expression $R \approx 1.1\,A^{1/3}$ fm fits the data on nuclei and even gives the reasonable value $R = 1.1$ fm if for a hadron one takes $A = 1$. Assuming that hadronic matter has the same distribution as the electric charge, the relevant quantity to be introduced in the short-wavelength condition is kR_t, where k is the wave number in the center-of-mass system, and R_t is the quadratic combination of the radii of the projectile and the target. For incoming protons to have $kR_t \gtrsim 10$ the laboratory momentum p_L has to be larger than ~ 5, ~ 1, and ~ 0.3 GeV/c when the targets are protons, ^{4}He nuclei, and ^{208}Pb, respectively. The same limits are applicable, within 20%, if the incoming hadrons are pions instead of protons.

The large-distance condition (Equation 3) is always satisfied because the distance D at which the scattered hadrons are observed behind the scatterer is at least of the order of 1 cm, so that $R/D \simeq 10^{-13}$. Since in all experiments performed so far kR is never larger than 10^4, the quantity kR^2/D is always smaller than 10^{-9}. Therefore hadron-hadron and hadron-nucleus scattering is analogous to Fraunhofer diffraction.

Typical diffraction patterns appear in the differential cross sections of Figure 3. As a first application of optical concepts we may consider the calculation of the profile functions. This already meets with a problem, since the complex scattering amplitude is needed in Equation 9, while the differential cross section gives only information on its modulus: $|f| = (d\sigma/d\Omega)^{1/2}$. The phase of the amplitude at $\mathbf{q} = 0$ may be obtained by measuring the interference between the nuclear and the Coulomb amplitude. In Section 4.4 we show that the amplitude $f(0)$ is essentially imaginary, i.e. that its phase is close to 90° in hadron-hadron collisions, at least for laboratory momenta larger than 10 GeV/c. For $\mathbf{q} \neq 0$ the phase of the nuclear scattering amplitude must be derived from some model. The simplest one corresponds to a purely absorbing interaction, i.e. to an imaginary phase shift $\Delta(\mathbf{a})$ and a real $\Gamma(\mathbf{a})$. This is an oversimplification; nevertheless, the fact that it is a reasonable one at very high energies justifies the present interest in optical diffractive models. In this case the amplitude of Equation 10 is imaginary at all momentum transfers and one speaks of "shadow scattering." This model is applicable when the energy is large enough and thus there are many open inelastic channels summing up to a large inelastic cross section [$\Gamma(0)$ is real and not too far from 1 in Equation 13]. In such circumstances the elastic scattering is mainly the shadow of the inelastic channels and the momentum-transfer dependence reflects the shape of the inter-acting hadrons. The profile function plotted in the inset of Figure 3C has been obtained within this model. This elicits both a question and a remark.

Question: What is the meaning of the parameter a in the case of hadron-hadron collisions? This may be clarified by writing the incoming hadron wave as a sum of partial waves of orbital angular momentum l and then by approximating the sum over l by an integral over the variable $a \simeq (l + 1/2)/k$. If the energy is

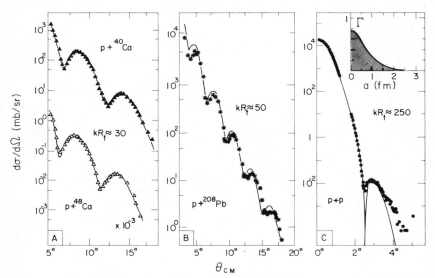

Figure 3 Typical diffraction patterns produced by hadronic waves: (*A*) elastic scattering of protons of 1.7 GeV/c on ^{40}Ca and ^{48}Ca (7). The fits are based on Glauber theory. (*B*) Elastic scattering of protons of 1.75 GeV/c on ^{208}Pb (8). The curve is computed by applying Glauber theory (10). (*C*) Elastic scattering measured at the CERN Intersecting Storage Rings (ISR) by the Aachen-CERN-Genova-Harvard-Torino Collaboration (11a). The equivalent laboratory momentum of the proton is $p_L = 1480$ GeV/c. The curve is a fit (11b) based on the optical model of Chou & Yang (33–35). The inset shows that even at these energies the proton is not a black disc.

large enough that a very large number of partial waves contributes, the wavelength is so small that a semiclassical description of the scattering process is applicable (4). The quantity a is nothing but the minimum distance of approach between the centers of the two colliding hadrons. For this reason it deserves the name of "impact parameter." Since the range R of the hadron-hadron forces is finite, their effect drops exponentially at large distances and the relevant angular momenta may increase in number as $l_{max} \lesssim kR \ln(kR)$. (We would have kR for a sharp edge). The short-wavelength condition becomes $l_{max} \gg 1$; the diffraction picture applies in its simplest form when many partial waves contribute. They are almost purely imaginary and coherence effects are maxima.

Remark: Experimentally it is found that the hadron-hadron differential cross section close to the forward direction is a Gaussian in the scattering angle (Figure 3C). Thus the scattering amplitude is well represented by the form

$$f(q) = k\sigma_t(i + \rho)\exp(-b|t|/2)/4\pi, \qquad\qquad 15.$$

where the ratio ρ between the real and the imaginary part is taken to be independent of t. The parameter b is the "slope" of the logarithm of the differential cross section plotted versus $|t| = \hbar^2 q^2$.[1] The profile function corresponding to Equation 15 is Gaussian and Figure 2D shows that its radius is

$$R = \hbar 2^{1/2} b^{1/2} \approx 0.3\, b^{1/2}\,\text{fm} \quad (b \text{ in GeV}^{-2}). \qquad\qquad 16.$$

For $R \approx 1$ fm this relation gives $b \approx 10$ GeV^{-2}.

2.4 Polarization Effects in Optics and Diffraction Dissociation

Our discussion has so far been limited to a scalar amplitude. It is, however, well known that when polarization is considered, new phenomena appear in optical diffraction. Let us for instance consider a screen with a "hole" that has different indexes of refraction for right and left circularly polarized beams (a tube with a sugar solution in it). In this case there are two profile functions $\Gamma_R(\mathbf{a})$ and $\Gamma_L(\mathbf{a})$ that form a two-by-two diagonal matrix in the representation that uses right- and left-circularly polarized states as base vectors. In a linear polarization basis the profile matrix is no more diagonal, so that the diffraction of a linearly polarized beam gives rise to two different waves. The first has the same polarization as the incident one and scattering amplitude proportional to the Fourier transform of $(\Gamma_L + \Gamma_R)$, while the second is polarized at $90°$ and is determined by the profile function $(\Gamma_L - \Gamma_R)$. This second component is not present in the incident wave and may be regarded as the production by diffraction of a new state degenerate in energy (frequency) from the initial state. We note that such a diffractive production becomes possible, since a new degree of freedom is introduced in the description of the incoming wave.

About twenty years ago Glauber (12), Feinberg & Pomeranchuk (13), Good &

---

[1] The four-momentum transfer-squared t is given by $t = 2m^2 c^4 - 2(E_1 E_2 - p_1 p_2 c^2 \cos\theta)$, where $(E_1, \mathbf{p}_1)$ and $(E_2, \mathbf{p}_2)$ are energies and momenta of the scattered hadron with mass m before and after the collision and θ is the scattering angle.

Walker (14), and others pointed out that similar effects must exist in the inter-action of hadrons with nuclei. In the Good & Walker approach, one views the state $|i\rangle$ of a hadron h_0 that moves through a nucleus as a linear combination of states $|n\rangle$ that have the same intrinsic quantum numbers as the hadron and are eigen-states for strong interactions in nuclear matter: $|i\rangle = \Sigma C_{ni}|n\rangle$. The S matrix is diagonal in the space spanned by the diffractive eigenstates $|n\rangle$ and "just behind" the nucleus the wave has the form $S(\mathbf{a})|i\rangle = \Sigma C_{ni}[1-\Gamma_n(\mathbf{a})]|n\rangle$. In general the profile functions Γ_n of the various states $|n\rangle$ are not all equal so that the diffracted wave does not coincide with the initial physical hadron. Let $\Gamma(\mathbf{a})$ be the profile function describing diffraction scattering of h_0. The leftover diffracted wave just behind the nucleus is then $\Sigma C_{ni}[\Gamma(\mathbf{a})-\Gamma_n(\mathbf{a})]$, which expresses the fact that each state $|n\rangle$ is present with an amplitude proportional to the difference between the profile function for elastic scattering of the hadron h_0 and the profile function referring to the eigenstate $|n\rangle$. Diffraction dissociation is thus expected to be small where these differences are small, for instance at the center of a heavy nucleus that is completely absorbing for most states $|n\rangle$. This already indicates that diffraction dissociation is expected to be more peripheral, in impact parameter space, than elastic shadow scattering.

The physical states $|f\rangle$ that are observed far away from the nucleus are linear combinations of the eigenstates $|n\rangle$ and in general do not appear as single-particle states, but rather as systems of two or more hadrons with energy M in their center of mass and of global laboratory momentum P. Nuclear coherence puts restriction on the possible systems h that may be diffractively produced. Indeed, they must have the same intrinsic quantum numbers as the incoming hadron h_0, i.e. the same charge, isotopic spin, nucleon number, strangeness, charge conjugation etc. However, they need not keep the spin and parity of h_0, because some angular momentum may be transferred to the internal motion of the hadron while the total angular momentum stays constant. A further limitation comes from the request of coherent production: the mass M should not be too different from the mass m of the incoming hadron. For zero-angle production the momentum transfer is $(p-P) \approx (M^2 - m^2)c^2/2p$, so that for a given incoming momentum p the minimum four-momentum transfer at which the mass M may be produced is

$$t_{min} = [(M^2 - m^2)c^2/2p]^2. \qquad\qquad 17.$$

In the transition the wave number k of the incident hadron varies by $\Delta k = (p-P)/\hbar$. The condition of coherent production amounts to the request that the product of Δk by the radius R of the diffracting nucleus has to be smaller than 1. Only in this case can the waves describing the hadrons h_0 and h stay in phase within the nucleus. In summary we have[2]

coherence condition: $M^2 - m^2 \lesssim 2p/R.$ \qquad\qquad 18.

Good & Walker's idea applies not only to hadron-nucleus collisions, but also to hadron-hadron collisions. In this case the basis $|n\rangle$ is defined as the set of states

---

[2] From now on we shall put $\hbar = c = 1$, so that energy, mass, and momentum transfer are all measured in GeV.

that do not mix in the interaction. The coherence condition is easier to satisfy since the radius R is smaller.

Evidence for diffraction dissociation of negative pions is shown in Figure 4. The data were collected at the CERN PS by the CERN-Milan-ETH (Zürich)-Imperial College collaboration (15–20) on the reaction $\pi^- + A \to (\pi^- \pi^+ \pi^-) + A$ around 15 GeV/c of laboratory momentum. It is seen that the mass distributions of the produced three-pion system peaks at low masses and that masses around 1.7 GeV are relatively more abundant with a beryllium target than with a tantalum one. This agrees with the coherence condition that is better satisfied for smaller nuclei. Moreover, the angular distributions show a very pronounced forward peak whose slope b increases as $A^{2/3}$, i.e. as R^2, as predicted by Equation 16. These data are discussed in more detail in Section 3.3.

3   DIFFRACTION IN HADRON-NUCLEUS COLLISIONS

We now consider hadron-nucleus scattering as a second step towards the identification of the general properties of diffraction phenomena in hadron-hadron collisions. Heavy nuclei are a useful tool because they have a well-defined radius R, which may be varied by changing the mass number, and because hadrons are much

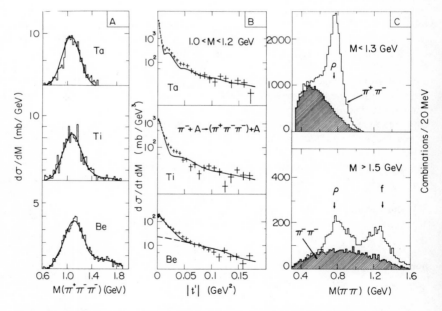

Figure 4   Diffraction dissociation of 15 GeV/c π^- on nuclei (15–20): (*A*) Mass distributions of the 3π system on tantalum, titanium, and beryllium nuclei. (*B*) Differential cross section on the same three targets as a function of $t' = t - t_{\min}$. The curves are fits based on Glauber theory (51). (*C*) Two-pion spectra for two samples of data having different 3π masses.

absorbed for impact parameters smaller than R. Moreover, a reliable theory exists to describe diffraction in hadron-nucleus scattering and what happens depends more on the size of the nucleus than on the details of the interaction with each nucleon. In the following we sketch the main lines of Glauber theory (21–24a), not only to discuss elastic hadron-nucleus collisions, but also to prepare the presentation of some geometrical models of hadron-hadron scattering.

3.1 Diffraction Theory of Coherent Hadron-Nucleus Scattering

Glauber theory describes the scattering of a hadronic wave by a system composed of various scattering centers. It is applicable to small-momentum transfer scattering. It is based on three main hypotheses: (a) the nuclear wave function has no time to change during the propagation of the hadron inside the nucleus; (b) the propagation of the hadronic wave within the nucleus follows the laws of geometrical optics; and (c) the phase shifts produced by the individual nucleons combine additively. We briefly consider the rationale for these hypotheses.

The first hypothesis is satisfied for laboratory momenta of the incident hadron larger than ~ 1 GeV/c, i.e. whenever the incident energy is much larger than the excitation energies of the nucleus (25).

According to the discussion of Section 2.1, the second condition is satisfied when $kR^2 \gg D$, where R is the interaction radius in a hadron-nucleon collision and D is the radius of the nucleus. This condition is also satisfied for $p_L \gtrsim 1$ GeV/c. [Gottfried has shown that some Fresnel effects are present in hadron-deuteron scattering below 1 GeV/c (26).]

Condition (c) does not represent a further assumption if the hadron-nucleon interaction can be described by a local potential (4), since the phase shift depends linearly on the potential and the potentials due to the different nucleons sum up. However, at sufficiently high energies the phases $\Delta_j(a)$ due to the interaction of the jth nucleon with the incoming hadron are complex and condition (c) becomes an independent hypothesis.

The S matrix for the hadron-nucleus interaction takes the form

$$S_A(\mathbf{a}; \mathbf{s}_1 \dots \mathbf{s}_A) = \prod_{j=1}^{A} \exp[i\Delta_j(\mathbf{a} - \mathbf{s}_j)],\qquad\qquad 19.$$

where $\mathbf{s}_j$ is the transverse position of the jth nucleon (Figure 5). The nucleus profile

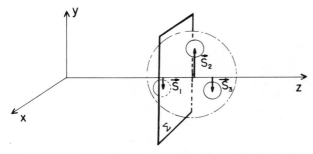

Figure 5   Definition of transverse positions in the Glauber approach.

function can then be written as the sum of A terms

$$\Gamma_A(\mathbf{a}; \mathbf{s}_1 \dots \mathbf{s}_A) = \sum_i \Gamma_i(\mathbf{a} - \mathbf{s}_i) - \sum_{i>j} \Gamma_i(\mathbf{a} - \mathbf{s}_i)\Gamma_j(\mathbf{a} - \mathbf{s}_j)$$
$$+ \sum_{i>j>k} \Gamma_i(\mathbf{a} - \mathbf{s}_i)\Gamma_j(\mathbf{a} - \mathbf{s}_j)\Gamma_k(\mathbf{a} - \mathbf{s}_k) - \cdots . \qquad 20.$$

This equation expresses the nucleus profile function in terms of the nucleon profiles and shows that when the interaction is weak ($\Gamma_i \approx 0$), the profile functions of such a "frozen" nucleus is equal to the sum of the profile functions of its constituent nucleons. This is not the case in general, however, and the other terms of this "multiple scattering series" must also be taken into account. For instance, the second term represents events in which the incoming hadron is scattered by two different nucleons in the nucleus, the third term represents triple scattering, and so on.

3.2   Elastic Scattering of Hadrons on Nuclei

For light nuclei the number of terms in the Glauber multiple expansion is small and may be computed explicitly by introducing a suitable parameterization of the hadron-nucleon profile. (A Gaussian profile is usually used, as implied by Equation 15.) In Figure 6A experimental data on pion-deuteron scattering (27–29) are compared with such calculations (29–31).[3] The forward peak is associated with the first term in the expansion. It represents single scattering of the pion either from the proton or from the neutron in the deuteron acting coherently. Its momentum-transfer dependence is essentially that of the deuteron form factor. For $|t| \gtrsim 0.5$ GeV^2, however, the second (and last) term in Equation 20 dominates; this corresponds to double scattering when the pion hits successively the neutron and the proton. It requires more localization in space and hence the t-dependence is less sharp than in the former case.

For intermediate and heavy nuclei the profile operator of Equation 20 contains many terms. The previous method becomes cumbersome and a simpler, yet reliable description of hadron-nucleus elastic scattering is obtained, neglecting correlations in the nuclear wave function, so that the ground state can be described by the nuclear densities $\rho_j(\mathbf{r})$ of the A nucleons ($\int d^3r \rho_j(\mathbf{r}) = 1$). Then the expectation value of the S matrix in the nucleus ground state $|0\rangle$ takes the form

$$\langle 0|S_A(\mathbf{a})|0 \rangle = \prod_{j=1}^A \int d^2s\, dz\, \rho_j(\mathbf{s}, z)[1 - \Gamma(\mathbf{a} - \mathbf{s})], \qquad 21.$$

where $\Gamma(\mathbf{a})$ is the hadron-nucleon profile. For $A \to \infty$ the product of these A factors may be written in the form $\exp(i\Delta_A)$, where

$$i\Delta_A(\mathbf{a}) = -\int d^2s\, \Gamma(\mathbf{a} - \mathbf{s}) \int_{-\infty}^{+\infty} dz\, \rho(\mathbf{s}, z) = -\int d^2s\, \Gamma(\mathbf{a} - \mathbf{s})\, T(\mathbf{s}). \qquad 22.$$

In this equation $\rho = \Sigma \rho_j$ is the total nuclear density ($\int d^3r \rho = A$) and $T(\mathbf{s})$ is the "thickness function," i.e. the integral of the density along a straight path at impact

---

[3] Proton-nucleus data have been reviewed by Saudinos & Wilkin (24a), and by Ciofi degli Atti (24b).

parameter **s**. Equations 21 and 22 are the starting point for most calculations of elastic scattering of high-energy hadrons on intermediate and heavy nuclei. For instance, the curves in Figure 3*A* have been computed with Equation 21 (7). The nuclear-matter distribution needed to fit the ^{40}Ca data comes out to be very similar to the charge distribution measured in electron-nucleus scattering. Similarly, the curve of Figure 3*B* is computed without adjustable parameters, using a self-consistent density distribution derived from Hartree-Fock calculations (10). In summary, the theory of hadron-nucleus diffraction is precise enough to provide information on the nuclear density of accuracy comparable to the one derived from electron scattering. This conclusion was already reached a few years ago (32).

To understand how an apparently smooth expression (Equation 22) gives rise to the many oscillations appearing in the differential cross section, it is useful to introduce the Fourier transform of the phase

$$\Delta_A(\mathbf{q}) = -(1/2\pi)\int d^2a \exp(i\mathbf{q}\cdot\mathbf{a})i\Delta_A(\mathbf{a}) = Af(\mathbf{q})F(\mathbf{q})/(ik); \qquad 23.$$

$f(\mathbf{q})$ is the hadron-nucleon scattering amplitude and $F(\mathbf{q})$ is the form factor of the nucleus so normalized that $F(0) = 1$. The hadron-nucleus scattering amplitude $f_A(\mathbf{q})$ is the Fourier transform of $\{1 - \exp[i\Delta_A(\mathbf{a})]\}$ and, expanding this exponential, it may be formally written as a series in $\Delta_A(\mathbf{q})$, namely

$$f_A(\mathbf{q}) = ik[\Delta_A(\mathbf{q}) - \Delta_A(\mathbf{q})\otimes\Delta_A(\mathbf{q})/2! + \Delta_A(\mathbf{q})\otimes\Delta_A(\mathbf{q})\otimes\Delta_A(\mathbf{q})/3!\dots]. \qquad 24.$$

The convolution of two functions in momentum space is defined as

$$N(\mathbf{q})\otimes M(\mathbf{q}) = (1/2\pi)\int d^2q'\, N(\mathbf{q})M(\mathbf{q}-\mathbf{q}'). \qquad 25.$$

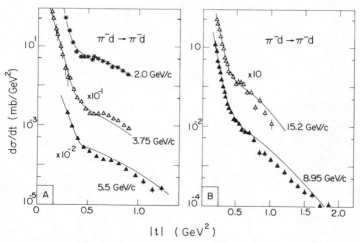

Figure 6 Data on pion-deuteron elastic scattering at different momenta (27–29) are compared with calculations based on Glauber theory (30).

If the interaction of the hadron with the nucleons is not very strong, the first term is sufficient, and the differential cross section is simply

$$(d\sigma/d\Omega)_{coher} = A^2 |f(q)|^2 F^2(q).$$ 26.

It has the typical A^2-dependence of a coherent process. The q-dependence of this expression is mainly determined by the nuclear form factor, i.e. by the size of the nucleus. The other terms in the expansion alternate in sign and give rise to the typical oscillatory behavior superimposed on the rapid decrease determined by the form factor. Note that if $\Delta_A(q)$ is approximated by $\exp(-cq^2)$, the n term of the development is proportional to $\exp(-cq^2/n)$, so that each successive term gives a flatter contribution that eventually dominates at larger angles.

An extension of Equation 22 has been used by Chou & Yang to interpret elastic scattering in very high-energy proton-proton collisions (33–36). This approach is based on the idea that hadrons are extended distributions of nuclear matter made up of an infinite number of very small constituents. The constituents of the two hadrons interact locally, i.e. their profile is a delta function. The phase $\Delta(\mathbf{a})$ of this model is immediately written down as an extension of Equation 22, which was obtained for a hadron interacting with a nucleus having thickness function $T(\mathbf{s})$. If T_A and T_B are the thickness functions of the interacting hadrons, the Chou & Yang phase is proportional to the integral $\int d^2s\, T_A(\mathbf{a}-\mathbf{s})T_B(\mathbf{s})$. The thickness function is then obtained from the charge distribution, i.e. from the measured electromagnetic form factors F_A and F_B of the two hadrons. With this hypothesis the Fourier transform of the phase, defined by Equation 23, becomes $\Delta(q) = C_{AB}F_A(q)F_B(q)$. C_{AB} measures the strength of the interaction and is the only free parameter of the model. In the case of proton-proton scattering the first term of the development (Equation 24) gives a cross section proportional to $|F_p(q)|^4$. The interference with the second term, negative and less steep, gives rise to a minimum at $|t| \approx 1.5$ GeV^2. The curve of Figure 3C is the result of such a calculation (11b). The data are very well reproduced. It must be stressed that the model was suggested to be valid at infinite energy, when a limiting angular distribution was supposed to be reached. As is discussed below, the results of the CERN Intersecting Storage Rings have shown that up to 2000 GeV/c, no limiting distribution is reached. One must conclude that the Chou & Yang approach is too simple; nonetheless, it has many appealing features.

3.3   Exclusive Diffraction Dissociation of Hadrons on Nuclei

3.3.1   PION-DIFFRACTION DISSOCIATION   Pion-diffraction dissociation is illustrated in Figure 4, which contains a sample of the data of the CERN-Milan-ETH-Imperial College collaboration. More data may be found in Lubatti's review article (37), where results on nuclear targets are presented in parallel with data on nucleons. Figure 4B shows that the slope of the forward cross section increases proportionally to the square of the nuclear radius. The mass spectrum of the three-pion system peaks around $M = 1.1$ GeV, and for light nuclei it shows another structure at $M \approx 1.7$ GeV (Figure 3A). These bumps are associated with the names of "A_1 and A_3 enhancements." The study of their properties has been one of the main

issues in the field of exclusive diffraction dissociation in the past few years. Some properties of these systems are illustrated in Figure 4C, where, from the data of the same collaboration summed over all targets, mass spectra for pairs of pions have been constructed. While the combination $\pi^-\pi^-$ has a structureless behavior, the peaks present in the $\pi^+\pi^-$ combination show that the three-pion system of mass smaller than 1.3 GeV/c^2 decays predominantly via the channel $\rho^0(770) + \pi^-$, while for masses larger than 1.5 GeV one finds both $\rho + \pi^-$ and f(1270) + π^- configurations.

Refined partial wave analyses of the three-pion system diffractively produced on hydrogen targets have been performed by the Illinois group (38) and are discussed in Section 5.2. The same analysis was used both in the experiment under discussion and in a more recent one performed at 23 GeV/c at the Brookhaven AGS by the Carnegie-Mellon–Northwestern–Rochester collaboration (39–41). In these experiments coherent production on nuclei was shown to be dominated by states in which the three pions have total angular momentum and parity $J^P = 0^-$ and 1^+ in the region of the A_1, and 2^- in the region of the A_3. Here we meet for the first time with a regularity that we shall encounter again in the following discussion: the pion, which has $J^P = 0^-$, diffractively dissociates in states belonging to the J^P series 0^-, 1^+, 2^- (unnatural parity states).

3.3.2 KAON DIFFRACTION DISSOCIATION   Kaon diffraction dissociation on nuclei has been studied first in heavy-liquid bubble chambers and later with counters by the CERN-Milan-ETH-Imperial College collaboration (17). The ($K\pi\pi$) mass spectrum, which presents similarities to the 3π spectrum, shows a prominent bump at $M(K\pi\pi) \approx 1.3$ GeV, referred to as the Q enhancement. At larger masses (~ 1.75 GeV), a small structure appears, usually named L enhancement, which is again not visible in heavy nuclei. This is due to the coherent condition. The Q enhancement has a strong contribution of 1^+ nature and decays mainly through the channel $K^*(890)\pi$ and ρK, thus paralleling the behavior of the A_1, which decays into $\rho\pi$. Again we see that a 0^- particle dissociates preferentially in a 1^+ state.

3.3.3 NUCLEON DIFFRACTION DISSOCIATION   Experiments on nucleon diffraction dissociation have been recently performed at the Brookhaven AGS. The Michigan group has studied the reaction $n + A \rightarrow (p\pi^-) + A$ (42) and the Carnegie-Mellon–Northwestern–Rochester collaboration has collected high statistics data at $p_L = 22.5$ GeV/c on the reaction $p + A \rightarrow (p\pi^+\pi^-) + A$ (40, 43). The angular distributions measured on C, Cu, and Pb are plotted in Figure 7 versus $t' = t - t_{min}$, together with the mass spectra of the ($p\pi^+\pi^-$) system. In this case too, there is a prominent bump at $M \approx 1.4$ GeV and, for carbon targets, a small enhancement around 1.7 GeV. The first is a typical threshold enhancement of spin-parity $1/2^+$. (It compares to what was found with the A_1 and Q). The second is very probably due to coherent production of the $5/2^+$ state $N(1690)$. Thus, for half-integer projectiles, diffraction dissociation favors transition of the type $1/2^+ \rightarrow 1/2^+$, $1/2^+ \rightarrow 5/2^+$.

Proton-diffraction dissociation on deuterons has been recently studied in a very elegant experiment by the Fermilab-Dubna-Rockefeller-Rochester collaboration working on the circulating proton beam of the 400 GeV/c Fermilab synchrotron

(44, 45). The technique, using a gas jet of hydrogen (or deuterium) as a target for protons circulating in the machine during the acceleration cycle, has been discussed in detail by Melissinos & Olsen (46). By measuring the kinetic energy and the angle of the proton (or deuteron), which for small momentum transfer t recoils at an angle close to 90°, one can determine both the value of t and the mass M of the produced system. Momentum-transfer distributions obtained at $p_L = 275$ GeV/c for three values of M are plotted in Figure 8A. The steep falloff is mainly determined by the deuteron form factor and is due to the fact that at very small momentum transfer, the double-scattering contribution is negligible. By analogy to Equation 26, the differential cross section for diffraction dissociation factorizes, namely

$$(d^2\sigma/dt\,dM)_{pd} = (\sigma_t^{pd}/\sigma_t^{pp})\,F^2(t)(d^2\sigma/dt\,dM)_{pp}, \qquad 27.$$

where $F(t)$ is the deuteron form factor. The ratio of the cross section is ~ 3.6, and substitutes in Equation 26 the factor $A^2 = 4$, so as to take into account shadowing effects. Equation 27 is used to extract proton-nucleon cross sections from proton-deuteron data. Proton-proton missing-mass spectra derived from proton-deuteron data at $p_L = 50$ and 275 GeV/c are compared in Figure 8B with previous proton-proton data at 20 GeV/c obtained at the AGS (47). The broad bump at $M^2 \approx 2$ GeV^2 is attributed to the production of the 1.4 GeV "enhancement" and the structure at $M^2 \approx 2.8$ GeV^2 is identified with the excitation of the well-known resonant state $N(1690)$.

In the above examples diffraction dissociation is identified by the presence of a

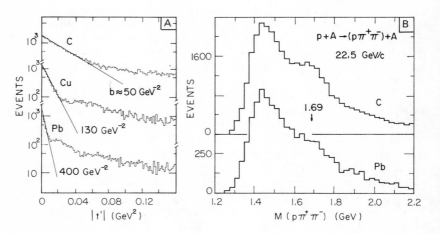

Figure 7   Proton-diffraction dissociation on nuclei measured by the Carnegie-Mellon–Northwestern–Rochester collaboration (40, 43). (*A*) the forward peak has a slope roughly proportional to the square of the nucleus radius. This is a clear indication of the diffractive nature of the process. (*B*) Mass spectra of the $(p\pi^+\pi^-)$ produced on carbon and lead targets.

large forward peak, the slope of which is related to the dimension of the nucleus. This small sample of data agrees with what is also found in hadron-hadron dissociation: not all energetically possible states are diffractively produced. Not only the intrinsic quantum numbers are preserved in the process, but also the Gribov-Morrison empirical rule seems to hold quite generally (48, 49). This rule is expressed by

$$P_f = P_i(-1)^{\Delta J};$$

28.

P_i and P_f represent the parity of the initial and final systems and ΔJ is the change of spin. In words, the change of parity is that which corresponds to the net gain in spin. This in turn corresponds to the minimum transfer of orbital angular momentum. For diffraction dissociation of the pseudoscalar mesons (π and K), the Gribov-Morrison rule allows the production of states with unnatural parity (0^-, 1^+, 2^-, 3^+, etc), while for diffraction dissociation of nucleons the allowed sequence is $1/2^+$, $3/2^-$, $5/2^+$, $7/2^-$, etc. For the collision of spinless particles at $t = 0$ this rule is equivalent to the conservation laws of parity and angular momentum (48). Goldhaber & Goldhaber have shown that to order A^{-1}, the same behavior is predicted for spinless hadrons diffracting on nuclei of nonzero spin (50). Away from

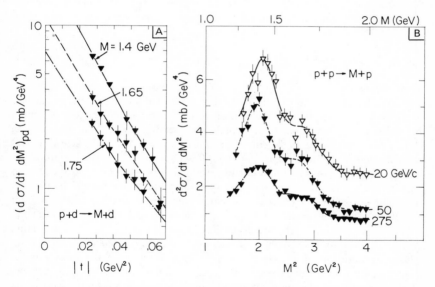

Figure 8 Proton-diffraction dissociation on deuterons at small momentum transfers (44–46). (*A*) Differential cross sections plotted versus $t' = t - t_{min}$ for various masses M of the produced system. (*B*) the black triangles represent the mass spectra in the reaction $p + p \rightarrow M + p$ measured at 50 and 275 GeV/c for $|t| \approx 0.04$ GeV². They were obtained by the USSR–USA collaboration working at Fermilab from the deuteron data by means of Equation 27. The open triangles were measured at the AGS (47). The curves are only to guide the eye.

the forward direction the rule implies some selective dynamical mechanism. The rule may often be enforced, since as the momentum transfer increases, the cross section is damped by the usual factor, which for coherent production on nuclei is proportional to $F^2(q)$ (Equation 26). As a consequence, a production amplitude, which is zero in the forward direction, will never have a chance to become appreciably large. However, this argument cannot be used to justify a rigorous rule. Thus Equation 28 must be considered an empirical relation whose dynamical origin has still to be understood. Indeed, some recent data even contradict it. For instance, the pion-dissociation experiment of the Carnegie-Mellon–Northwestern–Rochester collaboration on C, Al, Cu, and Ag targets indicates that the rule is not satisfied, since in the partial wave analysis of the data (41), evidence is found for coherent production of the $A_2(1310)$ meson ($J^P = 2^+$).

3.4   Cross Sections of Unstable Particles

Coherent diffraction on nuclei has been used as a tool for measuring the cross section of the produced states on the nucleons of the nucleus. In this application the nucleus acts both as a generator of particles and as a target. Let us first consider the elastic reaction $h + A \rightarrow h + A$ in the hypothesis that the thickness function $T(\mathbf{s})$ is slowly varying in comparison with the h-nucleon profile $\Gamma(\mathbf{s})$. The corresponding phase is obtained by introducing Equation 9a and the optical theorem in Equation 22:

$$i\Delta_A(\mathbf{a}) \approx -\sigma_t(1-i\rho) \int_{-\infty}^{+\infty} \mathrm{d}z\, \rho(\mathbf{a}, z)/2. \qquad 29.$$

When the real part of the h-nucleon scattering amplitude is zero ($\rho = 0$), this expression has a very intuitive meaning: the attenuation introduced by the nucleus in the probability of finding the hadron h at impact parameter $\mathbf{a}$ just behind the nucleus is proportional to the total cross section and to the thickness of nuclear matter traversed. In the diffractive process $h + A \rightarrow h' + A$, particle h survives up to a depth z in the nucleus and the diffractive state h' propagates from this point onward. The overall amplitude for production of the state h' off the nucleus contains the phase factor (Equation 29), describing the attenuation of the incident wave and in addition a phase factor containing the total cross section σ_M for h'-nucleon scattering together with the corresponding real-to-imaginary ratio ρ_M. Details may be found in the article by Kölbig & Margolis (51). The only point we intend to stress here is that by measuring the differential cross section for coherent production of a state of mass M as a function of the mass number A, it is possible to determine the total cross section of this unstable state on nucleons, because even though the state decays very quickly, it does not have time to vary appreciably while still inside the nucleus.

The curves of Figure 4B represent best fits to the data in which the total cross section σ_M was left as a free parameter (15–17). The same collaboration has performed phase-shift analysis of the produced three-pion system, so as to derive the total cross section for states of definite spin and parity (18). The results are summarized in Table 1, together with data from other experiments. It is most important

Table 1 Total cross sections of some unstable hadronic systems on nucleons obtained by measuring coherent production on nuclei

Reaction	p_L (GeV/c)	State	Mass (GeV)	σ_M (mb)	Ref.
$\pi^- A \rightarrow \pi^- \pi^- \pi^+ A$	15.1	$(\pi^- \pi^- \pi^+)$	$1.0 < M < 1.2$	23 ± 1.5	17
$\pi^- A \rightarrow \pi^- \pi^- \pi^+ A$	15.1	$(\pi^- \pi^- \pi^+)J^P = 0^-$	$1.0 < M < 1.2$	49 ± 8	18
$\pi^- A \rightarrow \pi^- \pi^- \pi^+ A$	15.1	$(\pi^- \pi^- \pi^+)J^P = 1^+$	$1.0 < M < 1.2$	15.8 ± 1.4	18
$\pi^- A \rightarrow 3\pi^- 2\pi^+ A$	15.1	$(\pi^- \pi^- \pi^- \pi^+ \pi^+)$	$1.5 < M < 1.9$	17 ± 8	15
$K^+ A \rightarrow K^+ \pi^+ \pi^- A$	13	$(K^+ \pi^+ \pi^-)$	$1.0 < M < 1.4$	26 ± 2	20
$pA \rightarrow p\pi^+ \pi^- A$	22.5	$(p\pi^+ \pi^-)$	$1.4 < M < 1.6$	24 ± 3	40
$pA \rightarrow p\pi^0 A$	22.7	$(p\pi^0)$	$1.3 < M < 2$	33 ± 7	52
$\gamma A \rightarrow \pi^+ \pi^- A$	< 7	ρ^0	0.77	26.7 ± 2	57
$\gamma A \rightarrow K^+ K^- A$	< 5	ϕ^0	1.02	12 ± 4	57

to remark that with the exception of the $0^-(\pi^- \pi^- \pi^+)$ state, all cross sections are of the same order of magnitude as the cross sections of stable hadrons. These experiments have taught us the very important fact that such states, usually seen as composite systems, propagate in the nucleus as single mesons. For detailed discussions of the meaning of this discovery we refer to the literature (53–56). For completeness in Table 1 we have also collected data obtained by studying the coherent photoproduction of vector mesons. In Section 5.7 we show that this is indeed a diffractive process. Also in this case we have to refer to the literature for more details (57, 58).

4 DIFFRACTION SCATTERING IN HADRON-HADRON COLLISIONS

4.1 *Characteristic Features of Hadronic Diffractive Phenomena*

Our previous discussion of diffraction phenomena for optical waves and for hadronic waves on nuclei allows us now to abstract those features that a particular final state produced in a hadron-hadron collision must present in order to be classified under the heading "diffractive phenomenon."

The range of strong interactions is of the order of 1 fm and the short-wavelength condition is satisfied if the laboratory momentum of a projectile hadron impinging on a nucleon is larger than ~ 5 GeV/c (Section 2.3). Figure 9A shows that at these momenta the inelastic cross sections of the usual six stable particles on protons are of the order of the area of a disc 1 fm in radius: $\sigma \approx 30$ mb (59). According to Equation 13, these large values may be obtained only if the hadron-hadron profile function is essentially real and, for small impact parameters, is not very far from the maximum allowed value: $\Gamma(0) \approx 1$. This is enough to "drive" shadow phenomena and one is left with the problem of recognizing them. Let us consider the inelastic two-body reaction

$$a + b \rightarrow c + d,$$

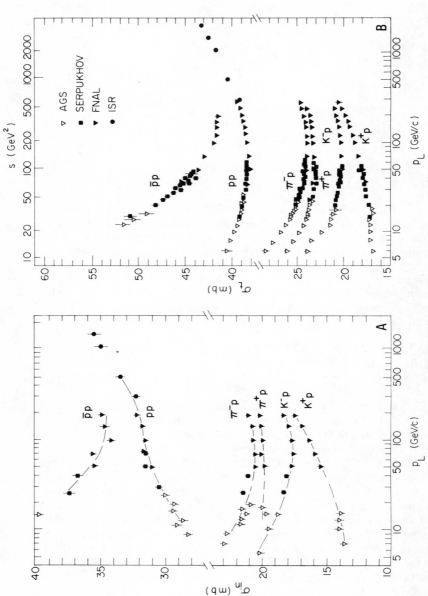

Figure 9 (A) Inelastic cross sections σ_{in} of six stable or metastable hadrons or protons. The usual convention is used to identify the experiments performed in the various laboratories. The curves are only to guide the eye. The values of σ_{in} are obtained by subtracting the elastic cross section σ_{el} from the measured values of σ_t. *(B)* Total cross sections σ_t of the same six hadrons on protons (59a–f, 66–67).

representing the interaction of the projectile hadron a with the target hadron b that produces two hadronic systems c and d, which in general are observed as multiparticle final states. The systems c and d are assumed to be well-separated and associated with well-defined quantum numbers. This reaction is labeled "diffractive process" if the following key features are observed.

1. The differential cross section has a pronounced forward peak whose slope is of the order of 5–10 GeV^{-2} (Equation 16). Note, however, that if for some dynamical reasons the process is peripheral in the impact parameter $\mathbf{a}$, the forward slope may be substantially larger. (Compare Figures 2A and 2B.)

2. The integrated cross section is a slowly varying function of the momentum of the projectile. This follows from the fact that the hadron-hadron profile function must vary little with the momentum, as indicated by the slow variation of the inelastic cross section (Figure 9A). What is meant by "slowly" cannot be stated in absolute terms, but may be defined only in comparison with the fast energy variation of cross sections that are certainly not of diffractive type. Typically they vary as inverse powers of s, the square of the center-of-mass energy. In this context a slow variation means, for instance, a logarithmic dependence on s.

3. The systems c and d have the same intrinsic quantum numbers as those of the hadrons a and b, respectively. Their spin and the parity may be different, but are a priori expected to follow the Gribov-Morrison rule of Equation 28. Note that the association of c with a and d with b is unambiguous because of the large forward peak required by condition 1.

4. The cross sections of processes initiated by a hadron and its antihadron behave similarly as a function both of momentum transfer and of laboratory momentum. This is expected from the fact that the range of strong interactions is practically universal and that the inelastic cross section of a hadron is numerically close to the cross section of its antiparticle (Figure 9A). Annihilations contribute a relatively small part of absorption.

5. The incoming momentum has to be large enough to allow for coherence over the dimensions of the hadron. According to Equation 18, this implies a threshold for the laboratory momentum p_L needed to coherently produce on the target of mass m a state of mass M:

$$p_L \gtrsim 2.5\,(M^2 - m^2), \qquad\qquad\qquad 30.$$

where momentum and masses are measured in GeV and we have taken $R = 1$ fm. As a numerical example, diffractive excitation of a 3-GeV system, even if allowed by other selection rules, will be depressed if p_L is less than about 25 GeV/c, while for $M = 10$ GeV, p_L must be of the order of 250 GeV/c.

In the case of diffractive elastic scattering one also expects the scattering amplitude to be mainly imaginary, at least in the forward direction. This happens if the interaction is essentially absorptive, so that the phase is imaginary and the profile function is real.

4.2   Exchange Picture of Diffraction

The optical picture of high-energy elastic scattering leads to a semiclassical approach that is very different from the usual quantum-field treatment so successful in electrodynamics. In this framework the basic process is of an exchange nature, the two interacting particles exchanging quanta of a field they are coupled to. This is the well-known picture of Coulomb scattering, where a photon is exchanged. With strong interactions this is the Yukawa theory, with pion exchange as the basic interaction between two colliding hadrons. The very nature of strong interactions does not allow, however, a perturbative expansion. The picture is even more complicated because of the great number of mesons that can be exchanged. Still, the Chew & Frautschi application of Regge's ideas (60) allows us to use a relatively simple exchange pciture for the description of hadron collisions, which is in fact the result of a tremendous averaging over elementary processes. It is out of the question to provide here an introduction to Regge models, for which we refer the reader to recent reviews (61, 62). We shall merely say that once a reaction can be labeled as proceeding through the exchange of a particular set of quantum numbers, the corresponding scattering amplitude can be approximated as

$$R(s,t) = \sum_i \gamma^i_{ab}(t)\gamma^i_{cd}(t)\eta_i(t)(s/s_0)^{\alpha_i(t)}, \qquad\qquad 31.$$

where s is the square of the c.m. energy ($s \approx 2p^a_t m_b$). Each term in the sum relates to the exchange of a Regge trajectory, which connects the spin J and the masses M of particles, having the quantum numbers exchanged in the process, through the relation $J = \alpha(M^2)$. It is this trajectory, continued from positive arguments (M^2) to negative ones (t), that fixes the phase and the energy dependence of the amplitude. The factors $\gamma^i_{ab}(t)$ and $\gamma^i_{cd}(t)$ may be interpreted as coupling constants of the exchanged trajectory with the upper line, describing the transition $a \to b$, and the lower line, describing $c \to d$. The function $\eta(t)$ is called the signature factor and has the explicit form

$$\eta(t) = \{\tau + \exp[-i\pi\alpha(t)]\}/\sin \pi\alpha(t). \qquad\qquad 32.$$

It brings in the amplitude poles due to the vanishing of the denominator for odd (even) integer values of $\alpha(t)$ according to the value $+1$ or -1 taken by the signature τ. In field theory the exchange of a particle gives rise to a pole in the amplitude. The signature factor is indeed producing this effect. The scale parameter s_0 is of the order of 1 GeV^2.

Introducing the relativistic amplitude

$$F(s,t) = 4\pi s f(q)/k, \qquad\qquad 33.$$

the full amplitude for the process $a + b \to c + d$ may be written

$$F(s,t) = R(s,t) + Z(s,t), \qquad\qquad 34.$$

where R sums the contributions from a very few Regge trajectories and Z stands for the corresponding remainder. One may now summarize over a decade of Reggeology, saying that whenever R is large, R alone, with the properties associated with Equation 31, meets an impressive amount of data and, as shown by the recent Fermilab results, is furthermore successful over a tremendous energy range. However, one still lacks a satisfactory form for the remainder Z. This is an important drawback, in particular whenever R is small. A specific example of a large $R(t)$ is provided by pion-nucleon charge exchange $(\pi^- p \to \pi^0 n)$, where the exchange trajectories have to carry one unit of charge and one unit of isospin and be even under G parity. One known meson meets all these properties, the ρ meson. The data may be fitted with a single term of Equation 31 in the ranges $5 < p_L < 100$ GeV/c and $|t| \lesssim 1$ GeV^2 with a trajectory of the form

$$\alpha(t) = \alpha_0 + \alpha' t, \qquad\qquad 35.$$

when $\alpha_0 \approx 0.5$ and $\alpha' \approx 0.7$ GeV^{-2} (Figure 10). This trajectory extrapolates well to the masses of the ρ and of the g mesons. This success is really impressive and is not isolated. One may also quote η production in πp collisions, K_S regeneration from K_L, and total cross-section differences. However, when the Regge pole amplitude is small, the remainder Z may not be neglected, as indicated for instance by the presence of a sizable polarization in $\pi^- p$ charge exchange (61, 64).

It is now tempting to assess the predicting power of a similar approach when it comes to elastic scattering, which is by all standards a large amplitude. In such a case one may tentatively introduce a leading Regge amplitude as

$$R^P(s,t) = \gamma_{aa}^P(t)\gamma_{bb}^P(t)\eta_P(t)(s/s_0)^{\alpha_P(t)}. \qquad\qquad 36.$$

The properties of the Regge trajectory that would give an a priori satisfactory form for a diffractive amplitude are readily obtained. The trajectory is associated with the exchange of the quantum numbers of the vacuum. It should be even under C- and G-conjugation and with positive signature. Since the optical theorem reads $s\sigma_t = \text{Im } F(s,0)$, by neglecting the remainder $Z(s,t)$ the total cross section becomes

$$\sigma_t^{ab} = \gamma_{aa}^P(0)\gamma_{bb}^P(0)\,\text{Im}\,|\eta_P(0)|s_0(s/s_0)^{\alpha_P(0)-1} + \sum_i \gamma_{aa}^i(0)\gamma_{bb}^i(0)\,\text{Im}\,|\eta_i(0)|s_0(s/s_0)^{\alpha_i(t)-1}. \quad 37.$$

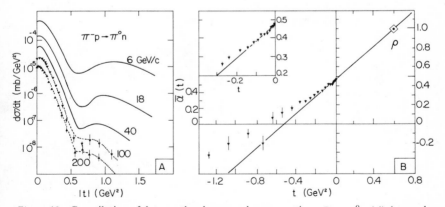

Figure 10 Compilation of data on the charge-exchange reaction $\pi^- p \to \pi^0 n$. (A) A sample of data on the differential cross section. The curves at lower energies are taken from Diddens review paper (63a). The points are some of the recent data of FNAL (63b). (B) Effective trajectory $\bar{\alpha}(t)$ derived from the FNAL data (63b). The straight line goes through the ρ and g meson masses. The inset shows the details close to $t = 0$.

The sum stands for the contribution of the nonleading trajectories, which have $\alpha_i(0) < \alpha_P(0)$. They could be neglected asymptotically ($s \to \infty$). The first term contributes a constant cross section if $\alpha_P(0) = 1$, which implies $\eta_P(0) = i$. This contribution is by construction the same for particle and antiparticle scattering off the same target (it is even under C). Since it dominates asymptotically, it satisfies the Pomeranchuk theorem, which states that the cross section of a particle and antiparticle off the same target are asymptotically equal (65). For this reason the trajectory with $\alpha_P(0) = 1$ is called a Pomeron.

Since the findings of the CERN Intersecting Storage Rings (66, 67) and of Fermilab (68, 69) we know that the total cross sections increase with s (Figure 9B). Thus, the leading term of the cross section should include at least extra logarithms in s that are not present in the Regge expansion of Equation 37. For the time being we consider the Pomeron amplitude only as a simple first approximation of the true amplitude describing shadow phenomena in hadron-hadron collisions. We then point out two consequences proper to the exchange picture of diffraction processes.

First, since $d\sigma/dt = |F|^2/(16\pi s^2)$, if Pomeron exchange with $\alpha(0) = 1$ dominates, by using the parameterization of Equation 35, one expects

$$d\sigma/dt = [\gamma_{aa}^P(t)\gamma_{bb}^P(t)]^2 \exp[2\alpha'_P t \ln(s/s_0)]/16\pi s_0^2. \qquad 38.$$

This expression implies that the slope b of the forward elastic scattering increases proportionally to $\ln s$, i.e. the diffractive peak shrinks with energy. If, as successfully done, the coupling constants γ_{aa}^P and γ_{bb}^P are parameterized by exponentials in t, Equation 38 also implies a profile function that is Gaussian in impact parameter (Figure 2D) with an interaction radius that increases logarithmically with energy: $R^2 = R_0^2 + 4\alpha'_P \ln(s/s_0)$ (Equation 16). Since the total cross section is constant, at the same time the central value of the profile must decrease: $\Gamma(0) \propto 1/\ln s$. Clearly, this prediction is very far from any black-disc analogy.

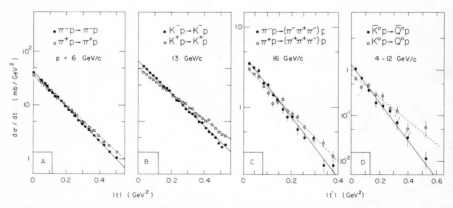

Figure 11 (A, B) Examples of forward hadron-proton elastic cross sections (70a, b). Note the crossover phenomenon. (C, D) The same effect appears in diffractive processes (70c, 124).

Second, the Pomeron amplitude contains the product $\gamma^P_{aa}\gamma^P_{bb}$, which implies factorization properties that are also not obvious in an optical picture. For instance, consider the proton-diffractive excitation initiated by a proton, a pion, or a kaon. The factorized Pomeron amplitude gives

$$\sigma(pp \to pN^*)/\sigma(pp \to pp) = \sigma(\pi p \to \pi N^*)/\sigma(\pi p \to \pi p) = \sigma(Kp \to KN^*)/\sigma(Kp \to Kp).$$

39.

Similarly,

$$[\sigma(pp \to pN^*)]^2 = 4\sigma(pp \to pp)\sigma(pp \to N^*N^*).$$

40.

Such relations should hold for differential cross sections.

In the following, these two predictions and the general features of diffraction processes discussed in the previous section are compared with the experimental data.

4.3   Elastic and Total Cross Sections

Examples of forward hadron-proton differential cross sections are plotted in Figures 11A and B. Besides the almost exponential behavior in t, we remark a "crossover" point at $|t| \simeq 0.2$ GeV^2 between the differential cross sections of particles and antiparticles. This effect may be interpreted geometrically as a consequence of the fact that antiparticles are more extended than particles in impact parameter space, so that both the forward cross section and the slope are larger for antiparticle-hadron than for particle-hadron interactions. From the exchange point of view, since the Pomeron is even under charge conjugation, its exchange gives equal particle and antiparticle cross sections. Thus the crossover phenomenon tells something about the other exchanged Regge trajectories. This point has been discussed extensively in the literature (71–73) and we do not dwell upon it here. The πp and Kp total cross sections are definitely different at $p_L \gtrsim 200$ GeV/c (Figure 9B). This implies different forward elastic cross sections, so that at $t = 0$ the Pomeron, if it acts as an isospin singlet, does not act as a unitary singlet [i.e. as a singlet under the SU(3) symmetry group of strong interactions]. This, however, seems to change with momentum transfer, since there are indications from Fermilab that the differential cross sections for πp and Kp eventually merge for $|t| \gtrsim 0.8$ GeV^2 (74a). This is connected with the long-debated Pomeron-f analogy, which we do not discuss here (74b,c).

Information on the secondary exchanges are also obtained by measuring the polarization parameter P in elastic scattering and the differences between particle and antiparticle cross sections. In the first type of experiment, hadrons are scattered on polarized protons. The technique introduced at the PS by the CERN-Pisa collaboration (75, 76) has been recently used at higher energy by the Saclay-Serpukhov-Dubna-Moscow collaboration working at Serpukhov (77). The energy dependence of the parameter P in $\pi^\pm p$ scattering at fixed t may be fitted by the interference of a Pomeron amplitude of slope $\alpha'_P = 0.27$ GeV^{-2} with a secondary "effective" trajectory of the form $\bar\alpha = 0.52 + 0.93t$ (77). The difference between particle and antiparticle cross sections gives also information on the effective intercept

$\bar{\alpha}_0$ of the secondary trajectories (Equation 37). The data plotted in Figure 12 give $\bar{\alpha}_0 \approx 0.5$, in agreement with the masses of the mesons, which can be exchanged, and with the polarization data. These two types of experiments are interesting to combine, since the latter one is only sensitive to the imaginary part of the non-flip secondary exchange contribution, while in the former one the real part of the spin-flip secondary exchange contribution mainly matters. In conclusion, the available data indicate that the secondary contributions have an effective intercept $\alpha_0 \approx 0.5$, so that they decrease about as $s^{1/2}$ with respect to the almost constant Pomeron cross section.

Proton-proton elastic scattering is very well known in large intervals of energy and momentum transfer. A compilation of data is shown in Figure 13. With increasing laboratory momentum a typical diffraction minimum develops around $|t| = 1.4 \text{ GeV}^2$. It occurs as the momentum increases between 100 and 200 GeV/c. The change between the kink appearing at lower momenta and the dip is presumably due to the gradual vanishing of the real part at these t values, as it loses the contribution of the secondary Regge trajectories. The CERN-Hamburg-Orsay-Vienna collaboration working at the CERN Intersecting Storage Rings has shown that the position of the diffraction minimum changes from $|t| = (1.44 \pm 0.02) \text{ GeV}^2$ to $(1.26 \pm 0.03) \text{ GeV}^2$ when the (c.m.) energy passes from 23 to 62 GeV, i.e. when p_L increases from 300 to 2000 GeV/c (79). This $(14 \pm 3)\%$ change of the position

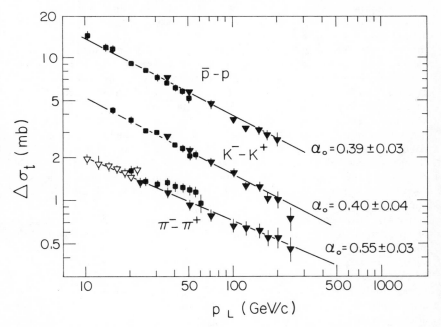

Figure 12   Energy dependence of the difference of antiparticle and particle cross sections on protons (59e).

of the minimum is equal, within the errors, to the $(11 \pm 2)\%$ increase of the total cross section and to the $(12 \pm 2)\%$ increase of the inelastic cross sections that are observed in the same energy interval (Figure 9). From an optical point of view such a coincidence is easily understood, assuming that the profile function is real and depends upon the incoming momentum only through a scale dilatation of its argument: $\Gamma(a, p_L) = \Gamma[a/R(p_L)]$. When the momentum increases, the value at $a = 0$ remains constant, while the radius $R(p_L)$ expands. This form of "geometrical scaling" (80, 81) implies that σ_t and σ_{in} are both proportional to $R^2(p_L)$, so that the data indicate that in the ISR energy range the radius of the profile function increases by $\sim 6\%$. Since b is proportional to R^2, the slope should then increase by 12%. The data plotted in Figure 14A give $(13 \pm 3)\%$.

The proton-proton forward peak shrinks with energy, and the slope b at $\langle t \rangle \approx 0.05$ GeV^2 increases logarithmically with s, as predicted by the Pomeron amplitude of Equation 38. A fit to the data of Figure 14A above $p_L \approx 30$ GeV/c gives for the

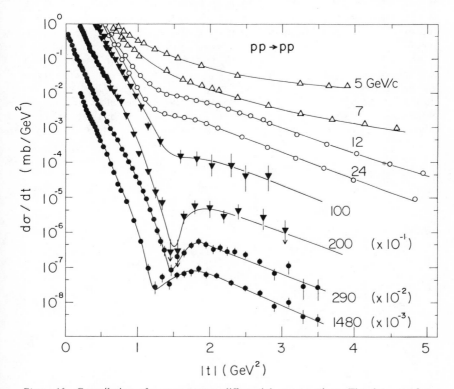

Figure 13 Compilation of proton-proton differential cross sections. The data are from Berkeley (5–7 GeV/c, 78a), PS (78b), FNAL (78c), and ISR (78d). The recent data of the CERN-Hamburg-Orsay-Vienna collaboration indicate that there is no other minimum up to $|t| \approx 8$ GeV^2 (79).

Pomeron slope $\alpha'_p \approx 0.30$ GeV^{-2}. This value agrees with the polarization measurement quoted above and is much smaller than ~ 0.8 GeV^{-2}, characteristic of all the other Regge trajectories.

Slopes computed at an average momentum transfer $\langle t \rangle = 0.2$ GeV^2 are also plotted in Figure 14. In the proton-proton channel the slope decreases, passing from $\langle t \rangle = 0.05$ GeV^2 to $\langle t \rangle = 0.2$ GeV^2. This reflects the sharp rise of the differential cross section for momentum transfers smaller than ~ 0.1 GeV^2 (84). The compilations of Figure 14 show that while the slope of $\pi^- p$ and $K^- p$ elastic scattering is almost energy-independent, the $\pi^+ p$ and the $K^+ p$ slopes increase with energy, i.e. their forward elastic peak shrinks. Indeed, one expects strong constructive interference between the first and the second term in Equation 37 for $K^- p$, but practically none for $K^+ p$.

Total cross sections for the usual six channels are plotted in Figure 9B. For a discussion of the relative values of these cross sections and their energy dependence in terms of the quark model we refer to the recent review by Wetherell (59a). A comparison between the energy dependences of the total cross sections (Figure 9) and the slopes (Figure 14) indicates that the channels whose total cross section varies more rapidly show also a faster increase of the slope. This is quantitatively shown in Figure 15, where the ratio b/σ_t is plotted versus s (85, 86). The fact that this ratio stays constant with energy is in agreement with the hypothesis that geometrical scaling is qualitatively valid in all six channels. The same conclusion may be reached by considering the energy dependence of the ratios σ_{el}/σ_t plotted in Figure 16. This figure demonstrates that at large enough energies these ratios are practically constant. It also shows that in all channels the ratio σ_{el}/σ_t is

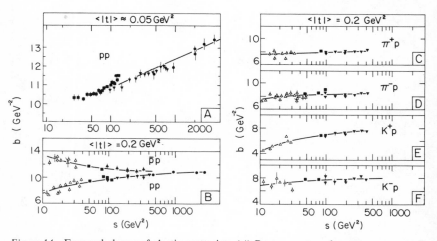

Figure 14 Forward slopes of elastic scattering. (*A*) Proton-proton slope at a very small average momentum transfers (82a–d, 84). The line represents the relation $b = (8.0 + 0.62 \ln s_{\text{GeV}^2})$. (*B–F*) Slopes computed for an average momentum transfer $\langle t \rangle \approx 0.2$ GeV^{-2} (82e–j).

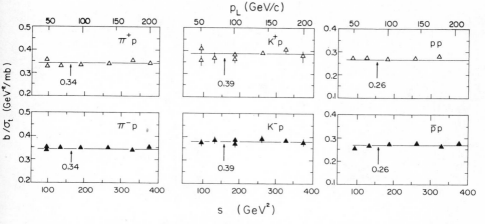

Figure 15   Ratios of the slope b to the total cross section σ_t as function of s, the square of the center-of-mass energy, and of p_L, the laboratory momentum (95, 86).

very far from 1/2, which is the value expected for a black disc (Equations 13 and 14). (For the proton-proton case, the fact that the profile function is very far from a black disc appears very clearly in the inset of Figure 3*C*.)

Geometrical scaling is a rather restrictive model in which σ_t, σ_{el}, and b are all proportional to each other. Actually, these three quantities must always satisfy a bound based on unitarity obtained by MacDowell & Martin (88):

$$\frac{\sigma_t^2}{18\pi\sigma_{el}} \leqq 2\left[\frac{d}{dt}\ln.\mathrm{Im}\,F(t)\right]_{t=0}. \qquad 41.$$

When the real part of the amplitude is zero ($\rho = 0$), the right-hand side of this inequality equals $b(0)$, i.e. the logarithmic slope of the cross section measured at $|t| = 0$. There is experimental evidence that the slope does not vary appreciably between $|t| = 0.05$ GeV^2 and $|t| = 0$, so that $b(0)$ can be obtained from the fit of Figure 14*A*. At $p_L = 300$ GeV/c the real part ρ is zero and the right-hand side of Equation 41 is numerically equal to (11.8 ± 0.25) GeV^{-2}. At the same momentum the left-hand side is equal to (10.0 ± 0.4) GeV^{-2}, so that the MacDowell-Martin bound is saturated within $\sim 20\%$. This could be expected, since the differential cross section is almost exponential in t and with a purely exponential behavior the identity $b = \sigma_t^2/16\pi\sigma_{el}$ holds for an imaginary amplitude.

4.4   Real Parts of the Strong Interaction Nuclear Amplitudes

The phase of the strong nuclear-scattering amplitude can be obtained by measuring its interference with the Coulomb-scattering amplitude of the colliding hadrons. The form of this Coulomb amplitude is well known, at least to first order in $\alpha = e^2/\hbar c$: $f_C = 2\alpha k/t$. For very small values of the four-momentum transfer ($|t| \approx 0.005$ GeV^2), this amplitude is of the same order of magnitude as a typical nuclear amplitude f

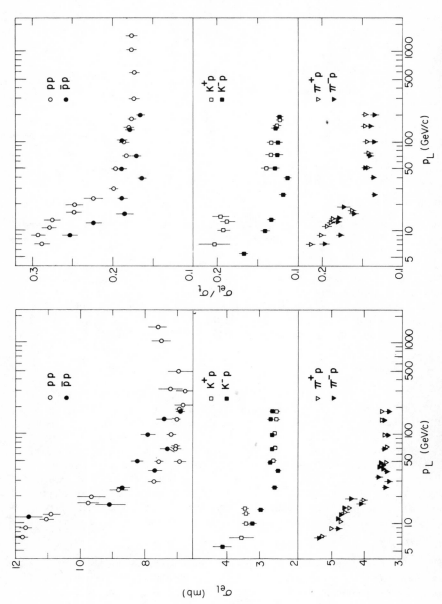

Figure 16 (*A*) Compilation of elastic cross-sections (87a–c). (*B*) Energy dependence of the ratio σ_{el}/σ_t.

(Equation 18) and the interference term gives a non-negligible contribution to the differential cross section: $d\sigma/d\Omega = |f|^2 + |f_C|^2 + 2f_C \, Re f$. The interference term is proportional to ρ, the ratio of the real to the imaginary part of the strong amplitude. Small corrections due to the hadron form factor and to the small imaginary part of the Coulomb amplitude are easily introduced (89), and the measurement of elastic scattering at very small values of $|t|$ gives information on ρ. The experimental data are plotted in Figure 17. It is seen that at high energies the parameter ρ is small, i.e. the nuclear amplitude is mainly imaginary, as expected for a diffractive process driven by a large absorption.

In optics the real part of the refractive index of a medium at a given frequency may be expressed as an integral over a function of the imaginary part of the refraction index extended to all frequencies. Similar dispersion relations, as a consequence of the causality principle, connect the real and imaginary parts of the nuclear scattering amplitude. Since the imaginary part of the forward amplitude is proportional to the total cross section, at any given energy the value of ρ may be expressed as an integral of the total cross section for particle and antiparticle over the whole energy range with special handling, if needed, of the unphysical (below elastic threshold) region (90, 91). In practice the value of ρ is mainly sensitive to the local derivative of the total cross section, since to a first approximation the result of the dispersion relation may be written in the form (92)

$$\rho \approx \frac{1}{\sigma_t}\left[tg\left(\frac{\pi}{2}\frac{d}{d \ln s}\right)\right]\sigma_t = \frac{1}{\sigma_t}\left[\frac{\pi}{2}\frac{d}{d \ln s} + \frac{1}{3}\left(\frac{\pi}{2}\right)^3\frac{d^3}{d \ln s^3} + \cdots\right]\sigma_t(s). \qquad 42.$$

This expression assumes that the scattering amplitude is even under crossing, as it should be asymptotically. It indicates, for instance, that if the cross section rises at the maximum rate allowed by the Froissart bound (93), i.e. as $\ln^2 s$, the parameter ρ approaches zero as $1/\ln s$ from positive values. This statement is indeed a much more general consequence of field theory when the total cross section increases asymptotically. It is the Khuri-Kinoshita theorem (94). Figure 17 shows that ρ indeed becomes positive in all channels in the range $100 \leq p_L \leq 200$ GeV/c in agreement with the observed rise in the total cross sections (Figure 9) and the Khuri-Kinoshita theorem. What happens to the real part at even larger energies is a very relevant question in the field of elastic diffraction because, as indicated by Equation 42, it could give a hint on the higher-energy dependence of the total cross section. At present the CERN-Rome collaboration is performing an ISR experiment that should provide values of ρ in proton-proton scattering up to $p_L \simeq 2000$ GeV/c.

4.5   Unitarity and Diffraction Scattering

Probability conservation implies the unitarity of the S matrix: $S^\dagger S = 1$. By introducing the transition matrix $(S = 1 + iT)$, the unitarity relation reads $i(T - T^\dagger) = T^\dagger T$. Sandwiching this between the initial state $|i\rangle$ of the two colliding hadrons and the outgoing elastic state $|f\rangle$, and using the usual normalization, one gets

$$4\pi \, Im f/k = \sum_{m \; el} \langle f|T^\dagger|m\rangle\langle m|T|i\rangle + \sum_{n \; in} \langle f|T^\dagger|n\rangle\langle n|T|i\rangle. \qquad 43.$$

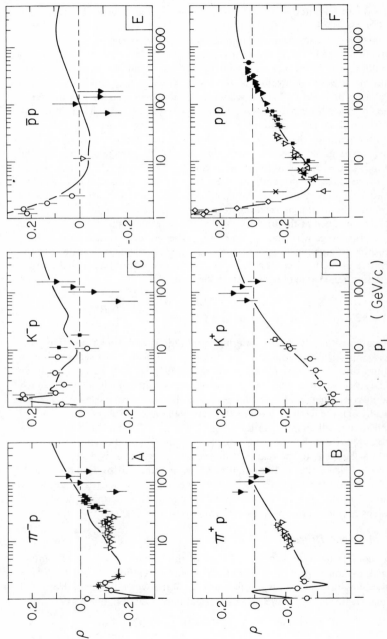

Figure 17 Ratio of the real to the imaginary part of the forward strong amplitude plotted versus the laboratory momentum for the usual six elastic channels. The symbols correspond to the following references: ○ = 95a–d; * = 95e; ✳ = 59c, 95f; ■ = 95g, 95h, 95i; □ = 95i; ◇ = 95j; × = 95k; ▲ = 95m; ▼ = 95n; and ● = 95o. The curves are from Hendrick & Lautrup (95p). They have been obtained by introducing in dispersion relations total cross sections that fit the data and increase asymptotically as ln *s* (95g).

The left-hand side is proportional to the imaginary part of the elastic scattering amplitude. The expression on the right-hand side has been obtained by introducing in $\langle f|T^{\dagger}T|i\rangle$ a complete set of states, and then distinguishing the elastic intermediate state $|m\rangle$ from the inelastic ones $|n\rangle$. Equation 43 is diagrammatically represented in Figure 18. It states that the imaginary part of the scattering amplitude receives contributions both from elastic and inelastic intermediate states. Following Van Hove (96), the two terms on the right of the Equation 43 are named elastic and inelastic overlap functions. They are functions of the momentum transfer t:

$$4\pi \operatorname{Im} f(t)/k = G_{el}(t) + G_{in}(t). \qquad\qquad 44.$$

For $t = 0$ this equation reduces to the optical theorem. Indeed, the overlap functions are so normalized that $G_{el}(0) = \sigma_{el}$ and $G_{in}(0) = \sigma_{in}$. If the real part of the elastic amplitudes is negligible, $G_{el}(t)$, which is expressed as an integral of the type $\int f^*(\mathbf{q}')f(\mathbf{q}-\mathbf{q}')d\mathbf{q}'$, contains only Im f, and Equation 44 becomes a nonlinear integral equation in $f(\mathbf{q})$, once the inelastic overlap function is supposed to be known. In this "shadow" approximation the knowledge of the inelastic transition matrix elements $\langle n|T|i\rangle$ allows the calculation of the elastic amplitude. This is the program of all s-channel approaches to high-energy elastic scattering: use a model for the inelastic processes to compute the inelastic overlap function and then solve Equation 44 for the elastic amplitude. The first step is the most difficult one because the model must specify both the modulus and the phase of the inelastic matrix elements. Physically this is due to the fact that the phase is related to the position in space where the particles are produced (97). Many attempts have been performed along these lines and the main results may be summarized as follows. If the production channels are described by an uncorrelated jet model (96–98) or by a multiperipheral model (99–101), it is impossible to reproduce the slope of the elastic amplitude. This difficulty may be cured by introducing correlations among the produced hadrons (99–101), a phenomenon now known to be present in production processes.

Above which laboratory momentum is the shadow approximation expected to be a good approximation? Since the integral of the overlap function $G_{in}(a)$ is proportional to the inelastic cross-section, it is natural to look to the energy dependence of σ_{in}. The data plotted in Figure 9A clearly indicate that one has to distinguish between two cases. In the pp and K^+p channels the inelastic cross section rises continuously, while in the others it decreases with the laboratory momentum up to at least 50 GeV/c.

This behavior is connected with the absence of direct formation resonances in the exotic channels pp and K^+p (87c). Thus these two reactions are the most suitable to study shadow effects at not-too-high energy. However, in these two cases the elastic cross section also decreases with energy for momenta smaller than 20–30 GeV/c

Figure 18   Diagrammatic representation of unitarity condition for elastic scattering.

(Figure 16A). This is due to the nondiffractive contributions, i.e. to the exchange of secondary Regge trajectories. A phenomenological fit to the cross-section data by Morrison (87c) indicates that the sum of these contributions decreases at s^{-1}. In conclusion, only above 20 GeV/c may one expect the shadow approximation to be a reasonable one. For the other nonexotic elastic channels the energy should be even higher. Equation 44 has often been used the other way around, i.e. to extract the inelastic overlap integral from the knowledge of $f(t)$. This is better done by going to impact-parameter space:

$$2 \operatorname{Re} \Gamma(\mathbf{a}) = |\Gamma(\mathbf{a})|^2 + G_{\text{in}}(\mathbf{a}). \qquad 45.$$

Due to angular-momentum conservation the knowledge of $\Gamma(a)$ determines the inelastic overlap function $G_{\text{in}}(a)$ at the same impact parameter. Note that the integral of G_{in} on all impact parameters gives the inelastic cross section (Equation 13). If the scattering amplitude is purely imaginary, one can write $\Gamma(a) = 1 - \exp[-\chi(a)]$ and Equation 45 gives immediately $G_{\text{in}}(a) = 1 - \exp[-2\chi(a)]$. Also, if the scattering amplitude has a non-negligible real part, Equation 45 may be solved, provided one knows the momentum transfer dependence $\rho(t)$ of the real part. Such a calculation has been recently performed, using dispersion relations to determine $\rho(t)$ (102). The computed proton-proton overlap integral at $p_L = 1480$ GeV is plotted in Figure 19A and its variation as a function of energy is shown in Figure 19B. This detailed analysis has shown not only that the rise of the total cross section is due to a peripheral increase of the overlap function (83), but also that above 50 GeV/c, $G_{\text{in}}(a)$ satisfies geometrical scaling. This is not the case for the profile function because the scattering amplitude has a non-negligible real part at lower energies. At high energies, when $\operatorname{Im} \Gamma \approx 0$, geometrical scaling of G_{in} implies the same property for $\operatorname{Re} \Gamma$. In conclusion, geometrical scaling, which is suggested for all channels by the data of Figure 15, is well satisfied by proton-proton elastic diffraction (103, 104). However,

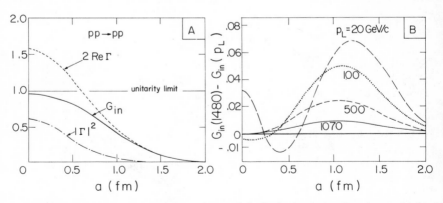

Figure 19   (*A*) Impact-parameter dependence of the inelastic-overlap function $G_{\text{in}}(a)$ for proton-proton scattering at $p_L = 1480$ GeV/c. The dashed and dash-dotted curves represent the other two terms appearing in the unitarity conditions of Equation 45 (102). (*B*) Energy variation of the inelastic overlap integral as computed by Grein et al (102).

a dynamical explanation for its origin is still missing. Recently, the energy-independence of the inelastic-overlap function for **a** = 0 has been related to the quark-glue picture of the protons (104). We come back to some of these points later, discussing present approaches to the Pomeron.

5   DIFFRACTION DISSOCIATION IN HADRON-HADRON COLLISIONS

5.1   *Missing-Mass Spectra*

Diffraction dissociation of protons in the reaction $a + p \rightarrow a + X$, where a is a hadron, was studied systematically at AGS and PS energies, using the process commonly called the missing-mass method. The experimental approach consists of measuring the momentum vector of the scattered hadron a by means of a magnetic spectrometer. The mass M of the excited system X can then be deduced from energy- and momentum-conservation laws.

The observed missing-mass distributions show peaks superimposed on a continuous background. Two typical examples of such data are displayed in Figure 20. The missing-mass spectrum obtained by Anderson et al (105) at low momentum transfer with π^- of 8 GeV/c is shown in Figure 20A. The clear signal at $M \approx 1.7$ GeV is attributed to the excitation of the well-known N(1690) baryon resonance. The $\Delta(1236)$ appears as a shoulder on the lower side of the missing-mass spectrum. The prominent bump at $M = 1.4$–1.5 GeV is interpreted as being due to an N(1400) state with some contribution from the N(1520) resonance. In Figure 20B, missing-mass spectra, obtained with incident protons of 24 GeV/c by the CERN-Rome

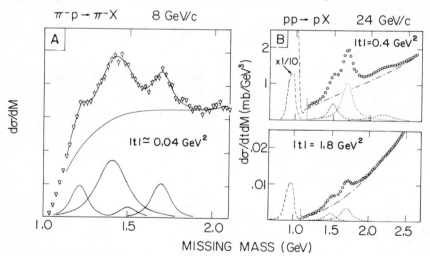

Figure 20   Missing-mass spectra for proton dissociation. (*A*) AGS data of (105). (*B*) PS data of (78b). The lines represent the results of fits obtained by adding the resonance contribution with Breit-Wigner shape to a smooth continuum.

group (78b) at larger momentum transfer, are shown. The excitation of the N(1520) and N(1690) resonances is clearly seen. At $-t = 0.4$ GeV^2 also the N(2190) gives an appreciable signal.

The identification of the peaks at $M \approx 1.5$ and 1.7 with the excitation of the N(1520) and N(1690) resonances having $J^P = 3/2^-$ and $5/2^+$, respectively, which are well-known from formation experiments, is supported by isospin and spin-parity analysis of bubble-chamber data, as discussed in the next section. We notice that for these resonances the Gribov-Morrison rule is satisfied (106).

The presence of a peak at $M \approx 1.4$ GeV, which rapidly disappears as the momentum transfer increases, is a common feature of all missing-mass experiments. However, this peak is not clearly identified with any known nucleon resonance. It appears in fact at a too-low mass value to allow a convincing identification with the $J^P = 1/2^+$ N(1470) resonance found in phase-shift analysis. At present the most likely interpretation (106, 107) of the 1.4-GeV peak is that it does not correspond to a resonance. It is thus indicated as a "kinematical enhancement."

The missing-mass method allows the study of diffraction dissociation up to high values of the momentum transfer; however, the lack of any information on the decay of the produced system represents a serious limitation.

5.2   Mass Distributions of Diffracted States

The mass distribution of a diffracted state can be studied either in bubble chambers or in counter experiments with a large-aperture multiparticle spectrometer by observing the decay products of the produced system.

We consider first *nucleon dissociation*. The excitation of the proton into the $n\pi^+$ system was studied at the Split Field Magnet of the CERN ISR at $(s)^{1/2} = 45$ GeV by the CERN-Hamburg-Orsay-Vienna collaboration (108). In Figure 21A the measured $n\pi^+$ mass distribution is shown. It exhibits peaks associated with the excitation

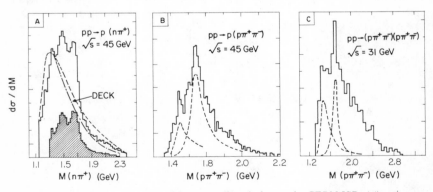

Figure 21   Invariant-mass spectra for proton dissociation at the CERN ISR: (A) $n\pi^+$ mass spectrum from (108). The lines are the result of a Deck-model calculation explained in Section 5.4; (B) $p\pi^+\pi^-$ mass spectrum from reference 109; (C) $p\pi^+\pi^-$ mass spectrum observed in double diffraction dissociation (110). The dashed lines in (B) and (C) represent resonance contributions.

of the N(1520) and N(1690) resonances. The signal due to these resonances is enhanced relative to the background when a lower cut is made on the momentum transfer of the incident proton to the neutron (shaded area of Figure 21A). The excitation of the proton into the $p\pi^+\pi^-$ system was studied at the same center-of-mass energy by the Aachen-CERN-UCLA-Riverside collaboration (109). The mass distribution of the $p\pi^+\pi^-$ system, reported in Figure 21B, shows two peaks that are again interpreted as due to the excitation of the N(1520) and N(1690) resonances, respectively. Diffractive dissociation of both incident protons was studied at the ISR in the channel $pp \rightarrow (p\pi^+\pi^-)(p\pi^+\pi^-)$ by the Pavia-Princeton collaboration (110). The mass spectrum of the $p\pi^+\pi^-$ system observed in this reaction (Figure 21C) is very similar to that obtained when the other proton does not dissociate (Figure 21B).

The quantum numbers of the state resulting from excitation of the proton were studied in several bubble-chamber experiments and have been summarized by Morrison (107). As an example we report in Figure 22 the results of the isospin analysis of the nucleon-pion system excited by incident π^+ of 8 and 16 GeV/c from the Aachen-Berlin-Bonn-CERN-Cracow collaboration (111). The simultaneous observation of the two reactions $\pi^+ p \rightarrow p\pi^+\pi^0$ and $\pi^+ p \rightarrow n\pi^+\pi^+$, where the nucleon-pion system appears in different charge states, allows one to separate the contributions of the isospin $I = 1/2$ and $I = 3/2$ states. As shown in Figure 22, the $I = 3/2$ state is dominated by the $\Delta(1236)$, whose cross section decreases quickly with energy, while the $I = 1/2$ state shows a broad enhancement around 1.4 GeV and a peak at 1.69 GeV. Spin-parity analysis (112a) of the nucleon-pion system, based on the study of their decay angular distribution in their rest frame, has

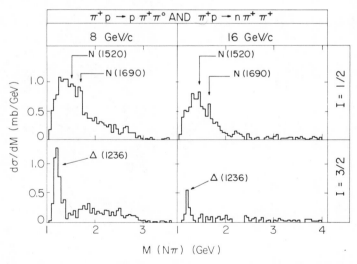

Figure 22   Results of isospin decomposition of the $N\pi$ system produced in $\pi^+ p$ interactions at 8 and 16 GeV/c (111).

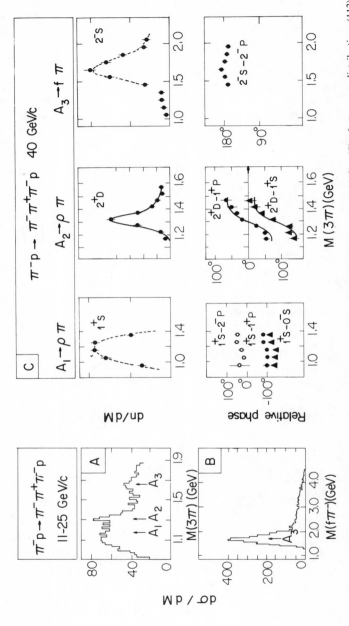

Figure 23 (A) $\pi^+\pi^-\pi^-$ mass spectrum from a world compilation of bubble-chamber data (113); (B) $f\pi^-$ mass distribution (113); (C) results of spin-parity analysis for A_1, A_2, and A_3. The mass distribution of the dominant spin-parity state is shown together with relative phases (115).

shown that the mass regions around 1.5 and 1.7 GeV are dominated by states with $J^P = 3/2^-$ and $5/2^+$, respectively. The identification with the N(1520) and N(1690) resonances is then clear, in agreement with the Gribov-Morrison rule. This rule is instead violated by some recent data showing a $1/2^-$ contribution in the $N\pi$ system around $M \approx 1.35$ GeV (112b).

The diffractive *pion dissociation* into a three-pion system has been the subject of thorough investigations in the past few years. The $\pi^+\pi^-\pi^-$ mass spectrum resulting from a world compilation by Ascoli et al (113) from bubble-chamber data on the reaction $\pi^- p \to \pi^+\pi^-\pi^- p$ at incident momenta between 11 and 25 GeV/c, is shown in Figure 23A. The spectrum is dominated by a broad bump in the mass interval $M(\pi^+\pi^-\pi^-) = 1.0$–1.4 GeV, which is due to A_1 and A_2 production. As discussed in Section 3.3, these two states are known to decay mainly into the $\rho\pi$ channel (Figure 4C). The peak at $M(\pi^+\pi^-\pi^-) \approx 1.65$ GeV represents the A_3 enhancement that essentially decays into $f\pi$. This statement is supported by the data of Figure 23B, which shows the shape of the 3π mass spectrum for events with at least one $\pi^+\pi^-$ combination in the f^0 mass region. Partial-wave analyses of the 3π system have provided a determination of the relevant quantum numbers (38, 113–115). The results of a partial-wave analysis for the reaction $\pi^- p \to \pi^-\pi^+\pi^- p$ at 40 GeV/c by the CERN-IHEP collaboration (115) are shown in Figure 23C. The mass distributions for the individual spin-parity states, in the $\rho\pi$ and $f\pi$ channels, are shown. The A_2 bump can be well fitted by a Breit-Wigner shape. The relative phase between the dominant partial wave and other partial waves that are not associated with a bump in the mass spectrum is displayed at the bottom of Figure 23C. For a resonance one expects such relative phases to increase by 90° over the full width of the resonance. This is indeed the case for the A_2, which thus shows up as a real resonance. For A_1 and A_3, however, no appreciable variation of the relative phase is observed. This indicates that these states cannot be considered as resonances.

Present evidence concerning the 3π system can be summarized by saying that the A_1 ($M \approx 1.15$ GeV) is a $J^P = 1^+$ state in the $\rho\pi L = 0$ partial wave, and the A_3 ($M \approx 1.65$ GeV) is a $J^P = 2^-$ state in the $f\pi L = 0$ partial wave. These states do not show the behavior typical of resonances, while the A_2 ($M \simeq 1.31$ GeV) appears as a genuine $\rho\pi$ resonance in the partial wave $L = 2$ with $J^P = 2^+$. As already remarked in Section 3.3, the A_1 and the A_3 states satisfy the Gribov-Morrison rule, while the A_2 does not.

We now consider kaon dissociation. The $K\pi$ mass spectrum, as observed in kaon-proton interactions, is dominated by the $K^*(890)$ and $K^*(1420)$ resonances, which have spin-parity 1^- and 2^+, respectively, and thus do not satisfy the Gribov-Morrison rule. They are not produced diffractively, as indicated by the rather steep energy dependence of their production cross section, to be shown in the next section. The $K\pi\pi$ mass spectrum (Figure 24A) shows a broad enhancement for $M(K\pi\pi) = 1.2$–1.5 GeV, which is mainly due to the production of the Q system (with perhaps some contribution from the $K^*(1420)$ meson) and, in addition, a less prominent bump at $M(K\pi\pi) \approx 1.75$ GeV (116), called L enhancement. An isospin analysis (117) of the Q system in the decay mode $K^*(890)\pi$ shows that this enhancement has $I = 1/2$ (Figure 24B). Spin-parity analyses of the Q and L systems have usually identified the Q as a $J^P = 1^+$ state that decays in s-waves in the $K^*(890)\pi$ and $K\rho$ channels, and the L as a $J^P = 2^-$ state with dominant s-wave decay into $K^*(1420)\pi$. However, as shown by the Aachen-Berlin-CERN-London Vienna collaboration (118a), the Q and L mass enhancements have a more complex structure, other partial waves being also present. It turns out that the Q and L systems, in spite of being complex structures, are a mixture of only unnatural spin-parity states (0^-, 1^+, 2^-, 3^+ ...), produced by exchange of a system that is mainly in a natural parity state. This is illustrated in Figures 24C and 24D. Recent results indicate the presence of two 1^+ states in the Q region (118b).

In reviewing the exclusive reactions that exhibit a diffractive behavior, we have met examples of the production of systems that do not correspond to resonances. This is the case for the N(1400) in the nucleon dissociation, the A_1 and A_3 in pion dissociation, and the Q in kaon dissociation. A common feature of all these systems is that they seem to be produced only in diffractive processes and have a rather small mass. A mechanism (the Deck effect) was proposed to describe the diffractive production of nonresonant systems. It is discussed in Section 5.4.

5.3 Differential and Total Cross Sections

A great deal of information is now available on the momentum transfer and energy dependence of diffractive-like processes. It is clearly impossible, due to lack of space, to cover the whole subject. We therefore limit ourselves to discussing the general features of the data and presenting a few typical examples of the most recent experimental results. For detailed discussions we refer the reader to the reviews of Lubatti (37), Morrison (107), Leith (119), and Kane (120).

In Figure 25A the momentum-transfer distributions of the diffractive processes

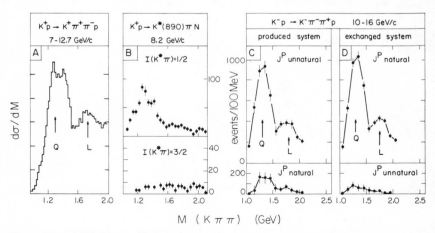

Figure 24 (*A*) World compilation of $K\pi\pi$ mass spectrum showing the Q and L enhancements (116); (*B*) isospin decomposition of the $K\pi\pi$ system (117); (*C*) results of spin-parity analysis of the $K\pi\pi$ system produced in the reaction $K^-p \rightarrow K^-\pi^-\pi^+p$ (118a); and (*D*) spin-parity of the system exchanged in the same reaction.

$pp \to pN^*$, measured at 24 GeV/c by the CERN-Rome collaboration (78b) in the low-t region are reported and compared to the shape of the elastic cross section. All reactions exhibit a forward peak, like elastic scattering, with an exponential slope b, which decreases with the mass of the excited system. In fact, for the N(1400) enhancement the t distribution is very steep ($b \approx 25 \text{ GeV}^{-2}$), more than for elastic scattering, while for the heavier nucleon isobars N(1520), N(1690), and N(2190) the t distribution is more gentle. This is a common feature of all diffractive processes. A similar dependence of the slope upon the mass of the diffractively produced system is also observed for pion dissociation into three pions and for kaon dissociation into the $K\pi\pi$ system. As an example of this feature of the data, we report in Figure 26A the t distributions for the process $pp \to p(n\pi^+)$ measured at the ISR by the

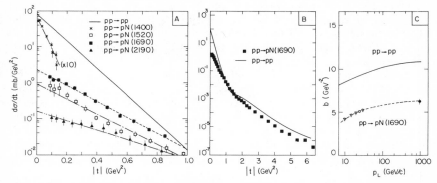

Figure 25   (A), (B) Momentum-transfer distributions of the exclusive reactions $pp \to pN^*$ at 24 GeV/c (78b); (C) compilation of the forward slope b for the reaction $pp \to pN(1690)$ as a function of energy (47, 78b, 109). The continuous curve represents elastic scattering data.

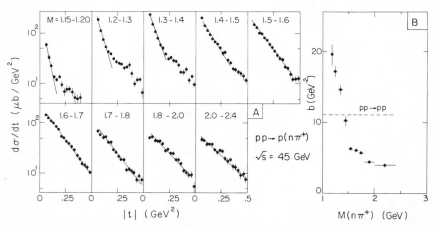

Figure 26   (A) Momentum-transfer distributions for the reaction $pp \to p(n\pi^+)$ at $(s)^{1/2} = 45$ GeV in different intervals of the $n\pi^+$ effective mass M (108); (B) dependence of the forward slope b on the $n\pi^+$ mass.

CERN-Hamburg-Orsay-Vienna collaboration (108). A strong correlation is clearly present between the slope parameter b, describing the low-t region, and the mass of the ($n\pi^+$) system (Figure 26B).

Concerning the large-t region outside the diffraction peak, only data at low energy are now available. The differential cross section for the reaction $pp \rightarrow pN(1690)$ measured by the CERN-Rome collaboration (78b) up to large momentum transfer ($-t \simeq 6$ GeV^2) is compared in Figure 25B with the shape of the elastic cross section. As shown in Figure 13, at low energies proton-proton elastic scattering starts to develop a structure that at ISR energies eventually becomes a very clear diffractive-like minimum. The shape of the momentum-transfer distribution for the reaction $pp \rightarrow pN(1690)$ at large t closely follows that of the elastic-scattering distribution, providing evidence for a common diffractive mechanism. Some evidence also exists for a shrinking of the forward peak in the process $pp \rightarrow pN(1690)$, in close analogy to the well-known shrinking of the elastic peak, as shown in Figure 25C, where data (47, 78b) at AGS-PS energies are compared to a recent ISR result (109).

The energy dependence of the total cross section for the two-body reactions $pp \rightarrow pN^*$ has been studied over a very wide energy range. The present experimental situation is illustrated in Figure 27, where data on the diffractive processes with excitation of the N(1400), N(1520), and N(1690) states are reported together with those referring to the reaction $pp \rightarrow p\Delta^+(1236)$, which involves change of isospin. The difference in energy dependence for the $I = 1/2$ states, all of which satisfy the Gribov-Morrison rule, and the $I = 3/2$ resonance is striking.

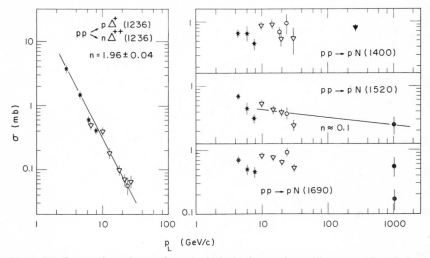

Figure 27 Energy dependence of two-body inelastic reaction with proton dissociation. Compilation of data from NIMROD (127b), AGS (47, 127a), PS (78b), FNAL (46), and ISR (108, 109) on the reactions $pp \rightarrow p\Delta$ (1236), $pp \rightarrow pN$ (1400), $pp \rightarrow pN$ (1520), and $pp \rightarrow pN$ (1690).

The energy dependence of two-body inelastic reactions is customarily represented by means of the power law $\sigma \propto 1/p_L^n$ (49). This is not just a convenient way of parameterizing the data. In fact, in the Regge picture, if the exchange of only one trajectory dominates, then the energy dependence neglecting logarithmic terms would be of the form $\sigma \propto s^{\alpha(0)-1}$, i.e. a simple power law (Equation 37).

In Table 2 we have collected the value of the exponent n for two-body inelastic reactions where a hadron dissociates without changing its intrinsic quantum numbers. We also state whether the Gribov-Morrison rule is satisfied or not. In addition, we give in Table 3 a few examples of reactions induced by pions where the

Table 2   Energy dependence of reactions $a+p \rightarrow b+p^a$

Reaction	Gribov-Morrison rule	n	Momentum range (GeV/c)	Ref.
$pp \rightarrow pN(1400)$	yes	≈ 0	5–280	this paper
$pp \rightarrow pN(1520)$	yes	≈ 0.1	5–1000	this paper
$pp \rightarrow pN(1690)$	yes	0–0.3	5–1000	this paper
$\pi^- p \rightarrow A_1^- p$	yes	0.40 ± 0.07	5–40	115
$\pi^- p \rightarrow A_2^- p$	no	0.51 ± 0.06	5–40	115
$\pi^- p \rightarrow A_3^- p$	yes	0.57 ± 0.21	10–40	115
$K^- p \rightarrow K^{*-}(890)p$	no	1.48 ± 0.04	3–40	121
$K^+ p \rightarrow K^{*+}(890)p$	no	1.90 ± 0.07	3–16	122
$K^- p \rightarrow Q^- p$	yes	0.4 ± 0.1	4–14	123
$K_L^0 p \rightarrow Q^0 p$	yes	0.59 ± 0.16	5–11	124
$K^+ p \rightarrow Q^+ p$	yes	0.67 ± 0.07	4–12	116
$K^- p \rightarrow K^{*-}(1420)p$	no	1.4 ± 0.1	4–40	121

[a] a and b have the same quantum numbers except for spin and parity. The value of the coefficient n from a fit $\sigma \propto p_L^{-n}$ is quoted, together with the corresponding momentum range.

Table 3   Energy dependence of some nondiffractive two-body inelastic reactions

Reaction	Type	n	Momentum range (GeV/c)	Ref.
$\pi^- p \rightarrow \rho^- p$	G-parity	1.87 ± 0.09	4–40	125
$\pi^+ p \rightarrow \rho^+ p$	change of	1.93 ± 0.03	2–15	126
$\pi^+ p \rightarrow B^+ p$	the meson	1.1 ± 0.2	4–18	126
$pp \rightarrow n\Delta^{++}(1236)$		1.96 ± 0.04	3–24	127a
$\pi^+ p \rightarrow \pi^0 \Delta^{++}(1236)$	isospin	1.85 ± 0.05	2–13	126
$\pi^+ p \rightarrow \eta \Delta^{++}(1236)$	exchange	1.8 ± 0.1	2–18	126
$\pi^+ p \rightarrow \omega \Delta^{++}(1236)$		1.86 ± 0.06	2–18	126
$\pi^- p \rightarrow K^0 \Lambda$		1.9 ± 0.2	2–6	128
$\pi^- p \rightarrow K^+ \Sigma^-$	strangeness	3.7 ± 0.4	2–4	128
$\pi^+ p \rightarrow K^+ \Sigma^+$	exchange	2.7 ± 0.1	1.5–5.4	126

produced meson has a G parity different from that of the pion, and examples of reactions involving change of isospin and also exchange of strangeness. We refer the reader to the review of Fox & Quigg (61) for a comprehensive treatment of two-body reactions together with their description in the Regge framework.

Considering proton dissociation into an $I = 1/2$ state, we see that in spite of present uncertainties in the high-energy data, which are mainly related to the difficulty of extracting the resonance contribution from the observed mass spectra, the wide range of energy that has been explored allows one to limit the value of n to the interval $n = 0$–0.3. This very slow energy dependence is quite similar to that of elastic scattering.

In Figure 28 we display the energy dependence of reactions where a pion dissociates. The reaction $\pi^- p \to \rho^- p$, which involves G parity exchange, has a steep energy dependence ($n \simeq 2$). On the contrary, for the reaction $\pi^- p \to A_1^- p$, $\pi^- p \to A_2^- p$, $\pi^- p \to A_3^- p$, where the intrinsic quantum numbers of the produced mesonic system are the same as those of the pion, and where only spin and parity may change, the cross section falls off rather slowly with energy, the fitted exponent n being about 0.5.

The Gribov-Morrison rule is satisfied for A_1 and A_3 production, but not for that of A_2. We notice that the t distribution for A_1 and A_3 shows clearly a diffractive peak, while for the production of the A_2 on protons (129) and coherently on nuclei (41), a forward dip is present. This is associated with important spin-flip contributions.

The energy dependence of some two-body reactions with kaon dissociation is illustrated in Figure 29. The cross section for production of the $K^*(890)$ and $K^*(1420)$ states, which do not satisfy the Gribov-Morrison rule, falls off rapidly

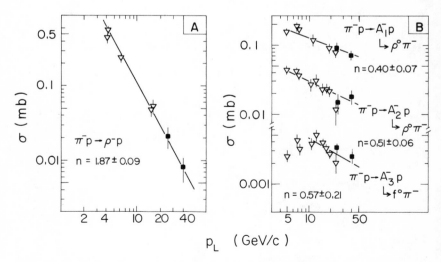

Figure 28   Energy dependence of two-body inelastic reactions with pion dissociation: $\pi^- p \to \rho^- p$, $\pi^- p \to A_1^- p$, $\pi^- p \to A_2^- p$, and $\pi^- p \to A_3^- p$ up to 40 GeV/c (115, 125).

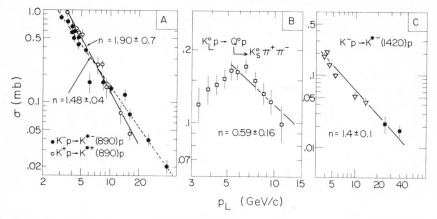

Figure 29 Energy dependence of two-body inelastic reactions with kaon dissociation into $K^*(890)$, $K^*(1420)$, and Q (121, 122, 124).

with energy ($n \approx 1.5$), while the production of the Q system has a rather slow energy dependence ($n \approx 0.5$).

We may therefore conclude that reactions proceeding without change of the intrinsic quantum numbers, and where spin and parity obey the Gribov-Morrison rule, represent a quite distinct class characterized by a low value of the exponent n (0–0.5). Similar values are also found for elastic cross sections in the low- and intermediate-energy region (Figure 16A). This fact points once more to a common mechanism for elastic and dissociation processes. We have seen that these diffractive reactions involve excitation of the nucleon into the $I = 1/2$ states N(1400), N(1520), and N(1690), of the pion into the A_1 and A_3 states and of the kaon into the Q system. The slightly higher value of the parameter n found for the pion and kaon reaction, with respect to the proton ones is probably related to nondiffractive contributions still present in the low-energy region where the former reactions have only been studied at present.

Another typical feature that diffractive reactions possess in common with elastic scattering is the crossover phenomenon in the angular distribution. This point has been discussed in Section 4.3 and illustrated in Figure 11, both for elastic and diffractive channels.

Before leaving the subject of the energy dependence of exclusive diffractive reactions we mention that diffractive-like behavior is exhibited by reactions as $pp \to p(n\pi^+)$, $\pi^- p \to \pi^- \pi^+ \pi^- p \ldots$ as illustrated in Figure 30. Here no selection is made for a resonance or mass enhancement in the final state, still, a clear trend is indicated for the cross section to flatten off at high energy.

5.4 *The Deck Effect*

A striking feature of all diffractive-excitation data is the prominent bump found at low mass values in the missing-mass distributions. As discussed in Section 5.2, the corresponding systems cannot be interpreted as genuine resonances. Moreover, these systems seem to be produced only in diffractive reactions. The Deck effect (130a) is a dynamical model that describes the diffractive production of such nonresonant systems. It was first introduced by Drell & Hilda (130b) to explain early data (131) on nucleon dissociation. In the special case of pion dissociation into the $\rho\pi$ system, the Deck mechanism is described by the graph of Figure 31, which also defines variables. The incident pion dissociates at the upper vertex and the virtual pion scatters off the nucleon. This mechanism is connected with diffractive excitation seen as a general consequence of the composite nature of hadrons. A pion component in the incident hadron may be perturbed by the proton "rushing by" at not too small an impact parameter. The associated Feynman amplitude F will contain as relevant terms the pion propagator and the pion-nucleon scattering amplitude, which is approximated by its on-mass-shell value $F_{\pi N}(s_1, t_1)$, so that one may write

$$F \propto F_{\pi N}(s_1, t_1)/(t_2 - m_\pi^2). \qquad\qquad 46.$$

At high energy s_1 is large, so that the pion-nucleon scattering is diffractive, i.e. it is characterized by a forward peak with low values of t_1 strongly preferred. Diffractive scattering tends to maximize s_1 at the expense of s_2, the $\rho\pi$ subenergy squared. The presence of the pion propagator is not enough to impose peripheralism with respect to t_2, since to a good approximation $(m_\pi^2 - t_2) \approx s_1 s_2/s$. As a result one gets the production of a $\rho\pi$ system peaking at low invariant mass, and mainly in an s-wave configuration. The energy dependence of the cross section, once s_2 is fixed, will be essentially determined by the elastic scattering at large s_1 and is therefore typical of diffractive processes. It is also clear that the process proceeds

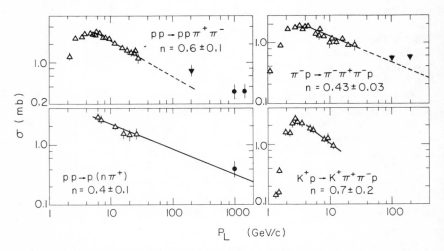

Figure 30   Energy dependence of exclusive reactions without identification of a specific final state. Compilation of data from (87b, d; 108, 109, 128).

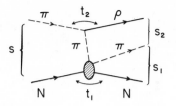

Figure 31 Deck-model graph for pion dissociation into the $\pi\rho$ system. The subenergies of the πN and $\pi\rho$ systems are indicated by s_1 and s_2, respectively. The four-momentum transfers squared of both the nucleon and of the ρ to the π meson are denoted by t_1 and t_2, respectively.

without exchange of intrinsic quantum numbers, the final mesonic system resulting from the dissociation of the incident pion. A refinement (132) of the original Deck calculation consists of introducing a Reggeized pion exchange as opposed to the simple pion propagator. This primarily includes an extra factor $s_2^{2\alpha_\pi(t_2)}$ that produces a further sharpening of the peak at low s_2, as demanded by the data, since the pion trajectory $\alpha_\pi(t_2)$ is always negative in the scattering region.

In order to illustrate how strong the Deck contribution is, we display in Figure 21A the results obtained by using the model in both its elementary (*slashed line*) and Reggeized form (*dotted-slashed line*), applied to the ($n\pi^+$) mass distribution observed at the ISR (108) in the reaction $pp \rightarrow p(n\pi^+)$.

A feature of the data not well described by the model is the strong slope-mass correlation that was discussed in Section 5.3 (see also Figure 26B). The calculated dependence of the slope upon the mass (133) is in fact much weaker than that experimentally observed. This can be remedied to a considerable extent by introducing absorptive effects (134), but one is soon back to an impact-parameter picture. Nevertheless, for a large fraction of the diffractive production of low-mass nonresonant systems, the Deck mechanism provides a satisfactory interpretation. However, the question whether in π and K dissociations genuine resonances with $J^P = 1^+$ are produced, superimposed to the large Deck background, is experimentally still open.

5.5   Tests of Factorization

The dominance of Pomeron exchange in diffractive processes at high energy implies factorization properties as mentioned in Section 4.2.

The validity of Equation 39 in the production of nucleon resonances with different projectiles was tested by Freund (135), using earlier AGS data. The results obtained by comparing more recent data on diffractive production of the $I = \frac{1}{2}$ nucleon resonances with π^- and p of 16 GeV/c are shown in Table 4a (105, 136–139, 160). The ratio R, as defined in Table 4a, is found to be constant within about 20%.

Many more data are available on exclusive diffractive processes without identification of specific final states. In Table 4b the ratio of the cross section for diffractive production of the nucleon-pion system to the elastic cross section is

Table 4 Tests of factorization for diffractive processes

(a) $R = \dfrac{\sigma(pp \to pX)}{\sigma(\pi^- p \to \pi^- X)}$				(b) $R = \dfrac{\sigma[ap \to a(N\pi)]}{\sigma[ap \to ap]}$			
X	P_L (GeV/c)	R	Ref.	a	P_L (GeV/c)	R	Ref.
p	16	1.99 ± 0.09	105	$\pi^\pm$	8	0.11 ± 0.02	136
N(1400)	16	2.3 ± 0.4	105	$\pi^\pm$	16	0.11 ± 0.02	136
N(1520)	16	2.3 ± 0.4	105	K^-	10	0.10 ± 0.02	137
N(1690)	16	1.88 ± 0.16	105	p	12	0.11 ± 0.02	138
N(2190)	16	2.3 ± 0.9	105	p	24	0.14 ± 0.02	138

(c) $R = \dfrac{\sigma[ap \to a(p\pi^+\pi^-)]}{\sigma[ap \to ap]}$				(d) $R = \dfrac{\sigma[\gamma p \to p(p\pi^+\pi^-)]}{\sigma[\gamma p \to \rho p]}$		
a	P_L (GeV/c)	R	Ref.	P_L(GeV/c)	R	Ref.
π^-	11	0.052 ± 0.005	139	6–10	0.053 ± 0.014	140
π^-	16	0.059 ± 0.003	139	10–14	0.035 ± 0.014	140
π^+	11	0.061 ± 0.006	139	14–18	0.055 ± 0.024	140
π^+	16	0.063 ± 0.003	139			
p	10	0.064 ± 0.007	140			
p	19	0.061 ± 0.008	140			
p	24.8	0.060 ± 0.009	140			

reported for different incident particles. Again, the ratio R, as defined in Table 4b, is found to be constant within 20%. Some more results concerning the diffractive production of the $(p\pi^+\pi^-)$ system are collected in Tables 4c and d. Here we quote also the values of R measured in photoproduction of the ρ meson, which is mainly diffractive, as is discussed in Section 5.8. It turns out that the R value for photoproduction is the same as that for hadron scattering.

From these and other data it may be concluded that the factorization property concerning total cross sections is always verified within the experimental errors, which are at most $\sim 20\%$. A quantitative test of the factorization property expressed by Equation 40 was also reported by the Pavia-Princeton collaboration (141), which compared the cross section for the double-diffraction-dissociation process $pp \to (p\pi^+\pi^-)(p\pi^+\pi^-)$ to the single-diffraction one $pp \to p(p\pi^+\pi^-)$.

In Section 4.2 factorization in diffraction was presented as a characteristic feature of the exchange picture, i.e. of the fact that the Pomeron pole dominates in the t channel. This kind of factorization does not arise naturally from an s-channel point of view (142), and at this stage experimental checks of its validity could be considered as supporting the fact that diffraction is better described by a factorizing Regge singularity. This statement is, however, incorrect. For many years it has been known that factorization tests at the 10–20% level are not conclusive if they refer to the full cross sections (135). This has been checked explicitly in various simple optical models. For instance, the collision of two hadrons of different radii made up of a not-too-large number of constituents (partons) was considered by Fishbane & Trefil (143). In this case Glauber theory predicts factorization to better than 20%, almost independently of the value of the parton-parton cross section, if the ratios of the radii and of the numbers of constituents of the two hadrons is within the range from 1/3 to 3. Similar conclusions are reached with other models. It even follows if hadrons are billiard balls, the radii of which are chosen to reproduce the observed total cross sections (144). These arguments show that factorization tests, to be really meaningful, should be performed on the differential cross sections at the level of precision of a few percent. The rising total cross sections, which exclude the association of the Pomeron with a simple pole, leads us to expect approximate factorization anyway. Concluding, we may say that approximate factorization appears to be a general property of diffractive excitation. It may be used to predict at the 10–20% level the value of yet-unmeasured cross sections.

5.6   Diffractive Excitation of Large-Mass Hadronic States

Diffractive excitation in exclusive channels with production of specific final states is now a well-established phenomenon with three typical properties: sharp forward peak, no quantum-numbers exchange, and weak energy-dependence. These properties can now be used as a criterion to identify a diffractive inelastic process in a more general sense, when a continuum of large-mass states is produced.

Evidence for an important diffractive component in the inclusive reaction

$$p + p \rightarrow p + X \qquad\qquad\qquad\qquad 47.$$

with excitation of large masses, was first established at the ISR by the CERN-Holland-Lancaster-Manchester collaboration (145). For general discussions on particle production at high energy we refer to the reviews of Bøggild & Ferbel (146), Whitmore (147), and Slansky (148).

Diffractive processes can be rather well identified with respect to the nondiffractive ones because of their specific kinematical configurations. When a proton in the final states of the reaction of Equation 47 is quasi-elastically scattered while the other one flares into a many-particle hadronic state X, a typical rapidity[4] configuration is obtained. This is illustrated in Figure 32A. The rapidity distribution of the final-state particles, represented by bars drawn at the corresponding rapidities,

---

[4] The rapidity y of a secondary produced in a high-energy collision is defined as $y = \{\ln[(E+p_l)/(E-p_l)]\}/2$, where E and p_l are the total energy and the longitudinal component of momentum, respectively. The maximum allowed value of the center-of-mass rapidity for a particle of mass m is $y_{max} \approx \ln(s/m^2)/2$. This expression gives also the rapidity of the incident-beam particle. For relativistic particles $y \approx \{\ln[(p+p_l)/(p-p_l)]\}/2 \approx -\ln tg(\theta/2)$. This last quantity, named "pseudorapidity," is usually denoted by η. The use of this variable has many practical advantages that have already been well discussed in the literature (146–148).

exhibits a gap between the rapidity of the quasi-elastically scattered proton, which is very close to that of the incident proton and those of all other particles. This is very different from what is now known to prevail for a typical many-particle configuration, with the full rapidity interval rather uniformly populated. A quasi-elastically scattered proton imposes such a rapidity gap. In effect, the secondaries resulting from the dissociation of an incident particle into a system of mass M appear on the rapidity plot as an elongated cluster centered at $y \approx \ln (s/M^2)/2$. The overall rapidity difference between the cluster and the quasi-elastically scattered proton is about $\ln (s/Mm)$, where m is the proton mass. However, even an isotropic cluster spreads out over at least two units of rapidity and thus, ascertaining the presence of a rapidity gap, does indeed require a high incident energy.

The kinematical configuration indicated in Figure 32A is then much more clearly identified at ISR energies, where the global-rapidity interval covers 8 units, than it is at AGS-PS energies, where it spans only 4 units. The occurrence of a large rapidity gap is readily associated with dominant Pomeron exchange, as depicted in Figure 32A'. The larger the rapidity gap, the more important is Pomeron exchange, as opposed to other exchange processes. Contributions from normal Regge trajectories are also present. However, once M is fixed, only Pomeron exchange will eventually remain at asymptotic energies.

Before discussing present experimental data on single-diffraction dissociation, we briefly mention two other important diffractive processes. The first one is double-diffractive dissociation. A typical corresponding rapidity distribution is shown in Figure 32B. A large rapidity gap separates the clustered distributions associated with the fragmentation of both diffractively excited protons. As mentioned in

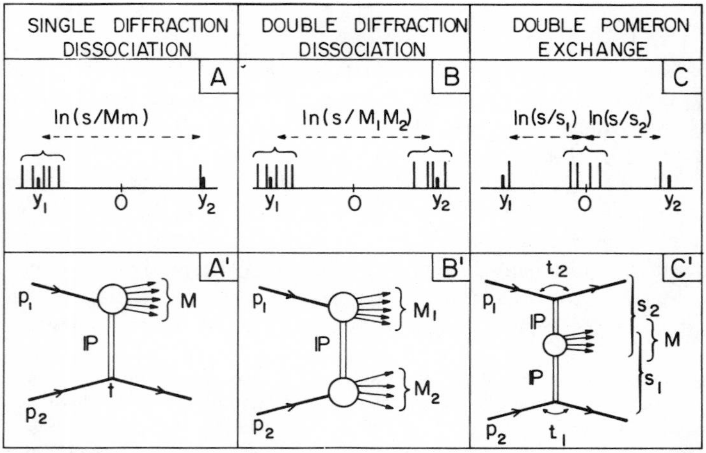

Figure 32 (A), (B), (C) Typical configurations in rapidity of single-diffraction dissociation, double-diffraction dissociation and double-Pomeron exchange. The corresponding Pomeron-exchange graphs are shown in A', B', and C', respectively.

Section 5.2, this process has been studied at the ISR only for the special case in which the protons are excited into rather low-mass states (110). No data are available at present on double-diffraction dissociation into large-mass states, a process that could be studied only with high difficulties at the ISR, even taking advantage of the large rapidity range available.

The second, presently very interesting, diffractive process is double-Pomeron exchange (Figure 32C'). The corresponding kinematical configuration now shows two rapidity gaps as illustrated in Figure 32C. This process is highly topical, as a diffractive process with no straightforward optical analogy, while having a predictable cross section in terms of the exchange picture of diffraction. It is therefore discussed at some length in Section 7.

Focusing now on single-diffraction dissociation (Equation 47), we recall that for any inclusive reaction two different sets of variables are commonly used. They are either the longitudinal and transverse component of momentum, p_L and p_T, respectively, of the quasi-elastically scattered proton or the four-momentum transfer–squared t and the invariant mass–squared M of the produced system X. As usual, $x = 2p_L/s^{1/2}$. At very large energies and for x close to one, the following relations hold

$$M^2 \approx s(1-x), t \approx -p_T^2/x. \qquad\qquad 48.$$

Data are usually presented in terms of the inclusive cross section, which is readily written in invariant form as

$$E\frac{d^3\sigma}{d^3p} \approx \frac{x}{\pi}\frac{d^2\sigma}{dx\,dp_T^2} \approx \frac{s}{\pi}\frac{d^2\sigma}{dM^2\,dt} \qquad\qquad 49.$$

using either set of variables.

Results on reaction 47, obtained by the CERN-Holland-Lancaster-Manchester collaboration (145) at two different ISR energies, are shown in Figure 33A. The inclusive cross section, for inelastically scattered protons at fixed p_T, exhibits a peak at $x \approx 1$. This behavior of the x distribution is remarkably different from that of all other secondaries produced in pp collisions, which show (146–148) a rapid falloff near the kinematical limit $x = 1$.

The data of Figure 33A can be interpreted in terms of two different processes, the first one corresponding to quasi-elastically scattered protons, to be described by the diffractive graph at Figure 32A', and the second one to a more common production process. The corresponding distributions are qualitatively illustrated in Figure 32A by the dashed and solid lines, respectively.

Events in the quasi-elastic peak ($0.96 \leq x \leq 1$) correspond to excitation of states with large masses, up to 5–10 GeV.

The fragmentation of the proton into such large-mass systems has been studied at the ISR. In Figure 33B results (149) are shown for the pseudorapidity distribution of secondaries resulting from the fragmentation of states of given mass M. In the experiment (149), the mass M is determined by measuring the momentum of the associated, quasi-elastically scattered protons. The clustering of secondaries in the rapidity plot and the associated gap are typical kinematical properties of single-diffraction dissociation.

It is interesting to remark that the multiplicity of the decay of a diffractively produced mass M, is very similar (146–148) to the one observed in hadronic collisions at a center-of-mass energy $s^{1/2} \approx M$.

From the ISR data of Figure 33A and similar FNAL data (146–148), it may be concluded that the inclusive cross section in the region of the quasi-elastic peak approximately scales with energy. This means that $E \, d^3\sigma/d^3p$, for fixed values of p_T and x, does not vary with s. Then according to Equation 49 also the distribution $s \, d^2\sigma/dM^2 \, dt$ will be energy-independent. Neglecting now any energy dependence of the diffractive excitation cross section at fixed mass, this implies that the mass distribution $d\sigma/dM^2$ should vary as $1/M^2$.

The data seem indeed to confirm this expectation, as shown by the compilation (46) reported in Figure 32C, where the inclusive cross section at fixed s and small t is displayed as a function of M^2. Part of these data were extracted from proton-deuteron scattering using the method discussed in Section 3.3.

The low-mass region ($M \lesssim 2$ GeV) is dominated by the N(1400) enhancement, which persists at very high energies with an almost constant cross section, as discussed in Section 5.3.

Above the resonance region, from $M \approx 2$ up to $M \approx 5$–6 GeV, the mass distribution closely follows the $1/M^2$ behavior. Expectations based on a diffractive-like

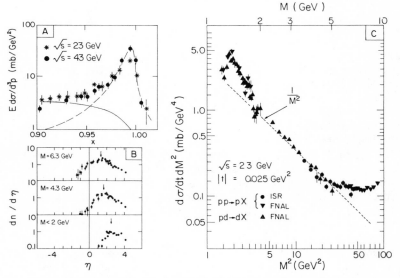

Figure 33 (*A*) ISR data (145) showing the x distribution of quasi-elastically scattered protons at $p_T = 0.525$ GeV/c. (*B*) Pseudorapidity distribution resulting from the fragmentation of large-mass M systems diffractively produced (149). The arrows indicate the center of the distribution as expected from kinematics. (*C*) Compilation of data (46) on M^2-dependence at fixed s and t.

mechanism are met. We remark, however, that the data from the Columbia-Stony Brook collaboration (150) seem to indicate a mass dependence steeper than $1/M^2$ in the low-mass region.

The momentum-transfer distribution in the region of the quasi-elastic peak has been extensively studied at FNAL and ISR. As an example we display in Figure 34*A* t distributions for two different values of M, measured at the ISR (151). These data were fitted by means of the function $\exp(bt + ct^2)$ in order to reproduce the curvature at large t.

We close this section with the following remark. If the inclusive cross section in the region of the diffractive peak actually scales with energy and if the diffractive production cross section at fixed mass is energy-independent, two properties suggested by present data, then by integrating the $1/M^2$ distribution up to a fixed x value, which corresponds to a limiting value of M^2 increasing as s, the total diffraction cross section σ_D should increase logarithmically with energy. Values of σ_D estimated by the CERN-Holland-Lancaster-Manchester collaboration (151) are shown in Figure 34*B*. While it is a definite experimental result that σ_D is comparable in magnitude to the elastic cross section $\sigma_{el} (\approx 7\text{--}8 \text{ mb})$, no definite statement can be made at present about the energy dependence.

From the measured value of σ_D, one can estimate the magnitude of the double diffraction dissociation cross section σ_{DD}, by using the factorization property $\sigma_{DD}\sigma_{el} = \sigma_D^2/4$. The value $\sigma_{DD} \approx 2$ mb is thus obtained.

5.7   Diffractive Effects in Photoproduction

There are great similarities between photoproduction and meson scattering. This is due to the fact that the relative weakness of the electromagnetic coupling

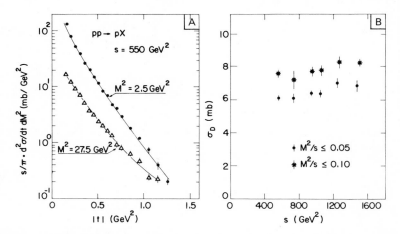

Figure 34   (*A*) ISR data (151) on the *t*-distribution of the diffractive reaction $pp \rightarrow pX$ for two different values of the mass M of the system X. (*B*) Energy dependence of the integrated single-diffraction dissociation cross section σ_D (151). The data suggest an increase of about 10% in the ISR energy range, equal to the increase of σ_{el}.

notwithstanding, the key features of the reaction are determined by the strong interactions of the hadrons that one finds in the final state. Regge models developed to describe the high-energy behavior of hadron-hadron scattering have been extended to meson photoproduction. The cross section for the photoproduction of an exclusive final state (say, $\gamma p \rightarrow \pi^+ n$) typically decreases with energy according to the relevant exchanged Regge trajectory. Of special interest in the context of diffraction scattering is the photoproduction of those mesons that have quantum numbers identical to those of the photon (spin 1, negative parity, odd under charge conjugation). In addition to the well-known vector mesons ρ, ω, and ϕ, the recently discovered J/ψ and ψ' resonances satisfy those requirements.

The photoproduction of vector mesons is usually described by the graph of Figure 35A, whereby the photon switches into a vector meson that diffractively scatters off the target. Because of the diffractive nature of high-energy meson-nucleon scattering we expect the photoproduction of vector mesons to exhibit typical diffractive-like behavior.

In the vector-meson dominance model a direct coupling, described by the parameter γ_V, is introduced between the photon and the vector-meson fields (152). This coupling can be expressed in terms of the partial width Γ_e for decay of the vector meson into a lepton pair ($e^+ e^-$ or $\mu^+ \mu^-$) as

$$\Gamma_e = \frac{\alpha^2}{12} \left(\frac{4\pi}{\gamma_V^2} \right) M_V, \qquad\qquad 50.$$

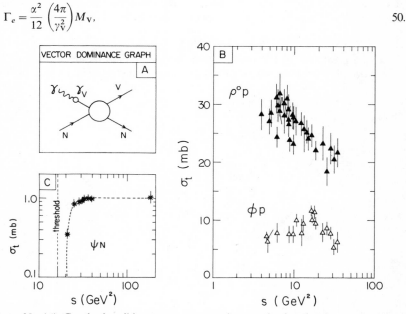

Figure 35 (A) Graph describing vector-meson photoproduction in the vector-meson dominance model. (B) Energy dependence of the $\rho^0 p$ and ϕp total cross sections as extracted from photoproduction data (153a). (C) Energy dependence of the $J/\psi(3100)$-nucleon total cross section (153b).

where α is the fine-structure constant and M_V is the vector meson mass. The partial width Γ_e has been determined experimentally in two ways; by measuring directly the branching ratio Γ_e/Γ_{tot} of vector mesons produced in hadronic interactions and by measuring the total cross section of the process $e^+e^- \to$ vector meson at the e^+e^- colliding-beam machines.

From the graph of Figure 35A one may express the cross section for photoproduction of vector mesons off nucleons, in terms of the elastic vector meson–nucleon scattering as

$$\sigma(\gamma N \to VN) = \frac{\alpha}{4} \left(\frac{\gamma_V}{4\pi} \right)^{-1} \sigma(VN \to VN), \qquad 51.$$

if we neglect off-shell effects.

The elastic cross section $\sigma(VN \to VN)$ can be written in terms of the vector meson-nucleon total cross section $\sigma_t(VN)$ and of the slope b, using the optical theorem and the simplifying assumption that the elastic amplitude is purely imaginary, which should hold at high energy:

$$\sigma(VN \to VN) = \frac{1}{16\pi b} \sigma_t^2(VN). \qquad 52.$$

By inserting Equation 52 in the expression (51), one may derive the vector meson–nucleon total cross section from photoproduction experiments, once γ_V is known.

Clearly, a photoproduction cross section that is constant or slowly varies with energy implies a similar trend for the vector meson–nucleon cross section.

A compilation of data on $\rho^0 p$ and ϕp total cross sections from Barger & Phillips (153a) is shown in Figure 35B. These cross-section values are typical of hadronic interactions and show a weak energy dependence. In Figure 35C current results concerning the $J/\psi(3100)$-nucleon cross section are reported (153b). This cross section quickly rises above threshold, reaching a plateau at the level of about 1 mb. The sharp rise and plateau behavior of vector meson–photoproduction cross sections are typical of diffractive processes.

Two important points connected with the previous discussion have been widely discussed in the literature. They are only mentioned here. The first one is the question of photon absorption in nuclear matter with its parallel to neutrino reactions (154). The second is the somewhat elusive evidence for a new selection rule that could be proper to diffractive processes, namely the conservation of the helicity during diffractive excitation (155). Strong evidence was first offered by vector meson–diffractive excitation showing only the transverse polarization imposed on the incoming photon.

5.8 Unitarity and Diffractive Dissociation

Diffractive production of particles is a specific "component" contribution to the inelastic cross section at high energies in view of the particular rapidity structure it imposes. In proton-proton collisions it corresponds to a cross section of ~ 8 mb. The nondiffractive or pionization component is approximately three times larger and shows a rather uniform rapidity distribution. Given these orders of magnitude, it is natural to consider the inelastic nondiffractive channels as origin of the absorption, whose shadow gives rise to both elastic scattering and diffraction dissociation. From this point of view it is in principle possible to calculate not only elastic scattering, but also diffraction dissociation sandwiching the unitarity relation between the initial state and an outgoing diffractive state (Section 4.5). Such calculations are much more difficult than in the elastic case, because one needs as input the amplitudes connecting multiparticle states. Interesting approaches based on the uncorrelated jet model and on the parton model were discussed by Bialas & Kotanski and by Resnick (156). However, progress in this field is slow, which should not be surprising, given the difficulties encountered in computing even the much simpler elastic shadow.

In connection with elastic unitarity (Equation 45), a much-discussed problem concerns the diffractive contribution $G_D(a)$ to the inelastic overlap function $G_{in}(a)$. Going back to the Good & Walker approach (Section 2.4), it is possible to derive an interesting bound on $G_D(a)$ (157). This bound is based on the hypothesis that diffractively produced states are physically distinguishable from pionization states, so that they do not interfere. The physical diffractive states $|h\rangle$ so defined may be decomposed in diffractive eigenstates $|n\rangle$, which by definition are not mixed by the interaction. Their profile functions $\Gamma_n(a)$, are supposed to be real. The profile function of the transition $|i\rangle \rightarrow |f\rangle$ (where $|i\rangle = \Sigma C_{ni}|n\rangle$ and $|f\rangle = \Sigma C_{nf}|n\rangle$) is simply $\Gamma_{fi} = \Sigma_n C_{nf}^* \Gamma_n C_{ni}$. Note that the C's form a unitary matrix. The elastic overlap integral is $\Gamma = \Sigma |C_{ni}|^2 \Gamma_n$ and its contribution to the unitarity condition (Equation 45) is $|\Gamma|^2$. The contribution to the same equation of all diffractive final states is $\Sigma_f |\Gamma_{fi}|^2 = \Sigma |C_{ni}|^2 \Gamma_n^2$. Since it comprises also the elastic contribution, in order to obtain the contribution of the diffractive final states one must subtract Γ^2:

$$G_D(a) = \sum_n |C_{ni}|^2 \Gamma_n^2 - \Gamma^2 \leqq \sum_n |C_{ni}|^2 \Gamma_n - \Gamma^2 = \Gamma(a) - \Gamma^2(a).$$ 53.

The inequality needed to obtain the Pumplin bound follows from the condition $\Gamma_n \leqq 1$, which expresses probability conservation for the diffractive eigenstates. By integrating Equation 53 over the impact parameter one gets the integral condition $\sigma_D < \sigma_t/2 - \sigma_{el}$. In the proton-proton case at ISR energies, the right-hand side equals ~ 13 mb, while the left-hand side is ~ 8 mb. The differential Pumplin bound is plotted in Figure 36A, together with the stronger one obtained by Caneschi et al, introducing an additional but physically reasonable hypothesis (158). Since to have a cross section as big as 8 mb this bound has to be almost saturated, the previous argument indicates very clearly that diffractive production is peripheral in impact parameter space. An explicit calculation of $G_D(a)$ was performed by Sakai & White (159). They used as input the experimental spectra of single particles produced at the ISR in diffraction processes, together with the hypothesis that helicity is conserved in the t channel, a hypothesis supported by some exclusive data. The computed $G_D(a)$ is plotted in Figure 36B and is very close to the upper bound of (158).

The peripheral nature of diffractive dissociation has been indicated by the data for many years, but its dynamical origin is not yet understood (120). However, it has a simple optical interpretation, because at small impact parameters, absorption effects play an important role. In this respect, elastic scattering behaves very differently from diffraction dissociation, despite having the same dynamics. Indeed, if for small impact parameters all Γ_n's are large, elastic scattering is large, while diffraction dissociation is small (Section 4.5). The peripherality of diffraction dissociation may also be related to the observed variation of the slope b with the mass M of the produced system (98b).

6   PRESENT APPROACHES TO THE POMERON

Our discussion has been so far as close as possible to the data and the following section should be looked upon as a parenthesis that may be skipped on a first reading. It only attempts to provide a hint at present lines of research at calculating diffraction amplitudes or, in other words, to calculate what is referred to as Pomeron exchange. What is seen is that present views favor a complicated Pomeron, even though a simple Regge pole may often offer a good enough approximation. Indeed, after this theoretical parenthesis we come back to it, despite its weakness but in view of its simplicity, to discuss further diffractive processes.

6.1   Outlook

We presented so far the optical approach and exchange approach as two different lines of research with quite different origins. They should nevertheless show a large amount of overlap. Indeed, the Regge-pole approach, as it applies to production

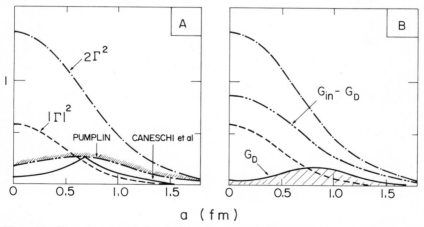

Figure 36   Impact-parameter description of proton-proton scattering at 1480 GeV/c. (*A*) The contributions to elastic unitarity are plotted together with the bounds due to Pumplin (157) and to Caneschi, Grassberger, Miettinen & Henyey (158) on the diffractive contribution G_D. (*B*) The shaded area represents the diffractive-overlap integral as computed by Sakai & White (159).

mechanisms, should allow for reliable calculations of all inelastic cross sections and, summing them over, one should be able to determine the shadow effects they are associated with. This was already discussed in Section 4.5. This should provide an exchange picture (field-theoretical) basis to the intuitive and experimentally well-founded properties from which the optical approach starts. If such a program looks a priori simple, it turns out to be extremely difficult to pursue in practice. This has to do in part with large interference effects that must be faced and are hard to control. In order to hint at the origin of such difficulties, one may assume that we know a proper procedure for handling all multiperipheral production amplitudes of the class of diagrams corresponding to Figure 37A. The corresponding shadow effect can in principle be calculated. However, the class of diagrams corresponding to Figure 37B, so far neglected, is different and yet interferes fully with that of Figure 37A. Taking it into account, one can modify in an important way whatever shadow effect was previously calculated. Despite much past effort to build up diffraction from inelastic amplitudes (160a), no general and satisfactory prescription is presently available. Nevertheless, it remains that in the framework of some simple field theories, such as ϕ^3 theory, the study of multiperipheral contributions to production amplitudes leads to an asymptotic energy behavior for the elastic scattering amplitude associated with a (Pomeron) Regge pole (160b). Such a constructed Regge-pole contribution is the shadow effect associated with a certain class of Feynman diagrams, as calculable in a particular theory, even if it is not yet the full-blown Pomeron looked for. It may therefore look like an a priori valuable input for a further-involved approach, which would include so far neglected inelastic processes. At present, theoretical approaches to the Pomeron start from such a "bare Pomeron" input, which one takes from some underlying field theory. One attempts to generate from it a better approximation of the actual diffractive amplitude. Indeed, a simple Pomeron pole, appealing as it may be, is not enough. We mentioned already the actually complicated nature of the Pomeron, as it gradually imposed itself from a phenomenological Regge approach and in particular after the discovery of rising cross sections.

For these reasons present approaches, starting from a bare Pomeron as previously defined, try to achieve building a full Pomeron amplitude that takes what one calls a scaling form:

$$F(s, t) = is(\ln s)^{\kappa} g[-t(\ln s)^{\nu}], \qquad\qquad 54.$$

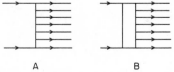

A   B

Figure 37   Two families of graphs that contribute to the calculation of the shadow effect brought about by the inelastic processes. (*A*) Standard multiperipheral amplitude. (*B*) Multi-peripheral amplitude with Pomeron exchange in the initial state.

as opposed to the simple Regge-pole form of Equation 36 (with $s_0 = 1$):

$$R(s, t) = i\beta(t)\eta(t)s \exp\{[\alpha(t) - 1] \ln s\}.\qquad\qquad 55.$$

Different approaches differ in their values for κ and ν and in their predictions for the function g. They nevertheless have the same global form (Equation 54). The total cross section grows as $(\ln s)^\kappa$. The slope parameter grows as $(\ln s)^\nu$. The impact-parameter picture for the asymptotic form 54, with an exponential for g is rather simple. The overall opacity grows as $(\ln s)^{\kappa - \nu}$, while the range grows as $(\ln s)^{\kappa/2}$ [the slope as $(\ln s)^\nu$]. Conversely, its angular-momentum plane description is complicated, with poles and cuts in the general case. It is interesting in any case that an approach that starts from the Regge-pole side eventually arrives at an amplitude amenable to a simple optical model picture, but a complicated Regge picture.

At present, reaching an asymptotic expression like Equation 54 from an input bare Pomeron is achieved in two different main ways. One uses an s-channel approach or a t-channel approach. What actually goes under such code names, which relates to the fact that particular attention is paid to unitarity constraints in either the s or the t channel, varies with time and schools. At present, one may say that the first one typically provides model amplitudes that saturate the Froissart bound ($\kappa = 2$) (93), while the second one flourishes under the name of Reggeon calculus or Reggeon field theory (160a, 160c, 161).

It is not attempted here to survey such approaches. Rather, we simply hint at the key features of either one.

6.2 The s-Channel Approach

In the s-channel approach, one typically starts with a Pomeron input, which alone would violate the Froissard bound. Its intercept is slightly greater than one $\alpha(0) = 1 + \varepsilon, \varepsilon > 0$. This is what is obtained in particular in the field-theory approach of Cheng & Wu (162). The input Regge amplitude (55) is written as

$$R(s, t) = -s\beta \exp\left(\frac{i\pi}{2}\alpha(0)\right)\exp\left(\frac{b}{2}t\right)s^{\alpha(0)-1},\qquad\qquad 56.$$

with $\beta = (b + 2\alpha' \ln s - i\pi\alpha')$, in the linear approximation $\alpha = \alpha(0) + \alpha' t$. The impact parameter amplitude is accordingly

$$R(a, s) = -\frac{\beta}{8\pi}\exp\left(-\frac{i\pi\alpha(0)}{2}\right)s^{\alpha(0)-1}\frac{1}{b}\exp\left(-\frac{a^2}{2b}\right).\qquad\qquad 57.$$

It would grow beyond bounds. Nevertheless, this may be remedied if it is used only as an eikonal phase so that one writes for the full amplitude $R_2 = [\exp(2iR) - 1]/2i$, which is by construction within the unitarity bounds (163). At high enough energies the input amplitude is large and provides full damping in the output amplitude. The increase with s, in Equation 57 can, however, be compensated for by an increase in a due to the last factor. Damping will occur, but only up to a value of a^2 increasing as $b \ln s$. Since b increases as $\ln s$, one eventually gets full opacity up to a value of a^2 (proportional to the cross section) increasing as $\ln^2 s$. The final result is of the form of Equation 54 with $\kappa = 2$. The approach thus outlined is of course too crude. More sophisticated methods have been devised. The procedures thus followed are, however, similar in spirit. One typically uses unitarity relations that involve themselves diffractive as well as nondiffractive amplitudes as a constraint on the input amplitude. The Pomeron contribution is then the solution of an integral equation. Various values for κ and v result, depending on the bare Pomeron intercept and on the conditions required (164, 165). The full Pomeron is always a complicated object. One may get a theoretically satisfactory asymptotic amplitude. However, the proton-proton profile function observed at ISR energies is very far from a black disc, and models of this type are certainly not applicable in this energy region. (See the inset of Figure 3C.)

6.3   *The t-Channel Approach*

The *t*-channel approach is of particular interest at present, since it relies on topical developments in field theory. Again one tries to achieve a richer amplitude from a Regge-pole input (160, 161). The first step is to translate the rules for Regge exchange (160), as obtained in an underlying field theory as ϕ^3 theory, into those of a field theory in two space and one time dimensions, where the two-momentum k is such that $-k^2 = t$ and the energy is $1 - \alpha(t)$. The Lagrangian contains a free term expressed in terms of the Pomeron field operator and a coupling term for which one finds good reason to limit oneself to a triple-Pomeron coupling. In this approach one takes into account interactions among input Pomerons — as they may enter unitarity from the *t*-channel point of view — as opposed to the multiple exchange of input Pomerons among scattering particles, which enters unitarity in the *s*-channel.

The solution of great interest corresponds of course to the output intercept-one limit, which, in view of the connection between $\alpha(t)$ and the energy, corresponds to the infrared limit of the theory. It is to obtain such an infrared behavior that one uses renormalization group techniques (161, 165). One then finds an asymptotic form of the type

$$F(s, t) \to s^{\alpha_{out}(t)}(\ln s)^{\kappa} g[-t(\ln s)^{v}],\tag{58}$$

where $\alpha_{out}(0) = 1$.

The Pomeron corresponds to some average field defined over the interacting hadrons in much the same way as a phonon field in a crystal. The behavior near $\alpha_{out} = 1$ corresponds to a phase transition limit where long-range interactions are associated with a massless excitation: the Pomeron [the equivalent of the energy is $1 - \alpha(t)$]. One may then hope that in much the same way as in the theory of phase transition, the critical exponents κ and v are representative of a whole class of underlying microscopic interactions, the Pomeron behavior would not depend on the details of hadronic interactions that are yet unknown. Equation 58 has a priori all required properties. There are, however, two main questions at stake. One is of a technical nature. It turns out that at present, κ and v can be calculated only within approximations that are probably not reliable. The second one is of a philosophical nature. The output value of $\alpha(0)$, which one wishes to be at 1, comes out at 1 only for a specified value of the input intercept (slightly above 1) as it appears in the Lagrangian one starts from. The fact that cross sections increase logarithmically, as opposed to asymptotically decreasing as inverse powers $[\alpha_{out}(0) < 1]$, appears therefore as an accident!

The Regge-calculus approach is, however, very far from having been fully explored and interesting developments should be expected. Other types of Lagrangians, with bare Pomerons originating from other types of underlying field theory, can be considered. We refer the reader to (161).

A conclusion at the present time is not possible. Rising cross sections have required asymptotic forms for the diffractive amplitude that are well within reach of available theoretical models. However, the basic property, the fact that up to logarithms, cross sections are apparently constant, has to be put into the theory. It does not follow, as one would certainly wish, from some fundamental property. At the same time, if cross sections can rise within present models, there is no understanding of why they rise so little. Within the ISR energy range, the s-channel approach would find it more natural to see the proton cross section rise a hundred times faster (full absorption), whereas the t-channel approach provides an acceptable asymptotic expression, but with values of κ and ν that have no reason to apply to present data. It should apply only for large values of ln s. What we see in present experiments is still very different from the output Pomeron arrived at. How much it resembles a simple-input Pomeron is unclear, yet we hope that this section will have convinced the reader that the question is very much alive.

7   DIFFRACTION EXCITATION SEEN AS POMERON EXCHANGE

7.1   *The Triple Regge Formalism*

As discussed in Section 5.6, single-diffractive excitation corresponds to an important and already rather well-explored inelastic mechanism. Its simple kinematical signature is the presence of a quasi-elastically scattered particle with (conservatively) $x > 0.96$. At lowest and highest ISR energies, this corresponds to excitation masses of 4.5 and 12 GeV, respectively. We now analyze the corresponding mechanism in the framework of the exchange picture of diffraction (Figure 32A'). Since the Pomeron-proton coupling is known from elastic scattering, within the factorization property associated with Pomeron exchange (thereupon assumed to be well represented in terms of an effective Regge pole), the whole process can be described in terms of the Pomeron-proton reaction process, as it appears at the upper vertex. To that extent, one may say that the ISR is equivalent to a Pomeron accelerator operating in an energy range comparable to that of Serpukhov for hadron-hadron interaction. Isolating kinematically the process corresponding to Figure 32A', when requiring a proton in the quasi-elastic peak of Figure 33A, one may actually study Pomeron-proton interactions up to a center-of-mass energy of 12 GeV.

The first relevant quantity in such a study is the Pomeron-proton total cross section. Its energy behavior can be easily defined within the triple Regge formalism

(146, 166–168). The by now well-known (146) calculation procedure is summarized by Figure 38. Diffractive excitation corresponds to trajectory α_i being identified with the Pomeron. The formalism includes, however, all Regge exchanges on the same footing. The discontinuity of a Reggeon-proton forward elastic amplitude, which, through the optical theorem, can be used to define a Pomeron-proton total cross section, is itself approximated in terms of its Regge behavior at large M, using Regge trajectories α_k. The inclusive distribution reads:

$$x\frac{d\sigma}{dx\,dp_T^2} = \sum_{ijk} G_{ijk}(t)\left(\frac{s}{M^2}\right)^{\alpha_i(t)+\alpha_j(t)-\alpha_k(0)}\left(\frac{s}{s_0}\right)^{\alpha_k(0)-1} \qquad 59.$$

$$= \sum_{ijk}\left(\frac{1}{1-x}\right)^{\alpha_i(t)+\alpha_j(t)-\alpha_k(0)} G_{ijk}(t)\left(\frac{s}{s_0}\right)^{\alpha_k(0)-1} \qquad 60.$$

One assumes large values for s/M^2 and M^2 to justify a Regge approximation based on a few terms. The function G, as it appears in Equation 59, can be expressed in terms of a coupling among three Reggeons, g (of Figure 38). Assuming, for instance, that trajectories i and j both correspond to the Pomeron, this reads

$$G_{PPk}(t) = \frac{1}{(4\pi)^2 s_0}[\gamma(t)]^2|\eta(t)|^2 g_{PPk}(0)\gamma_k(0)\,\mathrm{Im}\,[\eta_k(0)], \qquad 61.$$

where γ and η correspond respectively to the proton couplings and signature factors of the relevant Reggeons. With the same diffractive assumption, we can reunite the inclusive distribution in terms of a Pomeron-proton total cross section σ_P as:

$$x(d\sigma/dx\,dp_T^2) = s\,\frac{d\sigma}{dM^2\,dt} = [1/(4\pi)^2][\gamma(t)]^2|\eta(t)|^2\,(s/M^2)^{2\alpha_P(t)}(M^2/s)\sigma_P(M^2,t).$$

$$62.$$

We may then translate the key experimental results of Section 5.6, saying that $\sigma_P(M^2,t)$ practically goes to a constant at large mass (the Pomeron dominates among the trajectories α_k) in much the same way as what occurs in hadron-hadron scattering. From Equation 62 one may estimate it to be of the order of 1 mb (169). It should be stressed, however, that such a value is linked to our isolating only a factor $(s/M^2)^{2\alpha}$ in Equation 62, where the presence of s/M^2 translates only the scaling large-s and large-M^2 behavior. Extra constant terms could also have been included with a correlated change in σ_P. We should therefore only stress that the

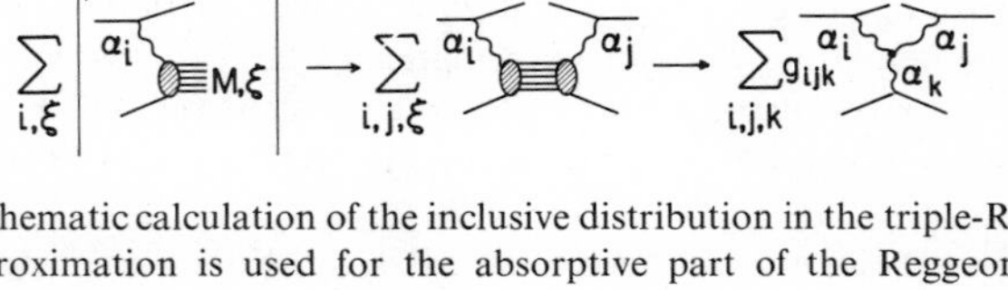

Figure 38 Schematic calculation of the inclusive distribution in the triple-Regge formalism. A Regge approximation is used for the absorptive part of the Reggeon-proton elastic amplitude.

Pomeron-proton cross section, after a sharp drop with increasing M^2, corresponding to the faster-than-M^2 fall in Figure 33C, practically levels for $M^2 > 5$ GeV^2. It shows a typical hadronic behavior and typical hadronic value.

The triple Regge formalism is also useful for isolating special kinematical dependences, such as those outlined in Figure 33A, relating them to specific Reggeon exchanges. One may for instance limit oneself to Pomeron P with intercept 1 and secondary Reggeon R with typical intercept 1/2, and to symmetric terms $i \equiv j$. The energy behavior of the inclusive cross section, together with its mass and x-dependences, are then summarized for small t in Table 5.

It is therefore Pomeron-exchange dominance in the Reggeon-proton amplitude that is connected with the scaling property of the quasi-elastic peak $[\alpha_k(0) = 1]$, whether we deal with Pomeron exchange (diffraction) or secondary Reggeon exchange (standard type of process) in the first place. With the PPP and RRP contributions, we obtain, respectively, the dashed and solid curves of Figure 33A.

7.2 The Pomeron-Proton Interaction

Continuing in our analysis of diffraction in terms of a Pomeron-proton interaction, next to the total Pomeron-proton cross section comes the inclusive distribution of the fragments of the excited proton. As previously mentioned, when of a low invariant mass they are separated from the quasi-elastically scattered proton by a large rapidity gap and they cluster over two units of rapidity. As the mass increases, however, they fill the kinematically available rapidity interval, which increases as $\ln M^2$. The inclusive distributions, as shown in Figure 33B, exhibit a rise of the rapidity distribution at maximum with increasing M^2, as the maximum shifts toward the center. It could eventually level off. The key point, however, is the spread of the rapidity distribution with increasing M^2. It is compatible with $\ln M^2$. Indeed, the proton fragments fill all of their kinematically allowed rapidity range, which, as previously discussed, increases as $\ln M^2$. All this is extremely similar to what is observed in hadron-hadron collisions. As already remarked, the associated charged multiplicity (the integral over the rapidity distributions in Figure 33B) also compares well with that observed in hadron-hadron collisions at $(s)^{1/2} = M$.

The kinematical spread of the secondaries has a striking effect when analyzed in terms of the full c.m. rapidity. The center of mass of the hadronic system of mass M shifts towards rapidity zero by an amount $\ln M$ from the incident proton rapidity $\ln (s)^{1/2}$. However, since secondaries are spread over an $\ln M$ segment on either side of

Table 5 Triple Regge contributions to inclusive spectra

Contribution	Energy Dependence	Mass Dependence	x Dependence	Scaling
PPP	s^0	M^{-2}	$(1-x)^{-1}$	yes
PPR	$s^{-1/2}$	M^{-3}	$(1-x)^{-3/2}$	no
RRP	s^0	M^0	$(1-x)^0$	yes
RRR	$s^{-1/2}$	M^{-1}	$(1-x)^{-1/2}$	no

their center-of-mass rapidity value, fragments still reach the boundary of the rapidity interval, whatever M is. This is compatible with Figure 33B. The use of pseudo-rapidity may, however, somewhat smear whatever is actually occurring at the kinematical boundary. At present there are many open questions that call for more data. The inclusive distribution may well not be symmetric at presently accessible masses, but show some skewing favoring the Pomeron side, in much the same way as what is observed in photoproduction. Data are compatible with that, but certainly not accurate enough yet. The inclusive distribution for protons may well show a leading-proton effect. There is no information about this yet. In the extreme case one may even envisage a leading proton of the quasi-elastic type. This would correspond to double-Pomeron exchange, as discussed later. So far, correlations among the diffractive-excitation fragments have not been studied. There is still much to be done to actually compare Pomeron interactions to proton or meson interactions with protons. Present information only indicates that there are many similarities. Following the original idea of Yang and co-workers (170, 171), one may thus see merely the fragmentation of the proton, viewed as a complicated object. At sufficiently high energies, it eventually occurs in a limiting way, irrespective of how strongly and with what it has been hit. A lot of what we attribute to the details of the Pomeron-proton interactions is probably far more general than what actually applies to diffraction proper. We probe the proton structure more than the details of the diffraction mechanism. The "Pomeron fragmentation" side (the more central region in Figure 33B) may, however, reveal specific features.

7.3   Double-Pomeron Exchange

Our present picture for diffraction scattering relies so strongly on the Pomeron-exchange approach that a detailed test of a rather specific property of this picture is of great interest. This is the presence of the double-Pomeron exchange mechanisms. If a Pomeron can be exchanged once, there is no a priori reason why it could not be exchanged twice or even several times. One is therefore led to search for processes where the dominant (or asymptotic) term in the amplitude would be represented by the graph of Figure 32C', which, in terms of rapidity, should lead to the rather special distribution of Figure 32C. There is of course no reason why the protons could not flare as well. Yet, in order to maximize the rapidity gaps so that double Pomeron exchange could be favored, one is presently led to focus on processes where the two protons are merely quasi-elastically scattered (with $x > 0.96$) and a few slow center-of-mass particles (typically a $\pi^+\pi^-$ system) are also produced.

The Pomeron-exchange picture of diffractive excitation calls for the occurrence of such a process with, in principle, a predictable cross section, estimates for which are at the 20-μb level. The kinematics of the process simplify in the asymptotic limit (large s/s_1 and s/s_2, large s_1 and s_2, and large M^2, using the notations spelled out by Figure 32C'). There are then two large rapidity gaps

$$Z_1 = \ln s/s_1, \, Z_2 = \ln s/s_2.$$ 63.

$$M^2 = s_1 s_2/s.$$ 64.

To the extent that the four units of rapidity available at PS (AGS) energies are enough to get strong evidence for single-diffractive excitation, one may hope that the eight units available at the ISR could be enough to secure evidence for double-Pomeron exchange. Background problems may, however, be very important. Furthermore, even at ISR energies, two large enough rapidity gaps ($Z > 3$, say) can be achieved only at the expense of taking a rather small M^2. Despite all that, it remains that double-Pomeron exchange has a particular interest in that the optical picture, which can be advocated for single and double diffraction, no longer readily applies. We have, rather, to deal with the hadronic polarization of the vacuum, following a process that the Pomeron-exchange picture of diffraction imposes. It is then worth searching for.

Granting that double-Pomeron exchange exists, one may attempt to obtain an estimate for the corresponding cross section and to spell out the expected kinematical oddities that could help in its identification. To that end, one may follow an approach that parallels that outlined when mentioning the triple Regge formalism. We then write, with the same notations, a double-inclusive distribution for the two quasi-elastically scattered protons as

$$\frac{x_1 x_2 \, d^2\sigma}{dx_1 \, dp_{T_1}^2 \, dx_2 \, dp_{T_2}^2} = \frac{1}{(4\pi)^2} \, \gamma^2(t_1)\gamma^2(t_2) \, |\eta(t_1)|^2 \, |\eta(t_2)|^2$$

$$\times \frac{\sigma_{PP}(M^2, t_1, t_2)}{(1-x_1)^{2\alpha(t_1)-1}(1-x_2)^{2\alpha(t_2)-1}},$$

65.

where we have used the asymptotic expressions

$$\frac{s}{s_i} = \frac{1}{1-x_i}, \quad M^2 = s(1-x_1)(1-x_2).$$

66.

Equation 65, of course, applies only to the double-Pomeron contribution. There are other (background) terms in the inclusive distribution. As previously done in Equation 62, we have isolated a total Pomeron-Pomeron cross section σ_{PP}, associated with the middle circle in the graph of Figure 32C' (172, 173). This is intuitively appealing, though it of course suffers from the same ambiguities as those already met in the definition of a Pomeron-proton total cross section. The key point is, however, to remain consistent and despite these ambiguities one may write a factorization relation, which is obtained from Equations 62 and 65 as

$$\frac{x_1 x_2 \, d^2\tilde{\sigma}}{dx_1 \, dp_{T_1}^2 \, dx_2 \, dp_{T_2}^2} = \frac{x_1 \, d\tilde{\sigma}}{dx_1 \, p_{T_1}^2} \, \frac{x_2 \, d\tilde{\sigma}}{dx_2 \, dp_{T_2}^2} \, \frac{1}{\tilde{\sigma}_t}.$$

67.

The tildes on top of the σ's are here to warn again that the relation applies only to the Pomeron contribution, to double-Pomeron exchange on the left-hand side, and to single-Pomeron exchange on the right-hand side. What belongs to Pomeron exchange (large M^2 limit) in single-diffractive excitation is, however, not precisely known yet (174). For these reasons, one may give only an order-of-magnitude estimate for the expected double-Pomeron exchange contribution in the large-M^2 limit. As anticipated, it is at the 20-μb level.

In the low-M^2 region, with typically a two-pion state produced, one may attempt a different estimate based on pion exchange in the central blob (172, 173). The expected cross section has the same order of magnitude. It is low for a background with hadronic processes, but large enough to be seen. The main question is of course the practical separation of the double-Pomeron contribution associated with Equation 65 from background terms, the main one being the large-mass, low-multiplicity tail of single diffraction excitation that could overcome the looked-for signal.

It is usually stressed that the double-Pomeron contribution is the only one to include a double-pole term as x_1 and x_2 approach one. Indeed, Equation 65, written for $\alpha(t) \approx 1$ gives

$$\frac{d\sigma}{dx_1 \, dx_2} \sim \frac{1}{(1-x_1)(1-x_2)}, \qquad\qquad 68.$$

when other terms, which can be tentatively associated with Regge intercept 1/2 in Equation 65, would behave as

$$\frac{d\sigma}{dx_1 \, dx_2} \approx \frac{1}{1-x_1} + \frac{1}{1-x_2}. \qquad\qquad 69.$$

This, however, holds only in the large-M^2 limit, where σ_{PP} or the corresponding term for PR interactions may be expected to be constant. As shown in Figure 39, which indicates where the kinematical limits are for fixed M^2 and s, following Equation 66, one however realizes that requiring two quasi-elastically scattered protons (with $x > 0.96$), severely limits the available mass range to rather low values. A constant σ_{PP} or σ_{PR} value is therefore certainly not going to be relevant at ISR energies. Since a variation with M^2 of the σ_{PP} term in Equation 65, adds, through Equation 66, an extra dependence on x_1 and x_2, the apparently neat separation between signal-and background, outlined by Equations 60 and 69, does not apply. At present, therefore, it appears necessary to fix the invariant mass of the central

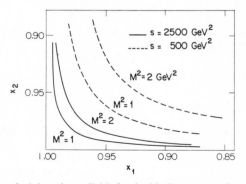

Figure 39 Kinematical domain available for double-Pomeron exchange at fixed invariant mass. The kinematical separation of the process may require x values as large as 0.96. Even at ISR energies, one is limited to low-mass values.

system and to study the x_1, x_2-dependence of the cross section, along hyperbolae on an x_1, x_2 plot, or along straight lines on a Z_1, Z_2 plot. One should then find a possible constant double-Pomeron signal against a likely dominant background symmetrical with respect to the $x_1 = x_2$ line, where it is minimum. Such a two-term parameterization, if tenable, may then be tested against experimental results at different energies using the different energy dependence of the signal and background at fixed M^2. They are s and $(s)^{1/2}$, respectively. One may then hope that such an a priori complicated and patient analysis may result in the identification of an unambiguous double-Pomeron contribution. As we hope to have sufficient stress here, the great interest for such a process justifies this endeavor. At present, several experiments have obtained candidate kinematical configurations that may partly include some double-Pomeron contribution. There is no evidence yet.

8   CONCLUSIONS

There is a long way into abstraction from old-fashioned optical diffraction to the energy dependence of a Pomeron-proton amplitude. Nonetheless, all steps follow each other in a rather simple way and an impressive array of experimental data can be correlated. Using optical concepts, one can achieve a successful description of most key features of a whole class of hadronic phenomena. In much the same way as optics can be developed into an important branch of physics without much reference to the details of electromagnetic interactions, one can now present a good facet of particle physics without having to go very far into the details of strong interactions among fundamental particles. It may be a deception that so much can be related to the mere fact that hadrons have a finite size, irrespective of their inner machinery. It should be realized, however, that if some general facts can thus be easily understood, many problems remain, which then clearly relate to the more detailed structure of hadrons. For instance, the fact that diffractive excitation accompanies diffractive scattering is readily related to the composite nature of the proton, though not yet in any precise way.

Two important discoveries at the ISR have deeply modified views that prevailed only three years ago. It was usually expected that at asymptotic energies, much greater than the rest masses of the particles involved, simplicity would prevail. Rising cross sections over the ISR energy range have shown that nothing seems to stabilize. Asymptopia was but an elusive concept! If present theories can easily accommodate rising cross sections, we have yet no explanation for the slow pace at which they grow. The Pomeron is now expected to be a complicated object at extremely high energies, but we have as yet no satisfactory explanation for the fact that it is empirically so simple at present energies. It was also usually assumed that diffraction excitation was a relatively unimportant phenomenon limited to low-mass hadronic states. It now appears as a very sizeable effect cross-sectionwise and there seems to be no bound on the mass of the diffractively excited object, provided the energy is high enough. When translating it easily in the framework of the triple Regge analysis, one partly eludes what it means in terms of the hadron structure.

There are many questions that should be explored experimentally better than they are at present. Does the elastic peak shrink at large momentum transfer? Are diffractive cross sections, for specific final states, constant over the ISR energy range? What are the correlations among the many particles into which a large-mass, diffractively excited system resolves itself? These are merely a few questions among many. The existence or nonexistence of double-Pomeron processes is also a very topical question in view of it being characteristic of the exchange picture of diffraction much more than any of the other processes.

The diffraction of hadronic waves is a very interesting facet of particle physics. It is simple by the basic concepts it uses, and rewarding by the many phenomena it correlates.

ACKNOWLEDGMENTS

We are grateful to A. B. Kaidalov for discussions about the Pomeron. We are very thankful to Mrs. C. Plumettaz and to Mr. M. Bellettieri for the very careful and patient work that they put into the drawing of the figures.

Literature Cited

1. Van Hove, L. 1967. In *High Energy Physics and Nuclear Structure*, ed. G. Alexander, pp. 259–73. Amsterdam: North-Holland
2. Gottfried, K. 1972. *Optical Concepts in High Energy Physics*. Geneva: CERN 72-20. 21 pp.
3. Cocconi, G. 1973. *Proc. R. Soc. Lond. A* 335:409
4. Perl, M. L. 1974. *High Energy Hadron Physics*. New York: Wiley
5. Born, M., Wolf, E. 1972. *Principles of Optics*. Oxford: Pergamon
6. Überall, H. 1971. *Electron Scattering from Complex Nuclei*. New York: Academic
7. Alkhazov, G. D. et al 1975. *Phys. Lett. B* 57:47
8. Bertini, R. et al 1973. *Phys. Lett. B* 45:119
9. Böhm, A. et al 1974. *Phys. Lett. B* 49:491
10. Auger, J. P., Lombard, R. J. 1973. *Phys. Lett. B* 45:115
11. Kac, M. 1973. *Nucl. Phys. B* 62:402
12. Glauber, R. J. 1955. *Phys. Rev.* 99:1515
13. Feinberg, E. L., Pomeranchuk, I. Ia. 1956. *Suppl. Nuovo Cimento* 3:652
14. Good, M. L., Walker, W. D. 1960. *Phys. Rev.* 120:1857
15. Bemporad, C. et al 1971. *Nucl. Phys. B* 33:397
16. Bemporad, C. et al 1972. *Nucl. Phys. B* 42:627
17. Mühlemann, P. et al 1973. *Nucl. Phys. B* 59:106
18. Beusch, W. et al 1975. *Phys. Lett. B* 55:97
19. Beusch, W. 1972. *Acta Phys. Pol. B* 3:679
20. Bellini, G. 1974. In *High Energy Collisions Involving Nuclei*, ed. G. Bellini et al, p. 317. Bologna, Italy: Compositori
21. Glauber, R. J. 1959. In *Lectures in Theoretical Physics*, ed. W. E. Brittin et al, 1:315. New York: Interscience
22. Glauber, R. J. 1967. In *High Energy Physics and Nuclear Structure*, ed. G. Alexander, p. 311. Amsterdam: North-Holland
23. Glauber, R. J. 1970. In *High Energy Physics and Nuclear Structure*, ed. S. Devons, p. 207. New York & London: Plenum
24a. Saudinos, J., Wilkin, C. 1974. *Ann. Rev. Nucl. Sci.* 24:341
24b. Ciofi degli Atti, C. 1975. In *Proc. Top. Meet. High-Energy Collisions Involving Nucl.*, ed. G. Bellini et al, p. 41. Bologna, Italy: Compositori
25. Foldy, L. L., Walecka, J. D. 1969. *Ann. Phys.* 54:447
26. Gottfried, K. 1971. *Ann. Phys.* 66:868
27. Hsiung, H. C. et al 1968. *Phys. Rev Lett.* 21:187
28. Fellinger, M. et al 1969. *Phys. Rev. Lett.* 22:1265
29. Bradamante, F. et al 1970. *Phys. Lett. B* 31:87
30. Alberi, G., Bertocchi, L. 1969. *Nuovo Cimento A* 63:285

31. Gurwitz, S. A., Alexander, Y., Rinat, A. S. 1974. *Non-Eikonal Approach to the Scattering of Energetic Hadrons from Nuclei.* Weizmann Rep. WIS-74/41 Ph, Weizmann Inst. Sci., Rehovot, Isr.
32. Glauber, R. J., Matthiae, G. 1970. *Nucl. Phys. B* 21:135
33. Wu, T. T., Yang, C. N. 1965. *Phys. Rev. B* 137:708
34. Byers, N., Yang, C. N. 1966. *Phys. Rev.* 142:976
35. Chou, T. T., Yang, C. N. 1968. *Phys. Rev.* 170:1591
36. Chou, T. T. 1975. *Phys. Rev. D* 11:3145
37. Lubatti, H. J. 1972. *Acta Phys. Pol. B* 3:721
38. Ascoli, G. et al 1970. *Phys. Rev. Lett.* 25:962
39. Russ, J. S. 1974. In *Experimental Meson Spectroscopy—1974,* ed. D. A. Garelick, pp. 237–58. New York: Am. Inst. Phys.
40. Gobbi, B. 1975. See Ref. 24b, p. 271
41. Kruse, U. E. et al 1974. *Phys. Rev. Lett.* 32:1328
42. O'Brien, D. D. et al 1974. *Nucl. Phys. B* 77:1
43. Edelstein, R. M. et al 1974. *The Coherent Dissociation Reaction* $p + A \to (p\pi^+\pi^-) + A$ *at 22.5 GeV/c.* Presented at Int. High Energy Phys. Conf., London
44. Akimov, Y. et al 1975. *Phys. Rev. Lett.* 35:763
45. Akimov, Y. et al 1975. *Phys. Rev. Lett.* 35:766
46. Melissinos, A. C., Olsen, S. L. 1975. *Phys. Rep. C* 17:77–132
47. Edelstein, R. M. et al 1972. *Phys. Rev. D* 5:1073
48. Gribov, V. N. 1967. *Yad. Fiz.* 5:197
49. Morrison, D. R. O. 1968. *Phys. Rev.* 165:1699
50. Goldhaber, A., Goldhaber, M. 1966. In *Preludes in Theoretical Physics,* ed. A. de Shalit et al, p. 313. Amsterdam: North-Holland
51. Kölbig, K. S., Margolis, B. 1968. *Nucl. Phys. B* 6:85
52. Gobbi, B. et al 1974. *Coherent Proton Dissociation on Nuclear Targets.* Unpublished
53. Rogers, C., Wilkin, C. 1972. *Nucl. Phys. B* 45:47
54. Van Hove, L. 1972. *Nucl. Phys. B* 46:75
55. Gottfried, K. 1972. *Acta Phys. Pol. B* 3:769
56. Voyvodic, L. 1976. *Phys. Rep. C.* In press
57a. Ting, S. C. C. 1968. In *Proc. Int. Conf. High Energy Phys., 19th, Vienna,* ed. J. Prentki, J. Steinberger, p. 43. Geneva: CERN
57b. Ting, S. C. C. 1970. In *Proc. Top. Semin.*

Interactions Elem. Part. Nucl., ed. G. Bellini et al. Trieste: INFN
58. Silverman, A. 1975. Review of high energy photoproduction. In *Proc. Int. Symp. Lepton Photon Interactions High Energy, 7th,* ed. W. T. Kirk, p. 355. Stanford: SLAC
59a. Wetherell, A. M. 1975. *Elastic and Total Cross-Sections and Implications.* In *Proc. EPS Int. Conf. High Energy Phys., Palermo, Italy, June 1975*
59b. Galbraith, W. et al 1965. *Phys. Rev. B* 138:913
59c. Foley, K. J. et al 1967. *Phys. Rev. Lett.* 19:857
59d. Denisov, S. P. et al 1973. *Nucl. Phys. B* 56:1
59e. Carroll, A. S. et al 1976. *Phys. Lett. B* 61:303
59f. CERN-Rome-Pisa-Stonybrook Collaboration 1976. *Phys. Lett. B* 62:460
60. Chew, G. F., Frautschi, S. 1962. *Phys. Rev. Lett.* 8:41
61. Fox, G. C., Quigg, C. 1973. *Ann. Rev. Nucl. Sci.* 23:219
62. Collins, P. D. B. 1971. *Phys. Rep. C* 1:103
63a. Diddens, A. N. 1974. In *Proc. Int. Conf. High Energy Phys., 17th,* ed. J. R. Smith, 1:41. Didcot, Engl: Rutherford Lab.
63b. Caltech-Berkeley Collaboration. Unpublished
64. Kane, G. L. 1973. In *Particles and Fields,* ed. H. H. Bingham et al, pp. 230–98. New York: Am. Inst. Phys.
65. Pomeranchuk, I. Ia. 1958. *Zh. Eksp. Teor. Fiz.* 34:725
66. Amaldi, U. et al 1973. *Phys. Lett. B* 44:112
67. Amendolia, S. R. et al 1973. *Phys. Lett. B* 44:1119
68a. Carroll, A. S. et al 1974. *Phys. Rev. Lett.* 33:928
68b. Carroll, A. S. et al 1974. *Phys. Rev. Lett.* 33:932
70a. Ambats, I. et al 1974. *Phys. Rev. D* 9:1179
70b. Brandenburg, G. et al 1975. *Phys. Lett. B* 58:367
70c. Beaupré, J. V. et al 1972. *Phys. Lett. B* 41:393
71. Berger, E. L., Fox, G. C. 1969. *Phys. Rev.* 188:2120
72. Davier, M., Harari, H. 1971. *Phys. Lett. B* 35:239
73. Berger, E. L. 1975. *Phys. Rev. D* 11:3214
74a. Akerlof, C. W. et al 1975. *Hadron-Proton Elastic Scattering at 50, 100 and 200 GeV/c Momentum.* Mich. prepr. VM HE 76-6
74b. Chew, G. F., Rosenzweig, C. 1975. *Phys.*

Lett. B 58:93
74c. Quigg, C., Rabinovici, E. 1975. Fermilab prepr. 75/81-THY
75. Borghini, M. et al 1970. Phys. Lett. B 31:405
76. Borghini, M. et al 1971. Phys. Lett. B 36:493
77. Gaidot, A. et al 1975. Phys. Lett. B 57:389
78a. Clyde, A. R. 1966. Univ. Calif. Rad. Lab. Rep. UCRL 16275
78b. Allaby, J. V. et al 1973. Nucl. Phys. B 52:316
78c. Akerlof, C. W. et al 1975. Phys. Lett. B 59:197
78d. Kwak, N. et al 1975. Phys. Lett. B 58:233
79. de Kerret, H. et al 1976. Phys. Lett. B. 62:363
80. Dias de Deus, J. 1973. Nucl. Phys. B 59:231
81. Barger, V. 1974. See Ref. 63a, 1:200
82a. Amaldi, U. et al 1971. Phys. Lett. B 44:116
82b. Beznogikh, G. et al 1973. Phys. Lett. B 43:85
82c. Bartenev, V. et al 1973. Phys. Rev. Lett. 31:1088
82d. CERN-Rome Collaboration. 1976. Unpublished
82e. Nurushev, S. 1974. See Ref. 63a, 1:25
82f. Derevshchikov, A. A. et al 1974. Phys. Lett. B 48:367
82g. Antipov, Yu. M. et al 1973. Nucl. Phys. B 57:333
82i. Fermilab Single Arm Spectrometer Group. 1975. Phys. Rev. Lett. 35:1406
82j. Akerlof, C. W. et al 1975. Phys. Rev. Lett. 35:1406
83. Amaldi, U. 1973. In Proc. Int. Conf. Elem. Part., 2nd, Aix-en-Provence. J. Phys. Paris C 34:241
84. Barbiellini, G. et al 1972. Phys. Lett. B 39:663
85. Grein, W., Kroll, P. 1975. Phys. Lett. B 58:79
86. Barger, V., Phillips, R. J. N. 1975. Rutherford Lab. Rep. RL75-176, T.145
87a. Carlson, P. J., Diddens, A. N., Giacomelli, G., Mönnig, F., Schopper, H. 1973. Landolt-Börnstein New Series, ed. K. H. Hellwege, Vol. 7. Berlin: Springer
87b. Particle Data Group. 1970. NN and ND Interactions—A Compilation. UCRL-20000NN. Univ. Calif. Rad. Lab., Berkeley, Calif.
87c. Morrison, D. R. O. 1974. Total Inelastic Cross-Sections. CERN/D, Pt. II/PHYS 74-38
87d. High-Energy Reactions Analysis Group. 1972. CERN/HERA 72-2, CERN/HERA 73-1

87e. Derevshchikov, A. A. et al 1973. See Ref. 83
88. MacDowell, S. W., Martin, A. 1964. Phys. Rev. B 135:960
89a. Locher, M. P. 1967. Nucl. Phys. B 2:525
89b. West, G. B., Yennie, D. R. 1968. Phys. Rev. 172:1413
90. Eden, R., Landshoff, P., Olive, D., Polkinghorne, J. 1966. The Analytic S-matrix. Cambridge: Cambridge Univ. Press
91. Queen, N. M., Violini, G. 1974. Dispersion Theory in High Energy Physics. London: Macmillan
92. Bronzan, J. B., Kane, G. L., Sukhatme, U. P. 1974. Phys. Lett. B 49:272
93. Froissart, M. 1961. Phys. Rev. 123:1053
94. Khuri, N. N., Kinoshita, T. 1965. Phys. Rev. B 137:720
95a. Baillon, P. et al 1975. Measurement of the Real Part of the Forward Scattering Amplitude. CERN 75-10. Geneva: CERN
95b. Jenni, P. et al 1976. Nucl. Phys. B. 105:1
95c. Baillon, P. et al 1976. Nucl. Phys. B 105:365
95d. Jenni, P. et al 1975. Nucl. Phys. B 94:1
95e. Govorun, N. N. et al 1973. E1-7552. Dubna: JINR
95f. Foley, K. J. et al 1969. Phys. Rev. 181:1775
95g. Apokhin, V. D. et al 1976. Elastic $\pi^+ p$-$K^{\pm} p$ and pp Scattering in the Coulomb-Nuclear Interference at Moments of 625 and 522 GeV/c. IHEP 76-6. (In Russian)
95h. Apokhin, V. D. et al 1975. Phys. Lett. B 56:391
95i. Carnegie, R. K. et al 1975. Phys. Lett. B 59:308
95j. Vorobyov, A. A. et al 1972. Phys. Lett. B 41:639
95k. Kirillova, L. F. et al 1966. Zh. Eksp. Teor. Fiz. 50:76
95l. Beznogikh, G. G. et al 1972. Phys. Lett. B 39:411
95m. Taylor, A. E. et al 1965. Phys. Lett. 14:54
95n. Bartenev, V. et al 1973. Phys. Rev. Lett. 31:1367; Ankenbrandt, C. et al 1975. Fermilab prepr. FERMILAB-Conf-75/61-EXP
95o. Amaldi, U. et al 1973. Phys. Lett. B 43:231
95p. Hendrick, R. E., Lautrup, B. 1975. Phys. Rev. D 11:529
95q. Hendrick, R. E. et al 1975. Phys. Rev. D 11:536
96. Van Hove, L. 1964. Rev. Mod. Phys. 36:657
97a. Michejda, L. 1968. Nucl. Phys. B 4:113
97b. Koba, Z., Namiki, M. 1968. Nucl. Phys. B 8:413

98a. Miettinen, H. I. 1973. See Ref. 83, 34: 263
98b. Miettinen, H. I. 1974. In *Neuvième Rencontre de Moriond. Interactions Hadronique à Haute Energie,* ed. J. Tran Thanh Van, 1:363. Orsay: CNRS
99. Hwa, R. C. 1973. *Phys. Rev. D* 8:1331
100. Hamer, C. J., Pierls, R. F. 1973. *Phys. Rev. D* 8:1358
101. Henyey, F. S. 1973. *Phys. Lett. B* 45:469
102. Grein, W., Guigas, R., Kroll, P. 1975. *Nucl. Phys. B* 89:93
103a. Martin, A. 1975. *Testing Geometrical Scaling.* Unpublished
103b. Barger, V., Luthe, J., Phillips, R. J. N. 1975. *Test of Geometrical Scaling and Generalizations.* Unpublished
104. Van Hove, L., Fiarowski, K. 1976. *Nucl. Phys. B* 107:211
105. Anderson, E. W. et al 1970. *Phys. Rev. Lett.* 25:699
106. Morrison, D. R. O. 1974. See Ref. 98b, p. 275
107. Morrison, D. R. O. 1974. In *Proc. Hawaii Top. Conf. Part. Phys., 5th,* ed. P. N. Dobson et al, p. 189. Honolulu: Univ. Press Hawaii
108. de Kerret, M. et al 1976. *Phys. Lett. B* 63:477, 483
109. Webb, R. et al 1975. *Phys. Lett. B* 55:331
110. Cavalli-Sforza, M. 1975. *Lett. Nuovo Cimento* 14:353
111. Boesebeck, K. et al 1971. *Nucl. Phys. B* 28:381
112a. Oh, Y. T. et al 1972. *Phys. Lett. B* 42:497
112b. Ochs, W. et al 1975. *Nucl. Phys. B* 86:253
113. Ascoli, G. et al 1973. *Phys. Rev. D* 7:669
114. Ascoli, G. et al 1972. In *Proc. Int. Conf. High Energy Phys., 16th,* ed. J. D. Jackson et al, 1:3. Batavia, Ill: Natl. Accel. Lab.
115. Antipov, Yu. M. et al 1973. *Nucl. Phys. B* 63:153
116. Bingham, H. H. et al 1972. *Nucl. Phys. B* 48:589
117. Leith, D. W. G. S. 1972. See Ref. 114, 3:321
118a. Deutschmann, M. et al 1974. *Phys. Lett. B* 49:388
118b. Brandenburg, G. W. et al 1976. *Phys. Rev. Lett.* 36:703, 705
119. Leith, D. W. G. S. 1973. In *Particles and Fields,* ed. H. H. Bingham et al, p. 326. New York: Am. Inst. Phys.
120. Kane, G. L. 1972. *Acta Phys. Pol. B* 3:845
121. Antipov, Yu. M. et al 1973. *Nucl. Phys. B* 63:202
122. Carney, J. N. et al 1972. *Phys. Lett. B* 42:124
123. Barloutand, R. et al 1973. *Study of the Kππ System Produced in K⁻p → K̄ππN Reactions at 143 GeV/c.* Saclay DPRPE 73-01
124. Brandenburg, G. W. et al 1972. *Nucl. Phys. B* 45:397
125. Antipov, Yu. M. et al 1973. *Nucl. Phys. B* 63:189
126. Particle Data Group. 1973. LBL-53. Lawrence Berkeley Lab., Berkeley, Calif.
127a. Ma, Z. M. 1970. *Phys. Rev. Lett.* 24:1031
127b. Blair, I. M. 1969. *Nuovo Cimento* 63:529
128. High Energy Reactions Analysis Group. 1972. CERN/HERA 72-1. Geneva: CERN
129. Gordon, H. A. et al 1974. *Phys. Rev. Lett.* 33:603
130a. Deck, R. T. 1964. *Phys. Rev. Lett.* 13:169
130b. Drell, S. D., Hilda, K. 1961. *Phys. Rev. Lett.* 7:199
131a. Chadwick, G. B. et al 1960. *Phys. Rev. Lett.* 4:611
131b. Cocconi, G. et al 1961. *Phys. Rev. Lett.* 7:450
132. Berger, E. L. 1968. *Phys. Rev.* 166:1525
133. Miettinen, H. I., Pirilä, P. 1972. *Phys. Lett. B* 40:127
134. Berger, E. L., Pirilä, P. 1975. *Phys. Lett. B* 59:361
135. Freund, P. G. O. 1968. *Phys. Rev. Lett.* 21:1375
136. Beaupré, J. V. et al 1973. *Nucl. Phys. B* 66:93
137. Graessler, H. et al 1972. *Nucl. Phys. B* 47:43
138. Böckmann, K. et al 1974. *Nucl. Phys. B* 81:45
139. Kittel, W., Ratti, S., Van Hove, L. 1971. *Nucl. Phys. B* 30:333
140. Liu, F. F. et al 1972. SLAC-PUB-1057. Stanford Linear Accel. Cent., Stanford, Calif.
141. Cavalli-Sforza, M. 1975. *Lett. Nuovo Cimento* 14:345
142. Van Hove, L. 1971. *Ann. Phys.* 66:449
143. Fishbane, P. M., Trefil, J. S. 1974. *Phys. Rev. Lett.* 32:396
144. Pumplin, J., Kane, G. L. 1974. *Phys. Rev. Lett.* 32:963
145a. Albrow, M. G. et al 1973. *Nucl. Phys. B* 51:388
145b. Albrow, M. G. et al 1974. *Nucl. Phys. B* 72:376
146. Bøggild, H., Ferbel, T. 1974. *Ann. Rev. Nucl. Sci.* 24:451

147. Whitmore, J. 1974. *Phys. Rep. C* 10: 273
148. Slansky, R. 1974. *Phys. Rep. C* 11:99
149. Albrow, M. G. et al 1974. *Phys. Lett. B* 51:424
150. Schamberger, R. D. et al 1975. *Phys. Rev. Lett.* 34:1121
151. Albrow, M. G. et al 1976. *Nucl. Phys. B*. 108:1
152. Sakurai, J. J. 1969. In *Lecture in Theoretical Physics: Elementary Particle Physics,* ed. K. T. Mahanthappa et al, p. 11A. New York: Gordon & Breach
153a. Barger, V., Phillips, R. J. N. 1975. *Nucl. Phys. B* 97:452
153b. Barger, V., Phillips, R. J. N. 1975. Univ. Wis. prepr. COO 881-462, Madison, Wis.
154. Bell, J. S. 1964. *Phys. Rev. Lett.* 13:57
155. Gilman, F. J. et al 1970. *Phys. Lett. B* 31:387
156a. Bialas, A., Kotanski, A. 1973. *Acta Phys. Pol. B* 4:659
156b. Resnick, L. 1974. *Can. J. Phys.* 53:2479
157. Pumplin, J. 1973. *Phys. Rev. D* 8:2899
158. Caneschi, L. et al 1975. *Phys. Lett. B* 56:359
159. Sakai, N., White, J. N. J. 1973. *Nucl. Phys. B* 59:511
160a. Zachariasen, F. 1971. *Phys. Rep. C* 2:1
160b. Gribov, V. N. 1968. *Sov. Phys. JETP* 26:414
160c. Gribov, V. N. 1968. *Sov. Phys. JETP* 28:784
161. Abarbanel, H. et al 1975. *Phys. Rep. C* 21:13
162. Cheng, H., Wu, T. T. 1970. *Phys. Rev. Lett.* 24:1456
163. Frautschi, S., Margolis, B. 1968. *Nuovo Cimento A* 56:1155
164a. Ball, J. S., Zachariasen, F. 1974. *Nucl. Phys. B* 72:149
164b. Ball, J. S., Zachariasen, F. 1974. *Nucl. Phys. B* 78:77
164c. Ball, J. S., Zachariasen, F. 1975. *Nucl. Phys. B* 85:317
165. Calucci, G., Jengo, R., Rebbi, C. *Nuovo Cimento A* 6:601
166. Fermilab-Rochester-Michigan Collaboration. Unpublished
167. Brower, R. C., de Tar, C. E., Weis, J. H. 1974. *Phys. Rep. C* 14:257
168. Roy, D. P., Roberts, R. G. 1974. *Nucl. Phys. B* 77:240
169. Kaidalov, A. B., Ter Martirosyan, K. A. 1974. *Nucl. Phys. B* 75:471
170. Yang, C. N. 1970. In *High Energy Collisions,* ed. J. A. Cole et al, p. 509. New York: Gordon & Breach
171. Benecke, J. et al 1970. *Phys. Rev.* 188:2159
172. Kaidalov, A. B., Khoze, V. A. 1975. *Diffractive Inelastic Processes at High Energies.* Unpublished
173. Chew, D. M., Chew, G. F. 1974. *Phys. Lett.* 53:191
174. Chen, M. S., Kane, G. L. 1976. *Phys. Lett. B* 60:192

Ann. Rev. Nucl. Sci. 1976. 26:457–509
Copyright © 1976 by Annual Reviews Inc. All rights reserved

EXCITATION OF GIANT MULTIPOLE RESONANCES THROUGH INELASTIC SCATTERING[1,2]

✻5579

Fred E. Bertrand

Oak Ridge National Laboratory, Oak Ridge, Tennessee 37830

CONTENTS

[1] The US Government retains the right to a nonexclusive royalty-free license in and to any copyright covering the article. All rights reserved.

[2] Research sponsored by the US Energy Research and Development Administration under contract with Union Carbide Corporation.

1   INTRODUCTION

An exciting development in nuclear physics during the past six years has been the observation of a new giant resonance. This new resonance is located at an excitation energy of $\approx 63\, A^{-1/3}$ MeV and has been identified as an isoscalar giant quadrupole ($E2$) resonance (GQR). Its observation in many nuclei implies that the GQR represents a general behavior of nuclei and as such may lend itself to a generalized theoretical description. Some evidence has also been obtained for the existence of isoscalar monopole ($E0$) and octupole ($E3$) resonances and an isovector quadrupole ($E2$) resonance. However, the isoscalar GQR is by far the most thoroughly studied and widely accepted of the new giant resonances.

Perhaps the best-known giant resonance in nuclei is the giant dipole resonance (GDR). The GDR is described in classical hydrodynamics as a class of nuclear motion in which the neutrons and protons within a nucleus move collectively against one another, providing a separation between the centers of mass and charge, thus creating a dipole moment. Although the existence of the GDR was established many years ago, the existence of giant resonances corresponding to other classes of nuclear collective motion has been proposed (1), but never experimentally identified.

1.1   *Scope of the Article*

Nearly all of the experimental evidence for the new resonances has been provided by inelastic scattering of medium-energy electrons and nuclear projectiles, rather than by the photonuclear reactions traditionally used to study the GDR. This article is intended to provide a summary of the experimental evidence for the new giant resonances and concentrates on the inelastic-scattering measurements. Particular emphasis is placed on topics for which there have been apparent discrepancies. Part of this review is devoted to a description of the systematic trends of the new resonances. Finally, a brief discussion of some recent microscopic calculations of excitation of giant resonances by inelastic scattering is given. Two reviews on the topic of the GQR have previously been published (2, 3).

Giant electric dipole measurements and calculations have been thoroughly reviewed before (4–6), as has recent work on excitation of the giant magnetic dipole resonance (7). These topics are not covered here.

1.2   *Complete Inelastic Spectrum*

It is instructive to consider the excitation of the new giant resonance in the context of a complete inelastic spectrum. Figure 1 shows such a spectrum at 27 degrees from ^{54}Fe bombarded by 61.7-MeV protons (8). This spectrum is complete in that nearly all protons emitted from the target were experimentally observed. The peaks at low excitation energies are produced by elastic scattering and by inelastic scattering to bound states of ^{54}Fe. For the spectrum shown, the integrated cross section (excluding elastic) from $E_x \approx 0$–12 MeV is less than 5% of the total (p, xp') cross section.

The large, broad peak dominating the high-excitation energy end of the spectrum is produced by the nuclear evaporation process. The region between the evaporation peak and the peaks from bound-state excitations is called the nuclear continuum. This region is often described in terms of an intranuclear-cascade model (9) or preequilibrium statistical model (10). The inelastic proton continuum is found to be featureless at angles greater than ≈ 40 degrees. However, as seen in Figure 1, at smaller angles a broad peak at an excitation energy of ≈ 16 MeV rises above the otherwise flat continuum. This structure is located approximately 2 MeV lower in excitation energy than the ^{54}Fe GDR and, as is discussed below, cannot be described as due to excitation of the GDR alone. It is broad structure of this type, located in the nuclear continuum, that has provided evidence for giant resonances other than dipole.

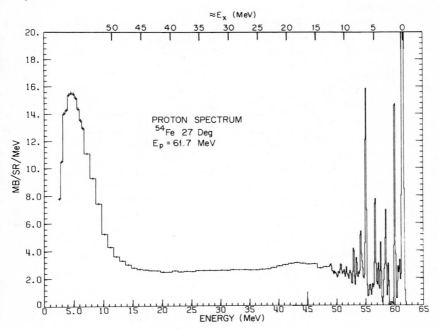

Figure 1   Proton spectrum at 27° from 62-MeV protons on ^{54}Fe (8). The energy of the outgoing proton (E_p) is plotted at the bottom of the figure, while the approximate excitation energy is plotted at the top. Data have been plotted in ≈ 1-MeV-wide bins up to $E_p \approx 49$ MeV, then plotted in 50-keV-wide bins. Protons below $E_p \approx 1.5$ MeV were not detected in the experiment. The small, broad peak near $E_x \approx 16$ MeV is identified as arising from excitation of the giant quadrupole and giant dipole resonances.

2   BRIEF THEORETICAL BACKGROUND

The discussion given here is intended to provide only a framework in which to consider and interpret the experimental observations described in the following sections. Detailed theoretical treatments of giant multipole resonances have been published previously (11–13).

2.1   *Giant Resonances and Collective Modes of Nuclear Excitation*

Over the years, the term *giant resonance* has come to be used nearly synonymously with the giant dipole resonance. However, such a correspondence is largely a product of history. In this discussion a giant resonance is regarded as a highly collective mode of nuclear excitation in which an appreciable fraction of the nucleons of the nucleus move together. These modes are generally not manifested as single, narrow states, but rather as a group of states. Since these modes usually occur at excitation energies where particle emission is possible, the states have an intrinsic spreading width. Those excitations corresponding to collective motion in which neutrons and protons move in phase are called isoscalar, while those in which neutrons and protons move out of phase are called isovector. The familiar low-lying 2^+ and 3^- levels are typical examples of isoscalar collective states, while the GDR is an example of isovector collective motion. [For a thorough discussion of collective modes, see (11).]

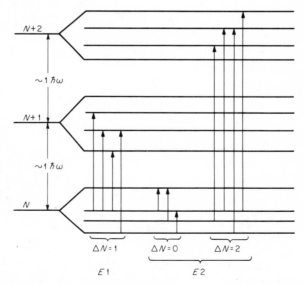

Figure 2   Schematic representation of $E1$ and $E2$ single-particle transitions between shell-model states of a hypothetical nucleus. Major shells are denoted as $N, N+1$, and $N+2$ and lie $\approx 1\hbar\omega$ or $\approx 41\, A^{-1/3}$ MeV apart.

Schematically represented in Figure 2 are single-particle transitions between shell-model states of a hypothetical nucleus. Collective transitions result from coherent superpositions of such single-particle transitions. Major shells are denoted as $N = 1$, $N = 2$, etc, and within each major shell are several subshells. The major shells are separated by $\approx 1\hbar\omega$ or $\approx 41/A^{1/3}$ MeV. The transitions shown represent some of the variety of vibrational (collective) modes that may occur by exciting a nucleon from one orbit to another via inelastic scattering, for example. Within a shell-model framework, giant resonances can be considered to result from transitions of nucleons from single-particle orbits in one major shell to another, under the influence of an interaction that orders these transitions into a coherent motion. The interaction operator for isoscalar excitations has the form $r^L Y_L^M$. (The interaction operator for inelastic scattering has similar characteristics.) This operator can excite a nucleon by at most $L\hbar\omega$, or, to state it differently, the nucleon can be promoted by at most L major shells. The number of shells is either odd or even in order to conserve parity. Thus one might expect a clustering of transitions every $1\hbar\omega$ in excitation energy.

From these arguments it follows that the GDR ($E1$) may be pictured as built up of transitions spanning $1\hbar\omega(\Delta N = 1)$, for example, between major shells N and $N+1$ on Figure 2. The GDR might then be expected to be located at an excitation energy of $\approx 41/A^{1/3}$ MeV; however, it is in fact located at $\approx 77/A^{1/3}$ MeV (6). This energy difference arises from the fact that the interaction between the nucleons in the nucleus is repulsive for the isovector mode, so that the excitation energy is pushed up from that expected. Conversely, the interaction is attractive for isoscalar modes, thus pushing the excitation energy down.

For $E2$ transitions we expect two different sets of transitions, as indicated on Figure 2. The first of these, with the lowest energy, is comprised of transitions within a major shell, the so-called $0\hbar\omega(\Delta N = 0)$ transitions. A second set is comprised of transitions between shells N and $N+2(\Delta N = 2)$ and would have energy of $2\hbar\omega$, pushed up or down for isoscalar or isovector modes. While the first class ($0\hbar\omega$) of $E2$ excitations is identified with the familiar low-lying 2^+ levels, the $2\hbar\omega$ class of transitions had not been identified before the observations described in this review.

2.2   Sum Rules

One way by which a transition is characterized as a collective excitation is by comparison of its measured electric transition strength [$B(EL)$] with the single-particle (or Weisskopf) estimate. The single-particle estimate provided by the interaction operator $r^L Y_L^M$ for excitation of a spin-zero nucleus to a state with spin L is given by the reduced transition probability,

$$B^{SP}(EL) = \frac{2L+1}{4\pi}\left(\frac{3}{L+3}R^L\right)^2 (T = S = 0),\qquad 1.$$

where R is the nuclear radius. A transition is generally considered collective if $B(EL)/B^{SP}(EL) \gtrsim 10$.

The measured transition strength may also be expressed in terms of a theoretical limit for the strength, a so-called sum rule. Giant resonances are so named because of the localization in excitation energy of a considerable fraction of the sum-rule strength. More detailed descriptions of sum rules have been given elsewhere (14, 15), and the application of sum rules to the problems discussed here has been previously described (12, 13, 16). The linearly energy-weighted sum-rule limit (EWSR) is especially useful, because nearly model-independent estimates of its value can be made. The EWSR limit is given by (15)

$$S_L = B(EL)(E_f - E_i) = \frac{L(2L+1)}{4\pi} \frac{\hbar^2}{2m} Ze^2 \langle r^{2L-2} \rangle, \qquad 2.$$

where E_f and E_i denote the energies of the final and initial states in the transition and $\langle r^L \rangle$ is the rms charge radius of the ground state. For $T = S = 0$ transitions and $\langle r^L \rangle = (3/L+3)R^L$ for a uniform mass distribution, the EWSR for a transition of multipole L can be written as

$$S_L = \frac{3A\hbar^2}{8\pi m} LR^{2L-2}, \qquad 3.$$

where m is the nucleon mass and A is the nuclear mass.

If the entire EWSR strength for a transition of multipole L were located at energy E (MeV), the transition rate would be

$$B(EL)/B^{SP}(EL) = S_L/B^{SP}(EL)E = \frac{A\hbar^2}{6mR^2E} L(L+3)^2$$
$$\approx 4.8 \ A^{1/3} L(L+3)/E, \text{ if } R = 1.2 \ A^{-1/3} \text{ fm.} \qquad 4.$$

For an 11-MeV $E2$ transition in ^{208}Pb that depletes 100% of the $T = 0$, $E2$ EWSR strength, $B(E2)$ is 130 single-particle units.

The determination of $B(EL)$ for a transition, and thus the fraction of the EWSR strength exhausted in the transition, is straightforward in the electromagnetic interaction of inelastic electron scattering. In inelastic hadron scattering, $B(EL)$ is not directly measured. Instead, a deformation parameter β_L may be obtained by comparison of the measured cross section to that calculated using the distorted-wave Born approximation (DWBA), in which the usual collective-model interaction is assumed:

$$\beta_L^2 = \frac{\sigma(L)_{\text{measured}}}{\sigma(L)_{\text{DWBA}}}. \qquad 5.$$

If β_L is assumed to be proportional to the mass multipole moment for a uniform distribution, then[3]

$$\beta_L^2 = B(EL)\left(\frac{4\pi}{3ZR^L}\right)^2. \qquad 6.$$

---

[3] It has to be recognized that the β_L in Equation 5 refers to the deformation of the optical potential, while that in Equation 6 is the deformation of the nuclear charge distribution. These two are not necessarily exactly the same, which gives rise to uncertainties in comparing excitation strengths obtained by electromagnetic interactions and inelastic scattering of nuclear projectiles.

The following expression for β_L^2 in terms of the EWSR may then be obtained:

$$\beta_L^2 = \frac{2\pi\hbar^2}{3m} \frac{L(2L+1)}{AR^2} \frac{1}{E} (T = S = 0) \approx \frac{60L(2L+1)}{A^{5/3}E}, \text{ if } R = 1.2 \, A^{-1/3} \text{ fm.} \qquad 7.$$

Again, taking the case of 100% of the $E2$ sum-rule strength located at 11 MeV in ^{208}Pb, $\beta_2 = 0.087$.

Similar sum rules hold for isovector excitations (12). The sum rules for $L = 1$ and $L = 0$ are special cases and are described in detail elsewhere [$L = 1$, (12); $L = 0$, (13)].

It has been known for some time that the first 2^+ levels in nuclei generally deplete less than 20% of the $E2$, $T = 0$ EWSR strength (15). However, few experiments have been reported in which the full $0\hbar\omega$ space has been examined. Table 1 shows the results from three measurements (17–20) on different nuclei, which covered the entire bound-state excitation region [i.e. excitation energies up to approximately the neutron separation energy (S_n)]. Only for ^{24}Mg is any appreciable fraction of the $L = 2$ EWSR strength found in the bound-state region. Indeed, less than one half of the possible sum-rule strength for any multipole is located in the bound states for the nuclei studied.

Since only two excitation modes, $0\hbar\omega$ and $2\hbar\omega$, are allowed for $E2$ transitions, the sum rule should be exhausted within these two sets of transitions. Thus, from Table 1 we conclude that if the estimate in Equation 3 for the EWSR is valid, considerable $E2$ collective strength must lie in the $2\hbar\omega$ transitions. There have been predictions that strength for the quadrupole mode of excitation would be found in the nuclear continuum near an excitation energy of $\approx 60 \times A^{-1/3}$ MeV (1).

2.3   Inelastic-Scattering Interaction

The methods used to analyze measurements of inelastic scattering to giant resonances are generally the same as those that have proven so successful [see, for example, (21) and (22)] for low-lying states. In order to describe the projectile-target coupling in inelastic hadron scattering, the optical potential used to describe the inelastic scattering is deformed. The optical potential used has the form

$$V = U_0 + U_1 \tau \cdot T, \qquad 8.$$

where τ and T are the isospin operators for the projectile and target, respectively.

Table 1   Percentage of isoscalar EWSR multipole strength depleted in bound states of ^{24}Mg, ^{40}Ca, and ^{208}Pb

Nucleus (Ref.)	Multipole								
	0	1	2	3	4	5	6	7	8
^{24}Mg (17)			40	10	~3				
^{40}Ca (18)	~0	~0	14	38	7	11	1	0.2	~0
^{208}Pb (19, 20)	~0	~0	20	47	14	3	3	2	1

Referring to Equation 8, the potential U_0 when deformed generates the isoscalar excitations ($T = 0$), while the second term, U_1, provides isovector ($T = 1$) excitations. For proton reactions, U_1 is often empirically determined by analysis of (p, n) scattering data to analog states (12). From a comparison of typical values for the potentials U_0 and U_1, we expect that for a given multipolarity, $T = 0$ states will be excited in inelastic proton scattering an order of magnitude more strongly than $T = 1$ states (12). (A comparison of the corresponding parts of the nucleon-nucleon effective interaction leads to a similar conclusion.) For the case of inelastic ^{3}He scattering, information about the potential U_1 is less well established. However, it is expected that $T = 1$ states will be excited even less strongly than $T = 0$ states. For the case of inelastic electron scattering, the electromagnetic excitation excites $T = 0$ and $T = 1$ states with equal strength.

Since the deuteron and alpha particle have isospin zero, excitation of a $T = 1$ state by these particles should be markedly reduced, and for scattering from self-conjugate nuclei ($N = Z$) there should be no isovector excitation. Charge-exchange reactions provide another isospin selection. The (n, p) and $(t, {}^3$He$)$ reactions on all nuclei and the (p, n) and $({}^3$He$, t)$ reactions on self-conjugate nuclei should excite only isovector states.

The excitation of isoscalar and isovector states in various reactions is summarized in Table 2. As will be seen later, comparison of various reactions leading to giant resonances in the same nucleus can be of great help in unraveling the isospin makeup of the resonances.

All of the inelastic-scattering data shown in this paper except the electron results were analyzed in terms of a macroscopic collective-model DWBA, and utilize prescriptions of the scattering process described in previous publications (12, 13). The calculations were performed using the computer codes DWUCK or JULIE, using optical-model parameters obtained from inelastic scattering. These parameters are generally described in the references cited and are not repeated in this paper. It is to be noted that for a given set of experimental data, use of different DWBA-analysis techniques and parameter sets can lead to considerable variation in the extracted EWSR values (16).

Table 2   Relative cross sections[a] of isoscalar and isovector excitations for various reactions

	Isoscalar	Isovector
(e, e')	1	1
(p, p')	1	$\approx 1/9$
$({}^3$He$, {}^3$He$')$	1	$\approx 1/30$
(α, α') (d, d')	1	≈ 0
(n, p) $(t, {}^3$He$)$	0	1
(p, n) $({}^3$He$, t)$ $N = Z$ nuclei	0	1

[a] Relative cross section normalized to 1 for the stronger excitation.

There are experimental methods other than those described above for excitation of giant multipole resonances, although less is said about these in this review. Over the years the photonuclear absorption process (as well as its inverse, nuclear capture reactions) has provided the bulk of the information on the GDR. For the study of quadrupole excitations, the capture process has the disadvantages that dipole excitation is favored over quadrupole excitation by a factor of 10 to 100 in cross section (23) and that generally only strength in the ground-state proton or alpha channel, i.e. (p, γ_0), (α, γ_0), is measured. However, the interpretation of capture angular distributions and polarization measurements is well understood so that precise measurements can yield accurate $E2$ strengths.

Another type of reaction that has been used to study giant resonances is the (π^-, γ) reaction. The limited number of such measurements that have been performed indicate (24) that isovector states will be preferentially excited. Thus there should be a correspondence between measurements of the (n, p) reaction and π^- absorption.

3   GIANT QUADRUPOLE RESONANCE

3.1   Early Experimental Evidence

The initial work leading to definitive identification of a GQR was done independently at three laboratories at about the same time. Electron-scattering measurements carried out at Darmstadt (25) and Tohoku University (26) and analysis of proton inelastic scattering measurements (27) from Oak Ridge National Laboratory provided the earliest evidence.

3.1.1   ELECTRON SCATTERING   The evidence from the Darmstadt group came from measurements (25) of inelastic scattering of 50- and 65-MeV electrons from Ce, La, and Pr. In the spectra from all three targets, a broad peak was observed at ≈ 15-MeV excitation and another at ≈ 12 MeV. The higher excitation energy agrees with the known GDR energy; however, no peak at 12 MeV was known from photonuclear measurements. The authors proposed that the 12-MeV peak was an $E2$ (or $E0$) excitation. [Further description of this work is given in (28, 29).]

The following year, a group at Tohoku University published (26) the inelastic electron spectra from ^{90}Zr shown in Figure 3. The spectra are for four different momentum transfers utilizing 150- to 250-MeV electron scattering. At the top of Figure 3a is the spectrum as obtained in photonuclear reactions on ^{90}Zr. In the electron-scattering spectra, a broad peak is observed that shifts in energy as a function of momentum transfer. At the higher-momentum transfers, the peak centroid has clearly shifted to a lower excitation energy. The authors decomposed the spectra into two peaks, as shown in the figure, and obtained the angular distributions shown in Figure 3b. The cross section for the 16.65-MeV peak (GDR energy) is well represented by an $E1$ calculation. However, the angular distribution for the 14.0-MeV peak is described by an $E2$ (or $E0$) calculation, but not by an $L = 1$ or $L \geq 3$ calculation. (The authors noted that the calculated $E2$ form factor cannot be distinguished from an $E0$ form factor at their electron energies. This

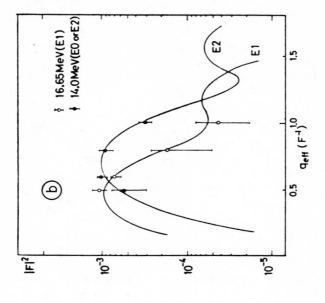

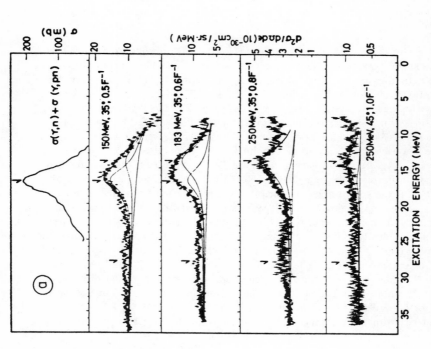

Figure 3 (*a*) Inelastic electron spectra from ^{90}Zr for various indicated momentum transfers (26). The top figure is the photonuclear excitation spectrum from ^{90}Zr, showing the giant dipole resonance. The electron spectra have been decomposed into two peaks, one at 16.65 MeV produced by excitation of the GDR and another at 14.0 MeV, which is identified as the GQR. (*b*) Angular distributions for the two peaks in the electron spectra shown in *a*. The calculated curve for $L = 2$ excitation is identical to the calculated curve for $L = 0$.

E2-E0 ambiguity is discussed in more detail later.) The measured cross section for the 14.0-MeV peak depletes ≈75% of the *E2* EWSR strength.

3.1.2 INELASTIC PROTON SCATTERING   In a large number of 60-MeV inelastic-proton spectra taken on several targets at Oak Ridge (8), a broad peak at high excitation energies was consistently observed at angles ≲40 degrees. The energy of the peak changed with target mass, and it was found (27) that the peak energy was always 2–3 MeV below the GDR energy in the nuclei examined. Figure 4 shows an example of the ^{56}Fe(*p, p'*) spectrum covering an excitation energy range of ≈40 MeV (8). It is clear that the broad peak observed at ≈16.5 MeV is centered 2–3 MeV below the GDR energy. Studies on several nuclei showed that the excitation energy of the observed peak varied as ≈63 $A^{-1/3}$ MeV (27), in good agreement with earlier predictions for the location of a GQR (1). These early data were soon substantiated (27, 30) by (*p, p'*) measurements on Al, Cu, In, and Pb.

The cross sections measured for the broad resonance peaks were too large to explain by depletion of 100% of the *E1* EWSR strength. However, inclusion of most of the *E2* EWSR strength with the *E1* strength provided good agreement with the measurements (27).

A previous observation of broad peaks in the inelastic proton continuum had been published in 1957–1958 (31). These data were taken on several nuclei ($A < 60$)

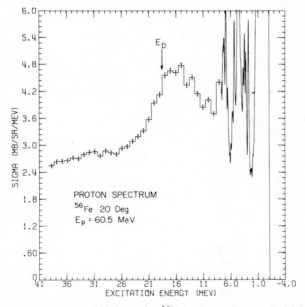

Figure 4 Proton spectrum at 20° from the ^{56}Fe(*p, p'*) reaction at 61 MeV (8). Above ≈8 MeV, the data have been plotted in ≈400-keV-wide bins to narrow the statistical fluctuations. The broad peak located at 16 MeV lies 2–3 MeV below the known excitation energy of the GDR, labeled E_D.

with 185-MeV incident protons. For each nucleus studied, a strong peak was found at 15–20 MeV of excitation. Comparison between 185- and 60-MeV data on ^{56}Fe showed (27) that the peak observed at 185 MeV is identical in excitation energy and shape to that shown in Figure 4 for 60-MeV protons. The peak observed in the 185-MeV data was assumed to be the GDR (31). However, we now know that the energies of the peaks and the cross sections for the peaks in the 185-MeV data cannot be explained on the basis of GDR excitation alone (12, 32).

3.2 Some Experimental Difficulties in Giant-Resonance Studies by Inelastic Scattering

Although the experimental techniques used to study high–excitation energy resonances are generally the same as those used for low-lying levels, inelastic scattering studies at high excitation energies do present some special problems. The resonance cross section is only a small fraction of the total inelastic cross section in the resonance region, and assumptions about the nature of the spectrum underlying the resonance peak must be made.

In order to obtain a cross section for the resonance, it is assumed that the unstructured nuclear continuum at excitation energies above the resonance (see Figure 1) extends under the resonance peak. Thus a shape and magnitude must be assumed for the underlying continuum. It is important to note that even small errors in the assumed shape and cross section of the underlying continuum can produce large errors in the extracted resonance cross sections. In the case of inelastic electron scattering, the ever-present radiative tail further complicates matters by producing nearly 90% of the raw counts obtained. However, this tail has a smooth shape and a calculable magnitude and can be subtracted from the raw data. [See (7) for a recent discussion of radiative tail corrections.] It is the author's judgment that a minimum uncertainty of $\pm 20\%$ must be assigned to resonance cross sections extracted from inelastic scattering results.

The manner in which the resonance cross sections are extracted has been described in many articles [e.g. see (33–36)]. The assumed shape and magnitude for the continuum underlying the resonance peak are usually determined by extrapolation of the slope of the higher–excitation energy continuum to a point near the neutron separation energy where the "true" continuum [from (p, pn) reactions] should begin. Although some angle-dependent calculations of the continuum are available (9), these calculations are not accurate enough to provide the continuum magnitude. However, the calculations do confirm the expected smoothness of the continuum spectrum.

As discussed in the preceding section, the GDR is expected to be excited in the (p, p') reaction. For the 60-MeV (p, p') data described in this paper, no attempt was made to subtract out a GDR component from the resonance peak. Rather, the resonance cross sections are determined over an energy interval broad enough to include both $E1$ and $E2$ contributions. These composite cross sections are then compared with calculated cross sections for an assumed $E1$ EWSR depletion plus the $E2$ component needed to fit the data. Other (p, p') data taken at 155 MeV (37) have been analyzed by subtraction of an assumed GDR cross section from the composite peak. In the case of inelastic alpha and deuteron scattering, there is no observable excitation of the $E1$ component (35, 36, 38, 39).

The high-excitation region of inelastic scattering spectra is particularly susceptible to spurious background. While such background at low excitation energies, where narrow, well-resolved peaks are observed, can generally be easily detected, background in the continuum spectrum is hard to detect. Many contaminants, such as lower-energy beam components, penetration of slits by the beam, and rescattering of small-angle elastically scattered particles, may have little effect on low-excitation spectra, but may show up in the nuclear continuum as broad peaks similar in character to the peaks from giant-resonance excitation. It is therefore of utmost importance to eliminate or reduce to negligible importance any spurious background. Observation of the incident beam by placing a detector in the reduced-intensity zero-degree beam [see, e.g. (40)] offers one method of detecting spurious background.

3.3 *Corroboration of Resonance Properties*

Following the initial observation of the GQR, many experiments utilizing a variety of reactions have been performed to study the resonance. In general, these results have tended to support fully the initial GQR proposals discussed in the previous section.

3.3.1 CONSISTENT EXPERIMENTAL OBSERVATION   Inelastic electron (41) and proton (37) measurements were soon extended to other nuclei with $A > 40$. In addition, studies of many nuclei by inelastic helium-3 (42–44), alpha (35, 39), and deuteron (36) scattering were undertaken.

Figures 5–7 show results from 60-MeV proton (33, 34, 35), 71-MeV helium-3 (42), and 96- to 115-MeV alpha-particle (35) inelastic scattering from several nuclei. In each of the nuclei studied there is a distinct peak in the continuum at an excitation of about $63 \times A^{-1/3}$ MeV. Consistent observation of the resonance has now been made in inelastic hadron scattering from an incident energy of 40 MeV to over 1.37 GeV (46). The available experimental results are tabulated in Section 5.

3.3.2 RESONANCE SPIN   Angular distributions measured for the resonance peak are in general well described by an $L = 2$ DWBA calculation. Figures 8 and 9 show several examples of the angular distributions from the (p, p') (33, 34, 45) and (α, α') (35) reactions, respectively. In the case of the (p, p') work, the measured cross sections should contain both the GDR and GQR contributions. The amount of the $E1$ EWSR strength depleted in the (p, p') cross section is assumed to be the same as that found in the same excitation energy region in total photonuclear cross-section measurements (47). For each nucleus studied, the calculated $E1$ cross section is not adequate to match the measurement. However, the addition of $E2$ strength in the amount shown on the figure provides good agreement with the data.

The agreement of the measured cross sections from the (α, α') results with $L = 2$ DWBA calculations is quite good. However, the (α, α') angular distributions cannot be fitted by an $L = 3$ calculation (35). Since the GDR should be excited to only a negligible extent by alpha particles, the amount of EWSR strength may be more

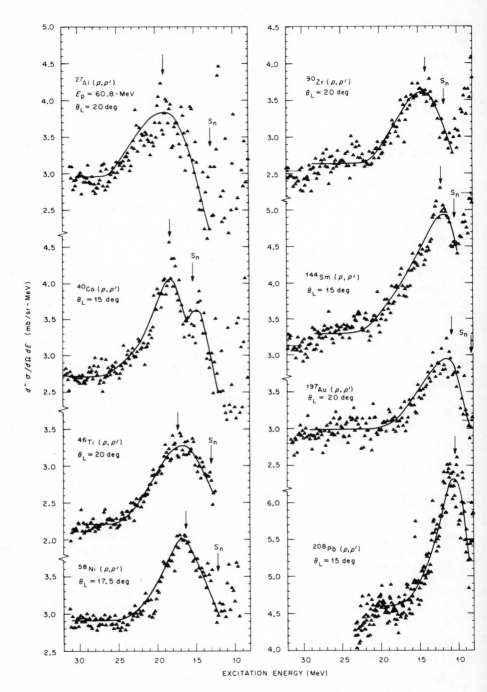

Figure 5 Inelastic proton spectra from 61-MeV protons incident on several targets (33, 34, 45). The neutron separation energy (S_n) and the energy $63 \times A^{-1/3}$ MeV are marked with arrows.

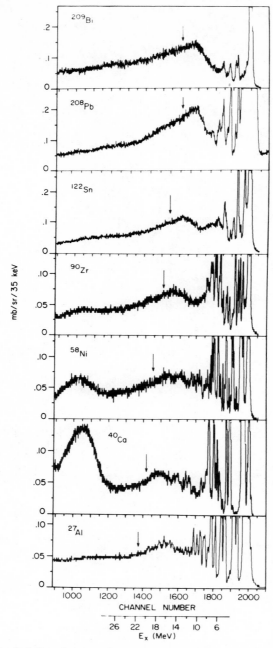

Figure 6 Inelastic helium-3 spectra from 71-MeV helium-3's incident on several nuclei, showing the systematic excitation of a giant-resonance peak (42). The peak in ^{58}Ni and ^{40}Ca at the high-excitation end of the spectrum is due to scattering from hydrogen in the target. The arrow marks the position of the GDR in each nucleus.

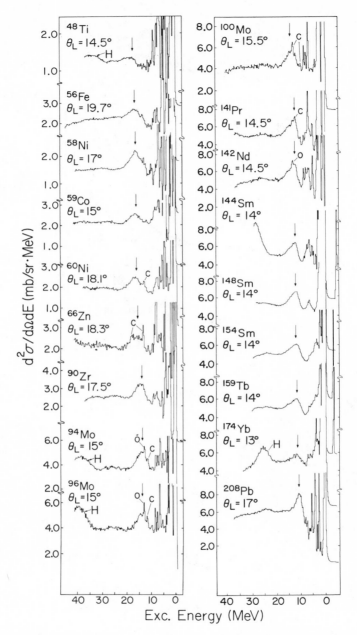

Figure 7 Inelastic spectra from 96- and 115-MeV alphas incident on several targets (35). The arrow indicates the excitation energy $63 \times A^{-1/3}$ MeV. Hydrogen, carbon, and oxygen contaminants are labeled.

Figure 8 Angular distributions for the giant-resonance peak measured in the (p, p') reaction at 60 MeV on ^{27}Al, ^{40}Ca, ^{58}Ni, ^{144}Sm, and ^{208}Pb [taken from (33, 34, 45)]. Calculated

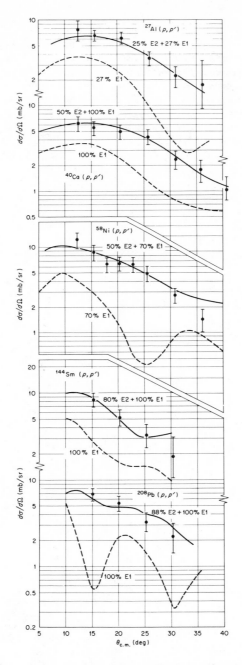

$L = 1$ (*dashed*) and $L = 2$ cross sections are normalized to percentage depletions of the respective sum rules shown in the figure. The $L = 1$ EWSR strength is assumed to be the same as that measured by total photonuclear reactions (47).

directly obtained than in the (p, p') reaction. The agreement in $E2$ sum-rule depletion between the (p, p') and (α, α') reactions is seen to be very good.

In the early inelastic electron–scattering work (26), it was noted that calculated form factors for quadrupole ($E2$) excitation and monopole ($E0$) excitation are identical for the momentum transfers used in the experiment (see Figure 3). Thus the possibility that the newly discovered resonance was a giant monopole resonance could not be ruled out with the (e, e') data. Calculations predicted (13) that angular distributions from inelastic proton scattering could distinguish between $E2$ or $E0$ excitation. Although the early (p, p') resonance cross sections (27) were shown (13) to have uncertainties too large to provide a clear distinction, more precise inelastic proton– (30, 33) and other inelastic-scattering measurements (36, 43) soon provided evidence that the resonance was not primarily a monopole excitation. Figure 10 shows the ^{58}Ni resonance angular distributions as measured using inelastic scattering of 70-MeV deuterons (36). The calculated angular distributions are normalized to the data at 15.5 degrees. It is easily seen that the data, while well reproduced by the $L = 2$ calculation, are not described by the $L = 0$ calculation.

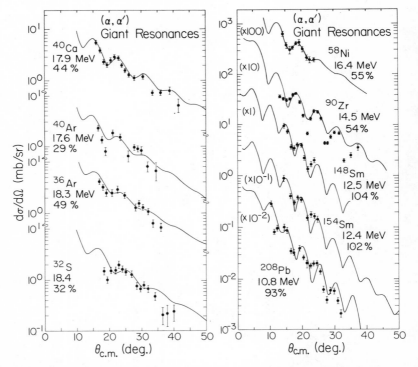

Figure 9 Angular distributions for the giant resonance peak from the (α, α') reaction at 96 and 115 MeV (35). The percentage of the $T = 0$, $L = 2$ EWSR strength depleted in the resonance is shown for each nucleus.

It is important to note that the $L = 0$ DWBA calculations are untested by comparison with experimental results. Unlike the $L = 2$ calculations, which can be compared with angular distributions for low-lying 2^+ levels, there are no known collective 0^+ states to which the $L = 0$ calculations can be compared. Thus all of the preceding comments dealing with observation or nonobservation of 0^+ resonance strength are based on an assumption of the validity of the model used.

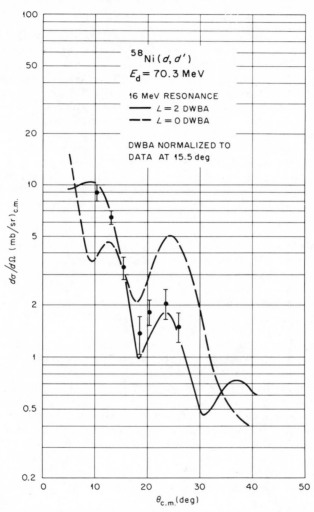

Figure 10   Angular distribution for the ^{58}Ni 16.5-MeV resonance peak measured in the (d, d') reaction at 70.3 MeV (36). The calculated curves for $L = 0$ and $L = 2$ are normalized to the data at 15.5°.

It has now been shown, using many inelastic reactions, that the $63 \times A^{-1/3}$-MeV resonance is predominantly a quadrupole excitation. However, small ($\lesssim 15\%$) contributions to the resonance cross section from excitations other than quadrupole cannot be ruled out because of the large uncertainties inherent in the extraction of resonance cross sections (recall Section 3.2). Indeed, as is discussed later, there is some experimental and calculational evidence to support the existence of such admixtures.

3.3.3   RESONANCE ISOSPIN   Identification of the GQR as an isoscalar excitation was provided by excitation of the resonance in the self-conjugate nucleus ^{40}Ca by isoscalar alpha particles (38). Alpha-particle scattering on a $T = 0$ nucleus should excite only isoscalar ($T = 0$) transitions, except for small contributions from Coulomb excitation. Since the GQR has $T = 0$, some difference should be expected in the $63\, A^{-1/3}$-MeV resonance structure as excited by alphas or deuterons and protons (see Section 2.1). Figure 11 shows a comparison between the (p, p') and (d, d') reactions on ^{58}Ni (36). A strong peak at ≈ 16 MeV due to excitation of the GQR is seen in both spectra. However, the resonance structure observed in the proton measurements extends several MeV beyond that for deuterons. The missing structure in the deuteron spectrum falls where the GDR ($T = 1$) would be expected. Thus, this comparison is consistent with the expectation that the $T = 0$ GQR is excited in both (p, p') and (d, d') reactions, while the $T = 1$ GDR is observed only in the (p, p') reaction.

3.3.4   ADDITIONAL EXPERIMENTAL EVIDENCE FOR THE GIANT QUADRUPOLE RESON- ANCE   Additional investigations of the GQR have been made using particle-capture reactions. For $A < 40$ nuclei, $E2$ strength has been definitely identified at high excitation energies, but no GQR peak has been found (48). In heavier nuclei, however, measurements of the ^{54}Fe$(\alpha, \gamma)^{58}$Ni reaction (49) have shown a GQR peak at ≈ 16 MeV. These results are in excellent agreement with inelastic-scattering results from ^{58}Ni. A study of the ^{36}Ar$(\alpha, \gamma)^{40}$Ca reaction (50) found $E2$ strength in the 20-MeV region of excitation, but not a well-defined GQR peak. However, the authors interpreted their data as being not inconsistent with the existence of a GQR at ≈ 18 MeV.

Most of the evidence for the new giant resonances discussed in this review has been provided by direct observation of peaks from giant resonance excitation. However, the existence of giant resonances may be deduced through other indirect methods, which do not involve direct experimental observation of a resonance structure. An important indirect method of resonance prediction is based on microscopic analysis of nucleon scattering to low-lying unnatural parity states (e.g. 2^- levels), in which the giant multipole resonances are treated as doorway states (51). Such calculations have yielded predictions for the energy and widths of 1^-, 2^+, and 3^- giant resonances. Comparison of these results with available data for direct resonance excitation shows remarkable agreement. Use of this indirect technique can identify the position of giant multipole resonances that may be too broad and/or too weak to be observed in inelastic scattering.

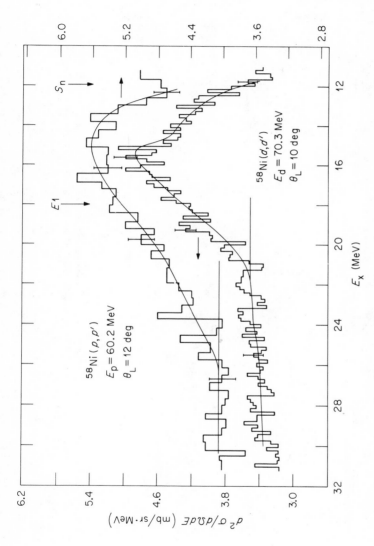

Figure 11 Comparison of cross sections in the nuclear continuum for proton and deuteron inelastic scattering from ^{58}Ni (36). $E1$ is the known energy of the giant dipole resonance; S_n is the neutron separation energy. The plotted uncertainties are statistical only. The smooth curves indicate assumed separations into the resonance and the underlying continuum.

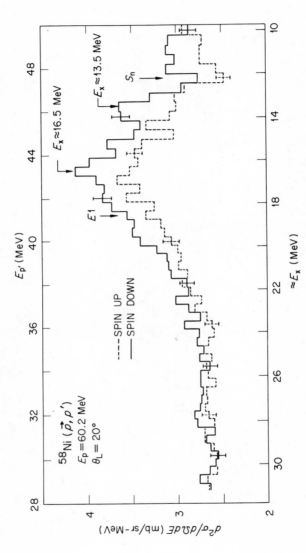

Figure 12 Polarized-beam cross sections from ^{58}Ni at 20 degrees plotted against outgoing proton energy (E_p) and approximate excitation energy (E_x) (34). S_n is the neutron separation energy; $E1$ is the known energy of the GDR. The GQR is located at 16.5 MeV, while another resonance, discussed in Section 3.8.2, is located at 13.5 MeV.

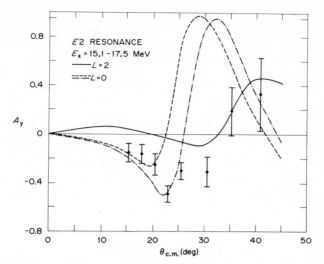

Figure 13 Analyzing powers for the giant resonance at $E_x \approx 16.5$ MeV in ^{58}Ni (52), compared with $L = 0$ and $L = 2$ calculations (13).

3.4 *Polarized-Proton Excitation of the Giant Quadrupole Resonance*

It was suggested (13) by DWBA calculations that the angular distribution of the analyzing power for 60-MeV polarized-proton excitation of the $63 \times A^{-1/3}$-MeV resonance should be markedly different for an $L = 0$ excitation compared with $L = 2$. To test this prediction, 60.2-MeV polarized protons were inelastically scattered from ^{58}Ni (34, 52).

Figure 12 shows 20° spectra obtained using incident spin-up and spin-down beams (34). In the region of the spectra where no structure was observed with unpolarized protons ($E_x \gtrsim 25$ MeV), the asymmetry[4] between the two spectra is essentially zero. However, in the region of the giant resonances a strong asymmetry is observed. An analyzing power for this excitation was extracted at each angle, using the relation

$$A_y(\theta) = \varepsilon/P_B, \qquad\qquad 8.$$

where P_B is the beam polarization and ε is the asymmetry. The analyzing power was obtained for the energy region $E_x \approx 15.1$–17.5 MeV by subtracting the underlying-continuum cross section and the GDR cross section [see (52) for details]. Figure 13 shows the results for the analyzing powers (52). The dashed curves represent DWBA calculations for two models of the $E0$ excitation (13), and the solid curve is the DWBA calculation for an $E2$ excitation (13). Clearly, the results are not well described by either calculation. It should be noted that the giant-

---

[4] Asymmetry $= \varepsilon = (\sigma\uparrow - \sigma\downarrow)/(\sigma\uparrow + \sigma\downarrow)$, where $\sigma\uparrow$ and $\sigma\downarrow$ are the measured cross sections for spin-up and spin-down protons, respectively.

resonance analyzing power has nearly the same angular dependence as that measured for the 1.45-MeV, $J^\pi = 2^+$ level in ^{58}Ni (52). Before definitive interpretation of the resonance multipolarity from polarized-proton experiments is possible, more theoretical and experimental work is needed.

3.5   *The Giant Quadrupole Resonance in Light Nuclei*

Although it has now been clearly demonstrated that a GQR peak exists in nuclei with $A \gtrsim 40$, the existence of such a peak in lighter nuclei is not well established. The available data on light nuclei fall into two distinct groups, that for several *sd*-shell nuclei and that for ^{12}C and ^{16}O.

3.5.1   *sd*-SHELL NUCLEI   Of the several topics discussed in this review, one that has stirred considerable controversy is the question of the existence of localized GQR strength in *sd*-shell nuclei (53). Much of the controversy can be eliminated by careful analysis of results and comparison of different experiments exciting giant resonances in these nuclei.

Early (p, p') results on ^{27}Al (27) showed a large resonance peak. The cross sections were found to be too large to be explained as an $L = 1$ (GDR) excitation alone. Subsequent calculations (12) showed that these data are consistent with $\approx 50\%$ depletion of the $T = 0$, $L = 2$ EWSR strength and 100% depletion of the $L = 1$ EWSR strength.

Following these observations, precise alpha-capture and polarized and unpolarized proton-capture measurements (23, 48) were made on a series of *sd*-shell nuclei. While the capture measurements clearly identified $E2$ strength in the 20-MeV excitation region, no evidence was found for any localization of this strength. It was thus felt that an inconsistency existed between the proton inelastic scattering and capture results (23).

Further data were provided by inelastic alpha-scattering measurements on several *sd*-shell nuclei (35, 39). Figure 14 shows results obtained with 96.6-MeV alpha particles on several targets. While a distinct peak is seen for ^{40}Ca, ^{40}Ar, and ^{36}Ar, no evidence for a resonance structure is present in the lighter nuclei. The dashed peaks in ^{27}Al and ^{16}O indicate the experimental resonance shape for an $L = 2$ excitation having a width of 6 MeV and exhausting 25% of the EWSR strength. The authors (39) note that a small fraction of the $E2$ EWSR strength spread over a larger energy region could be undetected in their data. These data are in apparent agreement with the capture result that no concentration of $E2$ strength occurs in nuclei near $A = 26$, and in disagreement with the ^{27}Al(p, p') data.

Figure 15 shows recently obtained (p, p') spectra in the resonance region of ^{28}Si, ^{27}Al, ^{26}Mg, and ^{24}Mg (54). Also shown for comparison are the 96.6-MeV (α, α') data on ^{27}Al. The proton data for ^{27}Al show the same broad resonance peak observed in earlier (p, p') data (27), which is clearly not observed in the alpha measurements. While definite and strong structure is seen near 20 MeV for the other nuclei, the structure is not contained in a single, broad peak as for $A \geqq 40$ nuclei. Indeed, the continuum structure is different for each of the nuclei studied.

The fact that the GDR should be excited in the (p, p') reaction, but not in the

(α, α') reaction, suggests that some strength observed in the (p, p') continuum may arise from $L = 1$ GDR excitation. Evidence for this is provided by the measured cross sections, plotted on Figure 16, for the structure seen between ≈ 15 MeV and 26 MeV in the (p, p') spectra. The calculated curves for the $L = 1$ excitation are based

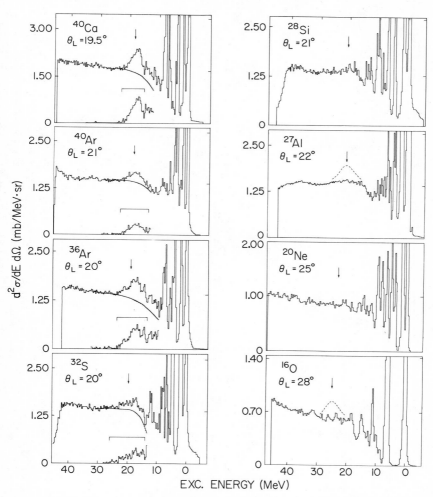

Figure 14   Spectra from the (α, α') reaction at 96.6 MeV (39). The data are shown at angles near the expected maximum for an $L = 2$ excitation. The smooth solid curve is the assumed continuum underlying the resonance. The GQR peak with background subtracted is shown at the bottom of the spectra for several nuclei. The arrows are located at 63 $A^{-1/3}$ MeV. The dashed peaks in ^{27}Al and ^{16}O indicate the expected resonance shape for an $E2$ peak having a 6-MeV width and exhausting 25% of the EWSR strength.

on two different models proposed previously (12). For each model the $L = 1$ calculation is normalized to the total photonuclear cross section (47, 55) measured within the energy limits shown on the plots. While the $L = 1$ strength calculated

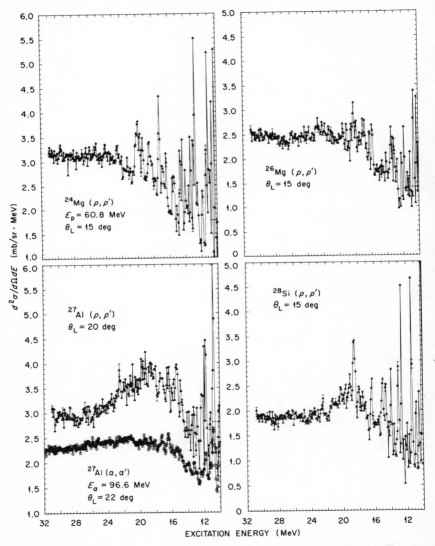

Figure 15 Spectra from the (p, p') reaction at 61 MeV for ^{24}Mg, ^{26}Mg, ^{27}Al, and ^{28}Si (54). The uncertainties shown on the data are statistical only. Energy resolution is ≈ 100 keV (FWHM). Also shown is the (α, α') spectrum from 96.6-MeV alphas on ^{27}Al. The (α, α') cross sections have been renormalized for ease of comparison with the (p, p') data.

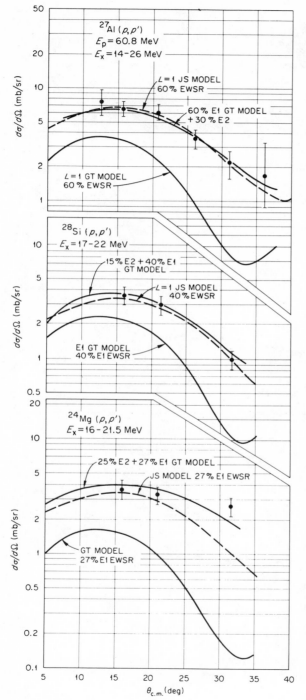

Figure 16 Angular distributions for the resonance structure measured in the (p, p') reaction on ^{27}Al, ^{28}Si, and ^{24}Mg (54). The excitation energy range integrated over is shown on each plot. The cross sections were obtained by subtracting a smooth continuum from beneath the peaks shown on Figure 15.

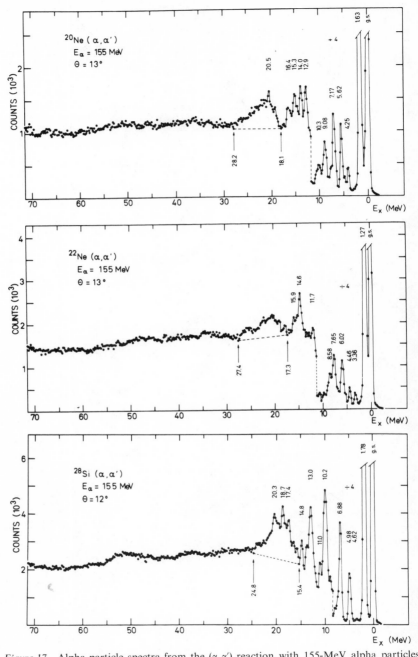

Figure 17 Alpha-particle spectra from the (α, α') reaction with 155-MeV alpha particles on ^{20}Ne, ^{22}Ne, and ^{28}Si (56). The dashed line represents the assumed shape of the underlying continuum.

with the Jensen-Steinwedel (JS) (12) model is adequate to account for the measured cross sections, use of the Goldhaber-Teller (GT) (12) model requires the addition of some $L = 2$ strength to agree with the data. However, no more than $\approx 30\%$ of the $L = 2$ EWSR limit, spread over many MeV (12 MeV for ^{27}Al), is needed to reproduce the measured cross sections.

Recently, (α, α') spectra have been measured (56) on some sd-shell nuclei, using higher-energy (155 MeV) alpha particles than previously used. Figure 17 shows spectra from ^{20}Ne, ^{22}Ne, and ^{28}Si. Unlike the lower-energy (α, α') data, a distinct peak is observed at $\approx 63 \ A^{-1/3}$ MeV. This seeming discrepancy is resolved when is is realized that the large peak in the higher-energy alpha data contains only $\approx 25\%$ of the $L = 2$ EWSR strength, a value in agreement with (p, p') results and not inconsistent with the limits of observation from the lower-energy (α, α') data. The large cross section in the high-energy (α, α') data for such a small EWSR strength depletion is provided by a more favorable $L = 2$ momentum transfer for these nuclei at 155 MeV than at 96 MeV.

The high-energy alpha results (56), that $\approx 25\%$ of the $L = 2$ EWSR limit is found in the sd-shell nuclei, would tend to support the GT model for the $E1$ (p, p') calculation (12). Other data on ^{27}Al obtained in the (d, d') reaction (57) yield $46 \pm 15\%$ of the $L = 2$ EWSR strength in a peak ≈ 8 MeV wide (FWHM).

In summary, we find that the available sd-shell data from inelastic scattering are generally consistent with one another and are not inconsistent with the capture results. The amount of GQR strength observed in these nuclei, 20–30% of the EWSR limit, is considerably less than the 80–90% observed in heavy nuclei. This difference may be partially explained by the larger depletion of $E2$ strength in the bound states for sd-shell nuclei than for heavier nuclei (see Table 1). In addition, the GQR strength in these nuclei may be fragmented and spread over a large energy region, as is the case for the GDR. Such fragmentation would make it difficult to observe resonance structure above the underlying continuum.

3.5.2 sp-SHELL NUCLEI

3.5.2.1 ^{16}O Inelastic scattering measurements to search for the GQR in ^{16}O have been performed with 104-MeV (58) and 146-MeV (59) alpha particles, 71-MeV ^{3}He (60), and 97- to 250-MeV electrons (61). In addition, measurements have been published using the capture reactions ^{12}C(α, γ_0) (62) and ^{15}N$(\vec{p}, \gamma_0)^{16}$O (63). The results from the $(\vec{p}, \gamma)$ reaction concern excitation of isovector GQR strength and are described in Section 3.10. The results from measurements using inelastic scattering of alphas and helium-3's are generally consistent with one another, but not with the (e, e') data.

Figure 18 shows the inelastic spectra from 104-MeV (58) and 146-MeV (59) alpha particles on ^{16}O. The energy resolution for the 104-MeV data is ≈ 150 keV, which is about three times better than that obtained at 146 MeV. Thus, what appears in the 146-MeV data as a somewhat structured broad bump centered at ≈ 22 MeV is resolved into a large number of peaks in the lower-energy data. From angular distributions from the 104-MeV data, $L = 2$ assignments were made (in addition to known low-lying 2^+ states) for states at 18.0, 18.5, 20.5, 20.9, and 21.85 MeV, with possible $L = 2$ assignments for states at 19.0, 22.5, 23.25, 23.85, and 24.4 MeV.

The total $L = 2$ cross section measured at 104 MeV in states with excitation energies between 18 and 25 MeV depletes $40^{+20}_{-10}\%$ of the $T = 0$, $L = 2$ EWSR strength (58). In the 146-MeV data, $67 \pm 25\%$ of the sum-rule strength is reported in the same region (59). The results from the $(^3\text{He}, {}^3\text{He}')$ reaction (54) show about 70% depletion of the EWSR strength. In the (e, e') measurements, only $25 \pm 10\%$ of the EWSR strength is reported (61) in the 18- to 25-MeV energy region, considerably less than in other inelastic-scattering measurements. However, the difficulty in subtracting the GDR from the GQR may be responsible for the lowest EWSR fraction reported for (e, e').

Further data (62) on ^{16}O have been obtained through the $^{12}\text{C}(\alpha, \gamma_0)$ reaction that cover the excitation energy region from 12–28 MeV. In this experiment, only the

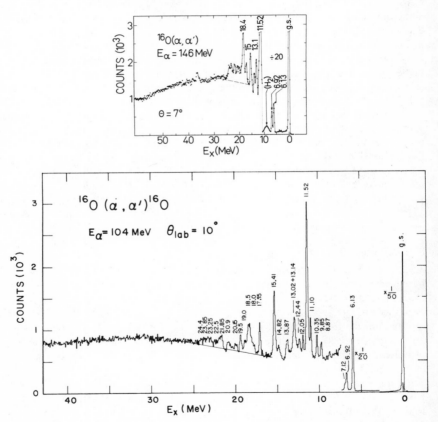

Figure 18 Spectra from the (α, α') reaction on ^{16}O using 146-MeV alphas [upper figure, (59)] and 104-MeV alphas [lower figure, (58)]. The energy resolution of the 104-MeV data is ≈ 150 keV (FWHM), which is about three times better than the resolution of the 146-MeV data. The solid curve is the assumed shape of the underlying continuum. Several of the excited states between 18.0 and 25 MeV are found to be 2^+ states.

ground-state (γ_0) channel is observed, and an estimate must be made of $E2$ strength in unobserved channels. The authors report a total estimated $T = 0$, $E2$ strength in ^{16}O (for all excitation energies up to ≈ 28 MeV) of about 70% of the EWSR limit. Since about 30% of this value lies in the excitation region below 18 MeV (59), the capture results imply that about 40% of the $E2$, $T = 0$ EWSR strength is in the 18- to 28-MeV region of excitation energy, in good agreement with the (α, α') results.

3.5.2.2 ^{12}C

Considerably less work has been done on the GQR in ^{12}C than on ^{16}O. Data have been obtained using inelastic scattering of 146-MeV alpha particles (56) and from the $^{11}B(p, \gamma_0)^{12}C$ reaction (64). In the latter measurements, the isospin of the observed $E2$ strength is not clearly determined, but is likely to be isovector. Some additional data have been reported using the (d, d') reaction (57).

The 146-MeV alpha data show several prominent peaks located at excitation energies above 15 MeV. Four peaks between 15.3 and 26.2 MeV are assigned as quadrupole and are found to deplete about 8% of the $L = 2$, $T = 0$ EWSR limit. Approximately another 8% of the sum-rule strength is reported to lie in the 26- to 30-MeV region (56). Thus, these experiments find less than 20% of the $L = 2$ sum-rule strength above 15 MeV in ^{12}C.

In the $^{12}C(d, d')$ reaction (57), a total of 40^{+15}_{-25}% of the $L = 2$, $T = 0$ EWSR strength is reported to lie in two broad peaks centered at ≈ 26 and ≈ 29 MeV.

3.6 Decay Modes of the Giant Quadrupole Resonance

Since the GQR is located at excitation energies above the particle separation thresholds, the resonance is expected to decay primarily by particle emission. Recent experiments (65) have measured these particle-decay branchings for the GQR in ^{40}Ca.

The GQR at ≈ 18 MeV in ^{40}Ca was excited by 115-MeV alpha particles, and the decay protons and alpha particles were detected in coincidence with the inelastically scattered alphas in the region of the GQR. Since the GDR is not significantly excited in the $^{40}Ca(\alpha, \alpha')$ reaction, no competition from GDR decay was present.

The GQR was found to proton decay 83^{+10}_{-40}% of the time. The proton-decay branches are ≈ 10% to the ^{39}K ground state and ≈ 17% to a group of states in ^{39}K at 2.6–3.0 MeV. Thus the majority of the proton decay goes to higher-lying levels of ^{39}K. No alpha decay to the ^{36}Ar ground state or first excited state was observed (<0.5%), and an upper limit of possible alpha decay to higher-lying states in ^{36}Ar was set at ≈ 10%.

These results help to explain why localized GQR strength is often not observed in capture reactions. In capture reactions [e.g. (α, γ)], usually only the gamma decay to the ground state, or perhaps to the first excited state, is observed. Since in ^{40}Ca these decay branches represent only a small fraction of the total GQR decay, the cross section for GQR observation in particle-capture measurements is quite small.

3.7 Broadening of the Giant Quadrupole Resonance in Deformed Nuclei

Since it is well established that the GDR is split in deformed nuclei [see e.g. (66)], one of the most interesting questions concerning the new resonances is whether the GQR is also split or broadened in deformed nuclei. The GDR splitting is attributed to different frequencies of dipole oscillation along the major and minor axes of a deformed nucleus (67). Since similar behavior has been predicted for an isovector $E2$ excitation (68), the shape of the $T = 0$ GQR might show a similar effect.

Inelastic-scattering experiments to study the GQR in deformed nuclei have been performed using incident 60-MeV protons (69), 50- and 64.3-MeV electrons (70), 80-MeV helium-3's (44), and 115-MeV alpha particles (71). Since the GDR is not significantly excited in the (α, α') reaction, the most definitive information has been obtained from those results.

Table 3 shows the measured GQR excitation energy and width for ^{144}Sm, ^{152}Sm, and ^{154}Sm, obtained from 115-MeV inelastic alpha scattering (71). It is clear that the width of the GQR increases by about 1 MeV between ^{144}Sm and ^{154}Sm. This broadening of the $T = 0$ GQR is discussed in (71) in terms of a quadrupole-quadrupole (QQ) nucleon-nucleon effective interaction. The authors find that while the usual treatment of the QQ interaction would predict about a 6-MeV GQR splitting, a more rigorous self-consistent treatment reduces the splitting to ≈ 2 MeV, in general agreement with the data.

The results from 60-MeV inelastic proton–scattering measurements (69), also on Sm nuclei, were more difficult to interpret, mainly because of excitation of the GDR. However, comparison of the (p, p') results with the (α, α') data shows that the two measurements yield consistent results. Shown in Figure 19 are the 15° ^{144}Sm (p, p') (72) and the 14° ^{144}Sm (α, α') (71) resonance peaks. In both cases the underlying continuum has been subtracted from the data. While the agreement between the two spectra at the low-excitation end of the resonance peak is excellent, the proton spectrum extends to higher excitation than the alpha-particle spectrum. The missing structure in the alpha spectrum falls near the GDR energy. Thus, as was the case with the previous comparison between (p, p') and (d, d') spectra (Figure 11), the extra cross section in the (p, p') spectrum is probably produced by excitation of the $T = 1$ GDR.

Inelastic electron–scattering measurements have been performed on the spherical nucleus ^{142}Nd and on the deformed nucleus ^{150}Nd (70). Data were taken at one angle, 93°, and for incident-electron energies of 50.0 and 64.3 MeV. For both targets the $E1$ contribution to the spectrum was removed by subtracting a GDR peak shape and cross section adopted from photonuclear measurements. For ^{142}Nd,

Table 3   Energies and widths for GQR in Sm nuclei[a]

	E_x (MeV)	Width (FWHM, MeV)
^{144}Sm	13.0 ± 0.3	3.9 ± 0.2
^{152}Sm	12.5 ± 0.2	4.3 ± 0.2
^{154}Sm	12.4 ± 0.3	4.7 ± 0.2

[a] Results from (71).

the remaining resonance peak has an energy of 12.0 ± 0.2 MeV with some additional structure centered at about 22 MeV. The 12-MeV peak is identified as the $T = 0$ GQR and has a width of ≈ 3.8 MeV. (The authors quote a value of 2.8; however, as pointed out in (71), the published spectrum indicates a 3.8-MeV width.) The GQR in ^{150}Nd is found at about 11.2 MeV (70), with some additional structure appearing at approximately 25 MeV. The width of the 11.2-MeV peak is about 5.0 MeV, about 1 MeV wider than for the spherical nucleus. The authors interpret the structure in both spectra at a higher excitation energy as being produced by excitation of the $T = 1$ GQR.

The widths of the GQR peaks for ^{142}Nd and ^{150}Nd are in good agreement with those for ^{144}Sm and ^{154}Sm, respectively, obtained from inelastic alpha-particle scattering. Thus there seems to be a consistent increase in GQR width of about 1 MeV for the deformed isotope, as compared to the spherical isotope for both the Sm and Nd nuclei. While it is fair to say that none of the present experimental results are totally conclusive, nevertheless the available data are consistent with a small broadening of the GQR due to nuclear deformation.

3.8   Fine Structure in the Giant Quadrupole Resonance

3.8.1 ^{208}Pb   For nuclei with $A \gtrsim 40$, the GQR is generally observed to be unstructured. An exception to this is the ^{208}Pb nucleus. In (e, e') (73) and (p, p') (31) measurements, several narrow (width < a few hundred keV) peaks were found to be

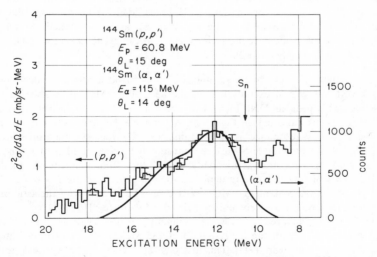

Figure 19   Comparison between the (p, p') (72) and (α, α') (35) reactions on ^{144}Sm. A smooth curve has been fitted to the (α, α') data from (35) for clarity of comparison. S_n is the neutron separation energy. The underlying continuum has been removed from the spectra in both cases. That portion of the (p, p') resonance that extends to excitation energies above the (α, α') spectrum is probably produced by excitation of the GDR, which is not excited by alpha particles.

superimposed on the broad GQR peak located at about 11 MeV. The fine-structure peaks observed in the (e, e') reaction were identified as $E2$ or $E0$ excitations. This observation of fine structure is especially interesting in that it represents selective excitation of a few 2^+ (or 0^+) states out of the hundreds of states in the high-excitation regions of ^{208}Pb.

More recent experiments on ^{208}Pb with electrons (74), alphas (35), protons (33, 37, 75), helium-3's (75), and the (γ, n) (76) reaction have confirmed the existence of fine structure and have attempted to identify multipolarities of the excited states. Of particular interest are the 70-MeV $(^3\text{He}, ^3\text{He}')$ and 45-MeV (p, p') measurements (75), in which energy resolutions of 45 keV (FWHM) and 35 keV, respectively, were obtained. Figure 20 shows the (p, p') spectra at two angles from these measurements, along with a more complete ^{208}Pb spectrum taken with 61-MeV incident protons (33). It is interesting to compare the high-resolution results with the 61-MeV spectrum taken with ≈ 100-keV resolution. The lower-energy data clearly show many fine-structure resonances not seen in the other data. The peaks marked $L = 1$ are identified as 1^- states, while the peaks at 9.35 and 10.3 MeV are assigned as octupole excitations. The angular distribution for the state at 9.11 MeV agrees well with an $L = 0$ calculation, which exhausts about 7% of the $T = 0$ EWSR strength.

The state at 9.35 MeV has been identified as $L = 2$ or 3 in other (p, p') work (33), as $L = 3$ in (α, α') (35), but as $L = 2$ or 0 in (e, e') (73) measurements. Similarly, the 10.3-MeV state was found to be $L = 2$ or 0 in (e, e') measurements (73). It is not clear whether the $L = 0$ state proposed (75) at 9.11 MeV is the same as the state seen at 8.9 MeV in electron scattering (74) and said to be $E0$. It should be noted that the fine structure seen in the (p, p'), (α, α'), and (e, e') reactions may not necessarily be the same, since the various reactions can excite isovector and isoscalar states differently (see Section 2).

3.8.2 fp-SHELL NUCLEI   In several nuclei of mass 56–60, a resonance peak located about 3 MeV lower in excitation than the $63 \times A^{-1/3}$-MeV resonance has been observed. This peak in ^{58}Ni, located at about 13.5 MeV and having a width of ≈ 2 MeV (34), is indicated on Figure 12. A probable $L = 2$ assignment was made for the 13.5-MeV resonance from these data (34, 52). Inelastic electron scattering on ^{58}Ni (77) indicated the 13.5-MeV resonance may be comprised of two peaks, one at 13.2 MeV and the other at 14.0 MeV. An $E2$ (or $E0$) assignment was made for both peaks, which were found to deplete a total of $12 \pm 1\%$ of the $T = 0$, $E2$ EWSR strength. A similar peak at 13.3 MeV, excited in ^{56}Fe by 155-MeV inelastic proton scattering (37), was tentatively assigned $L = 3$, while $(^3\text{He}, ^3\text{He}')$ (43) results for the same ^{56}Fe peak favored an $L = 2$ assignment. Further evidence for a peak at ≈ 13 MeV comes from the analysis (32) of 185-MeV inelastic proton–scattering data on ^{40}Ca [data from (31)]. This analysis reported the 13-MeV peak to have an $L = 3$ angular distribution and to deplete about 30% of the $T = 0$, $E3$ EWSR strength.

Hence, while it is rather well established that a secondary resonance, lower in excitation energy than the main GQR, exists in several fp-shell nuclei, a definitive determination of the multipolarity of this resonance is lacking.

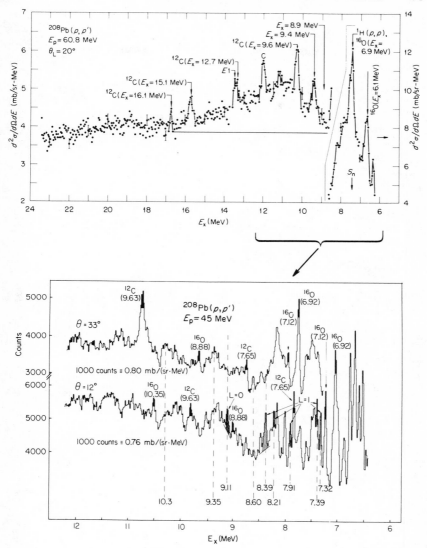

Figure 20   Inelastic proton spectra from ^{208}Pb. The upper figure shows the ^{208}Pb giant resonance spectrum from 60.8-MeV proton inelastic scattering (33). The energy resolution is ≈ 100 keV (FWHM). Several contaminant peaks from ^{12}C, ^{16}O, and H are marked. S_n is the neutron separation energy and $E1$ is the known energy of the GDR. The lower figure shows a portion of the giant resonance region in ^{208}Pb, studied with 35-keV (FWHM) energy resolution (75). A great deal of fine structure is observed in this spectrum that was not found in the upper spectrum.

Table 4 Isoscalar GQR energies, widths, and strengths

Nucleus	Resonance-excitation energy (MeV)	Resonance width (FWHM) (MeV)	EWSR[a] depletion (%)	Reaction	Incident-particle energy (MeV)	Ref.
^{12}C	b	2±1, 4±1	≈20	(α, α')	146	56
^{16}O	26±1, 29±1	d	40^{+15}_{-20}	(d, d')	70	57
	≈22[c]	d	40^{+20}_{-10}	(α, α')	104	58
	c	d	67±25	(α, α')	146	59
	c	d	≈70	$(^3He, {}^3He')$	71	60
	c		25±10	(e, e')	97–250	61
	c		≈40	$^{12}C(\alpha, \gamma_0)$	7–27.5	62
^{20}Ne	22.4	5.5	<30	(α, α')	146	56
^{22}Ne	22.0	6.0	<30	(α, α')	146	56
^{24}Mg	e	e	25±10	(p, p')	61	54
^{27}Al	≈20		30±10	(p, p')	61	54
^{28}Si	21.0±0.5	8±1	46±15	(d, d')	70	57
	19.7	5.1	<35	(α, α')	146	56
	≈19		15±10	(p, p')	61	54
^{32}S	18.4±0.6	7.1±0.5	32±15	(α, α')	96.6	39
^{36}Ar	18.3±0.2	5.6±0.3	49±15	(α, α')	96.6	39
^{40}Ar	17.6±0.2	4.7±0.3	29±10	(α, α')	96.6	39
^{40}Ca	18.0±0.3		50±12	(p, p')	60.8	45
	18.0±0.4		49±110	(p, p')	155	37
	18.1±0.3	3.5±0.3	44±10	(α, α')	96.6	35
	17.9±0.3	3.4±0.3	41±11	(α, α')	115	35
	16–22		37	(e, e'), $(^3He, {}^3He')$	150, 183, 250	89
^{44}Ca	≈17		≈75	(p, p')	70	42
	≈17		148	(p, p')	185	12,32
	≈17			(e, e')	124–250	41

Nuclide						
^{48}Ti	17.3±0.3	≈6	85±20	(^{3}He, ^{3}He')	80	43
	18.0±0.5	6.6±0.4		(α, α')	96	35
^{54}Fe	≈16			(p, p')	62	27
	≈16			(^{3}He, ^{3}He')	71	42
^{56}Fe	16.6±0.3	6.0±0.5	67±17	(^{3}He, ^{3}He')	80	43
	16.2	≈5.1	37±7	(p, p')	155	37
	16.1		80	(e, e')	124–250	78
^{90}Zr	16.7±0.3	5.7±0.5	59±15	(α, α')	96	35
	14.5±0.3	4.0±0.2	54±15	(α, α')	96	35
	14.0	4.8±0.6	56±17	(e, e')	150–250	26
^{94}Mo	14.4±0.4	5.2±0.5		(α, α')	115	35
^{96}Mo	14.4±0.6	4.8±0.6		(α, α')	115	35
^{100}Mo	13.6±0.4	5.1±0.5		(α, α')	96, 115	35
^{106}Pd	≈13			(e, e')	183	41
^{114}Cd	≈13			(e, e')	183	41
^{115}In	13.2±0.2	3.5	14±3	(p, p')	155	37
	13.7±0.6			(p, p')	66	27
^{116}Sn	12.0		120±36	(e, e')	150–250	77
^{120}Sn	13.5±1.0		93±20	(p, p')	62	27, 12
^{nat}Sn	12.7±0.2	3.3	17±4	(p, p')	155	37
^{122}Sn	≈13.5			(^{3}He, ^{3}He')	71	42
	≈13.5			(^{3}He, ^{3}He')	71	42
^{58}Ni	16.4±0.3	4.9±0.2	55±15	(α, α')	115	35
	16.0±0.5	4.5±0.3	50±10	(d, d')	70	36
	16.5±0.5	4.2±0.5	50±10	(p, p')	61	34
	16.0±0.5		40±15	(d, d')	46	36
	16.0			^{54}Fe(α, γ)		49
	16.3	≈4.5	57±6	(e, e')	150–250	77
^{59}Co	≈16			(^{3}He, ^{3}He')	70	42
	16.3	≈6		(α, α')	1370	46
	≈16			(^{3}He, ^{3}He')	80	43
	16.3±0.5	5.6±0.4	61±15	(α, α')	115	35

Table 4 *Continued*

Nucleus	Resonance-excitation energy (MeV)	Resonance width (FWHM) (MeV)	EWSR[a] depletion (%)	Reaction	Incident-particle energy (MeV)	Ref.
^{60}Ni	16.6±0.3	5.0±0.4	63±15	(α, α')	96, 115	35
	15.8			$(^3\text{He}, {}^3\text{He}')$	80	43
	≈16			(p, p')	62	27
^{nat}Cu	15.5±0.6			(p, p')	66	27
^{66}Zn	15.8±0.7	5.8±0.8		(α, α')	115	35
^{89}Y	13.8±0.2	3.2	24±5	(p, p')	155	37
^{139}La	≈12			(e, e')	50, 65	28, 29
^{nat}Ce	12.0		70±21	(e, e')	50, 65	25, 28, 29
^{141}Pr	≈12			(e, e')	50, 65	25, 29
^{142}Nd	13.3±0.4	4.0±0.4		(α, α')	115	35
	13.2±0.4	3.6±0.3	110±30	(α, α')	115	35
	12.0±0.2	3.8	65±15	(e, e')	50, 64	70
^{144}Sm	13.0±0.3	3.9±0.2	91±25	(α, α')	115	35
	12.8±0.3		80±20	(e, e')	67	69
	13.0	5.4	130±52	$(^3\text{He}, {}^3\text{He}')$	80	44
^{148}Sm	12.5±0.2	4.3±0.2	104±25	(α, α')	115	35
^{150}Nd	11.2±0.2	5.0±0.2	85±20	(e, e')	50, 64	70
^{152}Sm	11.5			(e, e')	150–250	41
^{154}Sm	12.4±0.3	4.7±0.3	102±25	(α, α')	115	35
	12.5±0.3		80±20	(p, p')	67	69
	12.2±0.5	5.5	150±60	$(^3\text{He}, {}^3\text{He}')$	80	44
^{159}Tb	12.0±0.6	4.7±0.4		(α, α')	115	35
	12.1±0.5	5.8	150±60	$(^3\text{He}, {}^3\text{He}')$	80	44
^{165}Ho	11.6±0.2	3.6	26±5	(p, p')	155	37
	12.2±0.5	6.1	150±60	$(^3\text{He}, {}^3\text{He}')$	80	44

^{169}Tm	12.0±0.5	6.2		150±60	(^{3}He, ^{3}He')	80	44
^{174}Tb	12.5±0.6	4.7±0.5			(α, α')	115	35
^{181}Ta	11.1±0.2	3.9		42±9	(p, p')	155	37
^{197}Au	11.5±0.2			90±20	(p, p')	61	34
	≈11	≈3		≈90	(^{3}He, ^{3}He')	71	42
	10.7±0.4			119±24	(e, e')	150–250	77
	10.8	2.9±0.2		77±18	(e, e')	90	74
^{208}Pb^f	11.0±0.3			90±20	(p, p')	61	34
	10.8±0.4	2.6±0.4		93±25	(α, α')	96	35
	10.6±0.4	2.6±0.3		92±25	(α, α')	115	35
	≈10.5			47±11	(e, e')	124–250	73
	10.5	2.8±0.3		95±35	(e, e')	80	74
^{208}Pb^f	10.5			39±8	(p, p')	155	37
	10.9±0.5	5.9		165±66	(^{3}He, ^{3}He')	80	44
	≈11			≈90	(^{3}He, ^{3}He')	71	42
	10.8			80±20	(e, e')	50	79
^{209}Bi	11.5±1.0	2.7±0.2		90±25	(p, p')	62	27
	≈10.5			38±8	(p, p')	155	37
^{238}U	≈11			85±20	(p, p')	66	96

^{a} These values were deduced from the resonance cross section with the assumption that the entire $63 \times A^{-1/3}$-MeV resonance is $L = 2$ [except for the (p, p') reaction, where an appropriate $L = 1$ cross section is assumed].

^{b} The GQR strength is observed to be split among several states (see Section 3.5.2.2).

^{c} Since the GQR strength is divided among many narrow states, this energy represents only the approximate center of the strength (see Section 3.5.2.1).

^{d} The GQR strength is split among several narrow states (see Section 3.5.2.1).

^{e} E2 strength may be fragmented in several peaks (see Figure 15).

^{f} Considerable resonance fine structure has been observed in ^{208}Pb. The values listed for width and energy are for the composite GDR upon which the fine structure is superimposed. The EWSR value shown is for all the GQR strength, including fine structure.

3.9   Systematics of the Isoscalar Giant Quadrupole Resonance

Shown in Table 4 are the energy, width, and EWSR depletion for the $T = 0$ GQR in each nucleus studied. The agreement among the various measurements is, in general, strikingly good. While there is some theoretical reason to expect that EWSR depletions obtained from inelastic scattering of nuclear projectiles and electron measurements may differ (16), there is little evidence to support that expectation in the data compiled in Table 4. Often the EWSR depletion is quoted without uncertainty, but in view of the inherent difficulties in obtaining the resonance cross sections in inelastic scattering, the uncertainty is probably at least $\pm 20\%$. One systematic discrepancy is observed in the EWSR strengths. For heavy nuclei ($A \gtrsim 120$), the EWSR strength deduced from 155-MeV proton scattering is 2–3 times less than that deduced from the lower-energy nuclear projectile scattering results and from some electron results.

In Figures 21–23, the resonance energy (MeV), width (MeV), and percentage depletion of the $T = 0$, E2 EWSR strength are plotted as a function of nuclear mass. Since there are far too many individual experimental numbers to plot, the average value or that which, in the present author's judgment, represents the "best" value has been plotted.

The variation of the GQR energy with nuclear mass is shown in Figure 21, along with a solid curve representing $E_x = 63 \times A^{-1/3}$ MeV. The agreement of the data

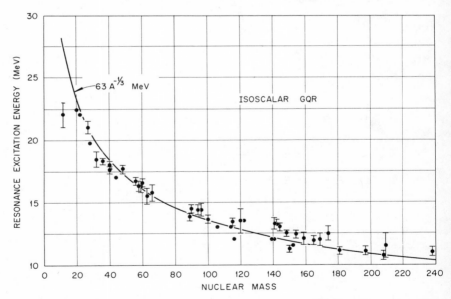

Figure 21   Excitation energy of the isoscalar GQR plotted against nuclear mass number. The data are averages or selections of results listed on Table 4. The solid curve represents the excitation energy $63\ A^{-1/3}$ MeV.

with the curve is excellent for nuclei with $40 \lesssim A \lesssim 120$. For nuclei in the *sd*-shell, there may be some evidence for a slight drop in energy; however, not enough data are available for a definite conclusion. For lighter nuclei, as was discussed earlier, the GQR strength seems to be fragmented into many narrow states, and thus it is not clear what is the best way to represent the energy of the GQR in these nuclei. For $A \gtrsim 120$, there is evidence that the measured GQR energies are higher than the $63 \times A^{-1/3}$ dependence would predict.

The trend of the GQR width with nuclear mass is shown in Figure 22. For those data reported without uncertainties, the assumption of an uncertainty of at least ± 0.2 MeV would be reasonable. The resonance width is narrowest for closed-shell nuclei and generally decreases with increasing nuclear mass. This trend is quite similar to that established for the GDR (5, 6).

Perhaps the most interesting of the systematics is that shown in Figure 23 for the percentage of the $T = 0$, $E2$ EWSR strength depleted in the GQR. (The plotted values do not include contributions to the $E2$ EWSR strength depleted in bound-state excitations.) A trend to larger sum-rule depletion with increasing nuclear mass is clearly evident. There may be several reasons for this. The amount of $T = 0$, $E2$ EWSR strength depleted in the low-lying states for nuclei with $A \lesssim 40$ is found to be considerably larger than for heavier nuclei (see Table 1). In addition, as discussed

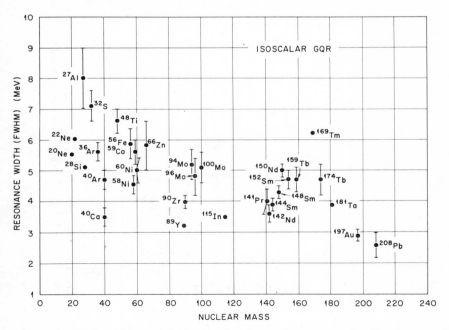

Figure 22 Widths of the $T = 0$ GQR plotted against nuclear mass number. The data are taken from Table 4. Uncertainties, when not given by the authors, are probably at least ± 200 keV.

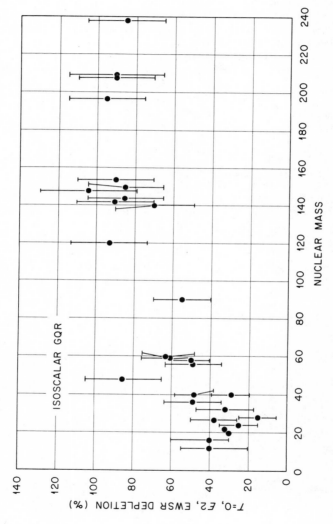

Figure 23   Percentage depletion of the $T = 0$, $E2$ EWSR strength in the GQR plotted against nuclear mass number. The data are averages or selection of results listed in Table 4. The values shown do not include the EWSR depletion in low-lying 2^+ states.

in Section 3, the $2\hbar\omega$, $E2$ strength is more likely to be fragmented in light nuclei, and thus difficult to observe above the ever-present inelastic nuclear continuum.

In summary, the present experimental evidence yields the following EWSR depletion in $2\hbar\omega$ states (i.e. in the GQR):

$A \lesssim 40$ $\lesssim 30\%$,

$40 \lesssim A \lesssim 100$ $\approx 60\%$,

$A \gtrsim 100$ $\approx 90\%$.

3.10   Isovector Giant Quadrupole Resonance

Experiments purporting to observe an isovector ($T = 1$) GQR are far fewer than for the isoscalar resonance. As described previously, isovector states are much less strongly excited in inelastic scattering of nuclear projectiles than are isoscalar states. On the other hand, the electromagnetic (e, e') interaction provides equal strength in both. Consideration of the strength, energy, and width of the $T = 1$ GQR's reported leads to the conclusion that they could not have been observed in any of the inelastic measurements using nuclear projectiles made to date. Indeed, the available evidence for the $T = 1$ GQR comes almost entirely from the (e, e') interaction, with some data from proton-capture results (63, 64).

In electron scattering, a very broad peak (8–10 MeV, FWHM) located at an excitation energy of 120–$130\ A^{-1/3}$ MeV has been observed in several nuclei. The

Table 5   Isovector GQR energies, widths, and strengths[a]

Nucleus	Resonance-excitation energy (MeV)	Resonance width (MeV)	EWSR depletion (%)	Electron energy (MeV)	Ref.
^{56}Fe	~ 32	~ 9		124–150	77
^{58}Ni	29		11 ± 1	150–250	76
	28.3 ± 0.3			150–250	95
^{60}Ni	28.5 ± 0.3			150–200	95
^{64}Ni	28.2 ± 0.3			150–200	95
^{90}Zr	27		≈ 25	150–250	26
^{116}Sn	25		58 ± 29	150–250	77
^{132}Nd	~ 22			50, 64	70
^{150}Nd	~ 24			50, 64	70
^{152}Sm	24		27	150–250	97
^{197}Au	23.0	7 ± 1	95 ± 31	90	74
	22.5		133 ± 27	150–250	77
^{208}Pb	22.5	5 ± 1	85 ± 28	90	74
	~ 22		60 ± 25	124–250	73
^{209}Bi	~ 24	~ 3.5		b	94
	~ 26.5			c	24

[a] All data from (e, e') measurements unless noted.
[b] Data from ^{208}Pb(p, γ) reaction for $E_p = 17.5$–25.0 MeV.
[c] From ^{209}Bi(π^-, γ) reaction.

angular distribution for the peak is generally described well by an $E2$ form factor, and the measured cross section depletes from 50 to 100% of the $T = 1$, $E2$ EWSR strength in the nuclei studied [see e.g. (70, 74, 78)].

The only independent substantiation of the (e, e') results comes from ^{208}Pb(p, γ) ^{209}Bi (62) and ^{209}Bi(π^-, γ) (24) reactions, in which a resonance suggested to be an isovector GQR is found near 25 MeV in ^{209}Bi.

In the light nucleus ^{16}O, evidence has been obtained through the ^{15}N($\vec{p}$, γ_0) reaction (63) for GQR strength that is probably isovector. These results, which are based only on observation of ground-state gamma transitions, are interpreted by the authors to yield $\approx 60\%$ of the $T = 1$, $E2$ EWSR strength.

A list of the proposed isovector GQR's is given in Table 5, and the peak energies are plotted in Figure 24 versus nuclear mass number. While the peak energies are on the average represented by $E_x \sim 120 \, A^{-1/3}$ MeV, there is a trend for the excitation energies to increase above that value with increasing nuclear mass number. There are too few measurements of the EWSR strength for this resonance to detect any trend with nuclear mass. It is clear that additional measurements of this resonance are needed to provide both confirmation and systematics.

4   OTHER NEW GIANT RESONANCES

While the GQR has been the most thoroughly studied of the new resonances, evidence has been found for other new resonances as well. Of these, the $E0$ and $M1$ have been the most discussed. As mentioned earlier, observation of giant magnetic dipole ($M1$) resonances through 180° electron scattering is a very active field and has been recently reviewed (7). It is worth noting that recent observation

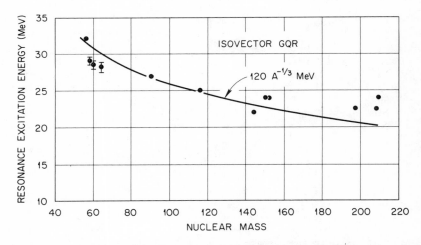

Figure 24   Excitation energy of the isovector GQR plotted against nuclear mass number. The data are taken from Table 5. The solid curve represents the excitation energy $120 \, A^{-1/3}$ MeV.

of the $M1$ resonance via (n, p) (80) and (p, n) (81) reactions may lead to another means of investigation of these resonances.

4.1 Giant Monopole (E0) Resonances

The possibility of observing an $E0$ giant resonance has been the topic of a large number of papers, since the earliest observation of the new giant resonances in 1971. This monopole resonance would represent the excitation of a collective breathing or compressional mode (13). This mode of excitation has a very special significance, since its observation could provide information on the compressibility of nuclear matter, which is unobtainable in other ways. Many theoretical calculations have predicted the existence of an $E0$ resonance near $2\hbar\omega$ [see e.g. (16, 82–85)], but to date no conclusive evidence for its observation has been provided.

The earliest candidate for the $E0$ resonance, the $63\ A^{-1/3}$-MeV resonance, has now been established as at least predominantly a GQR. However, due to the large uncertainties in the measured resonance angular distributions, some small contribution from other multipole resonances in this excitation region cannot be ruled out. As is described further in the next section, recent microscopic-model calculations (16) show that the available data for the $63 \times A^{-1/3}$-MeV resonance in ^{208}Pb are consistent with $L = 2 + 4 + 0$ excitations.

Another candidate for the $E0$ resonance was a state at 8.9 MeV in ^{208}Pb and at 9.2 MeV in ^{197}Au ($\approx 53\ A^{-1/3}$ MeV). These results (74), obtained with the (e, e') reaction, were interpreted as being consistent with $\approx 50\%$ depletion of the $L = 0$ EWSR limit. However, recent (e, e') measurements (79) have shown that this state could as well be interpreted as quadrupole. In addition, (γ, n) results (76) indicate a quadrupole state at this energy. Thus it appears that this candidate is also ruled out as the $E0$ giant resonance.

As was discussed in Section 3.7, a high-resolution study of the ^{208}Pb(p, p') and $(^3\text{He}, {}^3\text{He}')$ reactions has resulted in the observation of a peak at 9.11 MeV, the angular distribution for which is well described by an $L = 0$ calculation (75). The 9.11-MeV state depletes only $\approx 7\%$ of the $T = 0$, $L = 0$ EWSR strength and thus by itself would not be considered as the giant monopole resonance. Nevertheless, this observation is important since the 9.11-MeV state is the strongest monopole excitation so far observed (75).

Another candidate for the monopole resonance has been proposed from inelastic electron (86, 87) and deuteron (88) scattering. In these experiments, evidence is deduced for an $L = 0$ resonance located at the same energy as the GDR, $\approx 77 \times A^{-1/3}$ MeV, for the nuclei ^{40}Ca, ^{90}Zr, and ^{208}Pb. Since the proposed $E0$ resonance falls at the same energy as the GDR, a model-dependent subtraction of GDR strength must be made from the electron data. In addition, a GQR cross section is subtracted from the (e, e') composite resonance structure, and an underlying-continuum background is removed. When these subtractions are made, it is found that cross section remains in the resonance region near 13 MeV for ^{208}Pb and 17 MeV in ^{90}Zr. If the remaining strength is assumed to be $E0$, then for ^{208}Pb, $97^{+27}_{-14}\%$ or $10^{+20}_{-9}\%$ of the $T = 0$, $L = 0$ EWSR strength is depleted, depending on whether the Goldhaber-Teller or Jensen-Steinwedel model, respectively, is used for the GDR.

Similar results were obtained from inelastic deuteron scattering from ^{40}Ca, ^{90}Zr, and ^{208}Pb (88). It was observed in these data that the resonance peak centered at 63 $A^{-1/3}$ MeV was broader than the resonance peak excited in the same nuclei by the (α, α') reaction. When the (α, α') resonance shape was subtracted from the (d, d') data, a peak remained at 20 MeV in ^{40}Ca, 17 MeV in ^{90}Zr, and 13 MeV in ^{208}Pb. Angular distributions for these peaks were reported to be in agreement with calculations for an $L = 0$ transition. No $L = 0$ EWSR strength depletion has been extracted from the (d, d') data, although it is reported that the cross section accounts for a major part of the EWSR limit (88).

These observations provide a candidate for the elusive 0^+ giant resonance. However, as was the case for the GQR, these proposals must await the test of additional experiments.

4.2   Higher Multipole Resonances

There have been several reports of evidence for $E3$ giant resonances. Inelastic electron scattering on ^{208}Pb (73, 74, 86) and ^{197}Au (74) suggests an $L = 3$ resonance located at ≈ 17 MeV in both nuclei. The $L = 3$ EWSR strength deduced for ^{208}Pb is $\approx 90\%$ for a $T = 0$ mode or $\approx 60\%$ for $T = 1$. For ^{197}Au, the reported EWSR depletions are 45% and 30% for $T = 0$ and $T = 1$, respectively. Other electron data report some $L = 3$ strength in ^{40}Ca (89) and ^{116}Sn (78). The latter determinations were made by fitting the measured cross sections, integrated over a wide excitation-energy range (10–25 MeV), with $L = 1 + 2 + 3$ calculations.

Other candidates for an $E3$ resonance are the peaks observed at ≈ 13.5 MeV in several fp-shell nuclei and discussed in Section 3.8.2. However, as was pointed out, the multipolarity of this peak has not yet been firmly established.

Some evidence for an $L = 4$ resonance excitation has been provided through comparison of microscopic-model calculations with measurements on ^{208}Pb. These calculations are described in the next section.

5   MICROSCOPIC-MODEL CALCULATIONS OF EXCITATION OF GIANT RESONANCES

During the past 2–3 years there have been many theoretical calculations of the energies and excitation probabilities of high-excitation nuclear states. The number of these calculations is now so large that a separate review article is required for proper treatment of them. Since the present review is largely empirical in presentation, the results of one set of calculations that are particularly amenable to comparison with experimental results is described. An extensive but nonexhaustive list of references to other pertinent calculations is included (90).

Microscopic-model calculations for the excitation of nuclear states in the giant resonance region (8–20 MeV of excitation) of ^{208}Pb by 61-MeV protons and 115-MeV alphas have recently been published (16). The nuclear states utilized were those predicted by random phase approximation (RPA) nuclear structure calculations (91, 92). The calculations predicted over 400 normal parity, one-particle, one-hole excited states with spins from 0 to 5 for ^{208}Pb. Many of the states with high excitation energy were predicted to be of a collective nature, and thus may be those observed experimentally as giant resonances.

The inelastic scattering calculations were performed using the transition densities from the RPA calculations, together with an effective interaction between the

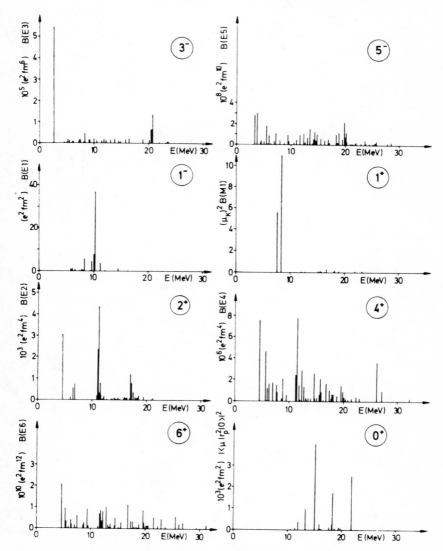

Figure 25   Spectrum of states of various J^π in ^{208}Pb predicted from the RPA calculations of (92).

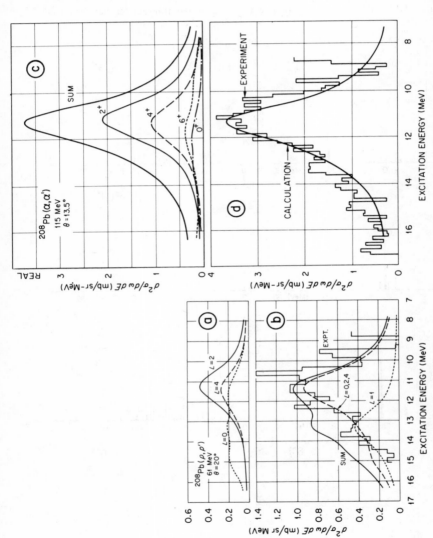

Figure 26 (*a*) Calculated 20° resonance spectra (16) for $L = 2$, $L = 4$, and $L = 0$ excitation by 61-MeV protons. (*b*) Comparison between calculated and measured (33) resonance spectra for the (*p, p'*) reaction on ^{208}Pb. The experimental spectrum is shown after removal of the underlying nuclear continuum. (*c*) Calculated 13.5° resonance spectra (16) for $L = 0$, 2, 4, and 6 excitation by 115-MeV alphas. (*d*) Comparison between calculated and measured (35) resonance spectra for the 115-MeV (α, α') reaction on ^{208}Pb. The experimental spectrum is shown after removal of the underlying nuclear continuum.

incident projectile and each target nucleon (16). The (p, p') and (α, α') cross sections for the ten 1^-, 2^+, 3^-, 4^+, and 5^- states that had the largest $B(EL)$ were then calculated in the distorted-wave Born approximation. A 2-MeV intrinsic spreading width was assumed for each state.

The excited states calculated for ^{208}Pb by the RPA method (92) are shown in Figure 25. For the 2^+ states the calculation predicts, in addition to the low-lying strength at about 4.5 MeV, strong excitation of states around 11 MeV and 19 MeV. These latter energies are in good agreement with the experimental results listed in Tables 4 and 5 for the $T = 0$ and $T = 1$ GQR in ^{208}Pb. It is also of interest to note the 3^- strength near 21 MeV, where experimental observation of $L = 3$ strength in ^{208}Pb has been reported in the (e, e') reaction. Of special interest is the $L = 4$ strength predicted to lie at the same energy as the 2^+ strength, ≈ 11 MeV. Several strong 0^+ levels are calculated; however, the strength is predicted to be spread over several MeV.

Figure 26 shows the results of the DWBA calculations applied to the states shown in Figure 25 and discussed above. The results were intended to cover only the excitation energy region from 8–17 MeV, and the experimental data are shown after the underlying nuclear continuum has been subtracted.

The calculations provide a good representation of the experimental data. It should be recalled that the strengths for the individual states selected from Figure 25 have been spread with Lorentzian shapes of 2-MeV width. As could be expected from Figure 25, the $L = 2$ and 4 strengths are concentrated in essentially the same region, while the $L = 0$ strength is more spread out. For the (α, α') calculation, a non-negligible $L = 6$ contribution to the data is also predicted. For alphas, the $L = 1$ component (GDR) does not contribute to the total cross section.

Perhaps the most intriguing feature of these calculations is the large $L = 4$ cross section predicted to occur at the same energy as the GQR. [A similar prediction was made earlier in a different calculation (93).] The calculated angular distributions are compared to (p, p') measurements on ^{208}Pb on Figure 27. While the sum of the calculated $L = 1$ and $L = 2$ cross sections is not as large as the measured cross section (as was the case for the macroscopic collective-model calculations shown in Figure 8), the addition of the $L = 4$ component provides good agreement with the data. The $L = 0$ contribution does not significantly improve the agreement. However, no definitive judgment of the need for $L = 4$ strength can be made on the basis of the angular-distribution shape. The same is true for the available (α, α') data. Thus a very important unanswered question is whether or not an $L = 4$ giant resonance is indeed contributing to the $63 \times A^{-1/3}$ resonance structure, as these calculations would predict. It is hoped that higher-energy data and studies on more nuclei will help provide this answer.

6   CONCLUDING REMARKS

It is now well established that giant resonances other than dipole exist as a general property of nuclei. The most thoroughly studied of the new resonances, the GQR, exhausts a major fraction of the $T = 0$, $L = 2$ EWSR strength and is located at

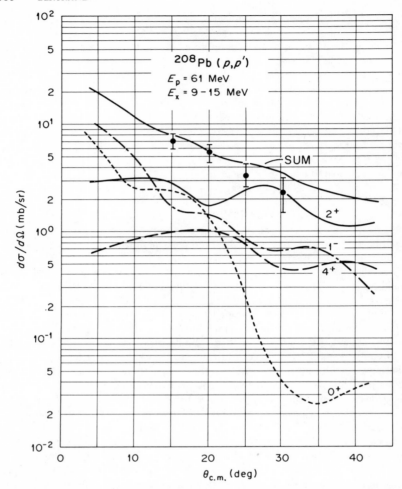

Figure 27 Calculated (16) angular distributions for various multipoles, compared with measurements (33) from the (*p, p'*) reaction on ^{208}Pb. The measured and calculated cross sections were obtained by integrating the cross section in the resonance region between 9 and 15 MeV.

$\approx 63 \times A^{-1/3}$ MeV of excitation. Some evidence for other new resonances—E0, E3, and E4—has been provided. Unfortunately, not all aspects of this field have been adequately covered in this review. In particular, the important areas of resonance-structure calculations, indirect methods of resonance observation, and the particle-capture work have been given too little discussion here to provide adequate description.

Hopefully, it has been established that there is generally good agreement among various measurements of the GQR. Often, as in the case of the sd-shell nuclei, agreement becomes apparent only upon consideration of the interaction properties of the various reactions, such as isospin selection or cross-section enhancement as a function of particle bombarding energy. In any case, many careful measurements have now provided a substantial set of data from which systematic trends of the GQR with nuclear mass have been established.

The field of inelastic excitation of giant multipole resonances is expanding rapidly, and it is clear that many new developments may have arisen by the time this review is published. Nevertheless, it is hoped that the present review provides both a summary of past work and a stimulus for new investigations.

ACKNOWLEDGMENTS

The author would like to recognize the researchers at the Oak Ridge National Laboratory who have been involved in the experimental and theoretical giant-resonance program and who have contributed to many of the ideas described in this review: Drs. E. E. Gross, D. J. Horen, D. C. Kocher, M. B. Lewis, E. Newman, and G. R. Satchler. The author thanks his many colleagues who helped make this review as up-to-date as possible by providing preprints and private communications about their work. The author particularly thanks Drs. Y. A. Ellis, E. E. Gross, D. C. Kocher, J. M. Moss, R. W. Peelle, G. R. Satchler, and D. H. Youngblood for critical review of the manuscript and S. J. Ball for preparation of the manuscript.

Literature Cited

1. Mottelson, B. R. 1970. *Proc. Solvay Conf. Phys., 15th, Brussels,* ed. I. Prigogine. New York: Gordon & Breach
2. Satchler, G. R. 1974. *Phys. Rev. C* 14: 97–127
3. Borzov, I. N., Kamerdzhiev, S. P. 1975. *FEI-580,* Phys.-Energ. Inst., Obninsk, USSR
4. Danos, M., Fuller, E. G. 1965. *Ann. Rev. Nucl. Sci.* 15: 29–66
5. Berman, B. L., Fultz, S. C. 1975. *Rev. Mod. Phys.* 47: 713–61
6. Hayward, E. 1970. *Natl. Bur. Stand. Monogr. No. 118.* Washington DC: GPO
7. Fagg, L. W. 1975. *Rev. Mod. Phys.* 47: 683–711
8. Bertrand, F. E., Peelle, R. W. 1973. *Phys. Rev. C* 8: 1045–64
9. Bertini, H. W., Harp, G. D., Bertrand, F. E. 1974. *Phys. Rev. C* 10: 2472–82 and references therein
10. Blann, M. 1975. *Ann. Rev. Nucl. Sci.* 25: 123–66
11. Bohr, A., Mottelson, B. 1975. *Nuclear Structure.* Vol. 2. New York: Benjamin
12. Satchler, G. R. 1972. *Nucl. Phys. A* 195: 1–25
13. Satchler, G. R. 1973. *Part. Nucl.* 5: 105–18
14. Lane, A. M. 1964. *Nuclear Theory.* New York: Benjamin
15. Nathan, O., Nilsson, S. G. 1965. In *Alpha-, Beta-, and Gamma-Ray Spectroscopy,* ed. K. Siegbahn, 1: 601–700. Amsterdam: North-Holland
16. Halbert, E. C., McGrory, J. B., Satchler, G. R., Speth, J. 1975. *Nucl. Phys. A* 245: 189–204
17. Singh, P. P., Yang, G. C., van der Woude, A., Drentje, A. 1976. *Phys. Rev. C* 13: 1376–87
18. Gruhn, C. R., Kuo, T. Y. T., Maggiore, C. J., McManus, H., Petrovich, F., Preedom, B. M. 1972. *Phys. Rev. C* 6: 915–44
19. Lewis, M. B., Bertrand, F. E., Fulmer, C. B. 1973. *Phys. Rev. C* 7: 1966–72
20. Wagner, W. T., Crawley, G. M., Hammerstein, G. R., McManus, H. 1975. *Phys. Rev. C* 12: 757–77
21. Bernstein, A. M. 1969. *Adv. Nucl. Phys.* 3: 325–476

22. Satchler, G. R. 1966. *Theor. Phys. C* 8:73–176
23. Kuhlmann, E., Ventura, E., Calarco, J. R., Mavis, D. G., Hanna, S. S. 1975. *Phys. Rev. C* 11:1525–36
24. Baer, H. W., Bistirlich, J. A., de Bottin, N., Cooper, S., Crowe, K. N., Truol, P., Vergados, D. 1974. *Phys. Rev. C* 10:267–69
25. Pitthan, R., Walcher, Th. 1971. *Phys. Lett. B* 36:563–64
26. Fukuda, S., Torizuka, Y. 1972. *Phys. Rev. Lett.* 29:1109–11
27. Lewis, M. B., Bertrand, F. E. 1972. *Nucl. Phys. A* 196:337–46
28. Pitthan, R., Walcher, Th. 1972. *Z. Naturforsch. Teil A* 27:1683–84
29. Pitthan, R. 1973. *Z. Phys.* 260:283–304
30. Lewis, M. B., Bertrand, F. E., Horen, D. J. 1973. *Phys. Rev. C* 8:398–400
31. Tyren, H., Maris, Th. A. J. 1957. *Nucl. Phys.* 4:637–42, 662–71; 1958 6:446–50, 82–86; 7:24–26
32. Lewis, M. B. 1972. *Phys. Rev. Lett.* 29:1257–60
33. Bertrand, F. E., Kocher, D. C. 1976. *Phys. Rev. C.* 13:2241–46
34. Kocher, D. C., Bertrand, F. E., Gross, E. E., Lord, R. S., Newman, E. 1973. *Phys. Rev. Lett.* 31:1070–73; erratum, 1974. *Phys. Rev. Lett.* 32:264
35. Moss, J. M., Rozsa, C. M., Bronson, J. D., Youngblood, D. H. 1974. *Phys. Lett. B* 53:51–53; Youngblood, D. H., Moss, J. M., Rozsa, C. M., Bronson, J. D., Bacher, A. D., Brown, D. R. 1976. *Phys. Rev. C* 13:994–1008
36. Chang, C. C., Bertrand, F. E., Kocher, D. C. 1975. *Phys. Rev. Lett.* 34:221–24
37. Marty, N., Morlet, M., Willis, A., Comparat, V., Frascaria, R. 1975. *Nucl. Phys. A* 238:93–110
38. Rutledge, L. L. Jr., Hiebert, J. C. 1974. *Phys. Rev. Lett.* 32:551–54
39. Moss, J. M., Rozsa, C. M., Youngblood, D. H., Bronson, J. D., Bacher, A. D. 1975. *Phys. Rev. Lett.* 34:748–51
40. Bertrand, F. E., Burrus, W. R., Hill, N. W., Love, T. A., Peelle, R. W. 1972. *Nucl. Instrum. Methods* 101:475–92
41. Torizuka, Y., Kojima, Y., Saito, T., Itoh, K., Nakada, A., Mitsunobu, S., Nagao, N., Hosoyama, K., Fukuda, S. 1972. *Proc. Int. Conf. Nucl. Struct. Stud. Using Electron Scattering Photoreaction,* Sendai, ed. K. Shoda, H. Ui, pp. 171–82; *Suppl. Res. Rep. Lab. Nucl. Sci.,* Vol. 15, Tohoku Univ.
42. Moalem, A., Benenson, W., Crawley, G. M. 1973. *Phys. Rev. Lett.* 31:482–85
43. Arvieux, J., Buenerd, M., Cole, A. J., de

Saintignon, P., Perrin, G., Horen, D. J. 1975. *Nucl. Phys. A* 247:238–50
44. Horen, D. J., Arvieux, J., Buenerd, M., Cole, J., Perrin, G., de Saintignon, P. 1975. *Phys. Rev. C* 11:1247–50
45. Bertrand, F. E., Kocher, D. C., Gross, E. E., Newman, E. Unpublished
46. Terrien, Y. 1976. Private communication
47. Wyckoff, J. M., Ziegler, B., Koch, H. W., Uhlig, R. 1965. *Phys. Rev. B* 137:576–94
48. Hanna, S. S. 1974. *Proc. Int. Conf. Nucl. Struct. and Spectrosc.,* ed. H. P. Blok, A. E. L. Dieperink, 2:249–85. Amsterdam: Scholar's
49. Meyer-Schutzmeister, L., Segel, R. E., Wharton, W., Debevec, P., Raghunathan, K. Submitted for publication
50. Watson, R. B., Branford, D., Black, J. L., Caelli, W. J. 1973. *Nucl. Phys. A* 203:209–20
51. Geramb, H. V. 1972. *Nucl. Phys. A* 183:582–92; Geramb, H. V., Sprickmann, B., Strobel, G. L. 1973. *Nucl. Phys. A* 199:545–59; Geramb, H. V., Eppel, D. 1973. *Z. Phys.* 261:177–86
52. Kocher, D. C., Bertrand, F. E., Gross, E. E., Newman, E. To be published
53. Search and Discovery, Odd Behavior of Giant Quadrupole Resonance. 1975. *Phys. Today* 28:18
54. Bertrand, F. E., Kocher, D. C., Gross, E. E. Unpublished
55. Ahrens, J., Borchert, H., Czock, K. H., Eppler, H. B., Gimm, H., Gundrum, H., Kroening, M., Riehn, P., Ram, G., Sita, Zieger, A., Ziegler, B. 1975. *Nucl. Phys. A* 251:479–92
56. Bauer, T., Breuer, H., Kiss, A., Knöpfle, K. T., Mayer-Böricke, C., Rogge, M., Wagner, G. J. 1976. Private communication; see also Institut für Kernphysik 1975. *Ann. Rep. KFA-IKP 10/76,* pp. 17–19, and Max-Planck Institut für Kernphysik. 1975. *Ann. Rep.*
57. Chang, C. C., Dideley, J. P., Kwiatkowski, K., Wu, J. R. 1976. Private communication
58. Harekeh, M. N., Arends, A. R., de Voigt, M. J. A., Drentje, A. G., van der Werf, S. Y., van der Woude, A. 1976. *Nucl. Phys. A* 265:189–212
59. Knöpfle, K. T., Wagner, G. J., Breuer, H., Rogge, M., Mayer-Böricke, C. 1975. *Phys. Rev. Lett.* 35:779–82
60. Moalem, A., Benenson, W., Crawley, G. M. 1974. *Nucl. Phys. A* 236:307–16
61. Hotta, A., Itoh, K., Saito, T. 1974. *Phys. Rev. Lett.* 33:790–94
62. Snover, K. A., Ebisawa, K., Brown, D. R., Paul, P. 1974. *Phys. Rev. Lett.* 32:317–20
63. Hanna, S. S., Glavish, H. F., Avida, R.,

Calarco, J. R., Kuhlmann, E., LaCanna, R. 1974. *Phys. Rev. Lett.* 32:114–17

64. Noe, J., Paul, P., Snover, K. A., Suffert, M., Warburton, E. K. 1975. Contribution to *Symp. Nucl. Struct., Balatonfured, Sept.*

65. Youngblood, D. H., Bacher, A. D., Bronson, J. D., Moss, J. M., Rozsa, C. M. 1976. Submitted for publication

66. Bergere, R., Beil, H., Carlos, P., Lepretre, A., Veyssiere, A. 1973. *Proc. Int. Conf. Photonucl. React. Appl., Asilomar,* ed. B. L. Berman, 1:525–34; CONF-730301, US AEC Off. Inform. Serv., Oak Ridge, Tenn.

67. Danos, M. 1958. *Nucl. Phys.* 5:23–32
68. Ligensa, R., Greiner, W. 1967. *Nucl. Phys. A* 92:673–95
69. Horen, D. J., Bertrand, F. E., Lewis, M. B. 1974. *Phys. Rev. C* 9:1607–10
70. Schwierczinski, A., Frey, R., Spamer, E., Theissen, H., Walcher, Th. 1975. *Phys. Lett. B* 55:171–74
71. Kishimoto, T., Moss, J. M., Youngblood, D. H., Bronson, J. D., Rozsa, C. M., Brown, D. R., Bacher, A. D. 1975. *Phys. Rev. Lett.* 35:552–55
72. Bertrand, F. E., Kocher, D. C. 1974. See Ref. 69
73. Nagao, M., Torizuka, Y. 1973. *Phys. Rev. Lett.* 30:1068–71
74. Pitthan, R., Buskirk, F. R., Dally, E. B., Dyer, J. N., Maruyama, X. K. 1974. *Phys. Rev. Lett.* 33:849–52; erratum, 1975. *Phys. Rev. Lett.* 34:848
75. Morsch, H. P., Decowski, P., Benenson, W. 1975. *MSUCL-197.* Mich. State Univ., East Lansing, Mich; 1976. *Phys. Rev. Lett.* 37:263–65
76. Sherman, N. K., Ferdinande, H. M., Lokan, K. H., Ross, C. K. 1975. *Phys. Rev. Lett.* 35:1215–19
77. Torizuka, Y., Kojima, Y., Saito, T., Itoh, K., Nakada, A. 1973. *Res. Rep. Lab. Nucl. Sci., Tohoku Univ.* 6:165–210
78. Torizuka, Y., Kojima, Y., Saito, T., Itoh, K., Nakada, A., Mitsunobu, S., Nagao, N., Hosoyama, K., Fukuda, S., Miura, H. 1973. See Ref. 66, 1:675–83
79. Schwierczinski, A., Frey, R., Richter, A., Spamer, E., Theissen, H., Titze, O., Walcher, Th., Krewald, S., Rosenfelder, R. 1975. *Phys. Rev. Lett.* 35:1244–47
80. Brady, F. P., King, N. S. P., McNaughton, M. W., Satchler, G. R. 1976. *Phys. Rev.*

Lett. 36:15–18
81. Doering, R. R., Galonsky, A., Patterson, D. M., Bertsch, G. F. 1975. *Phys. Rev. Lett.* 35:1691–93
82. Lipparini, E., Orlandini, G., Leonardi, R. 1975. *Phys. Rev. Lett.* 36:660–63
83. Krewald, S., Galonska, J. E., Faessler, A. 1975. *Phys. Lett. B* 55:267–69
84. Krewald, S., Speth, J. 1974. *Phys. Lett. B* 52:295–98
85. Bertsch, G., Tsai, S. F. 1975. *Phys. Rep. C* 18:125–58
86. Sasao, M. N., Torizuka, Y. 1976. Prepr.
87. Fukuda, S., Torizuka, Y. 1976. Prepr.
88. Marty, N., Morlet, M., Willis, A., Comparat, V., Frascaria, R., Kallne, J. 1975. *IPNO-PhN-75-11,* Inst. Phys. Nucl., Orsay, France. Unpublished
89. Torizuka, Y., Itoh, K., Shin, Y. M., Kawazoe, Y., Matsuzaki, H., Takeda, G. 1975. *Phys. Rev. C* 11:1174–78
90. Engel, Y. M., Brink, D. M., Goeke, K., Krieger, S. J., Vautherin, D. 1975. *Nucl. Phys. A* 249:215–38; Krewald, S., Speth, J. 1974. *Phys. Lett. B* 52:295–98; Speth, J., Zamick, L., Ring, P. 1974. *Nucl. Phys. A* 232:1–12; Zawischa, D., Speth, J. 1975. *Phys. Lett. B* 56:225–28; Bertsch, G. F., Tsai, S. F. 1975. *Phys. Rep. C* 18:125–58; Bertsch, G. F. 1973. *Phys. Rev. Lett.* 31:121–24; Krewald, S., Galonska, J. E., Faessler, A. 1975. *Phys. Lett. B* 55:267–69; Abgrall, Y., Caurier, E. 1975. *Phys. Lett. B* 56:229–31; Suzuki, T. 1973. *Nucl. Phys. A* 217:182–88; Hammerstein, G. R., McManus, H., Moalem, A., Kuo, T. T. S. 1974. *Phys. Lett. B* 49:235–38
91. Ring, P., Speth, J. 1973. *Phys. Lett. B* 44:477–80
92. Ring, P., Speth, J. 1974. *Nucl. Phys. A* 235:315–51
93. Hamamoto, I. 1972. *Nucl. Phys. A* 196:101–6
94. Snover, K. A., Ebisawa, K., Brown, D. R., Paul, P. 1974. *Phys. Rev. Lett.* 32:317–20
95. Gul'karov, I. S. 1974. *Sov. J. Nucl. Phys.* 18:267–69
96. Lewis, M. B., Horen, D. J. 1974. *Phys. Rev. C* 10:1099–1102
97. Torizuka, Y., Kojima, Y., Saito, T., Itoh, K., Nakada, A. 1973. *Res. Rep. Lab. Nucl. Sci., Tohoku Univ.* 6:1–29

AUTHOR INDEX

A

Abarbanel, H., 443-45
Abecasis, S. M., 291, 292
Abgrall, Y., 502
Abrams, G. S., 93, 122, 127-29, 131
Adler, H. H., 321
Adler, S. L., 202, 216
Ahrens, J., 482
Akerlof, C. W., 411, 412
Akimov, Y., 400, 401
Alberi, G., 396, 397
Albrow, M. G., 433, 435-37
Alburger, D. E., 253
Alder, K., 375
Alexander, E. C., 337
Alexander, Y., 396
Alff-Steinberger, C., 10, 15, 30
Alkhazov, G. D., 391, 397
Allaby, J. V., 411, 419, 420, 425, 426
Allegre, J., 328
Allegre, J. C., 324, 331
Alles, W., 3, 9
Alles-Borelli, V., 104, 108, 114, 131
Altarelli, G., 225
AMALDI, U., 385-456; 404, 408, 412, 413, 416, 418
Ambats, I., 408
Amendolia, S. R., 404, 408
Amman, F., 93
Anderson, E. W., 419, 431, 432
Anderson, R. L., 39
Andrews, R. A., 168, 169
Anholt, R., 364
Antipov, Yu. M., 412, 422, 423, 427-29
Apokhin, V. D., 413, 416
Appel, J. A., 40, 189
Appelquist, T., 137, 236
Arbuzov, B. A., 3, 8
Arends, A. R., 485, 486, 492
Arima, A., 283, 287
Armbruster, P., 364
Aronson, S. H., 21, 24
Arteaga-Romero, N., 99
Arvieux, J., 469, 474, 488, 490, 493-95
Ascoli, G., 399, 422, 423
Ashford, V. A., 34
Atwood, T. L., 140
Aubert, J. J., 93, 122, 228
Auerbach, E. H., 282
Auger, J. P., 391, 397
Augustin, J.-E., 92, 93, 103,

105-8, 112, 113, 118, 120-22, 125, 131, 136, 138, 139
Aulchenko, V. M., 131
Auslander, V. L., 92, 93, 118
Austern, N., 39
Avida, R., 485, 499, 500
Awschalom, M., 171
Ayers, D. S., 7

B

Bacci, C., 109-11, 131, 135, 138
Bacher, A. D., 468, 469, 472, 474, 480, 487-90, 492-95, 504
Baer, H. W., 465, 499, 500
Baglin, C., 39
Baier, V. N., 95, 99, 143
Bailey, J., 7
Baillon, P., 413, 416
Baird, J. K., 41
Baker, W. F., 175, 189
Balakin, V. E., 107-11, 119, 121, 131-33
Balats, M. Ya., 10, 30
Baldini-Celio, R., 112
Ball, J. S., 444
Baltay, C., 37, 38
Banerjee, B., 289, 305, 308, 312
Banner, M., 18, 31, 32
Baranger, M., 255
Barbarino, G., 130
Barbaro-Galtieri, A., 188
Barber, W. C., 92, 103, 104, 125
Barbiellini, G., 18, 32, 33, 111-13, 132, 412
Barbieri, R., 236
Bardeen, W. A., 213, 236
Bardon, M., 2
Barger, V., 411-13, 418, 438, 439
Barish, B. C., 188, 218
Barloutand, R., 427
Barmin, V. V., 6
Barnes, F. Q., 321
Barnes, S. W., 43
Barnett, M., 236
Barrett, B., 37
Barshay, S., 3, 8, 39, 40
Bartenev, V., 412, 413, 416
Bartlett, D. F., 39
Bartlett, J. F., 188
Bartoli, B., 104, 105, 135, 138
Basile, P., 26

Bassiere, H., 336
Bauer, E., 76
Bauer, T., 200, 484, 485, 487, 492
Beaupré, J. V., 408, 431, 432
Beck, R., 301
Beil, H., 487
Bell, J. S., 3, 7, 439
Bellini, G., 394, 403
Belyaev, S. T., 293
Bemis, C. E. Jr., 279, 281
Bemporad, C., 118, 120, 122, 394, 402, 403
Benaksas, D., 118, 120-22, 124, 132
Benecke, J., 448
Benenson, W., 469, 471, 485, 490-93, 495, 501
Bennett, S., 3, 11, 15, 20, 30, 31
Benvenuti, A., 188
Benz, P., 39
Berardo, P. A., 39
Berends, F. A., 103, 106-9
Berestetsky, W., 375
Berger, E. L., 409, 431
Bergere, R., 487
Berman, B. L., 458, 497
Berman, S., 219, 221, 226
Bernardini, C., 92, 133
Bernardini, M., 105, 130, 133, 135, 143
Bernstein, A. M., 463
Bernstein, J., 3, 7, 8
Beron, B. L., 105, 106, 108-10
Bertini, H. W., 459, 468
Bertini, R., 391
Bertocchi, L., 396, 397
BERTRAND, F. E., 457-509; 458, 459, 463, 465, 467-70, 473-80, 482, 483, 486, 488-95, 504
Bertsch, G. F., 501, 502
Bes, D. R., 292
Bethe, H. A., 242
Betz, W., 373, 381
Beusch, W., 394, 402, 403
Beuscher, H., 260
Beznogikh, G. G., 412, 413, 416
Bhargava, P. C., 311
Biafas, A., 440
Bigotte, G., 321
Billing, K. D., 39
Billinge, R., 161
Bingham, H. H., 130, 423, 424, 427

511

CUMULATIVE INDEXES

CONTRIBUTING AUTHORS VOLUMES 17-26

520

CHAPTER TITLES VOLUMES 17-26

522

DATE DUE

WITHDRAWN

GAYLORD | | | PRINTED IN U.S.A.